The Chemistry and Physics of Aerogels

Discover a rigorous treatment of aerogels processing and techniques for characterisation with this easy-to-use reference. This volume presents the basics of aerogel synthesis and gelation to open porous nanostructures, and the processing of wet gels such as ambient and supercritical drying leading to aerogels. It describes their essential properties with their measurement techniques and theoretical models used to analyse relations to their nanostructure. Linking the fundamentals and with practical applications, this is a useful toolkit for advanced undergraduates, and graduate students doing research in material and polymer science, physical chemistry and chemical and environmental engineering.

Lorenz Ratke is a senior scientist at the German Aerospace Center, DLR, Institute of Materials Research, where he served as department head until 2016. He is a member of the German Physical Society (DPG) and the German Society of Materials (DGM).

Pavel Gurikov is a professor at Hamburg University of Technology, Laboratory for Development and Modelling of Novel Nanoporous Materials. He is a member of the International Adsorption Society and the European Society for Engineering Education. In 2019 he received the Professor Siegfried Peter Prize for pioneering research work in high-pressure process engineering.

The Chemistry and Physics of Aerogels

Synthesis, Processing, and Properties

LORENZ RATKE
German Aerospace Center

PAVEL GURIKOV
Hamburg University of Technology

CAMBRIDGE
UNIVERSITY PRESS

University Printing House, Cambridge CB2 8BS, United Kingdom

One Liberty Plaza, 20th Floor, New York, NY 10006, USA

477 Williamstown Road, Port Melbourne, VIC 3207, Australia

314–321, 3rd Floor, Plot 3, Splendor Forum, Jasola District Centre, New Delhi – 110025, India

103 Penang Road, #05–06/07, Visioncrest Commercial, Singapore 238467

Cambridge University Press is part of the University of Cambridge.

It furthers the University's mission by disseminating knowledge in the pursuit of education, learning, and research at the highest international levels of excellence.

www.cambridge.org
Information on this title: www.cambridge.org/9781108478595
DOI: 10.1017/9781108778336

First published 2021

Printed in the United Kingdom by TJ Books Limited, Padstow Cornwall

A catalogue record for this publication is available from the British Library.

ISBN 978-1-108-47859-5 Hardback

To my wife Sylvia – Lorenz Ratke.

To the memory of F. N. Maskaev, who was more than a teacher to me.
– Pavel Gurikov.

Contents

Preface *page* xiii

1 **Introduction** 1
1.1 What Exactly Is an Aerogel? 1
1.2 Aerogel Classification 5
1.3 A Brief History of Aerogels 7

2 **Chemical Synthesis of Aerogels from Monomeric Precursors** 11
2.1 Silica Aerogels 11
2.1.1 Silica Aerogel Precursors 12
2.1.2 Monomeric Precurors 13
2.1.3 Hydrolysis and Polycondensation 14
2.1.4 Growth and Structure 20
2.1.5 Gelation 28
2.1.6 Ageing 29
2.1.7 Procedures and Results 31
2.1.8 Silica Aerogels with Lower Functional Silanes 32
2.2 Resorcinol–Formaldehyde Aerogels 34
2.2.1 Chemistry of Resorcinol and Formaldehyde 35
2.2.2 Reaction between Resorcinol and Formaldehyde 36
2.2.3 From Monomers to Polymers 39
2.2.4 Parameters of Synthesis 40
2.2.5 Thermodynamics of RF Solutions 54
2.3 Variables and Symbols 58

3 **Chemical Synthesis of Aerogels from Polymeric Precursors** 60
3.1 Synthesis of Cellulose Aerogels 60
3.1.1 Dissolution Agents for Cellulose 62
3.1.2 Gelation of Cellulose Solutions 68
3.1.3 Preparation of Bulky Cellulose Aerogels 72
3.2 Alginate: From Solution to Gel 76
3.2.1 Structure of Alginate 77
3.2.2 Degradation and Chemical Modifications 81
3.2.3 Gelation of Alginate 82
3.3 Variables and Symbols 91

4 Gelation 92
4.1 Viscosity of Gelling Solutions 92
4.1.1 Simple Models of Viscoelasticity 94
4.1.2 Evolution of Viscosity during Gelation 97
4.1.3 Tilting as a Simple Measure of Gelation 98
4.1.4 Rotating Bob Viscosimeter 100
4.1.5 Flow Damping 102
4.1.6 Light Transmission and Scattering 103
4.2 Theoretical Descriptions of Gelation 104
4.2.1 Percolation and Fractals 105
4.2.2 Diffusion-Limited Aggregation – DLA and DLCA 114
4.2.3 Mean Field Model of Smoluchowski 117
4.2.4 Scaling Solutions and Gelation 120
4.2.5 Polymerisation-Induced Phase Separation (PIPS) 124
4.3 Predictions of Gel Time 129
4.3.1 Gelation by Aggregation 129
4.3.2 Family–Viscek Scaling for Gel Times 132
4.3.3 Gel Time Prediction by PIPS 133
4.4 Variables and Symbols 138

5 Drying of Wet Gels 140
5.1 Ambient Drying 140
5.1.1 Some Observations during Drying 141
5.1.2 Evaporation Rate 144
5.1.3 Capillary Stress and Vapour Pressure 145
5.1.4 Flow of Fluid and Gas in the Network during Drying 147
5.2 Freeze Drying 149
5.2.1 Freezing of Single-Component Liquid 150
5.2.2 Freezing of a Multicomponent Liquid 152
5.2.3 Freezing of a Gel Liquid 157
5.2.4 Sublimation 162
5.3 Supercritical Drying 163
5.3.1 Supercritical Fluids 163
5.3.2 High-Temperature Supercritical Drying 167
5.3.3 Rapid Supercritical Extraction 171
5.3.4 Low-Temperature Supercritical Drying 173
5.4 Solvent Exchange and Gel Shrinkage 185
5.4.1 Solvent Selection 185
5.4.2 Gel Shrinkage: General Considerations 189
5.4.3 Gel–Solvent Interactions 193
5.4.4 Minimising Gel Shrinkage 196
5.5 Variables and Symbols 201

6 **Morphology of Aerogels** 203
6.1 Imaging Techniques 203
6.1.1 Scanning Electron Microscopy 203
6.1.2 Transmission Electron Microscopy 205
6.2 TEM Images of Aerogels 205
6.3 SEM Images of Particular Aerogels 209
6.4 SEM Images of Fibrillar Structures 212

7 **Density: Models and Measures** 214
7.1 Measurement of Envelope Density 215
7.2 Measurement of Skeletal Density 215
7.2.1 Measurement Principle 216
7.2.2 Effect of Adsorbates 217
7.2.3 Pressure Evolution 219
7.2.4 Effect of Test Gas on Skeletal Density 223
7.3 Porosity 225
7.4 Rule of Mixtures for the Envelope Density 226
7.5 Relation between Monomer Content and Final Density 227
7.5.1 Estimate of the Envelope Density of Silica Aerogels 228
7.5.2 Estimate of the Envelope Density of RF Aerogels 231
7.5.3 Density of Cellulose Aerogels 232
7.6 Variables and Symbols 235

8 **Specific Surface Area** 236
8.1 Definitions and Relations 236
8.2 Surface Area of Simple Shapes 237
8.3 Surface Area of Irregular-Shaped Bodies 241
8.4 Surface Area of Fibrillar Aerogels 243
8.5 Surface Area of Aerogel Composites 247
8.6 Measurement of Specific Surface Area 249
8.6.1 Langmuir Isotherm 251
8.6.2 A Bit of Thermodynamics of Adsorption 254
8.6.3 BET Isotherm 256
8.6.4 T-plot 261
8.6.5 Small Angle X-Ray Scattering (SAXS) 265
8.7 Variables and Symbols 265

9 **Pores and Pore Sizes** 268
9.1 Simple Geometrical Models 268
9.2 Stereological Pore Size Description 270
9.2.1 Pore Size Distribution: The BJH Method 273
9.2.2 Pore Size Distribution: Thermoporosimetry 278
9.3 Variables and Symbols 282

10	**Diffusion in Aerogels**	285
	10.1 A Phenomenological Approach to Diffusion	285
	10.2 Diffusion Coefficients	288
	10.2.1 Knudsen Diffusion	290
	10.3 Measurement of Gas Diffusion in Aerogels	292
	10.4 Variables and Symbols	295
11	**Permeability for Gases**	296
	11.1 An Experimental Approach	297
	11.1.1 A Dynamic Measurement Setup	298
	11.1.2 Stationary Measurement of Permeability	300
	11.2 Full Mathematical Model of Porous Media	301
	11.2.1 Characteristic Time of Permeation	303
	11.2.2 Stationary State	304
	11.2.3 Reynolds Number	304
	11.3 Flow through an Aerogel	305
	11.4 Knudsen Effect	307
	11.5 The Meaning of Permeability	307
	11.5.1 Parallel Cylindrical Pore Arrangement	308
	11.5.2 Karman–Kozeny Permeability	309
	11.6 Permeability of Some Aerogels	311
	11.7 Variables and Symbols	313
12	**Thermal Properties**	315
	12.1 Heat Conduction Equation	315
	12.2 Heat Conduction in Porous Materials: Aerogels	317
	12.2.1 Porous Media in General	318
	12.3 Thermal Conductivity of Aerogels	320
	12.3.1 Heat Conduction of the Solid Backbone	321
	12.3.2 Gas Phase Transport of Heat	325
	12.3.3 Radiative Heat Transport	331
	12.4 Thermal Diffusivity	332
	12.4.1 Specific Heat Capacity	333
	12.5 Measurement Techniques of Thermal Properties of Aerogels	336
	12.5.1 Stationary Measurement Methods	336
	12.5.2 In-Stationary Measurement Methods	339
	12.6 Application of Aerogels to Insulation Tasks	345
	12.6.1 Transfer of Heat from a Plate into the Surrounding Medium	345
	12.6.2 The Solution	346
	12.6.3 Stationary Case	347
	12.6.4 Instationary Case	349
	12.6.5 Heat Transfer from a Tube through an Insulation	351
	12.6.6 Stationary Solution for a Cylindrical Shell	352
	12.7 Variables and Symbols	354

13 **Mechanical Properties of Aerogels** 356
13.1 Mechanical Testing: A Brief Introduction 356
13.1.1 Bending 356
13.1.2 Compression Test 360
13.1.3 Tension and Brazilian Test 363
13.1.4 Flexibility 364
13.1.5 Compressibility 365
13.1.6 Young's Modulus from Sound Velocity Measurement 366
13.2 Stress–Strain Curves of Aerogels 368
13.3 Young's Modulus of Aerogels 370
13.3.1 The Gibson and Ashby (GA) Model for Porous Materials 371
13.3.2 Simple Extensions of the GA Model for Aerogels 372
13.3.3 Experimental Results 373
13.4 Yield and Fracture Strength of Aerogels 376
13.5 Modelling of the Mechanical Response of Aerogels 379
13.5.1 A Model for Compressive Stress–Strain Curves 380
13.6 Variables and Symbols 386

14 **How to Cook Aerogels: Recipes and Procedures** 388
14.1 Silica Aerogels 388
14.1.1 Classical Silica Aerogel 388
14.1.2 A Superflexible Silica Aerogel 389
14.2 Cellulose Aerogels 390
14.3 Alginate Aerogel Beads 391
14.4 Resorcinol-Formaldehyde Aerogels/Xerogels 392

Appendix A Thermodynamics and Phase Separation in Immiscibles 394
A.1 A Bit of Thermodynamics 394
A.1.1 Ideal and Regular Solutions 397
A.2 Phase Separation in a Regular Solution 399
A.2.1 Phase Separation by Nucleation and Growth 400
A.2.2 Phase Separation by Spinodal Decomposition 402

Appendix B Flory–Huggins Theory of Polymer Solutions 404

Appendix C A Brief Review on Scattering 411
C.1 Form Factor 414
C.2 Structure Factor 416
C.3 Specific Surface Area 420
C.4 Dynamic Light Scattering 421

Appendix D Mathematics of Polycondensation 424
D.1 A Simple Model of a Bimolecular Reaction 424

D.2 A More Complex Model of Polycondensation 425
D.2.1 A Variant of Smoluchowski's Aggregation Equation 425
D.2.2 Degree of Polymerisation 428
D.2.3 Flory–Schulz Distribution 429
D.2.4 Global Volume Fraction Polymers 429
D.2.5 The Maximum Volume Fraction Polymer 430

Appendix E Time-Dependent Heat Transfer through an Isolated Tube 432

References 439
Index 467

Preface

Aerogels are one focus area of nanostructured materials research since the mid-1970s and became especially popular in the last decade. Although proceedings of several aerogel conference series are available and even a handbook on aerogels was published some years ago, there is no text available still that treats the most important aspects of synthesis, processing and properties of aerogels in a concise manner such that a beginner in this research field can readlily become familiar with the concepts of making aerogels and mathematical models describing their properties.

This book emerged from scripts written for five summer schools on aerogels which were organised by the Aerogel Team at the German Aerospace Center, DLR, since 2008 in cooperation with the groups of Professor Irina Smirnova from the Hamburg University of Technology, Professor Nicola Hüsing from the Paris-Lodron University Salzburg, Dr Matthias Koebel, the Swiss Federal Laboratories for Materials Science and Technology (EMPA) Dübendorf and Dr Gudrun Reichenauer from the Zentrum für Angewandte Energietechnik (ZAE) Würzburg. These schools were open for master's, and PhD students, and postdoctorates and scientists from industry and academia. The focus of these summer schools was practical exercises in the lab to learn how aerogels are made and processed, how their properties are determined and how their microstructures can be described. For all summer schools, a script was prepared giving background information on aerogels and a detailed list of recipes to be used or cooked by the students in the lab, and test procedures were described for their characterisation.

Based on these scripts, this book is a basic course in the physical chemistry and processing of gels which can be converted to aerogels. We will use four different materials (silica, resorcinol-formaldehyde, cellulose and alginate) to study the mechanisms allowing them to form gels either from solutions of monomers or solutions of polymers, including a discussion of gelation mechanisms in dilute solutions such as diffusion limited aggregation or polymerisation-induced phase separation. The processes required to transform them into aerogels, such as ambient, freeze or supercritical drying, are described and analysed in detail based on thermodynamic considerations, solidification and crystal growth and critical state theory. Since in many cases the transformation of a wet gel to a dry aerogel requires an exchange of the pore fluid to one being miscible with the supercritical drying fluid, problems of solvent exchange and related issues such as shrinkage are discussed. The physical background to describe aerogel properties, such as density, specific surface area, porosity, diffusion, gas permeability, thermal and mechanical properties related to their nano- to mesoscopic microstructure will also be described in detail, and the models used are presented and explained.

We hope this book is helpful for master's and PhD students, postdoctorates as well as beginners in this fascinating field of bulk nanostructured materials: aerogels.

The authors are very thankful for the many discussions they had with Irina Smirnova, Matthias Koebel, Nicola Hüsing, Gudrun Reichenauer, Wim Malfait, Falk Liebner, Luisa Durães, Tatiana Budtova, Cedric Gommes, Aravind Ramaswamy, René Tannert, Marina Schwan, Maria Schestakow, Ameya Rege, Philipp Niemeyer, Markus Heyer, Barbara Milow, Kathirvel Ganesan, Fei Wang, Britta Nestler and Raman Subrahmanyam. We also thank all colleagues for carefully reading the script in various stages. They all helped us to understand aerogels and provided original data which we could use for illustration purposes.

1 Introduction

Aerogels are fascinating materials. Place a piece of a silica aerogel into someone's hands, and she or he is immediately fascinated, curious, how such a solid, stiff material can be that light, transparent, and withstand a burning flame of a welding torch and can still be held in the hand without any feeling of 'becoming hot'. The same experience is felt if, for instance, you take a cellulose aerogel into your hands: it is equally stiff, light, white and feels like a marshmallow without having a glueing touch to the fingers. A few questions immediately arise: how are such materials made, where is it applied and why are they called 'aerogels'?

In principle, the name says everything: the main component of this material is air which is embedded into a very filigree, solid network of very small particles or fibrils. This network can be designed in a way that the resulting aerogels have a density that is only three times that of air and if evacuated it would be lighter than air.

If an aerogel is given to a chemist, a physicist or a materials scientist, each of them would discover interesting properties which individually are also observed in other materials. For example, silica (SiO_2) aerogels have a high transparency that is close to that of glass and a refractive index close to that of air, a thermal conductivity lower than that of polystyrene or polyurethane foams, or very high specific surface areas, which are also found in activated carbon, a low sound transmission also to be found in soft polymeric foams. The unique feature of aerogels is the combination of these physical properties in one material which can have various chemical compositions ranging from organic to inorganic ones or hybrids of organic and inorganic materials.

1.1 What Exactly Is an Aerogel?

Aerogels belong to the more general class of porous materials. Materials are porous if they contain cavities, channels or interstices filled with air or a blowing agent (pentane or carbon dioxide, for instance). The pores may be regularly arranged as in honeycombs or zeolithes or metal organic frameworks (MOFs). However, the more common situation is an irregular pore structure, as obtained by cross-linking of polymer chains, aggregation or agglomeration of small particles or selective removal of elements of a solid (e.g., by etching or pyrolysis, chemical bleaching of one phase of a two-phase material). Porous solids can be further divided into closed-cell and open-cell ones. A typical foam as made, for instance, by blowing of polystyrene or polyurethane with

Table 1.1 Definition of terms used to characterise porous solids.

Term	Explanation
Porosity	Ratio of the pore volume to the volume occupied by the particles or fibrils
Pore shape	Ink bottle, cylindrical, funnel, or slit-shaped (see Figure 1.1)
Pore wall structure	Smooth or rough, closed or nanoporous
Pore accessibility	Closed, blind or through pores
Micropores	Size < 2 nm
Mesopores	2–50 nm
Macropores	> 50 nm
Density	Skeletal density: density of the solid network envelope density: mass per total volume (volume = solid phase + closed pores + open pores)

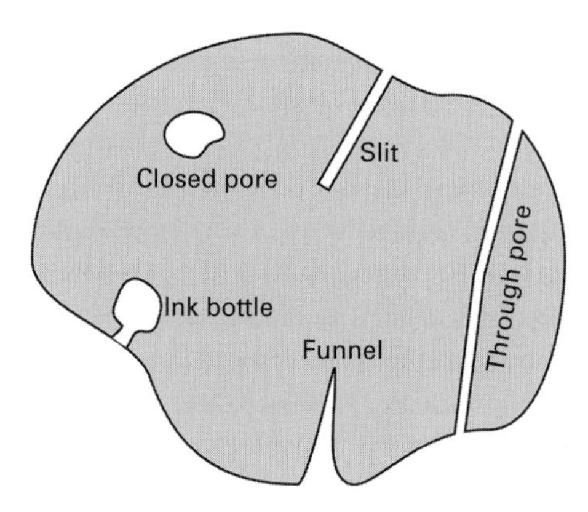

Figure 1.1 A schematic of a particle with various types of pores illustrating their definition and understanding.

a suitable gas consists of cell walls separating the pores. In an open porous foam, all pores are interconnected. An excellent overview about structure and properties of porous materials is provided in the textbook of Lorna Gibson and Michael Ashby [1]. The physical properties of a porous solid and its reactivity are effectively influenced by the kind, shape and size of the pores and of course the material it is made from. Some technical terms are defined in Table 1.1.

The term *aerogel* is by no means clearly defined in the literature. There are many definitions possible.

1. First, all materials prepared from wet gels by a special drying process, the supercritical drying technique, were called aerogels irrespective of their structural appearance and their properties. With the development of new drying techniques, this definition no longer appears appropriate.
2. According to an alternative definition, materials in which the typical structure of the pores and the network is largely maintained while the pore liquid of a gel is

replaced by air are called aerogels. However, it is not always clear to what extent the structure was maintained even in the case of supercritical drying. Rearrangement and shrinkage in the gel body during drying are normal phenomena!

3. Another definition used in this book would be that aerogels are open porous nanostructured solids obtained from a sol–gel process, irrespective of the way a wet gel is dried. This definition excludes many porous materials, which have pores in the micro- to hundred micrometer range and excludes porous materials made by some kind of foaming, templating techniques, formation of micelles and others. One can make the definition a bit more precise by requiring that an open porous material is only called an aerogel if the majority of pores are in the mesoporous range, meaning the pores are mostly below 50 nm.

The unique materials properties of aerogels result from the special arrangement of their solid network, which is schematically shown in Figure 1.2 for a SiO_2 aerogel.

The microstructures of aerogels are characterised by well-accessible branched mesopores and a solid network of nanoparticles or fibrils spanning the whole volume, meaning any branch of the network can be reached from any point. The pore structure of aerogels can be formed in different way: it can result from the nucleation or condensation of small (polymeric or colloidal) primary particles with a diameter of less than 10 nm or by the formation and aggregation of fibrils due to rearrangement of polymer strands or by phase separation in passing during polymerisation (polycondensation), a miscibility gap. Usually the formation especially of nanosized particles and their aggregation and network formation is controlled by a chemical process, the sol–gel process, and the underlying thermodynamics of the system (solvent and solute), leading for instance to a time-dependent miscibility gap and a possible phase separation process.

The sol–gel process is most easily explained by some terms related to it. In a *sol*, solid colloidal particles with diameters in the range of 1–1000 nm are dispersed in a liquid. A *gel* consists of a spongelike, three-dimensional solid network whose pores are filled with another substance (usually a liquid). The pore liquid mainly consists of water, but it also could be an alcohol or acetone or ionic liquids, salt-hydrate melts and others. The resulting ('wet') gels are therefore called aquagels, hydrogels, alcogels, acetogels etc. When the pore liquid is replaced by air without decisively altering the network structure or the volume of the gel body, aerogels are obtained. So-called cryogels or lyogels are formed when the pore liquid is removed by freeze-drying. A xerogel is formed upon evaporative drying of the wet gels, that is, by increase in temperature or decrease in pressure with concomitant large shrinkage and mostly destruction of the initially uniform gel body.

As previously mentioned, aerogels combine properties which make them attractive for applications: they are lightweight with densities in the range of 0.02–0.2 g/cm^3, have an extremely low thermal conductivity, typically in the range of 0.005–0.1 W/mK, a low speed of sound (typically around 100 m/s), huge specific internal surface area between 100 and 3000 m^2/g, can be brittle, but are elastically

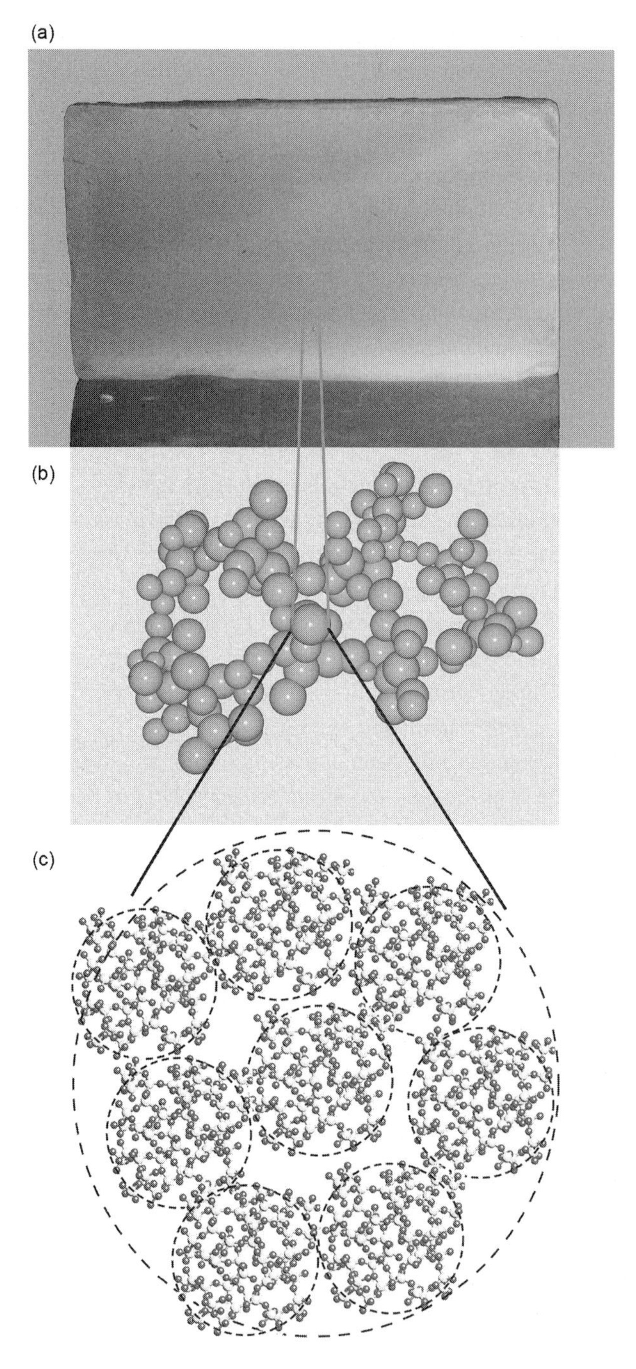

Figure 1.2 A photograph of a silica aerogel block. The typical appearance of such a tile illuminated from the front is a light blue, stemming from so-called Rayleigh scattering of the blue part of the light spectrum (looking throuh such a tile, it has a more yellow to orange appearance, since the blue light is scattered into the backward direction). Part (b) schematically illustrates an enlarged view of the particle skeleton building the solid network having itself a further substructure of very small amorphous silica particles in the range of a few nanometers. The structure of the amorphous silica particles (c) is taken from a figure of Lunt et al. [2], reproduced under Creative Commons licenses 4.0

deformable to a few percent, and certain aerogels can exhibit plastic deformation without breakage, can be made superflexible like soft rubber, have an adjustable Young's modulus, are typically electrically isolating but can be made electrically conductive like carbon aerogels, are excellent sound absorbers, are permeable to gases and humidity, can be functionalised in situ and then are especially useful as sensors or for catalysis or adsorptive filters. These properties are typically not all found in all types of aerogels, but they depend strongly on the type of aerogel and the way of chemical synthesis and drying procedure to get a suitable solid nanonetwork. Looking at either newspapers or many introductory remarks in scientific papers, aerogels are most often associated with lightweight or low density and especially with low thermal conductivity. But these properties are not always a preferred feature in several industrial applications, and it will be shown in this book that 'lighter' is not a property that goes along with, for instance, low thermal conductivity or high specific surface area. The core essence of aerogels is their tuneable properties, and these are found in a variety of shapes, from small beads, thin films to bulky pieces having lateral sizes in the metre range and thicknesses up to many centimetres.

1.2 Aerogel Classification

Aerogels can be divided into different classes according to their chemistry as inorganic or organic.

- Typical inorganic aerogels are all oxides, such as SiO_2, Al_2O_3, TiO_2, ZrO_2, niobates, tantalates, titanates or mixed oxides such as mullite, but also carbides such as SiC and SiOC or nitrides such as SiCN [3], and in the last decade even pure metal aerogels have been developed [4–7].
- The organic ones can be further subdivided into mainly five subclasses, biopolymeric, phenolic, protein based, polyole based and carbon aerogels derived from them (although they are then, strictly speaking, inorganic ones).
  - Biopolymeric aerogels are in the centre of research since just a decade. Typical aerogels developed are cellulosic ones, chitosan, alginates, starch, pectin, carrageenan and proteins.
  - The phenolic type aerogels are generally based, after their discovery by Pekala [8] on rescorcinol (a 1,3-dihydroxybenzene) or melamin, called RF and MF aerogels, but also mixed types such as RMF. At the turn of the millennium phenolic based aerogels were developed by Gross, Pekala and co-workers [9].
  - The protein-based aerogels came recently into the focus, although the corresponding hydrogels are known since centuries, for instance egg white.
  - The polyol-based aerogels are for instance all types of polyurethane aerogels, developed in the last decade, which are nowadays also industrially available [10].
  - Carbon aerogels are either made directly by pyrolysis of suitable precursors (RF, RMF, cellulose) or directly from graphene oxide as graphene aerogels.

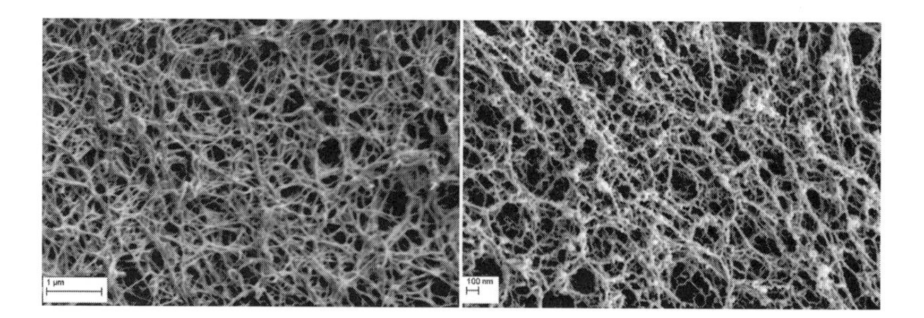

Figure 1.3 Scanning electron micrograph of a cellulose aerogel (left figure) exhibiting a nanofibrillar structure. Cellulose fibrils are interconnected yielding a three-dimensional network that also is able to transmit loads. The right figure shows a particulate aerogel prepared from resorcinol and formaldehyde. Courtesy of Ilknur Karadagli and Marina Schwan, DLR

Another way of classification would be based on the micro- or nanostructure. There are two different structures: a pearl-necklace-like chain of nanoparticles connected in 3D, as shown schematically in Figure 1.2 and in the right picture of Figure 1.3 or a network of nanofibrils in 3D, in which – in contrast to macroscopic fibre felts – the fibrils at their point of interconnection are chemically well bonded. An example is shown in the left picture of Figure 1.3. Almost all biopolymeric aerogels belong to the second class whereas the oxides, carbides, metallic and organic aerogels belong to the first class. There are, however, exceptions, such as whey protein aerogels [11].

Another classification would be the type of sol–gel processing. As mentioned previously, there are different ways of synthesis of aerogels.

1. One can start from monomers brought into solution, which then react with each other forming oligomers and particles generally via hydrolysis and polycondensation, which then connect in 3D to a gel network. This way of sol–gel processing, let us term it the classical routine, is the basis of all oxide aerogels and was already used by Kistler in his pioneering work on aerogels [12, 13].
2. Another way is to dissolve polymers, such as cellulose, carrageenan, pectins etc., to their polymeric level and rearranging the polymer strands in the highly diluted polymer solution to a new 3D structure of a gel by, for instance, cooling or neutralisation of the dissolving agent or for special processes even by heating. This unconventional sol–gel synthesis (sol–gel, since a solution is prepared and gelation induced) is typical for biopolymeric aerogels.

A recently published comprehensive review on all aspects of aerogels is provided in the *Handbook Aerogels* [14]. Older, but still enjoying excellent reviews, are the works provided by Hüsing and Schubert [15] and Pierre and Pajonk [16].

In this book, we concentrate on the chemical synthesis and processing of four different aerogels: two exhibiting a particulate microstructure (silica and RF) and two showing a fibrillar one (cellulose and alginate). These prototypes also serve to explain both the chemical synthesis routes valid for many other aerogels. Since the first critical step in making aerogels is gelation, the formation of a wet gel, we discuss the experimental techniques to determine the so-called gel point and models of gelation. As mentioned earlier, the drying of a wet gel leads to aerogels, and therefore all techniques described in the literature or used industrially to prepare aerogels from wet gels are thoroughly discussed. The properties of aerogels are sometimes unique and require a specialised knowledge on, for instance, gas transport in nano- to mesoporous solids, heat conduction in such materials or their mechanical response. We present an in-depth discussion of all aspects of the relation between microstructure and properties such as density, specific surface area, porosity, diffusion, permeability and thermal and mechanical response. Finally we present some well-proven and tested recipes to prepare aerogels in a chemical laboratory.

1.3 A Brief History of Aerogels

Before we proceed, it makes sense to briefly review the history of the discovery of aerogels. Looking at, for instance, the Web of Science or Scopus and asking to look for aerogels as a keyword, both easily produce a graph with the number of papers having the term *aerogel* either in its headline or as a keyword, and it readily is found that the number of papers on aerogels has increased exponentially in the last decade and a half. Thus it looks as if aerogels are a modern and recent invention. This is, however, not true. The first aerogels were prepared already in 1931 [12]. Steven S. Kistler, professor at the College of the Pacific in Stockton, California, was able to remove the liquid from a wet gel without damaging the solid component. He discovered the key aspects of aerogel production and coined the term 'aerogel':

> Obviously, if one wishes to produce an aerogel, he must replace the liquid with air by some means in which the surface of the liquid is never permitted to recede within the gel. If a liquid is held under pressure always greater than the vapour pressure, and the temperature is raised, it will be transformed at the critical temperature into a gas without two phases having been present at any time. [13]

It is not astonishing, that the first gels studied by Kistler were silica gels, since they were known for centuries to produces wet gels easily from a sol of aqueous sodium silicate. He made attempts to directly convert the wet gels with water as a pore fluid into dry aerogels by going beyond the supercritical point of water, 374.12°C and 22.1 MPa pressure, but he failed to do so, since supercritical water dissolves silica. He then tried to exchange water with an organic fluid that was miscible with water by first thoroughly washing the silica gels with water (to remove salts from the gel) and then exchanging the water for alcohol. Putting the wet gel (now an alcogel) in a high-pressure vessel and heating up beyond the critical point of the alcohol, which is for methanol 239.35°C with a pressure of 8.084 MPa, and allowing the alcohol to escape slowly from the vessel, he was able to prepare the first aerogel. The silica aerogels produced in this way were transparent (see Figure 1.3), had a low density and were highly porous. Kistler then thoroughly characterised his silica aerogels, and already prepared aerogels from many other materials, including alumina, tungsten oxide, ferric oxide, tin oxide, nickel tartarate, cellulose, cellulose nitrate, gelatine, agar, egg albumen and rubber. Organic aerogels came into focus especially in the last decade. Although Kistler already commercialised silica aerogels (he built a factory in Chicago with huge pressure vessels) and produced aerogel granulates, which were used in refrigerators but also as thixotropic agents in cosmetics and toothpaste, his invention of aerogels was forgotten, probably due to two other inventions, the easy and cheap production of fumed silica, which also is a thixotropic agent and widely used in industry today, and secondly the invention of polystyrene and poylurethane foams, which are also lightweight and can have a very low thermal conductivity depending on the foaming gas used. They are much better suited for large-scale industrial production than aerogels produced by Kistler's route.

Therefore, aerogels had been largely forgotten until Stanislaus Teichner at Universite Claud Bernard, Lyon, discovered another way to produce aerogels. His technique was a major breakthrough and paved the way for sol–gel chemistry to be applied to aerogels. In his process, he replaced the sodium silicate used by Kistler with an alkoxysilane, (tetramethylorthosilicate, or TMOS). Hydrolysing TMOS in a solution of water and methanol produced an alcogel in one step which under supercritical alcohol conditions allowed to produce high-quality silica aerogels and not only granules but also tiles of considerable size (the largest being made years later by Airglass in Sweden had a size of $90 \times 90 \times 6$ cm^3). In subsequent years, Teichner's group and others extended this approach to prepare aerogels of a wide variety of metal oxide aerogels [17].

After this discovery, new developments occurred rapidly, and the first aerogel conference was held in 1985 (and proceedings published in 1986, [18]), and new companies were founded producing aerogels. Since silica aerogels are transparent as glassy silica but are also open porous with porosities in the range of 90–99%, their refractive index is much less than that of silicate glasses. Typical refractive indices are in a range slightly above that of air ($n = 1.0003$). They are therefore well suited to fill the refractive index gap between air (index 1) and most solids and liquids starting at around 1.3 up to diamond, having a refractive index of 2.4. The refractive index of silica aerogels depends linearly on their density. Therefore, the first large-scale application and production of silica aerogels was for elementary particle physics detectors. A so-called Cherenkov counter is used to measure the speed of a beam of elementary particles. If an elementary particle (protons, electrons, positrons, anti-protons etc.) passes through a material at a velocity greater than that at which light can travel through such a medium (this is given by the vacuum speed of light divided by the refractive index of the material), then light is emitted in a cone whose opening angle allows determination of the speed at which the particle has passed the medium. Such a detector requires huge quantities of high-quality silica aerogel tiles. The first application was at the so-called TASSO detector at the DESY in Hamburg using 1700 L of silica aerogels (TASSO started to search for gluons, the bonding field particles between quarks, in 1979) and one at CERN in Geneva using a 1000 L detector. These aerogel tiles were produced at the University of Lund in Sweden. Soon afterwards, people from Lund founded the company Airglass Corp., which operated its business until around 2005 and produced silica tiles of around 10 × 10 inches with a thickness of around 1 inch and variable refractive index using the TMOS technology invented by Teichner and co-workers.

Since methanol is toxic and its usage underlies severe regulations, several groups developed processes to produce high-quality silica aerogels using tetraethylorthosilicate (TEOS) instead of TMOS and ethanol as a solvent. Replacing also the direct supercritical drying of the solvent (methanol, ethanol) by replacing first the liquid in the pores with carbon dioxide allowed use in much less hazardous conditions of supercritical drying, since CO_2 has a critical point of only 31.04°C at a pressure of 7.3 MPa. These developments made essentially in the Microstructured Materials Group at Berkeley Lab in the early 1980s promoted further academic and industrial research. In the late 1980s, researchers at Lawrence Livermore National Laboratory (LLNL) led by Larry Hrubesh prepared the world's lowest-density silica aerogel, which had a density of 0.003 g/cm^3, only three times that of air. BASF in Germany produced until 1996 a so-called BASOGEL made from sodium silicate and solvent exchange via supercritical drying with carbon dioxide. In 1992, Hoechst Corp. in Frankfurt, Germany, also began a program in low-cost granular aerogels also using sodium silicate as a starter. After a restructuring of Hoechst in the mid-1990s, Hoechst sold all patents on aerogels to Cabot (Boston, USA), and now Cabot produces on the former Hoechst site in Frankfurt, huge quantities of silica aerogel granulates essentially for thermal isolation purposes but also as a translucent material in roof windows.

Another breakthrough in the science of aerogels was also made at the LLNL in 1989 by Rick Pekala, who developed an organic aerogel made from resorcinol and formaldehyde. This aerogel was as light as silica aerogels, but opaque and with an even lower thermal conductivity. The essential issue, however, was that it could be converted by pyrolysis under a protective atmosphere to a carbon aerogel, which then is electrical conducting. Soon after this discovery, carbon aerogels were used as electrodes for batteries and supercapacitors. They outperform conventional carbon black. A company was founded around the turn of the millennium, which is now part of the Cooper Bussmann company. The company still produces carbon-aerogel-based capacitors. The development of aerogel-based batteries is still an active area of research.

Another important breakthrough in gel processing leading to aerogels was made at the University of New Mexico. Jeff Brinker and Doug Smith showed 1995 that one can eliminate the supercritical drying step by chemically modifying the inner surface of the gel prior to drying. Their idea was to attach hydrophobic trimethylchlorosilane groups to the inner surface of silica aerogels. This led to the discovery of the so-called springback effect, meaning on evaporative drying the gel first shrinks, but since the methyl groups at the pore walls of the gel repel each other, the dry gel elastically comes almost back to its original shape and then is an aerogel without supercritical drying.

In 2003, Aspen developed a technology to combine a wet silica gel with a fluffy fibre mat (glass or polymeric fibres) and supercritical drying of the composite gel to yield continuously produced aerogel fibre mats, which can be handled and used as the well-known rock or glass wool in thermal insulation applications, but with a much lower thermal conductivity of around 0.015 W/Km.

In the last decade and a half, organic and biopolymeric aerogels became popular and are now a focus of research. Polyurethane and polyimide aerogels were prepared having exceptional thermal and mechanical properties [19–21]. The starting point for biopolymers could have been the cellulose aerogels prepared by Tan and co-workers in 2001 [22]. Their work became well known and was mentioned even in newspapers, since they measured the impact strength of the cellulose aerogel sheets (5 mm thick) and could show that although their material had a high porosity, its strength exceeds that of resorcinol-formaldehyde (RF) aerogels. Since then, aerogels from alginates, carrageenan [23], starch, pectins and egg white have been synthesised mainly with a focus on food and medical applications [24–26].

The last decade is characterised by the development of routes to prepare so-called hybrid aerogels, meaning two different types of gels are mixed, such as silica with cellulose or pectins, or RF aerogels with silica, chitosan-silica etc. [25, 27–33] to optimise properties for applications.

2 Chemical Synthesis of Aerogels from Monomeric Precursors

A wide class of aerogels start from a solution of monomers. They are produced in three steps: formation of a solution to prepare a colloidal sol, gelation and drying. Prior to drying, typically always first a so-called ageing procedure is necessary which allows monomers and oligomers not yet built in the solid network to react with it and stiffen the network. Then, secondly, a washing procedure is necessary to both wash out the non-reacted chemicals and also to replace the gel fluid with a fluid that can be exchanged with that fluid used for supercritical drying, most often today, carbon dioxide. A new way, recently becoming again a focus of research, is performing gelation and drying in one step using a high-pressure vessel avoiding washing and a solvent exchange procedure. In the following two sections, we discuss the synthesis of silica aerogels, the most prominent, classical and standard one and an organic aerogel discovered 30 years ago, the resorcinol-formaldehyde aerogels.

2.1 Silica Aerogels

Silica aerogels are the most prominent of all aerogels. They were the first prepared by Samuel Kistler [12, 13], the first which were and are produced on an industrial scale either as granules [34] or as a fibre-reinforced felt [35] and they are still intensively investigated, since the process of preparation still has room for improvement, especially with respect to cost reduction.

Silica, SiO_2, is the most abundant material on earth: it is almost everywhere in the soil, deserts, beaches. It is mostly found in combination with other oxides. Almost any stone contains silica. How is an aerogel made out of silica? Formally, one could think of how expanding a piece of amorphous or crystalline silica by a factor of, let us say, 100 could yield an open porous (nano)structure. This indeed can be done in molecular dynamic simulations and yields structures which look very much like silica aerogels, as was shown by for instance Patil and others [36–38]. A real-world approach, however, does it chemically. Two different experimental routes have been established yielding aerogels in the last four decades: one starts with water glass, being already a colloidal solution of amorphous silica, and the second one uses alkoxides. Before we describe the reactions and mechanisms leading to silica gels and upon drying to aerogels, let us summarise in Table 2.1 the typical properties silica aerogels possess.

Table 2.1 Important structural properties of SiO_2 aerogels

Property	Units	Range	Typical value
Bulk density	g/cm^{-3}	0.003–0.500	0.100
Skeletal density	g/cm^{-3}	1.700–2.100	
Porosity	%	80–99.9	
Average pore diameter	nm	20–150	
Specific surface area	m^2/g	100–1600	600
Refractive index		1.003–1.24	1.02
Thermal conductivity	W/(m K) in air, 300 K	0.012–0.021	0.016
Modulus of elasticity	MPa	0.002–100	1
Sound velocity	m/s	<20–800	100
Acoustic impedance	$kg/(m^2\ s)$		10^4

The procedures leading to silica aerogels can and were also used to prepare (semi-)metal oxides, gels and aerogels. Therefore, the chemistry of silica aerogels outlined in this chapter is prototypic for many other aerogels. Excellent reviews on especially silica aerogels and certain aspects of their preparation, properties and applications were written in the past, and any reader is advised to look at them for more information and a huge list of relevant literature at the time they were published [14–16, 39–42].

2.1.1 Silica Aerogel Precursors

Two precursors for silica aerogels are used: sodium silicate and alkoxides. Before discussing both approaches, let us define a few terms:

- The term *silane* describes chemical compounds having the general formula Si_nH_{2n+2} and that are therefore formally similar to alkanes (in which Si is replaced by C). The simplest silane is SiH_4, but two others are important in aerogel synthesis: tetramethoxysilane $Si(OCH_3)_4$, TMOS and tetraethoxysilane $Si(OC_2H_5)_4$, TEOS. These are also called alkoxides, and in principle one can use alkoxides of higher alcohols such as tetrabuthoxysilane.
- The term *siloxane* always describes compounds in which there is an oxygen bridge between two Si atoms, such as ≡Si–O–Si≡.
- The expression *silanol* describes a ≡Si–OH group; a hydroxy group is bound to a silicon atom.
- The term *silicic acid* simply means $Si–(OH)_4$.

Sodium Silicate as a Precursor

A classical way to make silica gels and therefrom aerogels is the utilisation of water glass or more precisely sodium silicate. The general chemical composition of sodium silicates is $Na_{2x}Si_yO_{2y+x}$, which can be taken as a mixture of sodium oxide and silicon oxide. Typically, sodium silicate is characterised by the ratio of SiO_2 to Na_2O.

The classical way to make sodium silicate is to mix quartz sand with sodium carbonate and heat the mixture in a furnace at temperatures between 1100 and 1300°C, such that a sodium-silica glass results. Milling the glass to a powder and dissolution in water at 150°C and at higher pressure, around 5 bars, leads to a colloidal liquid, called water glass. This typically is alkaline in character and stable under neutral and basic conditions, since the colloidal silica particles are all negatively charged and therefore repel each other.

There are many variants of colloidal silica available and produced industrially in huge amounts. Water glass is strictly speaking not a monomeric precursor for aerogels, but already a nanosized particle-loaded dispersion, a *sol*. One of the most significant problems associated with utilising sodium silicate is the contamination of silica with residual sodium. There are several methods for separating sodium from sodium silicate-derived silica as already mentioned by Zsigmondy [43]. According to him, Graham [44] diluted sodium silicate and reacted it with an acid to make silica and an aqueous salt solution. The solution is then dialysed to get rid of the salt or cooked to increase the colloidal silica concentration. Depending on the dilution, the residual colloidal silica solution can be stable or gels after hours or weeks. The salts may also be removed by adding an organic solvent to precipitate salt crystals, which are removed by decanting or centrifugation. Another approach is to feed sodium silicate into an acid ion-exchange bed which exchanges the sodium ions with protons, providing an outlet stream of silicic acid. Gelation typically occurs in the ion-exchange resin at the reaction front, where the strong acid-base neutralisation occurs. Therefore, this technique is not simple to use, since gelation of silica occurs rapidly when the pH is lowered to about 6 (as discussed later). In any case, having succeeded in preparing a silica sol being almost free of sodium allows one to make a gel directly by pH reduction, adding an acid such as HCl, neutralising the charge on the nanoparticles and thereby allowing aggregation to take place and forming clusters, which again can aggregate eventually forming a gel. This gel then can be further treated to yield an aerogel by either evaporative or ambient drying (see Section 5.1) or supercritical drying (see Section 5.3); typically, before drying, any acid in the gel fluid has to be removed.

2.1.2 Monomeric Precurors

As a starting point to make silica gels, one could think to take silicic acid Si–$(OH)_4$. Silicic acid is a very weak acid which is prone to polymerisation. The solubility in water is not higher than 120 ppm at room temperature and pH values below 7. With increasing pH value in the basic regime, up to 1000 ppm are dissolvable. The solubility depends on temperature and increases from around 100 ppm at 20°C to 350 ppm at 100°C. The polymerisation of silicic acid readily happens, making dimers, trimers and rings. All these oligomers are in principle siloxanes such as $(HO)_3$Si–O–Si–$(OH)_3$ for the dimer and $(HO)_3$Si–O–Si$(OH)_2$–O–Si–$(OH)_3$ for the trimer. Imagine a molecule of silicic acid as a tetrahedron: in its centre is the silicon atom and at the four corners are OH groups. A dimer is then made of two tetrahedra oriented such that they are top-on-top and the two OH groups have created a siloxane bond and left a water molecule. For a trimer, several corner positions are available on a dimer, in principle six. A tetramer has even more possible positions, but one that is energetically favourable is already one which closes a ring of four tetrahedra. Therefore, the condensation of silicic acid does generally not lead to linear chains of ≡Si–O–Si≡, but to rings, and from those easily to larger rings and particles, being more or less compact. Important in these reactions is that always siloxane bonds are formed and reactive silanol groups are always at the inner and outer surface of oligomers (rings). Iler [45] made the destinction between an oligomer and a silica particle by defining that particles have in their interior the anhydrous form SiO_2 and only at their surface are silanol groups (in contrast to oligomers).

The polymerisation reaction of silicic acid is strictly speaking a polycondensation reaction. All silanols tend to produce an Si–O–Si bond, which has as a by-product, water. In the reaction, formally one might think that not only monomers react with each other but also dimers with monomers and dimers, trimers with dimers and monomers, etc. This does, however, not occur. Iler discusses this extensively and shows that always monomers are added to existing dimers, trimers or oligomers. This is the classical way of a step-growth polymerisation.

As mentioned in Chapter 1, Teichner and co-workers [17] developed an alternative route to prepare aerogels from metal or semi-metal alkoxides (in principle he was not the first using alkoxides. Grimaux described the process already in 1884 [46]). Alkoxides have the general formula $R\text{-}O^-$ in which R denotes an aliphatic rest, for instance $-CH_3$ or CH_2CH_3- for the methyl or ethyl group. In the case of interest here, the alkoxy group is bonded to a silicon atom, which always means four of the alkoxy groups are needed. Tetramethoxysilane (TMOS) and tetraethoxysilane (TEOS) are both liquids at room temperature. Whereas TMOS has a density slightly larger than water (1.032 g/mL), that of TEOS is slightly below (0.933 g/mL). Both are dissolvable in organic solvents, especially methanol for TMOS or ethanol or 2-propanol for TEOS. Addition of water induces a reaction called hydrolysis, meaning the alkoxy group is replaced by a hydroxy group, such that formally if the reaction would go to completeness silicic acid is produced. We come to this point in more detail later. Instead of four alkoxy groups at a silicon atom, there are other precursors which have been especially used in the last decade. These precursors have lower functionality, meaning one or two of the bonds at a silicon atom are blocked by simply hydrogen, a methyl, ethyl or vinyl group. Such precursors are for instance methyltrimethoxysilane (MTMS) or dimethyldimethoxysilane (DMDMS). Figure 2.1 shows structures of these molecules. Of course, one can replace in MTMS and DMDMS the methyoxy groups by ethoxy groups and the methyl groups by ethyl ones. Changing the size of the alkoxy group typically has an influence on the reaction rates, for instance the hydrolysis rate might be lower.

2.1.3 Hydrolysis and Polycondensation

The first step of the preparation of gels from alkoxide precursors is their dissolution in a mixture of water and an alcohol and adjusting the pH value of the solution by

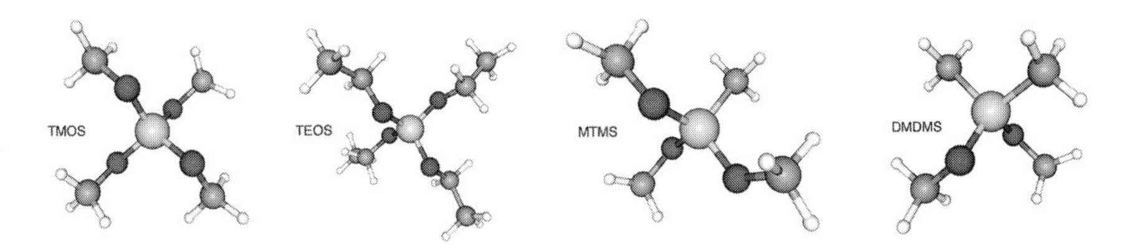

Figure 2.1 Three-dimensional pictures of the four important alkoxides used for aerogel synthesis. The central bigger atom in light grey always is silicon, surrounded always by four dark grey oxygen atoms in the case of TMOS and TEOS with either methyl or ethyl groups attached to them. In MTMS, three methoxy groups are attached and a methyl group directly at the central silicon atom. In DMDMS, only two methyoxy groups are bonded to silicon and two methyl groups directly at the silicon. The hydrogen atoms are coloured as almost white.

an acid or a base. Two effects occur then: hydrolysis and polycondensation leading to oligomers, particles, aggregates of particles and eventually a cluster spanning the whole volume (for more details on gel formation, see Chapter 4). The overall process was excellently visualised with a graphic going back to Iler [45] and is since then found almost everywhere in the literature on sol–gel processing. We reproduce it in Figure 2.2, and will then discuss the processes leading to gels.

The first process that always occurs once alkoxides come into contact with water is their hydrolysis, described by the following reaction scheme:

$$\equiv \text{Si}{-}\text{OR} + \text{H}_2\text{O} \quad \rightleftarrows \quad \equiv \text{Si}{-}\text{OH} + \text{ROH}. \tag{2.1}$$

The sequence from the left to the right is the hydrolysis, whereas the back reaction is called esterification. Once an $\equiv$ Si $-$ OH is formed, it can react with any other alkoxide described formally by the following reaction scheme:

$$\equiv \text{Si}{-}\text{OR} + \text{Si}{-}\text{OH} \equiv \quad \rightleftarrows \quad \equiv \text{Si}{-}\text{O}{-}\text{Si} \equiv + \text{ROH}, \tag{2.2}$$

called an alcohol condensation (the reverse reaction could be called alcoholysis). A third reaction does not lead to an alcohol, but water,

$$\equiv \text{Si}{-}\text{OH} + \text{Si}{-}\text{OH} \equiv \quad \rightleftarrows \quad \equiv \text{Si}{-}\text{O}{-}\text{Si} \equiv + \text{H}_2\text{O}, \tag{2.3}$$

and might be called a water condensation. The last two condensation reactions lead to a siloxane bond. These three fundamental reactions show that the essential step is the hydrolysis of the silicon alkoxide. There are many factors influencing these reactions:

- The pH value (also called in the literature the amount of catalyst)
- The type of molecular precursors used (type of alkoxide)
- The ratio of water-to-alkoxide (R_W)
- The solvent used and more specifically the dilution ratio (amount of water + alcohol divided by the amount of alkoxide)
- Temperature
- Additives, especially surfactants but also salts and other electrolytes

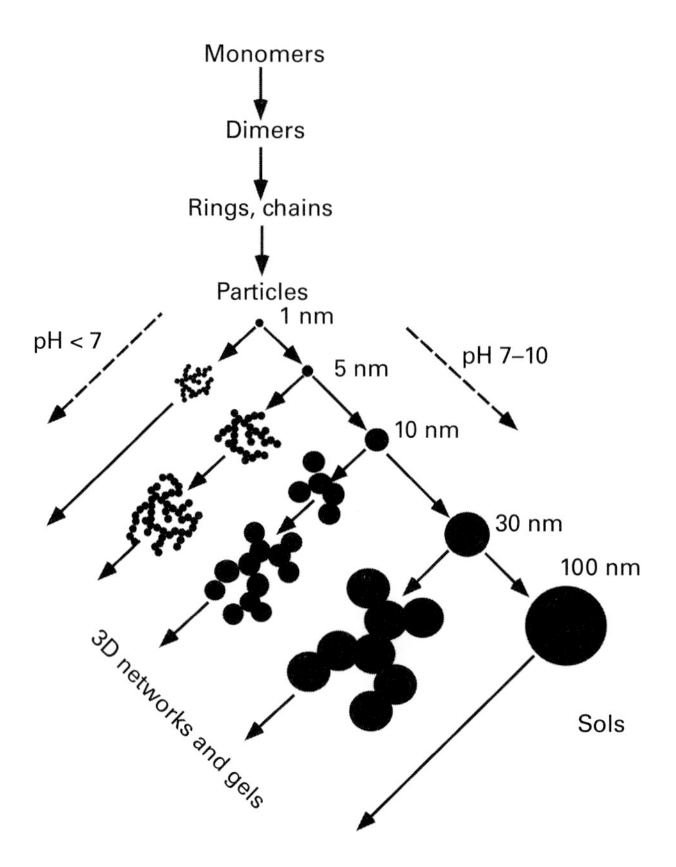

Figure 2.2 Scheme of the reactions leading under acidic conditions from monomers of, for instance, silicic acid or silicon alkoxides to oligomers, particles, open networks of particles or under base conditions to larger particles which also can aggregate to open porous networks or simply sols of very large particles in the range of 100 nm and above. In this sketch, it is assumed that the solution does not contain salts, since cations change the reaction processes. Redrawn from Iler's original picture [45] with permission from John Wiley & Sons

We are not going through all details but trying to concentrate on a few issues and give a few examples. The interested reader can find many more details in the cited book of Ralph K. Iler and the newer book of Jeff Brinker and George Scherer [3].

Let us first point to a simple issue, namely the solubility of alkoxides in water and alcohols. As mentioned previously, alkoxides are immiscible in water but are dissolvable to a large extent in alcohols. Therefore, typically a ternary solution is prepared (TMOS or TEOS with water and an alcohol, most often TMOS in methanol and TEOS in ethanol). Unfortunately, not that much thermodynamic data are available. Only a few research groups experimentally determined the ternary phase diagram and made an isothermal section through it. A new work of Wang and co-workers [47] shows in addition to the solubility lines given by Brinker and Scherer [3] that using pre-polymerised silica changes the solubility considerably. Figure 2.3 shows an isothermal section through the phase diagram of TEOS-ethanol-water. To imagine

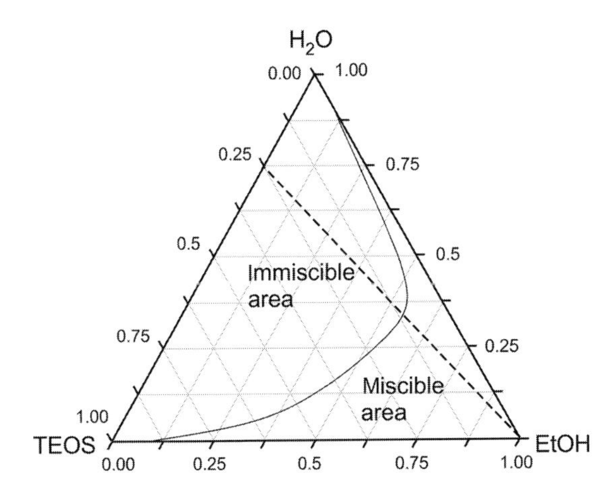

Figure 2.3 Isothermal section at 25°C of the ternary system TEOS–ethanol–water. A broad miscibility gap exists. To make a homogeneous solution of the three components, water and ethanol must be added in huge quantities.

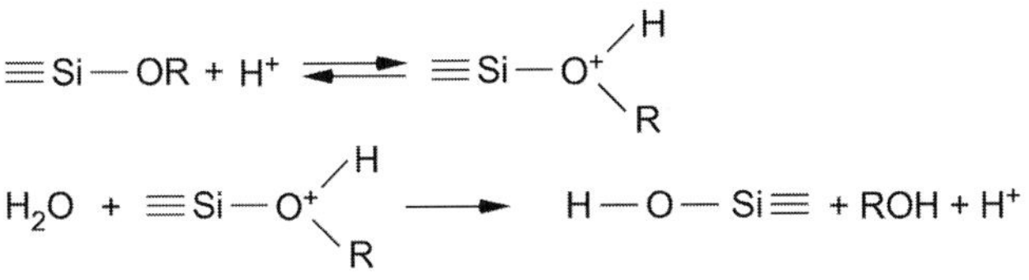

$$\equiv Si-OR + H^+ \rightleftharpoons \equiv Si-O^+(H)(R)$$

$$H_2O + \equiv Si-O^+(H)(R) \longrightarrow H-O-Si\equiv + ROH + H^+$$

Figure 2.4 Under acidic conditions, a silicon alkoxide is protonated and forms an intermediate. In the second step, water reacts with this intermediate leading to a replacement of the alkoxy group at the silicon and the formation of the alcohol. The proton (H^+) is again available for further reactions.

the meaning of the miscibility gap, draw a straight line from the right-lower corner to the TEOS-water side as sketched with the dashed line. Along this line, the molar ratio of TEOS and water is constant. Going down the line from the TEOS-H_2O side means a continuous increase in amount of ethanol, and the system stays inside the gap until it reaches the boundary line at approximately 0.33 mol fractions water, 0.54 mol fractions ethanol and the rest is TEOS. The shape of the gap reproduces that TEOS and water are immiscible and ethanol and water are largely miscible (they exactly build an azeotropic system with an azeotropic point close to pure ethanol). Therefore, preparing a homogeneous solution of TEOS and water requires use of an amount of ethanol such that the whole system is inside the miscible region.

As mentioned already, having a solution of water, alcohol and an alkoxide always initiates the hydrolysis reaction, meaning the alkoxy group on the silicon atom is replaced by an hydroxy group. Under acidic conditions, the reaction follows the scheme given in Figure 2.4, and under basic conditions it follows the scheme shown in Figure 2.5. Under acidic conditions, pH smaller than 7, a silicon alkoxide is protonated and forms an intermediate in which the oxygen atom bonded to silicon gets a positive charge and attached are a hydrogen atom and the alkoxy group. In the

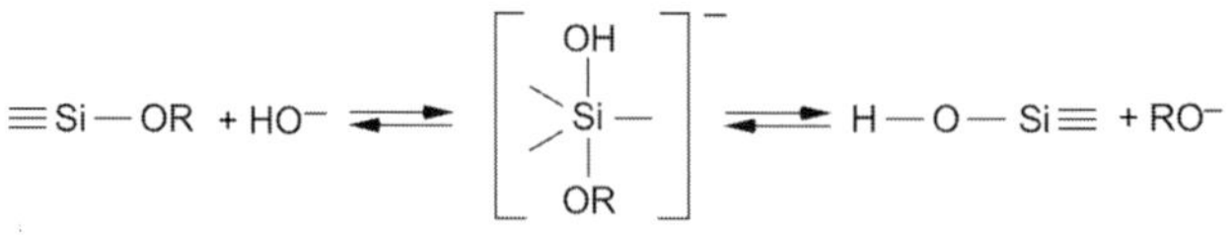

$$\equiv Si{-}OR + HO^- \rightleftharpoons \left[>Si(OH)(OR){-} \right]^- \rightleftharpoons H{-}O{-}Si\equiv + RO^-$$

Figure 2.5 Under basic conditions, the alkoxide reacts with the negatively charged hydroxy group and forms an intermediate, which decomposes into the hydrolysed alkoxide and a negatively charged alkoxy group.

second step, water reacts with this intermediate leading to a replacement of the alkoxy group at the silicon and the formation of the alcohol, and the proton (H^+) is again available for further reactions. Therefore, one can also talk about the acid as a catalyst, since it leaves the reaction unchanged. Under basic conditions, pH larger than 7, the alkoxide reacts with the negatively charged hydroxy group and forms an intermediate in which even the valency of silicon is increased. This intermediate decomposes into the hydrolysed alkoxide and a negatively charged alkoxy group.

The rate of the hydrolysis reaction depends on the concentration of the catalyst, both under acidic and basic conditions. The rate constant of the reaction decreases with increasing pH value and has a minimum around pH 7. This means that under acidic conditions, the hydrolysis rate decreases with increasing pH and increases above pH 7 the more basic it becomes. The reaction can be described as a first-order one [3, p. 123]. There are several parameters modifying the hydrolysis reaction. First there is steric effect, meaning the reaction rate depends on the length of the alkoxy group. Let Me denote the methyl group, Et the ethyl group and ^{i}Pr the isopropyl group. Then the reaction rate orders such as $Si(OMe)_4 > Si(OEt)_4 > Si(O^iPr)_4$. There is another electronic or inductive effect on the rate of hydrolysis. The electron density at the silicon atom varies, depending on the type of group attached to it. If an alkoxy group is there (≡Si–OR), the electron density at the silicon atom is larger compared to an OH group, and this is larger compared to ≡Si–O–Si≡. Under acidic conditions, this means that the reaction rate decreases the lower the electron density at the silicon atom is, and vice versa under basic conditions. As pointed out by Brinker and Scherer [3], this order has the following effect. With each hydrolysis step (meaning replacement of one OR by an OH group), the hydrolysis rate decreases under acidic conditions, whereas under basic conditions each subsequent hydrolysis step occurs more quickly in as much as hydrolysis (and condensation) proceeds. Another parameter affecting the hydrolysis reaction is of course the concentration of water, expressed as the molar ratio of water to $Si{-}(OR)_4$, R_w. In principle, since water is a by-product of the condensation reaction, a ratio $R_w = 1$ would be sufficient to replace all alkoxy groups by hydroxy ones, leading to silicic acid. A ratio of $R_w = 2$ is needed to completely transform all Si-alkoxides to SiO_2. In practice, a lower amount of alkoxide, $R_w < 1$, leads to more Si–OH instead of Si–O–Si. But even if $R_w > 2$, not only Si–O–Si is formed, but mixtures of the kind $[SiO_x(OH)_y(OR)_z]_n$ with $2x + y - z = 4$. This simply means that whatever one does during preparation of a gel from a mixture of TMOS or TEOS and water and ethanol, a silica network made via a sol–gel process will always contain a certain amount of alkoxy and silanol groups. The R_w ratio has been varied in many

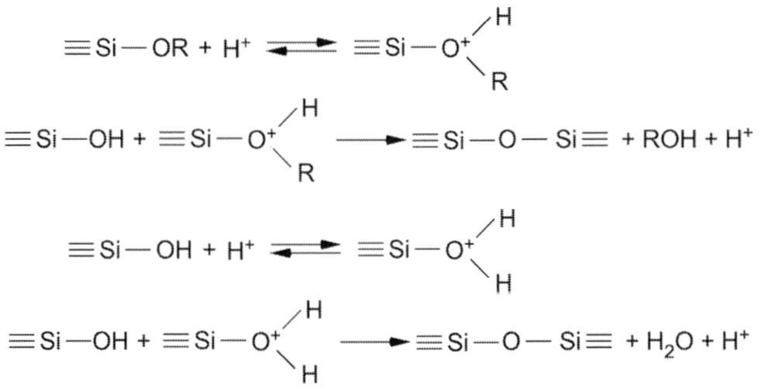

$$\equiv Si{-}OR + H^+ \rightleftharpoons \equiv Si{-}O^+{<}^{H}_{R}$$

$$\equiv Si{-}OH + \equiv Si{-}O^+{<}^{H}_{R} \longrightarrow \equiv Si{-}O{-}Si\equiv + ROH + H^+$$

$$\equiv Si{-}OH + H^+ \rightleftharpoons \equiv Si{-}O^+{<}^{H}_{H}$$

$$\equiv Si{-}OH + \equiv Si{-}O^+{<}^{H}_{H} \longrightarrow \equiv Si{-}O{-}Si\equiv + H_2O + H^+$$

Figure 2.6 Under acidic conditions, a silicon alkoxide or a partially hydrolysed one is protonated and forms an intermediate, which then reacts either with a silicon alkoxide or a partial hydrolysed one building a siloxane bond. Either water or the alcohol is a by-product.

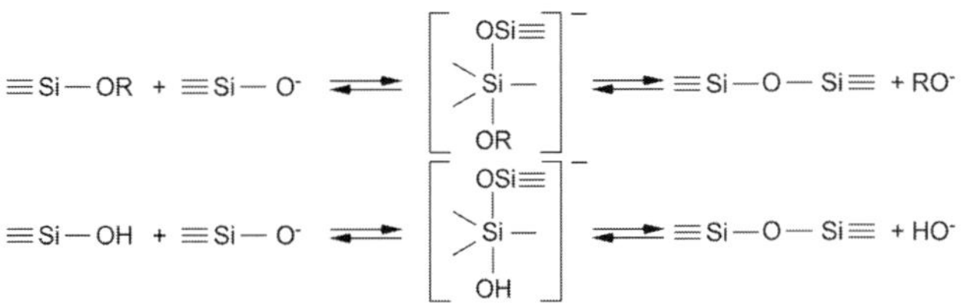

$$\equiv Si{-}OR + \equiv Si{-}O^- \rightleftharpoons \left[>Si{-}(OSi\equiv)(OR) \right]^- \rightleftharpoons \equiv Si{-}O{-}Si\equiv + RO^-$$

$$\equiv Si{-}OH + \equiv Si{-}O^- \rightleftharpoons \left[>Si{-}(OSi\equiv)(OH) \right]^- \rightleftharpoons \equiv Si{-}O{-}Si\equiv + HO^-$$

Figure 2.7 Under basic conditions, the silicon alkoxide or a partially hydrolysed specimen forms an intermediate, which decomposes into a siloxane-bonded silicon and a negatively charged alkoxy or hydroxy group.

investigations throughout the literature to a large extent, the gel time was measured and concentrations of the various species $Si\text{–}(OH)_n$, $Si\text{–}(OH)_n(OR)_m$ were measured by nuclear magnetic resonance (NMR). Their time depending variation showed that the larger R_w, the more rapid is the hydrolysis.

Having hydrolysed species, the formation of siloxane bonds (condensation) takes place and runs concurrently to hydrolysis, which is evident looking at Eqs. (2.2) and (2.3). The condensation reactions follow schemes given in Figures 2.6 and 2.7 under acidic and basic conditions. Under acidic conditions, a silicon alkoxide or a partially hydrolysed one is protonated and forms an intermediate in which the oxygen atom bonded to silicon gets a positive charge and attached to it a hydrogen atom and the alkoxy group. In the second step, the intermediate reacts either with a silicon alkoxide or a partially hydrolysed one and builds a siloxane bond with either water or an alcohol as a by-product. Also, for the condensation reaction, H^+ is unchanged during the reaction. Under basic conditions, pH larger than 7, the silicon alkoxide or a partially hydrolysed specimen reacts with a negatively charged silicon alkoxide and forms an intermediate in which even the valency of silicon is increased. This intermediate decomposes into siloxane-bonded silicon and a negatively charged alkoxy or hydroxy group.

The rate of the condensation reaction also depends on the pH value, and it was observed that the condensation is a bimolecular reaction of second order. With increas-

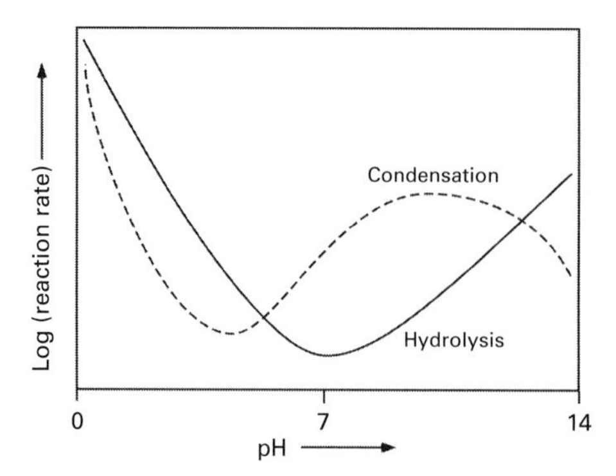

Figure 2.8 Schematic of the pH dependence of the reaction rates for hydrolysis and condensation of silicon alkoxides in water-alcohol solutions. Whereas the hydrolysis has a distinct minimum, the condensation reaction slows down at higher pH values due to dissolution of already formed siloxane bonds.

ing pH value, the rate constant decreases to a minimum around pH 4 and then increases again until pH 7. The condensation reaction under basic conditions is, as indicated in the preceding schemes, always accompanied by possible reverse reactions, meaning here that siloxane bonds formed can also be broken. This is especially important the larger the pH. At pH values above 8, the solubility of silica increases considerably until pH 14. This means the higher the pH, the lower the stability of siloxane bonds. Thus the overall condensation rate does not simply increase in as much as the pH is increased, but exhibits a maximum around pH 8–9. This contrasts to hydrolysis, which only has a minimum at pH 7 but increases towards lower or larger pH values. In total, one can draw a somewhat simplified diagram shown in Figure 2.8.

2.1.4 Growth and Structure

Once dimers, trimers and oligomers exist, the question arises, how do they grow with hydrolysis and condensation reactions happening simultaneously? The decisive parameter for growth to larger polymers or particles is again the pH value. Under acidic conditions, further condensation preferably occurs at the terminal silicon atom, since the acidity of any hydrolysed species increases with the number of siloxane bonds formed. A silicon atom surrounded by three siloxane bonds and one hydroxy group is more acidic than one with two or more hydroxy groups. The consequence is that once a silicon appears only bonded by siloxane bonds to neighbouring silicon atoms, any further condensation reaction proceeds away from this at those sites still having hydroxy groups, and the more reactive they are; see Figure 2.9.

Under basic conditions, condensation preferably occurs at the central silicon atoms. There is a nucleophilic attack, meaning first all hydroxy groups are replaced by siloxane bonds, the faster the less hydroxy groups are at a silicon atom. Figure 2.9 has to read just the opposite of the acid case: the darker a silicon atom is marked, the more active it is. The result is a clear difference of the microstructure obtained. Under basic

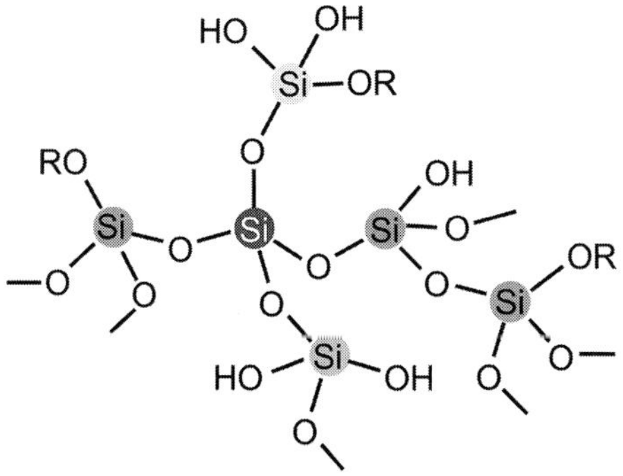

Figure 2.9 Schematic of the condensation reaction process under acidic conditions. The condensation takes place at those silicon atoms to which the most hydroxy groups are attached. The more siloxane bonds are formed, the less active is that site. The activity is schematically shown with grey colours, the darker the less active.

conditions, one therefore expects that compact structures occur, while under acidic conditions more open, branched polymeric structures will develop.

This difference was explained by Dale Schaefer [48] on the basis of small angle X-ray scattering results (see also [3]). In his picture, under basic conditions reactive monomers add to existing clusters (oligomers, particles) and make a new siloxane bond. This is in full agreement with the statements made by Iler on the condensation reaction of silicic acid, as mentioned earlier. This is the so-called monomer to cluster growth, or MC growth. Under strongly basic conditions in the presence of large amounts of water, the number of hydrolysed sites at silicon atoms increases. As the water/silicon ratio R_w drops, hydrolysis is less probable, and the reacting species have unhydrolysed or blocked sites that resist polymerisation. The growth can still be represented as a monomer-cluster addition, but not all sites are available. A few are blocked by alkoxy groups. Although in principle alkoxy groups could also undergo a condensation reaction, their reaction rate is much smaller than that of hydroxy sites. This behaviour was called a poisoned-MC growth [49], meaning not all impingements of molecules onto an oligomer lead directly to a siloxane bond, but several attempts, such as movements on the oligomer surface or back movements into the solution, are necessary before a siloxane bond is formed and thus the reaction completed. The result should be compact particles probably with a rough surface.

Schaefer as well as Brinker and Scherer argue that under acidic conditions, the rates of the hydrolysis and condensation are somehow reversed: hydrolysis increases with decreasing pH, and condensation does the opposite. Therefore, reactive monomers, meaning hydrolysed alkoxides, are produced in a huge concentration. This situation leads in their picture to a classic realisation of a so-called reaction limited cluster–cluster aggregation (RLCA) [49]. The structures observed are not compact, bulky particles with a smooth or rough surface, but polymeric ones. This is in full agreement with the previous statement that condensation proceeds at the terminal silicon atom. Therefore, polymeric-like structures appear. Let us briefly sketch a bit better the term RLCA. First, 'reaction limited' is contrasted by 'diffusion limited'. In any type of reaction, there are two extreme cases: the transport of reacting species through a medium (fluid, gas) determines the speed of reaction, and if this transport is by diffusion or convective diffusion, one can say the reaction is diffusion limited. This has a consequence: the speed of the reaction is then determined by concentration field of the interacting species and the local gradients. In the cases of, let us say, an oligomer or particle, its growth would be determined by the concentration gradient at the interface and how it develops with space and time [50]. In contrast to this situation, 'reaction limited' means that there is an energy barrier to overcome between two reacting species, before they build a new molecule or a molecule becomes part of another larger molecule or particle [50, 51]. This leads to the situation that the concentration field of the reacting species inside the fluid is not important and can be modelled as constant, and only the flux of reactive species across an interface, which is dictated by the energy barrier, is important. In reality, there are situations possible in which both extremes mix [52, 53]. Imagine now that not only monomers are in a solution and react with some oligomers (MC growth), but that the oligomers react with other oligomers, build at their point of contact a siloxane bond and such clusters of oligomers react further with other clusters, always sticking together at their point of contact. The result would then be a very loose structure.

A general scheme of reaction and growth during polycondensation is shown in Figure 2.10. A silicic acid or a partially hydrolysed alkoxysilane is drawn schematically as a tetrahedron. Connecting two tetrahedra at one of their tips shall indicate a siloxane bond. In as much as condensation proceeds, amorphous silica particles are formed which are loosely packed under acidic conditions and compact under basic ones. Note that in all cases the inner and outer surfaces of such a particle contain silanol and alkoxy groups as already mentioned. This is important for all aerogels. They always have at least hydroxy groups at their surface after drying and are then hydrophilic or they have to a varying degree alkoxy groups at their surface and are then hydrophobic. One should also note that these alkoxy groups in dry aerogels undergo hydrolysis under atmospheric conditions by water vapour. The process takes time, sometimes years, but it happens. Quite often, therefore, the silanol or alkoxy groups are burned away by a thermal treatment of the aerogels (typically in a furnace at around 300–400°C for a few hours).

Although the interpretation given in the work of Schaefer [48] and in the textbook of Brinker and Scherer [3] is intriguing, it is difficult to map mathematically the chemistry and physics of the transformation from a solution to a gel in terms of all the aforementioned parameters and to make predictions. The real problem with all aerogels making solutions is that there are chemical reactions leading from monomers to oligomers running simultaneously with condensation reactions (and dissolution); the evolving entities (polymers of more or less compact appearance) undergo stochastic or nonstochastic motions inside the solution forming clusters by some type of process, and the clusters and oligomers can aggregate. The processing parameters such as the pH value, the R_w ratio and the concentration of alcohol affect to a large extent all four

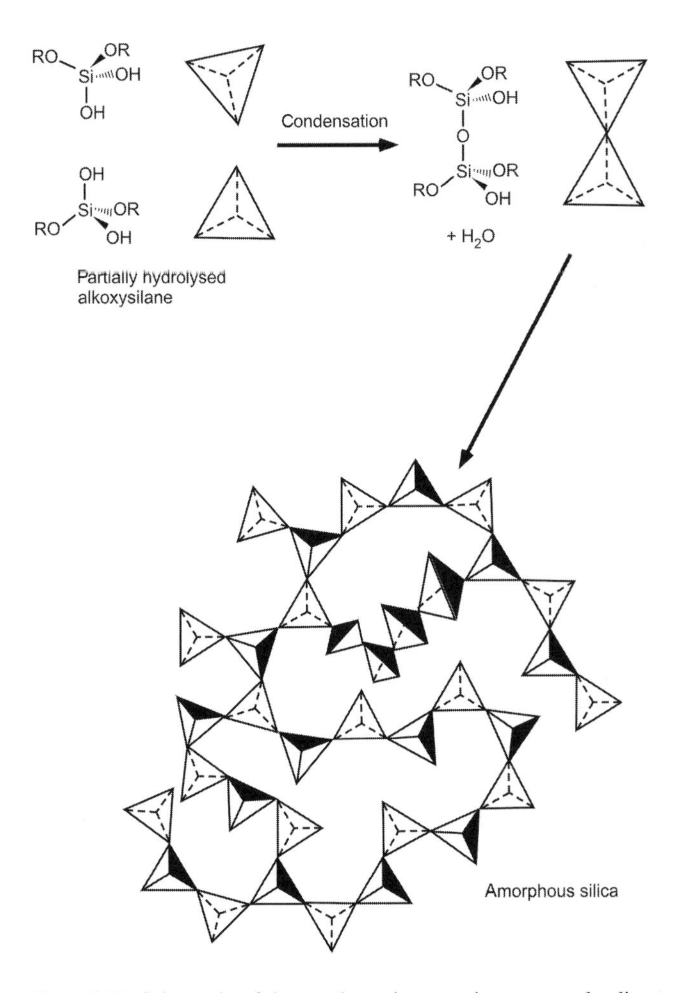

Figure 2.10 Schematic of the condensation reaction process leading to amorphous silica particles. The silicic acid or partially hydrolysed alkoxysilane is drawn schematically as a tetrahedron. Connecting two tetrahedra at one of their tips indicates a siloxane bond. In as much as condensation proceeds, amorphous silica particles are formed, which are loosely packed under acidic conditions and compact under basic ones. Note that in all cases, the inner and outer surface, of such a particle contain silanol and alkoxy groups.

fundamental processes: hydrolysis, condensation and their reverse processes, oligomer to polymer formation and aggregation to a gel.

In the following we would like to model these effects with a few equations and leave the processes leading to gelation open, since they are treated in more detail in Chapter 4.

Let us imagine an oligomer of silica including siloxane bonds, hydroxy and alkoxy groups and let this be embedded in a solvent having in the case of basic condensation a concentration of $\equiv$Si–O$^-$ ions, denoted as x_{SiO} determined by the pH value. Under acidic conditions, the concentration of protonated siloxanes is relevant, denoted as

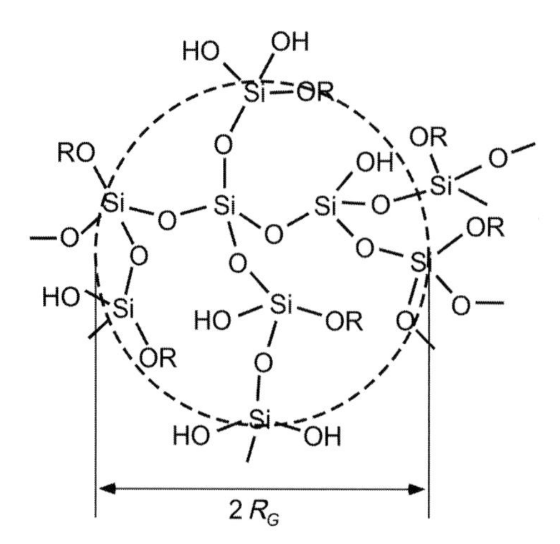

Figure 2.11 Schematic of an oligomer with reactive sites, including hydroxy and alkoxy groups at the silicon atoms. One can construct a radius of gyration and handle the oligomer as if it would be a particle with a certain radius R_G.

x_H, which is essentially the pH value.[1] Then one can imagine the further growth of the oligomers is due to their incorporation into the interface and reacting under basic conditions inside the oligomer nucleophilic with silicon atoms having preferably one hydroxy group. Let us first try to model the growth under basic conditions. In order to model growth, we construct inside an oligomer a radius of gyration. This is illustrated in Figure 2.11. The radius of gyration R_G is quite often used in polymer physics [54]. Its square, R_g^2, is the second moment around the centre of mass of all connected molecules. The centre of mass is defined as

$$\mathbf{r}_c = \frac{1}{N+1}\sum_{i=0}^{N} \mathbf{r}_i \tag{2.4}$$

with $\mathbf{r}_i$ the position vectors of all atoms in the molecule. Then the radius of gyration is calculated as

$$R_G^2 = \frac{1}{N+1}\sum_{i=0}^{N} \langle(\mathbf{r}_i - \mathbf{r}_c)^2\rangle. \tag{2.5}$$

Multiplying with the mass of a molecule, we would have exactly the mechanical expression of a moment of inertia for a rotation around the centre of mass. $\langle\cdots\rangle$ indicates averaging over all atoms. The rate of growth is directly proportional to the local concentration of the $\equiv$Si–O$^-$ ions times the number of active sites x_{as}

[1] Let pK_a denote the strength of the acid, and C_a the concentration added to a solution in moles. The molar concentration of H^+ is determined from the equation $x_H^2 + K_a x_H - K_a C_a = 0$. Here $K_a = 10^{-pK_a}$. Knowing the proton concentration then gives the pH value as pH = $\log_{10} x_H$, if the activity and the molar concentration are equal.

(hydroxy groups) being there. Assume that the concentrations of active sites, or better the concentration of hydroxy groups on alkoxides, follows a first-order reaction. The concentration of hydrolysed sites on existing alkoxy monomers changes with time proportional to the concentration of alkoxys there, but reduced by the number of already hydrolysed positions on the alkoxys (which we call here active sites and denote them by x_{as}). Therefore the rate of change dx_{as}/dt can be written as

$$\frac{dx_{as}}{dt} = k_H(T,\mathrm{pH})(x_{alk} - x_{as}). \tag{2.6}$$

This equation is easily solved by direct integration and yields

$$x_{as} = x_{alk}\left(1 - \exp(-k_H t)\right). \tag{2.7}$$

It says that initially there are no active sites and finally all sites on the silicon alkoxide are hydrolysed and thus are active. The time dependence is determined fully by the reaction rate constant k_H, which depends on the pH value (and temperature). The growth of oligomers and polymers by a condensation reaction is a bimolecular one and thus the growth of radius can be described by the following relation [51, 52]:

$$\frac{dR_G}{dt} = k_g(T,\mathrm{pH})x_{SiO}^2 x_{as}(t). \tag{2.8}$$

Inserting the result from Eq. (2.7), one obtains for the radius a superposition of a linear growth term and an exponential one (we omit the explicit expression here). Depending on the ratio of the rate constants, there might be an initial parabolic-like increase in particle radius always followed by a linear increase in size. If such a simple picture would be correct, one would obtain compact particles whose growth is dictated by the concurrent hydrolysis and condensation reactions. The expression derived has a shortcoming: the particles would grow, and their growth would never stop. This is definitely not real and is due to the single particle approximation used. One can correct Eq. (2.8) taking into account that the number of monomers is limited. Let ϕ_0 be the maximum volume fraction of silica which would result if all alkoxides would be fully reacted to SiO_2 (this can be calculated from the composition of the solution used; see Chapter 7). The actual volume fraction $\phi(t)$ is a product of the number density $n(t)$ of oligomers in the solution and the volume of the growing particle $R_G(t)$ as $\phi(t) = n(t)4\pi/3R_G(t)^3$. We then can correct the growth equation in a simple way:

$$\frac{dR_G}{dt} = k_g(T,\mathrm{pH})x_{SiO}^2 x_{as}(t)(\phi_0 - \phi(t)). \tag{2.9}$$

Initially the volume fraction of silica polymeric particles is zero and the growth rate large. In as much as the volume fraction increases, the additional term reduces the growth rate. The only bad issue in this equation is the unknown number density of polymeric particles or oligomers that can grow, $n(t)$. Especially for growth under basic conditions, it is quite often assumed in the literature, that viable oligomers are a result of a nucleation process; see [3, 45]. Without further asking how a nucleation process proceeds in silica solutions, we take this idea and propose that under base conditions, out of a sudden a burst, oligomeric particles appear, which can be described by a

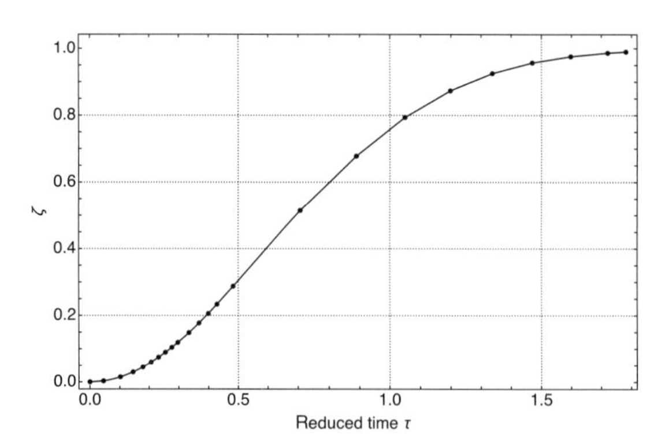

Figure 2.12 Relative radius $\zeta = R_G(t)/R_G^{\max}$ as a function of the time integral of the right side of Eq. (2.11). The normalised time τ is defined as $\tau = k_H t$. The parameter "$A*$" in Eq. (2.12) is set to 1.

density n_0 and a radius $R_G^0 = R_G(0)$, such that initially $\phi(0) = n_0 4\pi/3 R_G(0)^3$. Then one can rewrite Eq. (2.9) as

$$\frac{dR_G}{dt} = k_g(T, pH) x_{SiO}^2 x_{alk} (1 - \exp(-k_H t)) \left(\phi_0 - n_0 \frac{4\pi}{3} R_G(t)^3 \right). \quad (2.10)$$

If we also rewrite the expression for the maximum volume fraction as $\phi_0 = n_0 4\pi/3(R_G^{\max})^3$, and introducing the new variable $\zeta = R_G(t)/R_G^{\max}$ one yields an equation

$$\frac{d\zeta(t)}{dt} = A^*(1 - \exp(-k_H t))(1 - \zeta(t)^3) \quad (2.11)$$

with A^* the abbreviation of all other constants. The equation can be solved, but gives only an implicit expression for the radius as a function of time, namely

$$-\frac{1}{6} \ln \left(\frac{(1-\zeta)^2}{1 - \zeta + \zeta^2} \right) + \frac{1}{\sqrt{3}} \left(\arctan \left(\frac{2\zeta + 1}{\sqrt{3}} \right) - \frac{\pi}{6} \right) = A^* \left(\tau - (1 - e^{-\tau}) \right) \quad (2.12)$$

with $\tau = k_H t$. Numerically evaluating the result yields the functional form depicted in Figure 2.12. The figure shows that initially there is a parabolic growth of the radius of gyration with reduced time. This is understandable by looking at the right side of Eq. (2.12) and expanding it to second order around zero. At small times, it evolves like τ^2. The left side of Eq. (2.12) is at small ζ proportional to ζ itself. Thus the change fo the radius of gyration with time is like $R_g(t) \propto \tau^2$. At larger times, the radius increase levels off and approaches the final size, which is dictated by the volume fraction and the number density of nucleated oligomers. Although this picture is somehow fine, it cannot predict the final particle size. There are factors not taken into account: with increasing pH value, the reverse condensation reaction, meaning the dissolution of oligomers becomes important, and there also is an aggregation of oligomers and particles leading to a faster increase of size and finally that curvature induced solubility changes will modify the morphology of particles and their aggregates.

The polycondensation process under acidic conditions is different from that under basic ones. Instead of compact particles, rather ramified or polymeric structures are built. In order to construct a simple model, we take into account the fractal nature of the emerging particles. Fractals are explained in more detail in Chapter 4. Here it should be sufficient to note that fractal objects are porous structures which are in principle self-similar, meaning the structure is the same independent of the magnification they are looked at [55–57]. Real objects are only fractals in a certain size range. There is a difference between so-called mass fractals and surface fractals. In the first case, an object in 2D or 3D has a ramified porous structure; in the second case, the object might be compact, but the surface only is ramified or rough. Both explanations become clearer using the following definitions. Let N_m be the number of monomers in an object whose radius of gyration is R_G. In a fractal object, this number can be calculated as

$$N_m = k_1 \left(\frac{R_G}{a} \right)^{d_f}, \tag{2.13}$$

with a the radius of the monomers, k_1 a numerical constant of the order of unity and d_f the fractal dimension. This equation also gives a rule, how to determine the fractal dimension: measure the number of monomers (or particles) as a function of radius from an arbitrarily chosen centre and plot $\log N_m$ as a function of $\log R_G$. The slope determines the fractal dimension. The radius of the monomer can be calculated from the molar volume Ω_m and Avogadro's number N_A. The volume of a fractal particle V_p is then calculated as

$$V_p = \frac{\Omega_m}{N_A} N_m \propto R_G^{d_f}. \tag{2.14}$$

If the object would not be a fractal, then $d_f = d$, the Euclidian embedding dimension. In 3D, this means that the volume would be proportional to R_G^3. This also tells that for fractal objects, $d_f < d$. If the surface area is also a fractal, the number of monomers at the surface can be calculated in a similar way:

$$N_s = k_2 \left(\frac{R_G}{a} \right)^{d_s}, \tag{2.15}$$

defining the surface fractal dimension d_s. The surface area is then

$$A_p = a^2 N_s \propto R_G^{d_s}. \tag{2.16}$$

Again, if the object would not have a ramified surface but has a smooth one, then $A_p \propto R_G^2$ and also the surface fractal dimension will be smaller than that of Euclidian dimension of the surface. Unfortunately the fractal dimension of real objects[2] cannot be predicted by any analytical theory, but has to be determined using suitable mathematical or numerical models. For a deeper discussion, see also the book of

[2] For many mathematical fractals, the fractal dimension can be calculated exactly; see Chapter 4 for more details on fractals.

Tamas Viscek [49]. The volume of a particle (being a polymer or oligomer) changes with time due to the attachment of activated alkoxides (fully or partially hydrolysed; see Figure 2.6) at the surface:

$$\frac{dV_p}{dt} = a^3 N_s \Gamma(T) x_{as}(pH) \propto R_G^{d_s} \Gamma(T) x_{as}(pH). \tag{2.17}$$

In this equation, $\Gamma(T)$ is the attachment frequency of activated molecules which react with surface molecules to form a siloxane bond. $x_{as}(pH)$ is the molar concentration of activated molecules (protonated ones), and thus this concentration is directly proportional to the concentration of H^+ and thus the pH value. Inserting Eq. (2.14) gives an expression in terms of the radius of gyration that can be integrated to yield

$$R_G(t)^{d_f - d_s} \propto \Gamma(T) x_{as}(pH) \cdot t. \tag{2.18}$$

If the oligomer, polymer or particle would be a compact object and not a fractal, the growth would be linear in time. Due to the exponent $d_f - d_s$, there is a slight deviation from linearity. Interesting in this equation is another aspect. If the polymer grows until the container size is reached, one could argue that then gelation has happened (spanning cluster). This would mean

$$t_{gel} \propto \frac{1}{(\Gamma(T) x_{as}(pH))^{d_f - d_s}} \tag{2.19}$$

and thus the smaller the pH value, the larger the concentration of activated molecules and thus the shorter the gel time. Since one can assume that $\Gamma(T)$ follows an Arrhenius type relation, with an activation energy for the reaction at the surface, the larger the temperature, the shorter the gelation time, as it should be. One must, however, take into account that during growth, both under acidic and basic conditions all particles move with respect to each other, e.g., by Brownian motion, and thus might collide and form small clusters, which themselves will change their morphology due to the curvature-induced solubility (Ostwald ripening or dissolution and re-precipitation). Under basic conditions, this is more pronounced, since the solubility of silica increases with increasing pH value.

2.1.5 Gelation

Particles as defined earlier as well as oligomers are not fixed in space. They move either due to thermally activated Brownian motion, sedimentation, if their density is different from the embedding solvent or by solutal-driven flows inside a container driven by local density differences in the solvent. Any such motion will lead to encounters between particles or oligomers, and then they might become bound to each other and move further as a particle cluster. Details are discussed in Chapter 4. Here it is sufficient to notice that gelation in silica is generally thought to be due to an aggregation process of nanosized particles, which might have a fractal substructure or are compact.

2.1.6 Ageing

There is another process also running concurrently with hydrolysation and polycondensation: ageing. Although conventionally this is discussed for gels as a way to further modify the gel structure, e.g., stiffening, reducing syneresis etc., it was already pointed out by Iler that the size-dependent solubility of silica leads to rearrangement of aggregated silica particles, provided they are sufficiently small [45]. When defining the term 'ageing' needs to first think about the solubility of silica particles upon their curvature and second to visualise how this might change the structure of aggregated particles.

The curvature dependence of solubility stems from the fact that the free energy of a particle is not only due to the bulk free energy, but there is a surface energy term. The chemical potential of silica in a solution therefore gets a new term $\mu_p = \mu_p^0 + \sigma_{sl}\mathcal{H}$ in which σ_{sl} denotes the surface tension of the silica particle in the solution, $\mathcal{H}$ is the mean curvature and μ_p^0 the reference state for a planar surface. The curvature of an arbitrarily shaped body can be constructed at an arbitrary point of its surface by two circles (of different radii) being perpendicular to each other and approximating at best the curve at that point. Their two radii R_1 and R_2 define the mean curvature $H = 1/R_1 + 1/R_2$. The mean curvature can be positive or negative (imagine a saddle).[3] This means, the chemical potential can be higher or lower depending on the position on a surface compared with the reference state. For dilute solutions, one can express the chemical potential in terms of concentration of solute and obtains for the solubility the expression already given by Iler:

$$c_{SiO_2} = c_{SiO_2}^0 e^{\frac{\sigma_{sl}\Omega_m \mathcal{H}}{k_B T}}. \quad (2.20)$$

If the solid-liquid interface can be described as spherical with a radius R, then $\mathcal{H} = 2/R$. For more details, see the book of Ratke and Voorhees on growth and coarsening [50]. Iler gave some values which allow us to calculate the radius effect of solubility for amorphous silica. Typically, the solubility of particles increases from values around 100 ppm for particles with 10 nm diameter to 200 ppm for particles with 2 nm diameter.

The effect of this size dependency of solubility is manifold. First, if there is a dispersion of particle sizes in a solution, the size distribution will change, the average particle size increases and the small particles dissolve at the expense of larger ones to which their excess solute is transported by diffusion processes, for instance [50]. Second, if particles are in contact with each other as in aggregates, the radius of curvature changes from negative at the contacting neck to positive on the way to

[3] Let a surface be described by a function g(x,y,z) = C, which is transformable to z = f(x,y). Define then $p = \partial z/\partial x, q = \partial z/\partial y, r = \partial^2 z/\partial x^2, s = \partial^2 z/(\partial y \partial x, t = \partial^2 z/\partial y^2$. Then according to standard textbooks on differential geometry, the average curvature is calculated by $\mathcal{H} = (r(1+q^2 - 2pqs + t(1+p^2))/(1+p^2+q^2)^{3/2}$. For a sphere with the general equation $x^2 + y^2 + z^2 = R^2$, one then obtains after lengthy calculations for the average curvature the result $\mathcal{H} = 2/R$. This could have also be obtained, for spheres, from $\mathcal{H} = \partial A/\partial V$. Note: for a cylinder with radius R, the average curvature is half the value of that of a sphere.

the equatorial plane of the particles. This means the higher solubility there leads to a dissolution of silica and deposits it on the neck area. Provided there is enough time, a loose aggregate of particles can transform to even a fibril or fibrillar network, as already pointed out by Iler [45] and examples of such a more or less fibrillar morphology are presented in Chapter 6 see Figures 6.2 and 6.5. Third, one should note that the transformation of a chain of loosely connected particles to a fibril must not necessarily be the thermodynamic state with the smallest free energy. As pointed out by Nichols and Mullins [58, 59], depending on the type of mass transport, a cylinder with wiggles on the surface can become unstable and decompose into particles whose size depends on the counteracting effects of capillarity and transport by, for example, volume diffusion in the surrounding matrix or surface diffusion along the surface of the perturbed cylinder. Therefore, even for prolonged ageing of a silica aggregate in a solution, a fibrillar state must not be the final one, but intermediates might be stable at least on typical experimental time scales of a few hours or days. A fourth point has to be taken into account: even after the gel point, there are always monomers and oligomers of different degrees of polymerisation in the solution which will deposit onto the existing gel network and thereby modify its morphology. Depending on their concentration, their action might be more powerful than the Ostwald ripening of aggregates.

The changes in morphology have been carefully investigated and the term 'ageing' was cast into the discussion of such effects. There are numerous investigations available in the literature concerning ageing of silica. Especially extensive studies were performed by Mari-Ann Einarsrud and co-workers [60–62]. They studied the effect of solvent composition (water, TEOS or alcohols), time and temperature of ageing (leaving the gel in such solutions) on properties of the resulting dry gels, such as shrinkage, density, mechanical strength, surface area etc. They were able to show that adding new monomers to an already formed network of silica particles from a TEOS solution led to a precipitation on the already hydrolysed and condensed network, with the effect that the shear modulus increased by a factor of 15 and the shrinkage upon ambient drying was reduced by 30%. A similar observation was made later by Smitha and co-workers [63], who also showed that adding a mixture of TEOS with ethanol led to an increase in specific surface area, while the bulk density and the shrinkage decreased. Reichenauer [64] showed that ageing of silica gels in water at temperatures between 60–80°C increased their mechanical stability, decreased their specific surface area and the amount of microporosity, made the particle size more uniform and drastically reduced shrinkage, in full accordance with the observations made by Einarsrud [60]. In a recent study, Iswar and co-workers [65] systematically investigated the effect of ageing on properties of silica aerogels. The gels were aged for various times and temperatures in their wet gel fluid, hydrophobised in hexamethyldisiloxane and dried either at ambient pressure or after solvent exchange using supercritical carbon dioxide. They observed that the specific surface area decreases with increasing ageing time and temperature and attributed this to Ostwald ripening processes. Interestingly, they observed a decrease of bulk density with prolonged ageing time and temperature for ambient dried gels. Supercritically dried ones did not show such an effect. Therefore, they conclude that ageing increases the ability of the gel network to withstand irreversible pore collapse during ambient pressure drying.

2.1.7 Procedures and Results

The wonderful scheme of silica gel evolution under acidic or basic conditions, developed by Iler and shown in Figure 2.2, directly gives advice on how to prepare gels with a desired structure. Instead of preparing gels at a fixed pH, it readily was realised that a two-step routine is optimal: first prepare very fine nanostructured sol under acidic conditions (small particles and clusters of more or less polymeric appearance) and then change the pH to basic conditions, using, for instance, ammonia to initiate fast condensation and gelation. Such a two-step procedure and recipe is described in Chapter 14.

Silica aerogels are produced industrially in huge quantities. Granulates of silica are produced using sodium silicate solutions as a starting material and inducing gelation by adding acidified water (typically with HCl). Cakes of the wet gel are broken into small pieces, washed with solutions also containing trimethylchlorosilane (TMCS) to make the surface of the gel hydrophobic and to utilise the so-called springback effect, which allows drying the wet gel under ambient conditions with minimal shrinkage (for more details, see Section 5.1 on ambient drying). For more details and their applications, see either the webpage of Cabot [34] or the article by Cabot in the *Aerogels Handbook* [14].

In the last decade of the last century, Aspen developed technologies to produce fibre felt composites with aerogels [35]. In order to make a composite, a lofty batting is infiltrated with a pre-polymerised silica solution (typically a TEOS solution acidified and sufficiently hydrolysed). Just before infiltration, the solution is mixed with ammonia to induce the condensation and also a more or less rapid gelation (within a few minutes). Originally Aspen produced sheets with 5 m length, a metre wide and around 10–20 mm thickness. At the turn of the millennium, Aspen was able to produce such fibre-reinforced battings almost continuously, by combining the infiltration and gelation step on a conveyor belt and wrapping the wet gel fibre batting. The wet gels are washed with trimethylchlorosilane and dried supercritically with carbon dioxide. For reinforcement glass, polymeric and carbon fibre felts are used, leading to different properties (see also the article of Aspen in the *Aerogels Handbook* [14]).

Silylation is today most often not performed using TMCS but using hexamethyldisilazane (HMDS). HMDS consists of two silicon atoms bonded via an NH group together. The other bonds of the silicon atoms are made by methyl groups. The overall formula of HMDS is therefore $[(CH_3)_3Si]_2NH$. HMDS is a colourless liquid with a melting point of $-78^\circ C$ and a boiling point of $126^\circ C$ (be careful the flame point is at 15°C). Its odour is that of an amine. In the presence of water, it hydrolyses in an exothermic reaction producing in an intermediate step ammonia and trimethysilanol, which is unstable and reacts to form hexamethyldisiloxane. This decomposition of HMDS can be used to make silica hydrophobic, since the trimethysilanol replaces the -OH group with an -Si–$(CH_3)_3$ group. The interesting aspect using HMDS is that one can avoid using ammonia itself to finish the aforementioned two-step procedure, but can catch two flies with one swatter: making the solution basic and performing the hydrophobisation.

2.1.8 Silica Aerogels with Lower Functional Silanes

Whenever mechanical properties of aerogels are discussed, the standard statement found in the literature even today is that they are brittle and have therefore bad mechanical properties. Although such a statement is generally wrong, as will be shown in greater detail in Chapter 13, it nevertheless initiated research trying to make especially silica aerogels somehow flexible or bendable. The first results on such a chemical approach to flexibility was reported by Rao et al. in 2006, in which he used a trifunctional precursor, MTMS, and could show that aerogels result which are fully bendable and recover completely elastically after compression [66]. The idea behind it was to block one bond at a silicon atom by an inert functional group, e.g., methyl or ethyl, such that hydrolysis and condensation will result in structures more polymeric chain-like, leading to some intrinsic flexibility. Guo et al. prepared flexible aerogels from a mixture of tetramethoxysilane and vinyltrimethoxysilane to which a special organic sulfide was added (bis[3-(triethoxysilyl)propyl]disulfide) such that disulfide bridges could enhance the elastic recovery after compression [67]. Aravind and co-workers [68] developed nanostructured flexible aerogels derived from trifunctional alkoxides, MTMS with 3-(2,3-epoxypropoxy)propyltrimethoxysilane (GPTMS) as a co-precursor. Since aerogel forming solutions using trifunctional alkoxides exhibit a tendency for phase separation, it is common to suppress this using surfactants, especially trimethylammonium chloride (CTAC). Optimising the synthesis procedure allowed preparation of nanostructured flexible aerogels, with a huge amount of mesoporosity and pores in the range of 10–50 nm, but also larger ones could be observed in SEM figures. The materials recover elastically after compression up to 30% strain. Figure 2.13 shows an SEM picture as an example of such a material. After the discovery of Rao [66], many other recipes for superflexible aerogels have been developed. Two recent reviews of various reinforcing strategies developed in the last few decades to improve or modify the mechanical response of silica aerogels discuss these recipes [41, 42]. Especially the research group around Nakanishi and Kanamori [69–74] developed fascinating new synthesis procedures and a whole bundle of different chemical building blocks to prepare superflexible and also superhydrophobic silica-based aerogels. They mixed tetrafunctional silanes with tri- and difunctional ones, especially DMDMS and MTMS, or used only tri- and difunctional silanes. The procedures to make such superflexible aerogels are very similar to the procedures described earlier: Mixing of water with acetic acid and urea

Figure 2.13 SEM picture of an MTMS–GPTMS aerogel after [68]. The microstructure shows nanosized particles connected in 3D with pores in the range up to 100 nm and more.

Figure 2.14 SEM pictures of an MTMS–DMDMS aerogel prepared according to the route given in Chapter 14. It demonstrates that such superflexible superhydrophobic aerogels are mainly macroporous materials. The particle size is in the micron range as well as the pore size.

and adding then MTMS together with a solution of a surfactant (most often CTAC) yields hydrolysis of the OH groups at the MTMS molecules. After stirring, one can, but must not, add for instance, DMDMS and let it gel at higher temperature in a furnace (typically between 70–90°C). The gel can be aged and has to be washed with water and ethanol to get rid especially of the surfactant. The resulting wet gel can be dried under ambient conditions. Figure 2.14 show the typical microstructure of an MTMS-DMDMS aerogel made according to the recipe given in Chapter 14.

These types of aerogels always exhibit an open macroporous structure with particles in the range of a micron up to a few tens of microns in diameter and a varying amount mesoporosity. Both, the intrinsic flexibility of the more chain-like molecules and the larger pores give them an extraordinary flexibility. Depending on the chemical details, how they were made, they can recover from compression even up to 70% fully elastically. One of their features, making them attractive, is their superhydrophobicity, which stems from the huge amount of methyl or ethyl groups at their surface. Hydrophobicity means vice versa oleophilicity and thus these materials can be used to separate oil from oil–water mixtures. For some recent studies, see also [75–77].

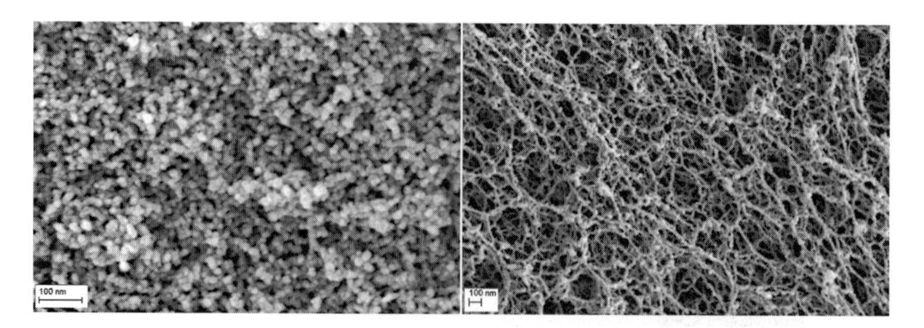

Figure 2.15 Scanning electron microscopy pictures of RF aerogels prepared after the recipe of Pekala [8]. Depending on the synthesis conditions, one can observe a particulate structure (left) in which polymeric particles of around 10 nm in diameter are connected in 3D to form an open porous network, or a string-of-pearls-like structure can be achieved (right), giving the impression of fibrils being interwoven in 3D. Parameters of the left figure: $R/F = 0.5$, $R/W = 0.014, R/C = 200$. Parameters of the right figure: $R/F = 0.5, R/W = 0.014$, $R/C = 50$, pH adjusted. Courtesy M. Schwan, DLR

2.2 Resorcinol–Formaldehyde Aerogels

In 1989, Pekala and co-workers [8] demonstrated that organic aerogels can be prepared by a base-catalysed polycondensation of resorcinol and formaldehyde (RF aerogels) using a sol–gel process followed by drying of the solvent using supercritical carbon dioxide. Such aerogels were later shown to display an extremely low thermal conductivity [78]. Resorcinol–formaldehyde aerogels belong to a wider class of organic aerogels such as melamin–formaldehyde, cresol–formaldehyde, phenol–furfural, polyurethanes and polyimides.

RF aerogels have extraordinary properties such as a porosity of 90–98%, a specific surface area up to 1500 m^2/g, a very low thermal conductivity down to 0.012 W/(K·m), an envelope density between 0.05–0.3 g/cm^3 and a high stiffness and compression strength compared to silica aerogels. As their inorganic counterparts, however, RF aerogels are typically brittle in nature, which renders them difficult to handle and process [79]. RF aerogels are not inflammable, non-toxic and decompose at higher temperature without fume formation. They can also be converted to carbon aerogels by carbonisation under a protective atmosphere. Carbon aerogels are materials for supercapacitors and batteries, if the nanostructure is designed properly. Carbon-based supercapacitors are already on the market and there still is an enormous research activity to tailor the nanostructure of RF and carbon aerogels for optimal performance in batteries and supercaps.

Although we present some morphologies of RF aerogels in more depth in Chapter 6 on various morphologies that can be observed, we will show here as an appetiser a few typical microstructures of RF aerogels. Figure 2.15 shows two microstructures that can be observed in RF aerogels, if resorcinol and formaldehyde react under basic conditions. A completely different structure results if a strong acid is used as shown in Figure 2.16. The colour of RF aerogels varies considerably with the synthesis conditions, especially the pH value, as wonderfully written by Job in her PhD

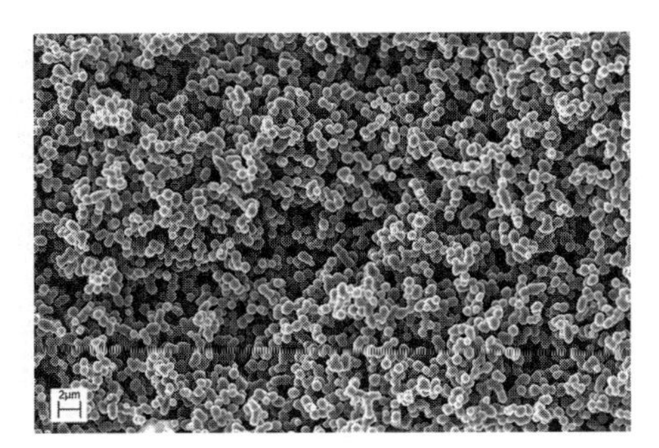

Figure 2.16 RF aerogels show almost always a particulate structure with spherical RF particles, if resorcinol and formaldehyde react under strong acidic conditions. The aerogel shown was prepared using hydrochloric acid [80].

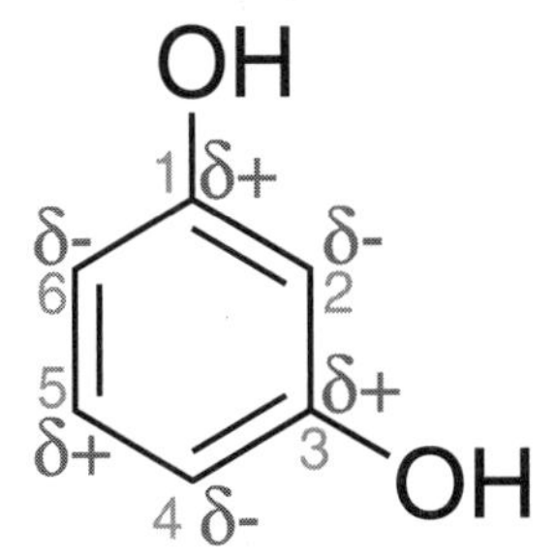

Figure 2.17 Scheme of a resorcinol molecule. At positions 1 and 3, OH attached to the benzene ring induces a shift of electronic charges along the ring such that positions 2, 4 and 6 bear negative charges. These positions are therefore nucleophilic in character. Position 2 is called ortho, whereas positions 4 and 6 are called para positions.

thesis [81]. If the pH value is smaller than 4, they have an opaque orange appearance; if it is between 4–5.5, they are light beige, between a pH of 5.5–6.5 they are opaque brown and above they are red to violet but translucent. The aerogels change their colour, since they all contain an aromatic ring which is chromophore.

The chapter is designed as follows. We first briefly review the chemistry of resorcinol and formaldehyde. Then discuss reactions between them under base and acidic conditions, and we summarise the effects of various experimental parameters, like temperature, pH value or catalyst concentration, dilution ratio (e.g., amount of water) and ratio between formaldehyde and resorcinol. We then discuss how polymeric networks and particles might be formed.

2.2.1 Chemistry of Resorcinol and Formaldehyde

Resorcinol and resorcin are trivial names for 1,3-dihydroxybenzene. The hydroxy groups render the aromatic more electron-rich than phenol or benzene (see Figure 2.17). The hydroxy groups induce a shift of the electron density along the ring. The positions 1, 3 and 5 bear a positive charge and positions 2, 4 and 6 a negative one.

The higher electron densities at the positions 2, 4 and 6 make these positions reactive to any molecule that likes to take up electron charges (called an electrophilic molecule). In a recent molecular dynamic study of crystalline resorcinol, Chatchawalsaisin et al. [82] confirmed this charge distribution, especially pointing out that the carbon atoms at positions 2, 4 and 6 bear a negative charge. Most reactive are positions 4 and 6. Position 2 is sterically hindered for reactions due to the neighbouring OH groups. Resorcinol is at room temperature a crystalline material occurring in two polymorphs. It melts at 108.8°C but has a sufficient vapour pressure even in the solid state: it smells (our nose is very sensitive). Resorcinol is soluble in water, but is also solvable in ethanol, acetone, benzene or acetonitril. In all cases, the solubility of resorcinol increases with temperature from around 10 wt% at 50°C to 90 wt% at 80°C. An overview of properties, compounds and materials made with resorcinol is given in the book of Durairaj [83].

Formaldehyde is the simplest aldehyde. It consists of the aldehyde group –CHO, in which the oxygen atom is double-bonded to carbon. In formaldehyde, the remaining substituent at the aldehyde functionality is hydrogen, giving CH_2O as the formula. Formaldehyde is a colourless gas at room temperature. It is very well soluble in water, in which it forms a hydrate called methanediol. Since formaldehyde easily polymerises to paraformaldehyde having the repeating unit -CH_2O- and this polymerisation is self-initiated, a commercially available solution of water and formaldehyde contains not more than 37 wt% formaldehyde and is in addition protected as acetals by about 10 wt% methanol against oxidation and formation of paraformaldehyde. Such solutions are typically used to produce aerogels. Formaldehyde is a very reactive compound due to the strong dipole character of the C=O bond, such that at the oxygen atom there is a negative charge and a positive one at the carbon.

2.2.2 Reaction between Resorcinol and Formaldehyde

The reaction between resorcinol and formaldehyde in water depends strongly on the pH value of the solution. If one mixes resorcinol with formaldehyde, the solution automatically has a pH value between 3.9–4.1, decreasing with the amount of formaldehyde. Even in the non-catalysed solution, resorcinol and formaldehyde react with each other with an activation energy of around 80 kJ/mol [84]. The classical way to make RF aerogels is to use sodium carbonate as a so-called catalyst[4] in low concentrations such that the pH value is in the base regime, meaning pH values are larger than 4. Acid-catalysed synthesis of RF aerogels was developed much later by Barbieri et al. [85]. In both regimes, the chemical reactions are different and lead to different microstructures as shown previously. We therefore discuss first the reaction in a base medium and then reactions under acid conditions.

[4] The expression 'catalyst' for sodium carbonate is not really correct, because it leaves the reaction not unchanged. It serves in reality as a base and works as a hydrogencarbonate buffer. More correctly, it would be called an activator. In the aerogel literature, however, it is common to name it a catalyst and so we do here.

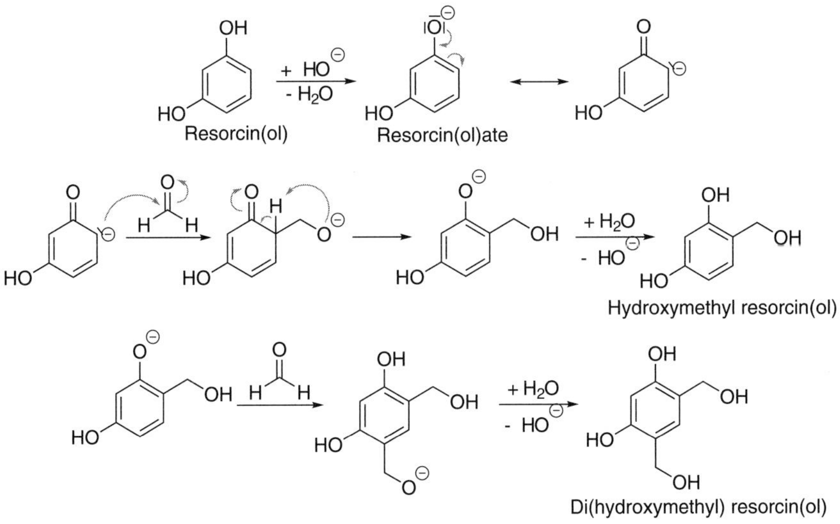

Figure 2.18 Scheme of the reaction steps in a resorcinol–formaldehyde in a basic aqueous solution. First the resorcinol is activated such that at position 2 or 4 in the ring, a negatively charged CH exists which can react with the electrophilic formaldehyde to build hydroxymethylresorcinol. If there is an excess of formaldehyde, the other position (2 or 4) is also electrophilic substituted with CH_2OH.

Reaction between Resorcinol and Formaldehyde under Base Conditions

The first step of a reaction between formaldehyde and resorcinol under base conditions is a deprotonisation of one hydroxy group, meaning from one OH group a proton atom is taken away by an excess OH ion, an electron charge transfer takes place such that the free oxygen is now double-bonded to the carbon atom deleting by that step a carbon double-bonding in the ring and a negatively charged CH is left. This activation of the benzene ring is promoted by base catalyst like all carbonates and depends strongly on the valence and size of the cation. Once the resorcinol is activated, an electrophilic substitution by formaldehyde takes place, leading to hydroxymethylresorcin. The hydroxymethylresorcin reacts with another formaldehyde to dihydroxymethylresorcin, if there is an excess of formaldehyde. All steps are shown together in Figure 2.18. In principle, the reaction is a bit more complicated since concurrently to the formation of a hydroxymethylresorcinol runs another reaction, which leads to a so-called ortho-quinon methide. Quinones are oxidised dihydroxy benzenes. For instance, in a benzene ring two –OH groups are oxidsed to form –C=O groups, and this is of course connected with a rearrangement of the double bonds between the carbon atoms in the ring. In the case of resorcinol, the ortho-quinone methide is an unstable intermediate product. A scheme of the reaction is shown in Figure 2.19. These quinones are very reactive and can directly bond to another activated resorcinol, building thereby a dimer with a methylene ($–CH_2–$) or a dimethylene ether ($–CH_2–O–CH_2$) bridge between the resorcinol rings. A detailed chemical study of

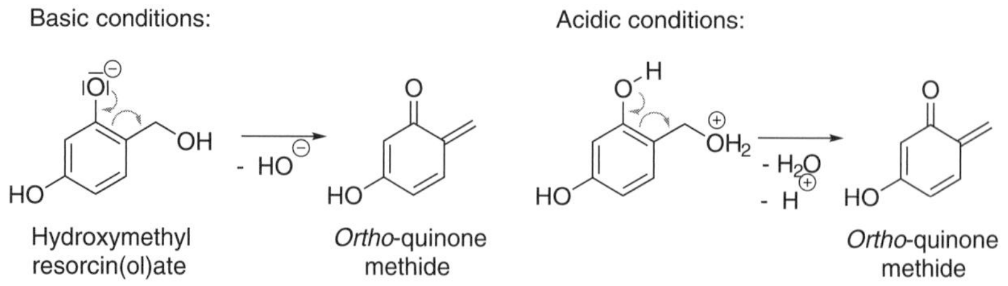

Figure 2.19 Scheme of the formation of a so-called ortho-quinonemethide stemming form the reaction of an activated resorcinol ring with formaldehyde. A charge transfer from the activated oxygen leads under an excess of OH^- ions to a double-bonded oxygen in the ring and a methylene group aside.

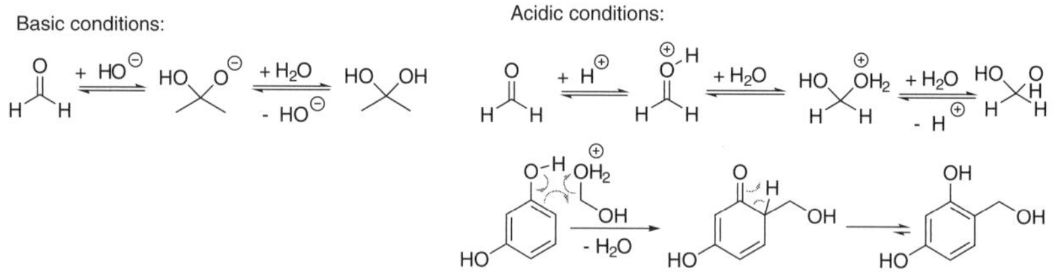

Figure 2.20 The upper line shows the scheme of formaldehyde activation under acidic conditions. The reaction between water and formaldehyde always leads to methanediol. Methanediol is protonated such that it can make an electrophilic attack on resorcinol shown in the lower line of pictures leading to a protonated hydroxymethylresorcinol.

this reaction was recently given by Li and co-workers [86] and can also be read in the review of Mulik and Leventis [87]. In both cases, the overall reaction could be written in a simplified manner. Naming the activated hydroxymethlyresorcin HMR and the resorcinol as R, we have 2 HMR $\rightarrow$ R-CH_2-R + CH_2O + H_2O or 2 HMR $\rightarrow$ R-CH_2-O-CH_2-R + H_2O. These are classical polycondensation reactions with water as the by-product. The influence of catalyst, formaldehyde and water concentration will be described later.

Reaction between Resorcinol and Formaldehyde under Acidic Conditions

In acidic conditions, meaning pH values below 3, there is no way to activate the resorcinol molecule, since there is an excess of protons (H^+) and to activate the positions 2, 4 or 6 of the ring, an electrophil would be necessary. However, in acidic conditions, instead of the resorcinol the formaldehyde can be activated, which was already mentioned. Methanediol or formaldehyde monohydrate (an older name of this is methylene glycol) with the formula HO–CH_2–OH is always present in water formaldehyde solutions. This hydrated formaldehyde is protonated, becoming thereby electrophilic and thus can react with resorcinol at positions 2, 4, and 6 to make a protonated hydroxymethylresorcinol. The scheme is shown in Figure 2.20. The protonated hydroxymethylresorcinol can directly react with resorcinol to make a dimer

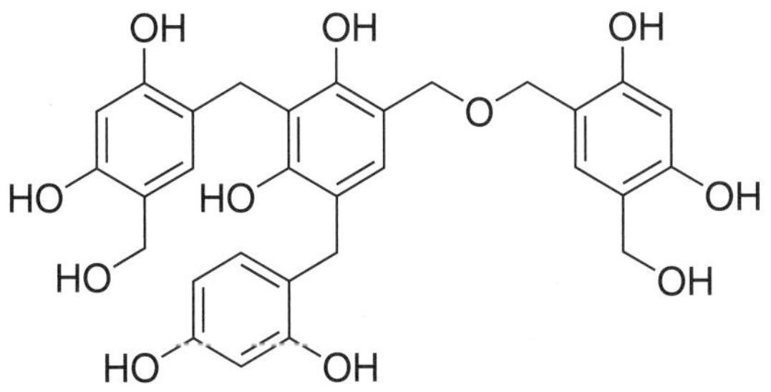

Figure 2.21 Scheme of a so-called resol structure, which typically is observed under an excess of formaldehyde. The benzene rings are connected by methylene groups and CH_2OH groups are attached for further reactions.

with a methylene bridge. Dimethylene ether bridges are never formed under acidic conditions. Also, in this case the reaction between resorcinol and protonated methanediol leaves a water molecule and as such it also is a polycondensation reaction.

2.2.3 From Monomers to Polymers

Contrary to resins made from resorcinol (see [83]), the synthesis of aerogels demands an excess of formaldehyde ($R/F < 1$) to be sure to promote more than just one substitution at the aromatic resorcinol and consequently more than just one networking condensation reaction. The synthesis of a dimer treated previously also shows how the process is going on. Dimers react with activated monomers or with other dimers. Once trimers are available, they can react with monomers, dimers and trimers, and thus it easily is possible to make oligomers. The activation at the aromatic ring always is fertilised by the catalyst. There are, as discussed, two bridges between the rings possible, methylene -CH_2- or dimethylene ether bridges. In the case of an excess of formaldehyde and base conditions, typically so-called resol structures are produced. A sketch of such a structure is shown in Figure 2.21.

If there are either acid conditions or an R/F ratio >1, so-called novolak structures are developed. These are less branched, more chain-like, but the essential difference from resols is that they always only contain methylene bridges. A sketch is shown in Figure 2.22. Especially with low formaldehyde concentrations, fusible novolak structures are possible, which have the appearance of just a chain polymer. This is shown in Figure 2.22. Due to their branching and network formation, resol structures are never fusible and behave like their well-known phenolic counterpart, the bakelite. The way from monomers to oligomers or polymers is a polycondensation reaction. As outlined earlier, the balance equations in base and acid catalysis always have as a by-product water. The polycondensation reaction can be described as a second-order reaction, meaning the reaction rate is proportional to the product of the resorcinol and the formaldehyde concentration. A mathematical model of polycondensation and its consequences are presented in the Appendix D. In the context here, it is sufficient to note that polycondensation reactions are slow reactions, and the degree of

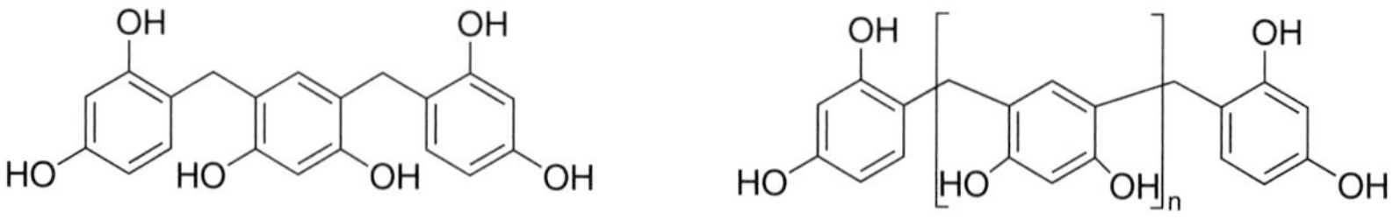

Figure 2.22 The left figure shows a scheme of a so-called novolak structure, which typically is observed if the molar concentration of formaldehyde is less than that of resorcinol. The benzene rings are connected by methylene groups. The figure to the right shows a chain-like novolak structure, which also is achieved at R/F < 1. Such a polymer is fusible [83].

polymerisation, DP, meaning the number of monomer units (here resorcinol rings) in a linear or branched structure is usually small, since Carother's law describes that it depends inversely on the conversion rate p like $DP = 1/(1-p)$. This means that it needs, for instance, a 99% conversion rate to achieve a DP of 100. A conversion rate of 95% yields a DP of 20. A resorcinol molecule has a diameter of roughly 1.12 nm. At a DP of 20, a plane polymer could have a size of around 22 nm, and at a DP of 100, the size could be 112 nm, assuming a dense plane molecule packing. This would show up in scattering methods, discussed later, as disk-shaped particles or simply branched polymers.

2.2.4 Parameters of Synthesis

The sol–gel reaction is influenced by many parameters and thereby the resulting properties of the aerogels. Here a list of parameters:

- Molar ratios
  - Resorcinol to formaldehyde (R/F)
  - Resorcinol to catalyst (R/C)
  - Resorcinol to water (R/W)
- Reaction conditions
  - pH value
  - Catalyst
  - Temperature

We go through this list point by point.

Effects of R to F and R to C Ratio

First of all, parameters affecting the reaction between resorcinol and formaldehyde is of course the concentration of formaldehyde, more specifically the ratio between formaldehyde and resorcinol, R/F, defined quite generally as

$$R/F \text{ ratio} = \frac{n_{\text{resorcinol}}}{n_{\text{formaldehyde}}} \quad (2.21)$$

with $n_{\text{resorcinol}}, n_{\text{formaldehyde}}$ as the moles of resorcinol and formaldehyde. Since in resorcinol the positions 4 and 6 are the most active ones stoichiometrically, two formaldehyde molecules per resorcinol are sufficient to fully bond a resorcinol

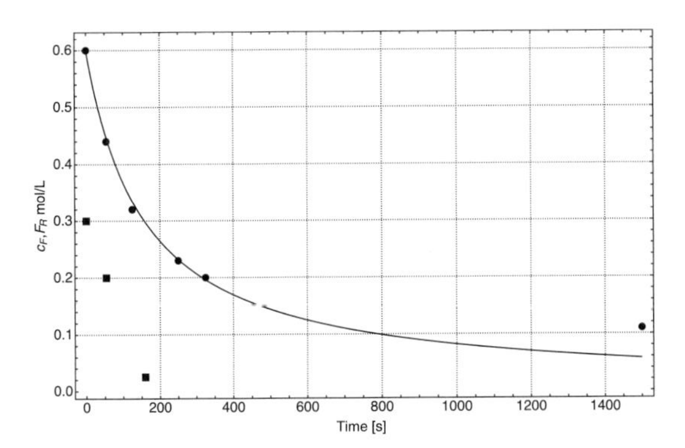

Figure 2.23 Experimental determination of the free monomer concentrations, R (squares) and F (circles), in an RF reaction as a function of the polymerisation time at 85°C; R/C ratio 200. Data extracted from Pekala and Kong's original paper [88]. The line through the formaldehyde concentration is fitted with a second-order reaction model of Eq. (2.32).

molecule into a more or less branched polymer. This would mean a ratio R/F of 1:2 would be the preferred one. But smaller and larger ones are also possible, and indeed it was observed that gels also form at different ratios. In most studies on RF aerogels, however, a R/F ratio of 1:2 is used.

An excess of formaldehyde always accelerates the reaction, which can be followed by measurement of the resorcinol and formaldehyde concentration as a function of time. A first measurement of the reaction kinetics was performed already by Pekala and co-workers [88]. They observed that the formaldehyde concentration seems to decrease with time slowly in a two-step procedure, whereas the resorcinol concentration drops almost immediately. Their result is shown in Figure 2.23 with a fit of a theoretical model discussed later. They discuss the observation that approximately 60% of the original formaldehyde is consumed after 200 minutes. In this time, also the resorcinol is almost fully reacted, which at least means it is transformed to hydromethylresorcinol and probably to larger polymers. In a recent study of Schwan [89], the formaldehyde concentration in RF solutions was measured as a function of time for four different R/C ratios in a highly diluted system ($R/W = 0.008$) at stochiometric composition $R/F = 1{:}2$. Her result is shown in Figure 2.24. This figure shows that the formaldehyde concentration decreases continuously with time and the experimental data can be fitted by a second-order reaction scheme.

The reaction of the resorcinol anion with formaldehyde to yield hydromethylresorcinol was measured by Yamamoto [90] with liquid chromatography. They describe the reaction as a substitution one with a rate equation

$$r_{sub} = k_{sub} c_R c_F, \tag{2.22}$$

with c_R, c_F the concentration of resorcinol and formaldehyde per unit volume and r_{sub} the reaction rate and k_{sub} the rate constant. Let us take a closer look at this equation, whose chemical background was not explained in the original paper of Yamamoto [90]. Assume the resorcinol (R) and formaldehyde (F) react in the presence of a

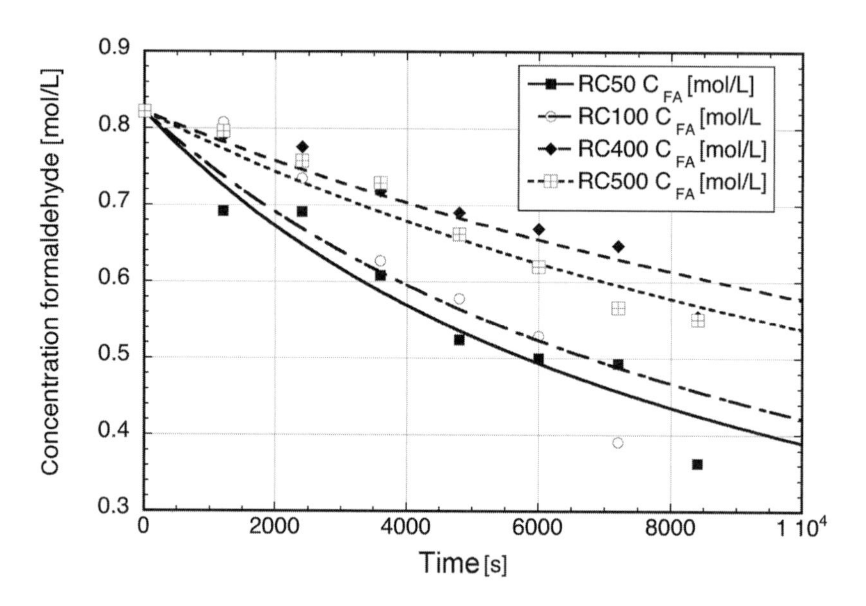

Figure 2.24 Formaldehyde concentration as it varies with time for a solution with R/F = 0.5, R/W = 0.008 and four different R/C ratios. The fits are made with a second-order kinetic equation.

catalyst to form hydroxymethylated resorcinol (HMR), R + F → HMR. Then the rate of resorcinol consumption can indeed be written as

$$-\frac{dc_R}{dt} = k_{sub}c_R(t)c_F(t). \tag{2.23}$$

The concentration of R and F are, however, not independent. We generally have from mass conservation

$$c_F(t) = c_R(t) + (c_F(0) - c_R(0)) = c_R(t) + (c_F^0 - c_R^0), \tag{2.24}$$

with $c_F(0) = c_F^0, c_R(0) = c_R^0$ the initial concentrations. Using this, we arrive at

$$-\frac{dc_R}{dt} = k_{sub}c_R(t)(c_R(t) + (c_F^0 - c_R^0) = k_{sub}\left(c_R^2 + (c_F^0 - c_R^0)c_R(t)\right). \tag{2.25}$$

This is a mixture of a second- and a first-order reaction. The equation can be integrated to give

$$c_R(t) = c_R^0 \frac{(c_F^0 - c_R^0)}{c_F(0)\exp\left((c_F^0 - c_R^0)k_{sub}t\right) - c_R^0}, \tag{2.26}$$

and equally we have for the formaldehyde concentration

$$c_F(t) = c_R^0)\frac{(c_F^0 - c_R^0)}{c_F^0\exp\left((c_F^0 - c_R^0)k_{sub}t\right) - c_R^0} + (c_F^0 - c_R^0) \tag{2.27}$$

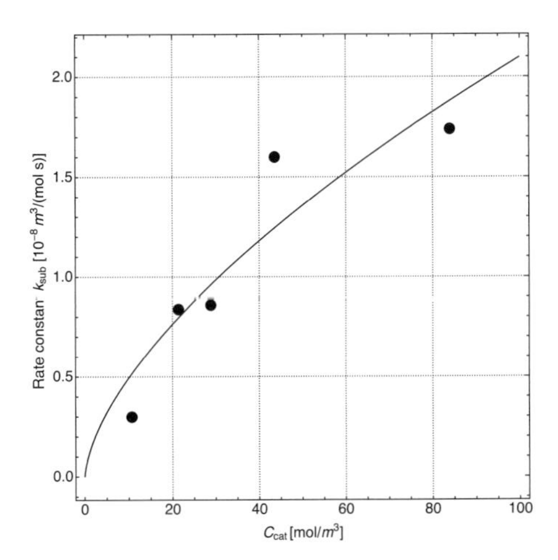

Figure 2.25 Rate constant of the substitution reaction defined in Eq. (2.22). The catalyst concentration, sodium carbonate, varies between R/C 25 and 200 and the R/W ratio was fixed at 0.5 (mol/mol). The reaction temperature is 25°C. Redrawn from the original data published by [90] with permission from Elsevier

The interpretation is straightforward: the consumption of R is roughly exponential in time and thus agrees qualitatively with the result of Pekala. From Eq. (2.24), it follows that formaldehyde follows the same trend as resorcinol. The reaction rate constant will, however, be influenced by the catalyst concentration. Yamamoto's experimental result is shown in Figure 2.25. Assuming the kinetic equation reflects reality, it also tells that at large R/F ratios the concentration of resorcinol is only dictated by the rate constant and the initial formaldehyde concentration $c_R(t) = c_R^0 \exp(-c_F^0 k_{sub} t)$. The simple approach in Eq. (2.23) does, however, only reflect reality in the initial state of the reaction. Eq. (2.27) would predict that the substitution reaction stops, once all R molecules are converted to HMR, and this means there would be an excess of formaldehyde left, namely of the molar concentration of R initially. This points to the fact that the simple reaction rate equation suggested by Yamamoto is not sufficient, since only one position at the resorcinol molecule is substituted. One has to modify the rate equations to take a substitution at two positions, say 4 or 6 at the aromatic ring into account (and principally also at position 2). Instead of the simple reaction scheme given previously we could set up a reaction scheme like

$$\mathrm{R} + \mathrm{F} \rightarrow \mathrm{HMR}_1 \tag{2.28}$$

$$\mathrm{HMR}_1 + \mathrm{F} \rightarrow \mathrm{HMR}_2 \tag{2.29}$$

$$\mathrm{HMR}_2 + \mathrm{F} \rightarrow \mathrm{HMR}_3. \tag{2.30}$$

In this HMR_k scheme, the index $k = 1, 2, 3$ indicates the substitution of a formaldehyde at three positions. Although one can write down the kinetic equations, we omit this, since there is no analytical solution of a system of second-order rate equations known [91, 92]. Paventi has, however, shown that a particular solution is possible for the scheme even if n consecutive reactions would occur [92]. His special solution is

$$c_F(t) = \frac{s \cdot c_R^0 - c_F^0}{\frac{s \cdot c_R^0}{c_F^0} \exp(s \cdot c_R^0 - c_F^0 kt) - 1}, \tag{2.31}$$

with s the number of active sites on the resorcinol molecule (formally 3, but in reality only 2). In the case studied in many RF aerogel papers, we have $2c_R^0 = c_F^0$, or an R/F ratio of 0.5. Taking the limit $sc_R^0 \rightarrow c_F^0$ yields a surprisingly simple result:

$$c_F(t) = \frac{c_F^0}{1 + c_F^0 kt}, \tag{2.32}$$

with an equation looking like a simple second-order reaction kinetic. Performing a fit to the formaldehyde data of Pekala yields the result shown in Figure 2.23, which is surprisingly good. The resorcinol concentration should follow the same functional behaviour, namely $c_R(t) = c_R^0/(1+c_F^0 kt)$. It does not, as already mentioned by Pekala. It decreases more drastically than hyperbolically with time, as the three data points in the figure show. This behaviour points to the fact, that free resorcinol or free anions are consumed readily into branched polymers, faster than the methylation by formaldehyde.

The R/F ratio is important, but more important is the catalyst concentration, since besides a few studies, most researchers leave R/F at 0.5. The catalyst concentration usually is given as a molar ratio:

$$R/C \text{ ratio} = \frac{n_{\text{resorcinol}}}{n_{\text{catalyst}}}. \tag{2.33}$$

In all cases an increase in catalyst concentration increases the reaction rate, which either leads to a faster decrease of formaldehyde concentration or an increase in the rate constant of the substitution reaction. The importance of the catalyst stems from its action on resorcinol: it promotes the formation of the resorcinol anion by hydrogen abstraction, as shown in Figure 2.18. At the beginning of the reaction, both uncharged resorcinol molecules and always a small percentage of charged resorcinol anions exist. Resorcinol anions are more reactive towards the addition of formaldehyde than the uncharged resorcinol molecules, as explained previously. Instead of sodium carbonate, which is readily available and cheap, several other catalysts were tested [93]. In most cases, just their effect on the final aerogel microstructure was looked at and a few properties measured, such as the specific surface area. Unfortunately, the chemical kinetics was not studied and compared with that of sodium carbonate. The microstructure was analysed by scanning electron micrographs of the dried materials. The variation of the microstructure using different catalysts was analysed by small angle X-ray scattering (SAXS). Although this is a way to somehow understand the effect of a catalyst, it is problematic, since between mixing of the components to a solution, the reaction of the dissolved molecules, gelation, solvent exchange and drying might lead to changes

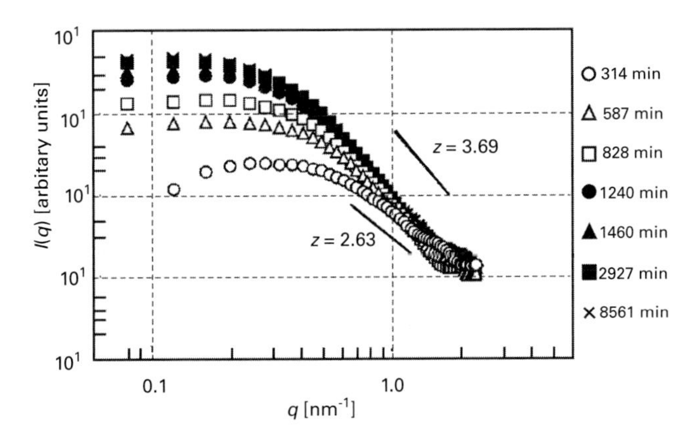

Figure 2.26 SAXS curves of an R and F solution during curing to a wet gel after [94]. $R/C = 25, R/W = 0.125, R/F = 0.5$ at room temperature. Reprinted with permission from Elsevier

in the morphology, which are difficult to track back unambiguously to, for instance, the R/C ratio and more specifically the kind of catalyst used. There are only a few investigations looking with SAXS or dynamic light scattering (DLS) or chemical methods (e.g., liquid chromatography) into the solution during transformation to first a sol and finally a gel.

The earliest report on SAXS measurement during the transformation from a solution to a gel was performed by Tamon and Ishikaza [94]. They prepared RF aerogels according to the original recipe of Pekala, but let them gel at room temperature, such that it was easy to follow the evolution of structure by SAXS. As explained in more detail in Appendix C on scattering, the scattered intensity follows at small scattering vectors a Guinier behaviour, characterised by an exponential decay of the scattering intensity:

$$I(q) = I_0 \exp\left(-\frac{1}{3}R_G^2 q^2\right), \tag{2.34}$$

in which R_G is the so-called radius of gyration of the scattering particles and I_0 is the beam intensity. A plot of $\ln I(q)$ against q^2 should give a straight line, which allows to determine the radius of gyration. The time evolution of R_G should tell something about particle growth. In the so-called Porod regime, the intensity follows a power-law behaviour:

$$I(q) \propto q^z, \tag{2.35}$$

with the exponent z taking in principle various values. For cylindrical objects $z = -1$, for disc-shaped ones $z = -2$ and for spherical particles $z = -4$. Fractional values can be understood as mass or surface fractal objects (for more details, see Chapter 4).

An important observation of Tamon and co-workers is that during gelation the shape of the scattering curve changes drastically. An example of a data set is shown in Figure 2.26. At short times, the scattering curve exhibits a maximum at intermediate

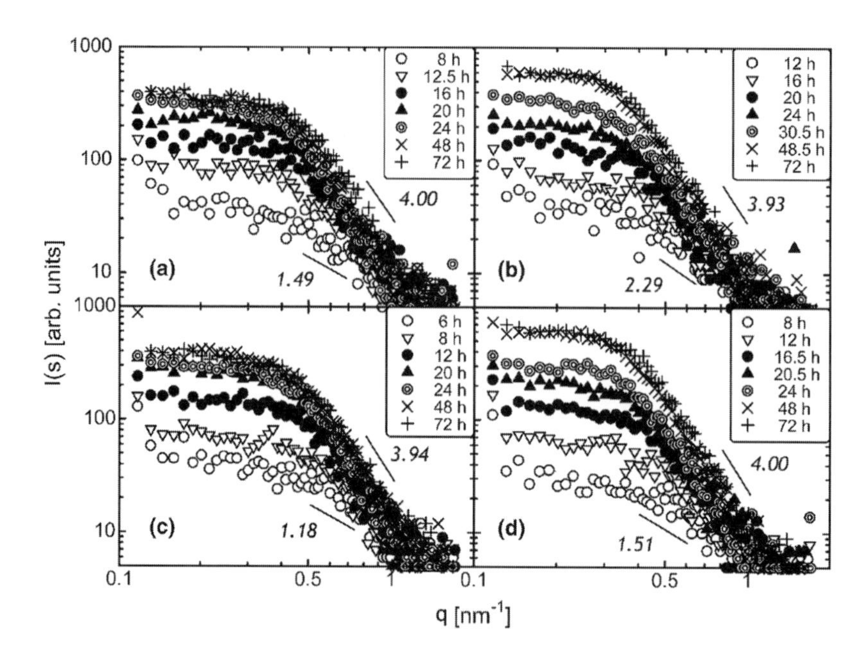

Figure 2.27 SAXS curves measured during the transformation of RF solutions to RF gel at 298 K with different catalyst species at R/C = 50: (a) K_2CO_3, (b) $KHCO_3$, (c) Na_2CO_3 and (d) $NaHCO_3$. Reprinted from [95] with permission from Elsevier

q-values and then drops off in a power-law fashion with an exponent around 2.6. With increasing time, the scattered intensity increases, meaning also more scattering objects are there, and a plateau develops at low q-values. The transition from the plateau into the power-law region is the Guinier region. The slope of the power-law decay at larger q-values changes with time. Tamon et al. evaluated the scattering curves and showed that z increases from values around 2.5 to almost 4. In the solutions with a smaller amount of catalyst (R/C = 200), they observed at long times a z-value larger then 4 and attribute this to a surface fractal or rough particles constituting the gel.

The idea to follow the sol–gel transition with SAXS was taken up by Horikawa and co-workers in 2004 [95]. In principle, they performed the same experiments as Tamon, but they changed the R/C ratio from 50 to 1000 and used four different catalysts (K_2CO_3, $KHCO_3$, Na_2CO_3 and $NaHCO_3$). The R/W ratio was 0.037 and the R/F ratio 0.5, and the solutions were allowed to gel at room temperature. Therefore, their results are very well comparable with Tamon's. Figure 2.27 shows four SAXS patterns obtained with different catalysts. In all cases, the shape of the SAXS curves change considerably during the transformation from a solution to a gel. Initially there is almost no clear power-law dependence of the scattering intensity as a function of q. This, however develops in all cases to almost the same shape: namely a plateau with a Guinier decay followed by a Porod decay with an exponent very close to 4. Especially the Porod regime is also observed in supercritically dried RF aerogels and was taken as a clear indicator that the aerogels are constituted by smooth spherical particles [96]. Many studies of RF aerogels have confirmed the original observation and analysis of Schaefer and Pekala made already in 1993, such as [85, 97–99].

In a study on structure formation during gelation, Gommes and co-workers [100] also used SAXS to interpret the structures observed, but they followed a different analysis method and came to important conclusions concerning the formation of a gel network. They synthesised RF aerogels with an R/F ratio of 0.5, a dilution ratio of M = 0.175 (see Eq. (2.38)) and an initial pH value of 3.7. They observed that at early reaction times below 15 min the SAXS curves exhibit a decay at low angles with a slight curvature, with a flat background at larger angles. Plotting at this time $I(q) \cdot q^2$, named a Kratky–Porod plot, shows a clear maximum at very small q-values, which is indicative of a branched polymer. The maximum shifts slightly towards smaller angles with time. At intermediate reaction times, 20–35 min, first the scattering intensity increases by an order of magnitude, which reflects that more scattering objects are in the solution. After half an hour, annealing the scattering intensity exhibits a power-law scaling with an exponent of −2, which is replaced after longer times by a power-law decay with an exponent of −3. Gommes and co-workers argue that the exponent of −2 is a strong indication of critical opalescence in a solution transforming to a gel. Such opalescence indicates that a phase separation takes place starting with the formation of nanosized droplets everywhere in the sample. This is similar to the development of water drops in a supercooled, supersaturated water vapour atmosphere below the dew point. They analysed the scattering intensity with an expression derived by Ornstein, Zernicke and Debye, namely

$$I(q) = \frac{I_0(t)}{1 + \xi^2 q^2}, \tag{2.36}$$

in which ξ is the correlation length, being a measure of the size of the scatterers. The analysis of the data allowed plotting I_0 as a function of time. Figure 2.28 shows one result. The sharp increase in scattering intensity at around 20 min is due to a huge amount of small scatterers appearing in the solution, indicating, for example, the formation of polymer-rich droplets. This effect will be analysed and described later in Chapter 4 on gelation. The formation of a bunch of droplets could be, for instance nucleation events occurring once a miscibility gap is reached, or better, the system developing a gap and after a certain time the whole system moves from a one-phase field into a two-phase field. We will discuss this Section 2.2.5 on the thermodynamics of RF solutions.

Instead of X-ray small angle scattering, Gaca and Sefcik used DLS to study the processes in RF solutions until gelation [101]. They prepared solutions with an R/F ratio of 0.5 and R/C ratios from 50–600 using sodium carbonate as a catalyst. The initial pH value decreased with the increasing R/C ratio from 7.51 (R/C 50) to 6.11 at R/C 600. The ratio of R/W given in mol/mol was fixed at 0.0164 in the series of R/C from 50 to 600. They also prepared solutions with R/W ranging from 0.0328 to R/W = 0.082. The solutions were put into suitable vials for DLS. The solutions were cured at 55 and 80°C but the measurements of DLS always

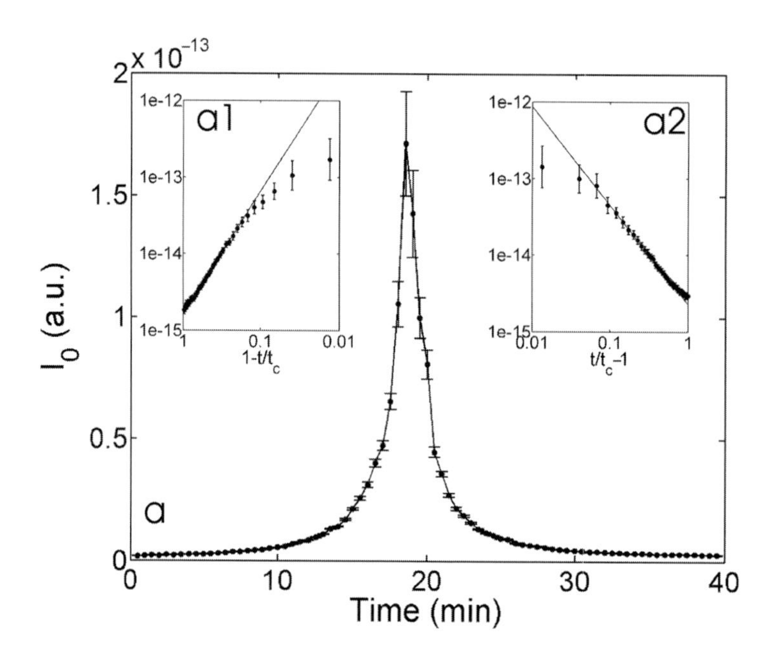

Figure 2.28 Increase in scattering intensity in an RF solution undergoing a phase separation after [100]. The sharp increase in scattering intensity at around 20 min indicates that a huge amount of small scatterers appear in the solution, indicating for example, the formation of polymer-rich droplets. Reproduced under Creative Commons license

performed at room temperature. The gelation time increases linearly with R/C and also R/W. As explained in Appendix C on scattering, in DLS one measures the autocorrelation function, which is has a simple shape in a monodisperse solution, namely an exponential decay $\exp(-\Gamma t)$, and in a polydisperse solution it is the sum of such exponentials. The decay rate Γ is the product of the diffusion coefficient and the square of the scattering vector q. Having determined the decay rate by fitting the measured autocorrelation function with an exponential allows using the Stokes–Einstein relation (see Eq. (10.15)) to calculate the hydrodynamic radius of the particles or particle clusters. Figure 2.29 shows some measured autocorrelation functions as they change with reaction or curing time. Figure 2.30 shows the variation of the hydrodynamic radius with R/W ratio as a function of reaction time. The hydrodynamic radius starts with values of around 1 nm and increases to values of 4–5 nm. In the figure, we have drawn in three simple models for the time dependence of the hydrodynamic radius. The solid line corresponds to a simple linear growth with reaction time, the dashed mimics a diffusive growth, the radius would then depend on the square root of reaction time and the dotted line reflects a cube root growth of the particles. A square root growth would be expected if diffusion of monomers and oligomers is the time-dependent and thus the slowest step. A cube root growth would mean a curvature-dependent solubility determines the change of the morphology (e.g., a coarsening of a spinodally demixed structure; see Section 4.2.5). Clearly, according to the data of Gaca and Sefcik, growth is

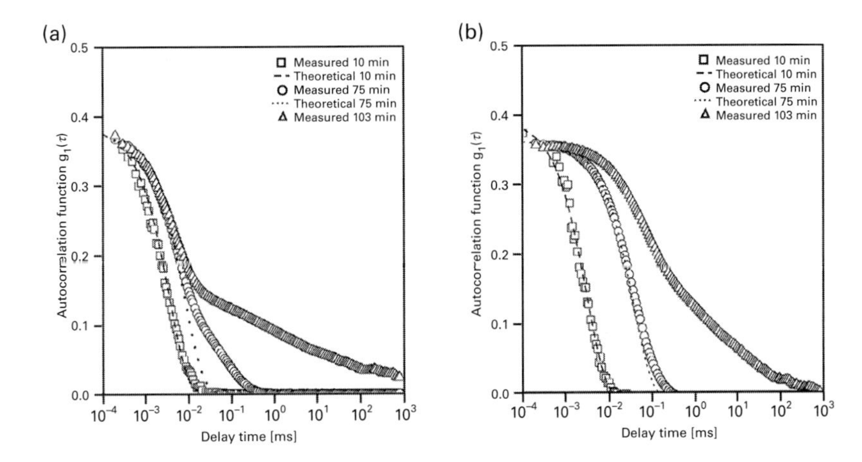

Figure 2.29 Measured autocorrelation functions. In (a), the R/W ratio is 0.1, R/C fixed at 50. In (b), the R/W is at 0.5. Curing is in both cases at 55°C. In both figures, the dashed lines are calculated from a simple exponential decay. Reprinted from [101] with permission from Elsevier

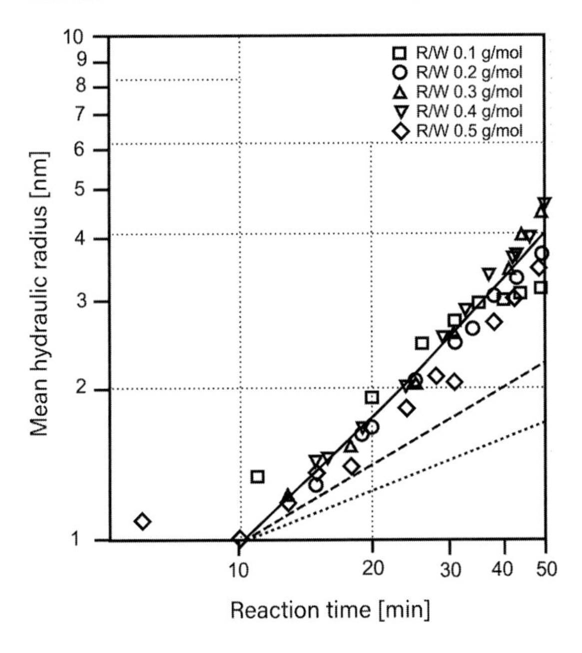

Figure 2.30 The hydrodynamic radius increases with reaction time but is almost independent of the R/W ratio [101]. Note that both scales are logarithmic ones. Redrawn from [101] with permission from Elsevier

linear with reaction time, and thus growth is determined by a first order kinetic [51]. For all conditions investigated by Gaca and Sefcik, the size of primary clusters varies on changing concentrations of reactants the R/C ratio. The authors conclude that their results are consistent with a process of nanoscale demixing, controlled by

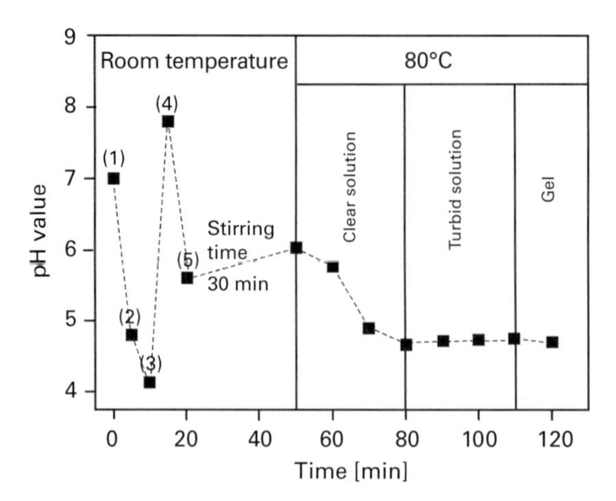

Figure 2.31 Variation of the pH value in an RF solution with R/C = 50, R/F = 0.5 and R/W = 0.008 after [89].

solubility/miscibility equilibria. They suggest that certain intermediates produced in resorcinol–formaldehyde polycondensation reaction reach a limit of their solubility in the solution at a given temperature. Then primary particles or clusters appear, but the process of polycondensation and shift of the phase equilibrium goes on. We discuss this briefly in Section 2.2.5 on thermodynamics and in more detail in Chapter 4 on gelation.

Effect of pH Value

The difficulty in all investigations with RF solutions is, that the catalyst and the amount of catalyst fixes an initial pH value and in principle a solution can then be left alone and will yield a gel. However, after adding a catalyst, the pH value can be adjusted in both directions, more acid- or base-like, by using, e.g., nitric acid or sodium hydroxide. We will in the following discuss a few exemplary investigations which only adjusted the pH to a certain value. Brandt investigated in his PhD thesis the whole range from acid conditions using acetic acid and sodium carbonate with different R/C ratios [102]. Schwan discusses in her PhD thesis in detail the pH change during processing using sodium carbonate and sodium hydroxide with their effect on the microstructure [89]. Ludwig and co-workers [103] presented a detailed investigation of citric acid catalysed aerogels, and Mulik and co-workers invented a new synthesis routine to produce nanostructured RF aerogels under strong acidic conditions.

The pH value is not a simple constant during synthesis of RF aerogels. Schwan [89] measured the pH-value in every step of preparation. Her result is shown in Figure 2.31. The preparation of the solution starts with de-ionised water (1) and thus a pH value of 7. Adding resorcinol in the second step decreases the pH to 4.6–4.8, since resorcinol reacts with water and makes it slightly acidic. Adding in step (3) formaldehyde again decreases the pH value to a value slightly above 4, since formaldehyde in water also has an acid character. Adding sodium carbonate increases the pH to a value between 7.5 and 7.7 (step (4)). In the last step, nitric acid is used to adjust the pH value to 5.5. Then a stirring period of 30 min follows with a small increase in pH. Leaving the solution at rest but heating it to 80°C, it stays clear until after 80 min it becomes milky and turbid, followed by gelation after 110 min. Once the solution turned opaque, the pH value stays almost constant slightly below 5.

The pH value of a solution also depends on the catalyst used. Conventionally in preparation of resorcinol resins, sodium hydroxide is used, whereas for RF aerogels sodium carbonate is most common. The behaviour of the pH value is different in RF solutions. Schwan measured that for an RF solution catalysed with NaOH and Na_2CO_3. She observed that mixing both catalysts in a solution with $R/F = 0.5$, $R/C = 50$ and $R/W = 0.008$ leads to the following behaviour: the pH value of around 5.5 for both catalysts used changes as a function of stirring time. However, with NaOH, the pH value remains constant at 5.5 all the time it increases to a value of 6 if sodium carbonate is used. Heating the solution to 80°C leads in both cases to a decrease. In the cases of the carbonate, it eventually reaches 4.5. In the NaOH case, it falls down to 3.5. The increase upon the addition of sodium carbonate is an effect of its buffer capacity and has also the side effect, that the gels contain small bubbles, especially at the gel surface.

As mentioned, an extensive study of the effect of pH value and R/C ratio was performed by Brandt [102], who used RF aerogels as precursors for carbon aerogels. In his study, he fixed the R/F ratio to 0.5 and varied both the mass fraction, (see Eq. (2.38)) between 5 and 55 and the R/C ratio (see Eq. (2.33)) from values between 5 and 2000. For base catalysed aerogels, he used sodium carbonate as a catalyst, and for acid catalysed aerogels he used acetic acid as a catalyst. The solutions were gelled 1 day at RT, 1 day at 50°C and 3 days at 90°C and then washed with acetone and dried 3 days in a drying cabinet at 50°C. These ambient dried RF aerogels were characterised with standard methods. In this context, it is interesting to note that gels could not be obtained for all combinations of R/C and mass ratio M; see Eq. (2.38). He collected these results in an interesting diagram shown in Figure 2.32. Interestingly, at high R/C ratios (low catalyst concentration) and also low amount of resorcinol (M-ratio), precipitates form which also can settle (the skeletal density of RF particles is around 1.5 g/cm^3 and thus higher than the density of water). In a small region adjacent to the sedimentation region, inhomogeneous gels are obtained, which probably is due to a combined effect of sedimentation of precipitates and their agglomeration to larger particles.

The effect of the pH value on gelation time was determined by Job in her PhD thesis [81]. She was able to investigate at constant R/F ratio of 0.5 and a mass ratio of 17.5% the gel time as a function of pH value and could reproduce an old figure shown in Durairaj for resorcinol formaldehyde resins [83], namely that the gel time is short at low and high pH, but has in between a maximum. Her result is shown in Figure 2.33.

For purely acid-catalysed aerogels, Milow et al. [104] determined the gelation time as dependent on the pH value. They used an R/F ratio of 0.77 and varied the R/W and R/C ratio using citric acid. Citric acid does not allow varying the gel time over a large scale, but in the given scale it varies almost linearly with pH as shown in Figure 2.34.

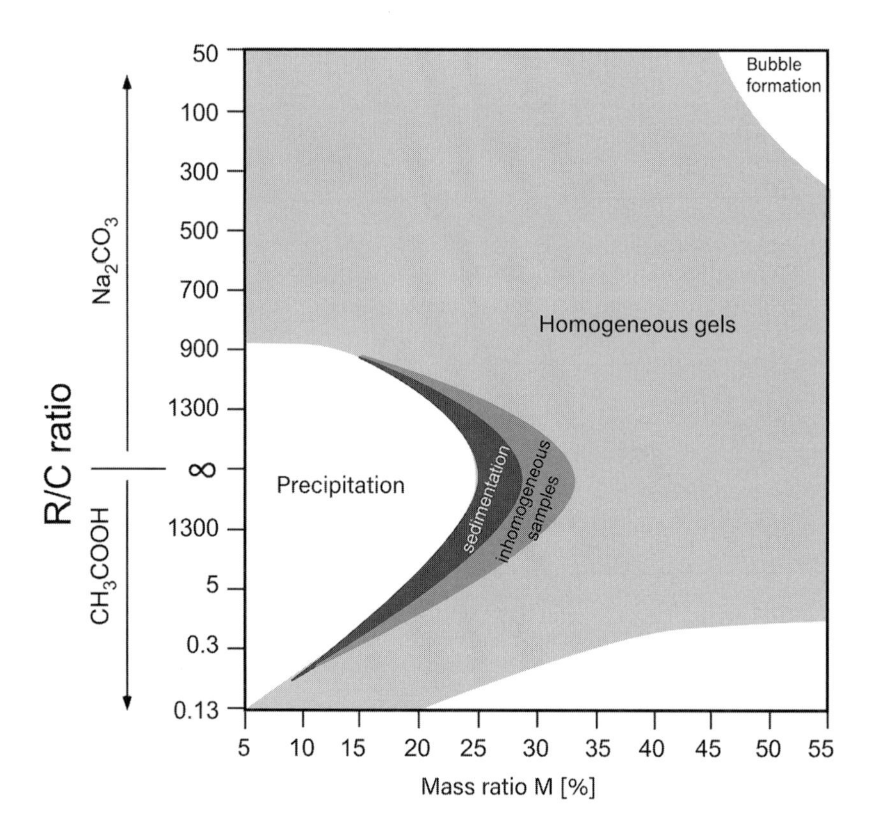

Figure 2.32 Region allowing to produce homogeneous aerogel after [102] as dependent on the acid- or base-catalyst ratio R/C and the mass fraction.

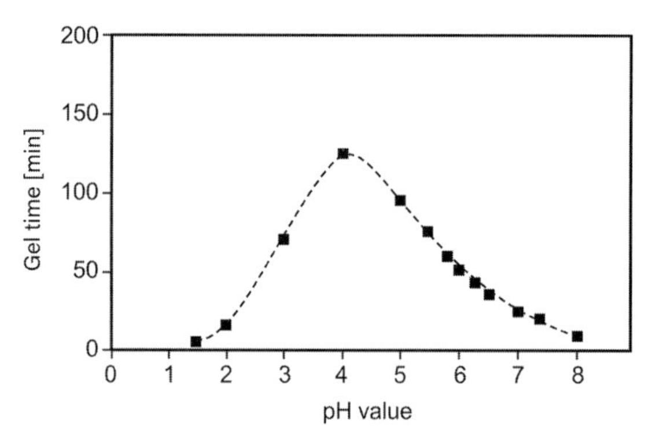

Figure 2.33 Gel time in RF aerogels prepared with $R/C = 0.5$, $M = 17.5\%$ as dependent on pH value adjusted with an acid. Redrawn with the data presented by Nathalie Job in her PhD thesis [81]

Effect of Dilution: R/W Ratio

In most cases, resorcinol is dissolved in water and then formaldehyde is added as well as the catalyst. The chemical reactions in such a solution are of course determined kinetically also by the amount of water used. This typically is expressed as the ratio R/W. At a fixed amount of resorcinol, a decreasing R/W ratio remarks a

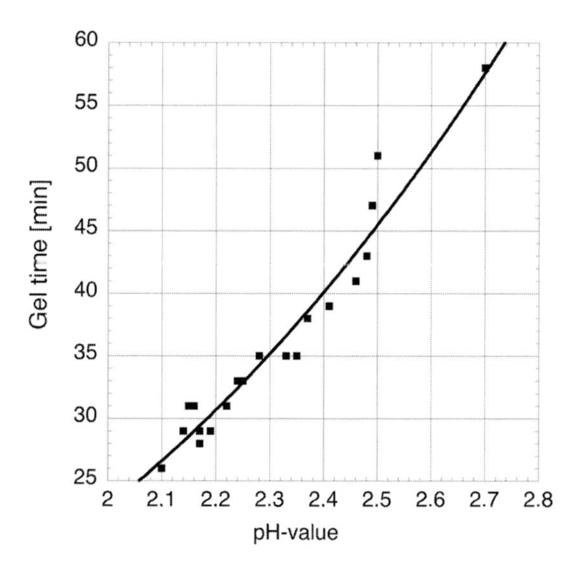

Figure 2.34 With increasing the pH value of RF solutions using citric acid the gel time increases linearly with it [104]. The pH-value variation is a larger variation in R/C from 24 to 1500.

higher dilution, and thus for any reactant it takes more time to diffuse to each other and react if the solution is diluted more and more. The dilution ratio determines many properties of aerogels, the most simple being the envelope density, as will be described and calculated in detail in Chapter 7. The dilution ratio also affects the microstructure and especially the particle size and the porosity. The effect of R/W on the formation of RF aerogels was carefully investigated by Schwan [105, 106] for base-catalysed aerogels and for acid catalysed ones by Milow et al. [104].

$$R/W \quad \text{ratio} = \frac{n_{\text{resorcinol}}}{n_{\text{water}}} \tag{2.37}$$

which will be a ratio of moles resorcinol to water. Instead of this ratio, a so-called mass ratio is often used

$$M = \frac{m_{\text{resorcinol}} + m_{\text{formaldehyde}}}{m_{\text{resorcinol}} + m_{\text{formaldehyde}} + m_{\text{catalyst}} + m_{\text{water}}}. \tag{2.38}$$

For citric acid-catalysed aerogels, the gel time looks as if it depends parabolically on the R/W ratio. A result is shown in Figure 2.35. The gel time increases with the molar fraction water as expected. Since no gelation model is currently available that theoretically predicts gel times, one can only state that inasmuch as the solution is diluted, the diffusion distance between molecules increases, thus the chemical reaction needs more time to start, and of course, once oligomers and particles have developed, they need more time to move to each other. One could also argue that if a phase separation process as discussed briefly in Section 2.2.5 would be responsible, then the dilution would shift the state point of the system to the water side, and during the kinetics of the phase diagram evolution it will take a longer time for the evolving miscibility gap to pass this state point.

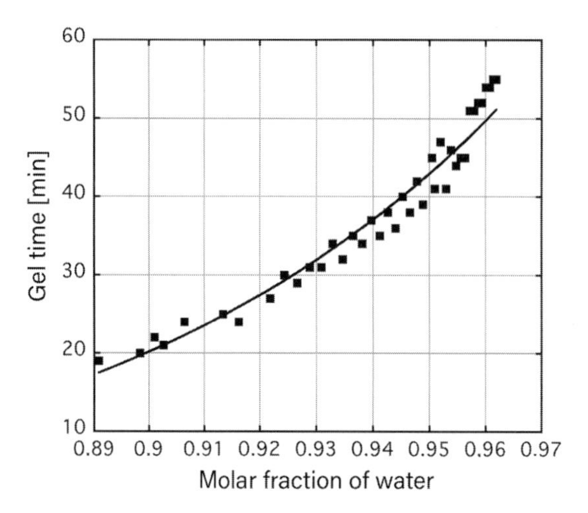

Figure 2.35 The R/W ratio increases or the dilution of the solution with water the gel time increases in a simple parabolic manner.

A different approach to look into the effect of dilution ratio was made by Schwan in a series of papers [105, 106]. She showed that on extreme dilution, gels can be produced with such a porosity that they become flexible, meaning one can squeeze them and they come back almost to their original shape. This flexibilisation is very prominent after supercritical drying, whereas ambient drying yields soft but more cork-like mechanical behaviour. The microstructure of these RF aerogels still is a particular one, with chains of bigger particles. The pore size is rather large such that on deformation the chains can bend freely without restrictions by neighbouring particles.

Influence of Temperature

As usual for all chemical kinetics, the reaction of R and F is strongly temperature dependent. The chemical reaction kinetics, however, have not been studied in such detail, that the temperature dependence and the activation energy of all chemical process steps are known. Typically, the gelation time is measured instead as it varies with temperature or the viscosity. Both are presented and discussed in Chapter 4.

2.2.5 Thermodynamics of RF Solutions

Thermodynamics is concerned with equilibrium of phases, caloric properties of materials, their specific heat, heat of melting or vaporisation, relations between enthalpic and entropic properties, functions describing the state of a material (gas, liquid, solid) as dependent on pressure and temperature etc. The bad news for RF aerogels is that there are no experimental investigations available on RF solutions. We are therefore left with, in a certain sense, speculations or at least reasonable arguments, and we will try to make some good guesses for RF solutions with the help of well-known theoretical models for polymer solutions. The most often used model for solutions of polymers in a solvent is the Flory–Huggins (FH) model. This model is a variant of the

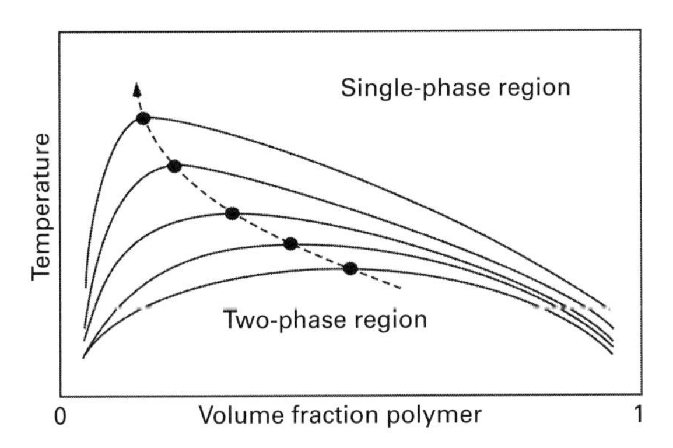

Figure 2.36 Schematic of a phase diagram of a solvent mixed with a polymer, exhibiting a miscibility gap. For different degrees of polymerisation the maximum of the binodal line separating the single-phase field (solution) from the two-phase field (two liquids) shift to the left, the solvent side.

more general thermodynamics of solutions. In Appendix A, we discuss the origin of a miscibility gap, the explanation of terms 'binodal' and 'spinodal' and the process of phase separation. The interested reader is advised to take a look at the appendix and of course standard textbooks on physical chemistry and thermodynamics [54, 107–109]. The Flory–Huggins model is explained in more detail in Appendix B.

Let us summarise one principle result of the FH model: it predicts a phase diagram for a solvent and a polymer. Such a phase diagram is the two-dimensional representation of phase fields and a schematic shown in Figure 2.36. On the ordinate, the temperature is plotted and in the abscissa the concentration of the polymer; most often in polymer physics, one uses the volume fraction polymer. A phase diagram then shows regions in which a single phase exists and another region in which two phases are in equilibrium. The boundary line between the fully dissolved state and a field of two liquid phases is called a binodal line. It covers the miscibility gap. If the temperature and concentration point is in the two-phase field, equilibrium thermodynamics require that two phases coexist. If, for instance, in such a situation the temperature is changed (generally increased) the state point can move into the single-phase region and the two-phase system is transformed into a one-phase system (see Appendix A). Flory and Huggins have shown, that the shape of the binodal line depends on the degree of polymerisation, DP. The higher the DP, the more skew-symmetric is the binodal line, and since this line has a maximum, that maximum shifts towards the solvent side. One could guess from this behaviour, especially with the experiments described earler, that RF solutions may exhibit a miscibility gap. The curves shown in Figure 2.36 have a so-called upper critical solution temperature (UCST) point, meaning above that point the two components are miscible, below not. There are, of course, systems in which the curves are just inverted and there is a lower critical solution temperature (LCST) point. This especially occurs in systems in which the two components (solvent and solute) form weak bonds, e.g., hydrogen bonds.

There is, however, a principle problem with the polymer solution model of Flory–Huggins and possibly all other polymer solution models and the application to RF aerogels: the phase diagram is of irreversible nature in RF gels. Any thermodynamic model assumes that the phase boundaries are equilibrium lines, meaning that once a phase separation has occurred, a change in temperature or concentration can lead from a two-phase region to a single-phase one and vice versa. In RF gels, this is never the case. Once the polycondensation reaction has formed polymers of a given DP, they cannot be dissolved anymore in the solution. Therefore, the application of thermodynamic models to RF aerogels is outside equilibrium thermodynamics. We think that in principle one can nevertheless utilise the Flory–Huggins model to understand certain issues of structure formation in RF gels.

The application of the classical FH model to RF solutions and their conversion to an RF aerogel looks more complicated compared with conventional polymer-solvent systems, since one has to consider several effects:

1. The degree of polymerisation depends on time, the pH value and the amount of catalyst. Resorcinol reacts with formaldehyde, and these activated hydroxymethylresorcinol react, leading to the polymer. Thus $DP = DP(t)$. The time dependence is defined by the polycondensation reaction as described in Appendix D.
2. There is a time dependence of the volume fraction of polymers $\phi(t)$, and thus the state point is not fixed; see also Appendix D.

Therefore, the miscibility gap varies its shape with time and eventually the system passes either into the region between the binodal, decomposes in a spinodal fashion or does both in a sequence.

Before we go on, we have to clarify: what if RF solvent systems have a lower or an upper critical point? Both systems behave in one sense completely different, and this allows us to experimentally discriminate between them. Figure 2.37 shows schematically both systems and a state point of an RF mixture indicated by the black circle. If one would perform a sudden temperature change in a UCST system, the state point would move into the two-phase regime, whereas in the LCST system the

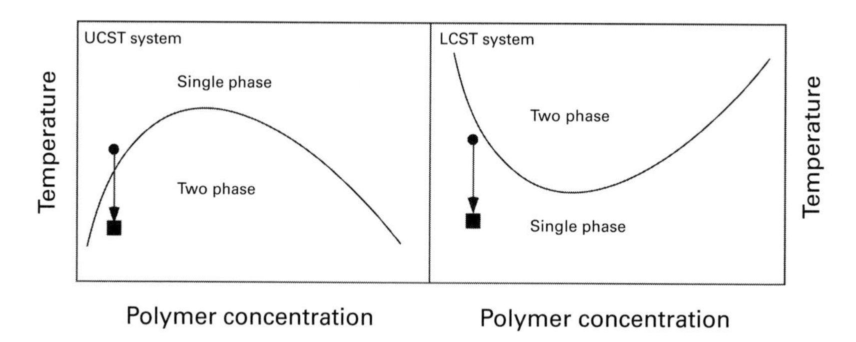

Figure 2.37 Scheme of a RF solution with an evolving miscibility gap having an upper or a lower critical point. The black circle represents for a given time the state of the system being close to the binodal line. A quench in a UCST system moves the state point into the two-phase region, whereas the same quench in an LCST system leaves the state point in the single-phase region. Reprinted from [110] with permission from Springer-Nature

state point stays in the single-phase regime. Experimentally, it is exactly the latter that is observed. Whenever a solution that stayed at a certain temperature (RT and higher) is transferred to a fridge, it can stay there without gelation for quite a while. If the RF water system would exhibit an upper critical point, this would not be possible. A quench to a lower temperature would immediately lead to a phase separation and eventually to gelation. Therefore, from an experimenter's point of view, the RF water system must be one with a lower critical point. Let us therefore assume that RF solutions belong to the class of systems with a lower critical point. Inasmuch as curing proceeds at a given curing temperature, the miscibility gap develops and the state point (polymer volume fraction and temperature) moves into the two-phase field, shown schematically in Figure 2.38. Inside the two-phase region, polymer-rich

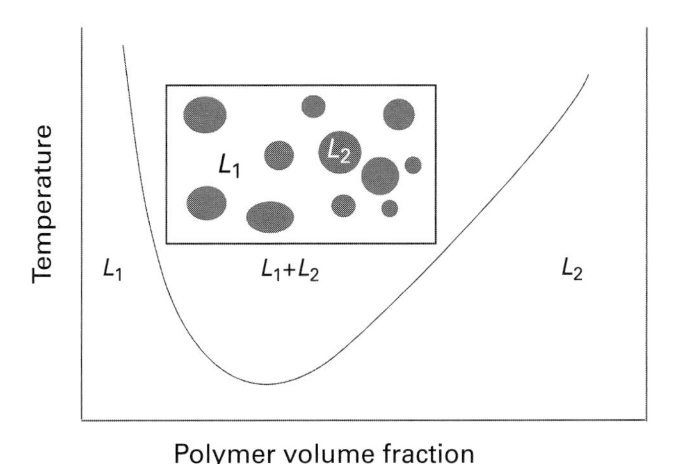

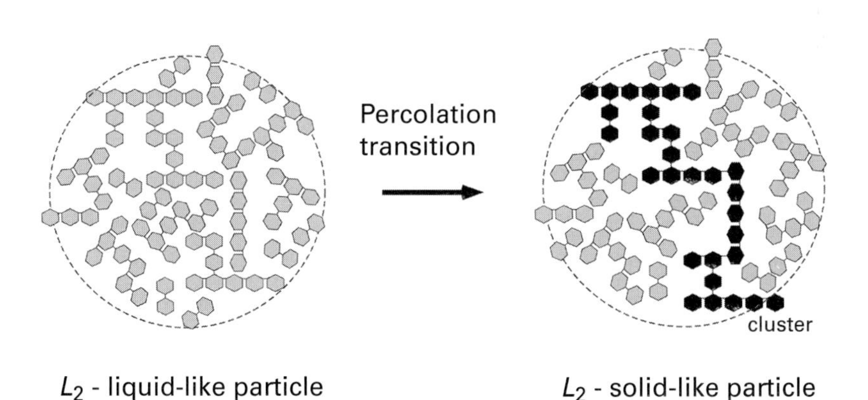

Figure 2.38 Schematic of a RF–polymer solvent system (water) with a lower critical point showing the two-phase region and both single-phase regions to the right and left of the critical point. If the system enters the miscibility gap, liquid droplets rich in polymers appear, and these droplets could make a percolation transition (they gel) and then exhibit a solid-like behaviour, although they are still rich in liquid and the surface also has liquid properties. The lower part is reprinted from Fei Wang et al. [110] with permission from Springer-Nature

droplets appear, and their volume fraction will change as well as their size with time. The solution turns to a suspension of droplets in a polymer-depleted solvent. We now can ask the question, what happens with the liquid droplets of the second phase? Looking at the phase diagram, it is clear that they have a volume fraction of polymers above 30–50% (see also Appendix A). Whatever model of percolation we apply (site or bond percolation, lattice or continuum), this means the amount of solid-like oligomers in the droplets is beyond the percolation threshold and thus, once these droplets form, they transform into a wet gel particle, meaning they immediately make a percolation transition. The phase separation leads to solid-like gel particles in a matrix liquid containing still polymers of varying degrees of polymerisation (for an introduction into percolation, see [111] and Chapter 4, which show that for continuum or off-lattice percolation, the percolation threshold can be rather low, below 15%). A scheme of the phase separation and a possible percolation transformation in a system with a lower critical point is also depicted in Figure 2.38 adopted from Fei Wang et al. [110]. We can conclude that the application of the FH theory with a lower critical point to RF aerogels would always lead to a situation that a phase separation in the metastable region of the phase diagram happens first. Spinodal decomposition is possible especially in LCST systems, since the spinodal already is extremely steep at the solvent boundary for even not too high DPs (see Appendix B).

2.3 Variables and Symbols

Table 2.2 lists some of the variable and symbols of importance in this chapter.

Table 2.2 List of variables and symbols in this chapter.

Symbol	Meaning	Definition	Units
a	Radius of a monomer	–	m
$c_F(t)$	Concentration formaldehyde in solution	–	moles/m^3
$c_R(t)$	Concentration resorcinol in solution	–	moles/m^3
c_{SiO_2}	Concentration silica in solution	–	moles/m^3
d_f	Mass fractal dimension	–	–
d_s	Surface fractal dimension	–	–
f_ϕ	Phase fraction of ϕ in a solution	–	–
$I(q)$	Scattering intensity	–	
k_B	Boltzmann coefficient	1.38×10^{-23}	J/K
k_H	Reaction rate constant	–	1/s
k_{sub}	Reaction rate constant	–	1/s
$\mathcal{H}$	Mean curvature	–	1/m
N	Number of atoms in a molecule/polymer	–	
N_A	Avogadro's constant	$6.02\ 10^{23}$	mol^{-1}
N_m	Number of monomers in an object of radius R_G	–	

Table 2.2 Continued

Symbol	Meaning	Definition	Units
N_s	Number of surface sites on a particle	–	
n_0	Number density	–	1/m^3
$\vec{q}$	Wave vector	$\lvert\vec{q}\rvert = 2\pi/\lambda$	1/ m
$\vec{r}_c$	Location of the centre of mass	–	m
$\vec{r}_i$	Location of a molecule in a polymer	–	m
R_G	Radius of gyration	–	m
R_G^{max}	Maximal radius of gyration	–	m
t_{gel}	Gel time	–	s
V_p	Particle volume	–	m^3
$x_{species}$	Molar fraction of *species*	–	
z	Exponent in power law	–	–
λ	Wavelength	–	m
$\Gamma(T)$	Attachment frequency	–	1/s
μ_p	Chemical potential	–	
Ω_m	Atomic or molar volume	–	m^3
ϕ	Volume fraction	–	–
ϕ_s	Volume fraction solid	–	–
ϕ_p	Pore volume fraction	–	–
ρ	Density	–	kg/m^3
ρ_e	Envelope density	–	kg/m^3
ρ_s	Skeletal density	–	kg/m^3
σ	Interfacial energy	–	J/m^2
σ_{sl}	Interfacial energy between a silica particle and solution	–	J/m^2
τ	Dimensionless time	–	–
ξ	Correlation length	–	m
ζ	Dimensionless length	$R_G(t)/R_G^{max}$	–

3 Chemical Synthesis of Aerogels from Polymeric Precursors

In this chapter, we present the synthesis of aerogels in which a solution of polymers dissolved in a suitable solvent is the starting point. In contrast to the aerogels of the preceding chapter, no polymerisation reaction is first necessary, but instead a given polymer has to be dissolved and then by suitable processing rearranged to form an open porous nanostructured network. We will discuss two types of aerogels made from cellulose and in the context of alginate as a model biopolymer as well as a few other biopolymers.

3.1 Synthesis of Cellulose Aerogels

Cellulose is probably the most abundant natural polymer on our planet. It is produced by nature in teratons per year, mainly by plants but also some bacteria and fungi. In plants, cellulose is in fibrillar form and coexists with hemicellulose and lignin. The combination of cellulose and lignin gives plants and their cell walls mechanical stability. The cellulose fibrils embedded in lignin (a phenolic polymer) give excellent tensile strength, whereas the lignin increases the compressive strength. For commercial applications, cellulose is obtained from cotton, bast fibers, flax, hemp, sisal and jute or wood. The extraction of cellulose from plants needs special chemical processes to separate it from lignin and hemicellulose. The pure cellulose is a biopolymer and exists in various modifications. The macromolecule consists of repeating cellobiose units, shown in Figure 3.1. Speaking chemically more exact, cellulose is a chain of 1-4-linked β-D-glucopyranose. The figure shows the link between two rings made by an oxygen atom. The repeating rings in the cellobiose units are also called anhydrous glucose units (AGUs), which are connected by the oxygen bridge. In the cellulose polymer, each AGU has three OH groups. Two of them are directly located at the carbon atoms of the AGU, and one is outwards via a hydroxymethyl group. The cellobiose unit is not a flat molecule, in which all atoms are in a plane, but around the oxygen bridge the molecule is twisted.

Combining cellobiose units gives a cellulose polymer. The chain of cellulose is stiffened by various hydrogen bonds, intramolecular bonds and intermolecular bonds. There is ample literature on how these bonds are arranged. To get at least a simple picture of a few bonds, Figure 3.2 shows two cellulose chains with interchain

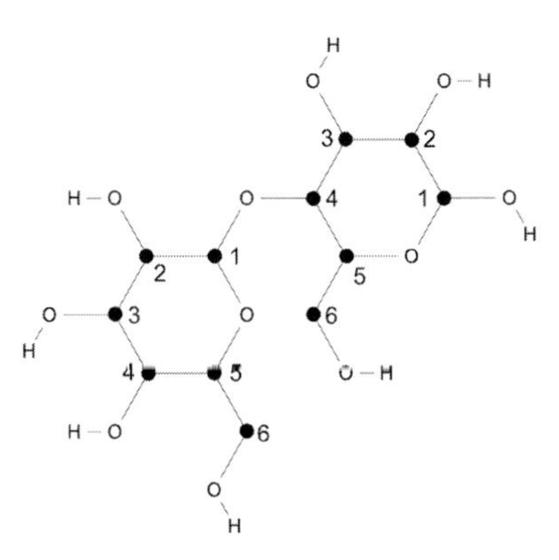

Figure 3.1 Chemical structure of a cellobiose unit: two β-D-glucose rings are connected via an oxygen bridge at positions 1 and 4 in the ring. The full circles indicate the carbon atoms and the numbering of their position in the ring. Note first that not all hydrogen bonded at the carbon atoms is drawn and, second, that the basic unit of a cellulose polymer contains in its ring oxygen atoms and of course in the -OH groups. This is important with respect to pyrolysis of cellulose to produce carbon or graphite.

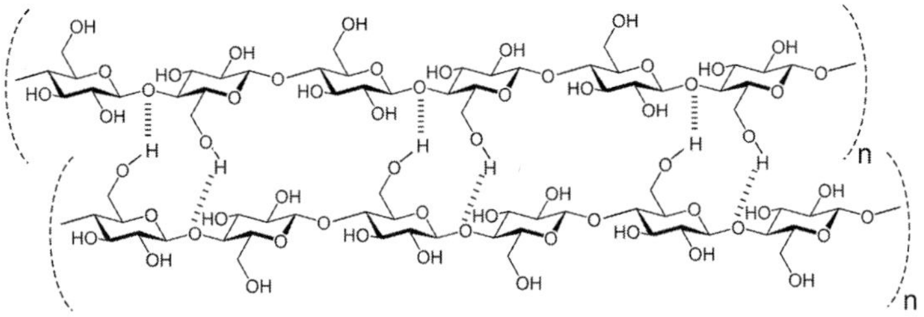

Figure 3.2 Schematic view of two parallel strands of a cellulose polymer. The hydrogen bonds between the chains are shown as a bar with stripes.

hydrogen bonds. The order of the chains shown in Figure 3.2 can vary. There are four modifications known: cellulose I, more exactly I_α and I_β, and II–IV [112]. All plants produce so-called cellulose I, which means that the polymers are arranged parallel in sheets and have the same direction. Regeneration after dissolution to the polymeric level leads to fibres consisting of pure cellulose, but generally as a type II cellulose, meaning the polymers are arranged anti-parallel in sheets (rayon and viscose process [113, 114]). Cellulose aerogels are mainly made from regenerated cellulose. Natural cellulose exhibits different degrees of polymerisation (DP). Cellulose from wood pulp has between 6000 and 10,000 monomer units (AGU), and cotton between 10,000 and 15,000 [112, 114].

All natural cellulose fibres consist of areas of crystalline order intermixed with amorphous ones as shown schematically in Figure 3.3. Typically the amount of amorphous regions varies in plant cellulose between 40 and 60% [115]. Although cellulose is hydrophilic and considerable swelling occurs in water, it is insoluble in water and most organic solvents. Since cellulose is chemically a very stable material,

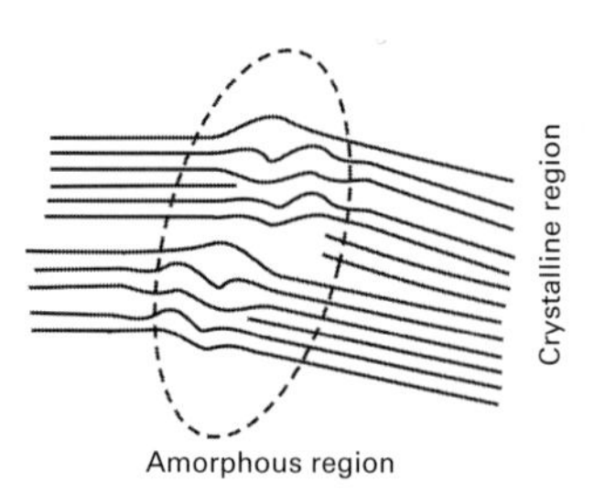

Figure 3.3 A schematic picture of the fringe-fibril model developed by Hearle [115]. Areas of crystalline order of the polymer chains are interrupted by areas with no order (amorphous). Especially in the amorphous regions, water as well as other chemicals can attack and eventually break the hydrogen bonds.

the production of aerogels from cellulose needs a technology or processing route to disintegrate the cellulose into the polymer strands (best without degradation of the polymer or reduction of the degree of polymerisation) and then to rebuild them into a suitable low-density, open porous gel that can be dried to obtain a mesoporous microstructure.

3.1.1 Dissolution Agents for Cellulose

If cellulose is a very stable macromolecule, the question arises, how can it be dissolved? An answer might formally come from a more fundamental question: What drives dissolution of a polymer in a solvent? Thermodynamically the difference between the dissolved state of a polymer and the undissolved one is a difference in Gibbs free enthalpy ΔG. The Gibbs free enthalpy is built up by two terms: an enthalpic (H) and an entropic (S) one, $G = H - T \cdot S$. The entropic term characterises the order or the number of states that can be taken by a system, and the enthalpy describes in a simplified picture the sum of all bond energies (internal energy). The enthalpy and entropy depend on the composition of the system. If G_1 would be the free Gibbs enthalpy of the system if its components would be completely dissolved into each other, making a homogeneous mixture and G_2 that of the same system in which both components are not miscible and separated, then the sign of $\Delta G = G_1 - G_2$ determines if the components are dissolvable or not. If $\Delta G > 0$, the components are not miscible; if $\Delta G < 0$, the components are miscible in each other and make a solution.

Therefore, if one would like to dissolve a monomer, such as a salt or sugar, there is much entropy gain by dissolution, meaning the distribution of all monomers in a solvent such as water increases the disorder and thus the entropy. In addition, the enthalpy changes, since crystalline bonds in a salt or a sugar crystal are broken and new bonds between the solvent and the solute replace them. If there is also a gain in enthalpy, the total free energy or enthalpy reduces, leading to the situation that the dissolved state is energetically favourable compared to the undissolved state. If instead a polymer is brought into contact with a solvent (polystyrol in cyclohexane or acetone), the gain in entropy is less, compared to a monomer, since the total number of configurations a polymer can take in a solvent is less than that of monomers (if the polymer keeps its degree of polymerisation, or DP). In addition, if the solvent is not able to break the bonds between the polymers, there is also no gain in enthalpy.

The first formal theory of polymer dissolution and the interaction of polymers with a solvent was developed by Flory and co-workers almost 80 years ago [116]; for a more in-depth discussion, see Chapter 4 and Appendix B. Cellulose in its native form with often a very long polymer chain or a high DP has a therefore low solubility since there is a lack of entropic gain for dissolution. But this is only one side of the coin. A solvent for cellulose must be able to break the hydrogen bond network between the polymer chains. A typical hydrogen bond has an energy of around 20 kJ/mol [117], which is not a big value. A solvent for cellulose should have a higher binding energy to, for instance, all the OH-groups between and along the cellulose polymer strand and thus it would be energetically favourable to replace the hydrogen bonds by bonds to the solvent. Olsson pointed out, [118] that for dissolution of cellulose it is not sufficient to think about breakage of the hydrogen bond network, but to take into account that cellulose has an amphiphilic nature [119, 120]. The CH-groups being essentially perpendicular to the plane of the polymer chain provide in their picture a hydrophobic interaction between the polymer chains, whereas the OH-groups provide the hydrophilic hydrogen bonds. Both must be attacked by a suitable solvent for cellulose.

There are two ways to dissolve cellulose: first, make a cellulose derivative, and second, try chemicals that dissolve it without any further modification. Derivatisation of cellulose is an old technique. Cellulose can be converted to cellulose acetates, esters and ethers (methylcelluose is the basis of many wallpaper pastes). The derivatisation to an acetate is, as an example, achieved by mixing cellulose fibres with glacial acetic acid and acetic anhydride, a colourless liquid, with sulfuric acid as a catalyst yielding cellulose triacetate. The reaction is stopped by addition of water, which also partially hydrolyses the triacetate. Cellulose acetate is a crystal clear, tough, and flexible plastic and is the most stable cellulose derivative. Methylation of cellulose is performed by heating cellulose in a sodium hydroxide solution and then treating it with methyl chloride, leading to a substitution of the –OH groups with a methoxy group. The degree of substitution can be adjusted.

Another classical way to treat cellulose was discovered in the nineteenth century and is the basis of the viscose process leading to rayon filaments. The process is called mercerisation. Cellulose is soaked in a strong alkali solution such that the crystal structure changes from cellulose I to cellulose II. For cellulose to dissolve in alkaline aqueous media, it needs to be cooled below room temperature. In a classical paper, Hirosi et al. [121] were able to present a section of the ternary phase diagram of the system cellulose–NaOH–H_2O; see Figure 3.4. They found that sodium hydroxide–water mixtures can dissolve low molecular weight cellulose at temperatures below 0°C. The alkali–cellulose solution is mixed with toxic carbon disulfide to form cellulose xanthate. The resulting very viscous liquid, viscose, is extruded into a sulfuric acid bath through a spinneret to make rayon. The sulfuric acid converts the viscose back into cellulose [113, 114].

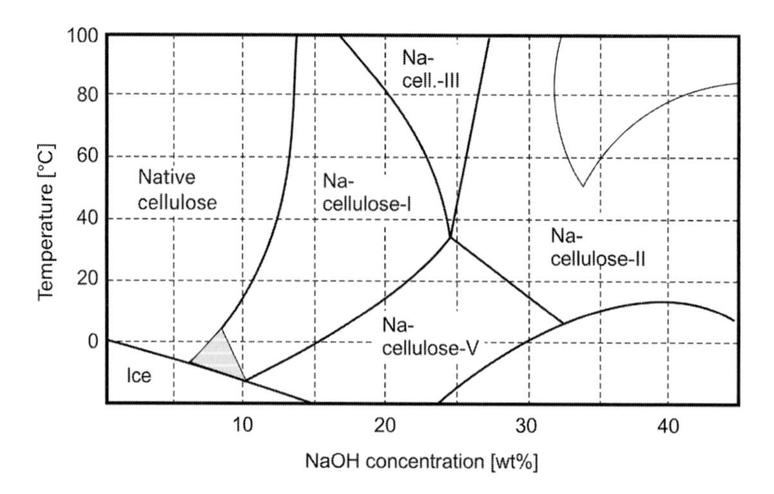

Figure 3.4 Section through the ternary phase diagram of cellulose, water and sodium hydroxide. The shaded region is the concentration and temperature range in which cellulose is dissolvable in NaOH–water solutions. Redrawn from the original paper of Hirosi et al. from 1939 [121]. The other areas denote various modifications with so-called alkali cellulose.

Other non-derivatising solvents were developed in the last decades like metal-inorganic complexes, salt melt hydrates and ionic liquids, to name a few (for more details, see [114, 118]). For aerogels, three dissolving agents are widely used:

- NaOH or LiOH water solutions with additions of small amounts of urea, thiourea, polyethylene glycol (PEG), ZnO and others
- Ionic liquids, most frequently based on N-methylmorpholine-N-oxide (NMMO) with stabilisers against oxidation
- Salt hydrate melts, such as $ZnCl_2{\cdot}3H_2O$ or $Ca(SCN)_2{\cdot}4H_2O$ (also called rhodanide)

Recently Budtova and Navard [120] reviewed the dissolution of cellulose in sodium hydroxide water solutions. They develop hypotheses of the chemical processes of swelling and dissolution and the effects of various additives which enhance the dissolution power of NaOH. Why do NaOH–water solutions dissolve cellulose at all? The basic idea is that the sodium cation has a solvation shell which can interact with the hydroxyl groups of the cellulose chains and thereby breaks the hydrogen bonds; see [120]. Dissolution in NaOH–water solutions means that a lot of water is present around the cellulose chains. The water molecules and the hydroxyl groups polarise each other and may thereby induce orientation correlations of species close to the cellulose chain, such that the alkali cations can interact with the bonds. The mutual polarisation can explain the influence of temperature on dissolution ability since low temperatures foster polarisation and orientation of molecules as opposed by thermal energy, which destroys it. This picture was questioned by Medronho, Lindman and co-workers [119, 122] bringing back into the discussion the amphiphilic nature of cellulose.

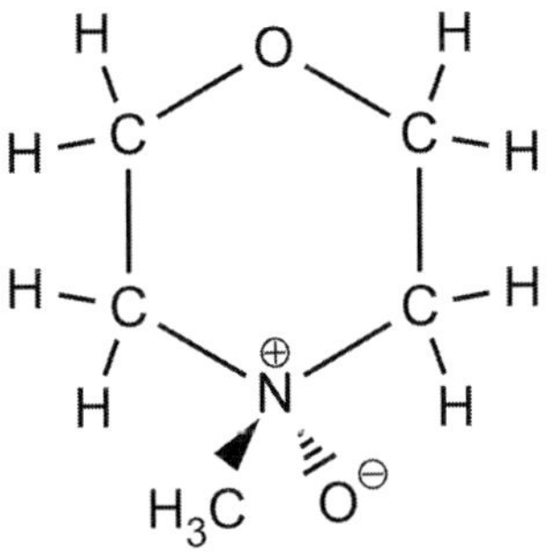

Figure 3.5 Chemical structure and conformation of NMMO, an ionic liquid used to dissolve cellulose. The charges on the nitrogen atom and the attached oxygen atom shall show that the molecule has a strong dipole moment, which allows strong interaction with cellulose.

Ionic liquids are modern dissolution agents for cellulose. Over the last three decades, ionic liquids were developed to industrially produce cellulose fibres and yarns (Lyocell or Tencel fibres). Ionic liquids consist of organic cations and organic or inorganic anions. They typically have a low melting point, a high viscosity and a good thermal stability. The most prominent ionic liquid used in aerogel production is NMMO, although many others have been studied in the last decade. Figure 3.5 shows the chemical structure of NMMO. Interesting here is a ring of six atoms in which on opposing edges an oxygen and a nitrogen are placed. At the nitrogen atom is a methyl group bonded and an oxygen atom. The N–O bond bears a strong dipole moment.

NMMO as well as other ionic liquids is complete soluble in water. The high polarity of the N-O bond allows easy formation of hydrogen bonds with nearby hydroxyl groups such as are present in water or cellulose. The procedure for dissolving cellulose in NMMO usually starts with a suspension of cellulose in NMMO to which a large excess of water is well mixed. The excess water provides low viscosity and thereby excellent mixing. The excess water is removed from the solution by evaporation above 100°C and reduced pressure until the cellulose is completely dissolved. According to the phase diagrams of NMMO and cellulose up to approximately 14 wt% cellulose can be dissolved in a mixture of 10% water and 76% NMMO [123, 124]. NMMO is also a strong oxidant and sensitive to all forms of impurities in the cellulose pulp. Therefore, typically stabilising agents are used, such as propyl gallate.

A different class of dissolving agent for cellulose is so-called salt hydrate melts. If one dissolves an inorganic salt in water and increases the salt content up to the coordination number of the salt cation a salt-cation-water complex may be formed which has a uniform melting temperature (such as a congruent melting phase). Typically, just adding salt at room temperature to water does not help; most often the temperature has to be increased. For example, if one uses as a salt rhodanide, which is $Ca(SCN)_2$, one has to increase the temperature up to 120°C, whereas using and $ZnCl_2$ a lower temperature like 70–80 °C is sufficient. Figure 3.6 shows the phase diagram of zinc chloride water as determined by Wilcox and co-workers [125]. Wilcox and co-workers discuss the idea that the salt–water complex behaves like an ionic liquid,

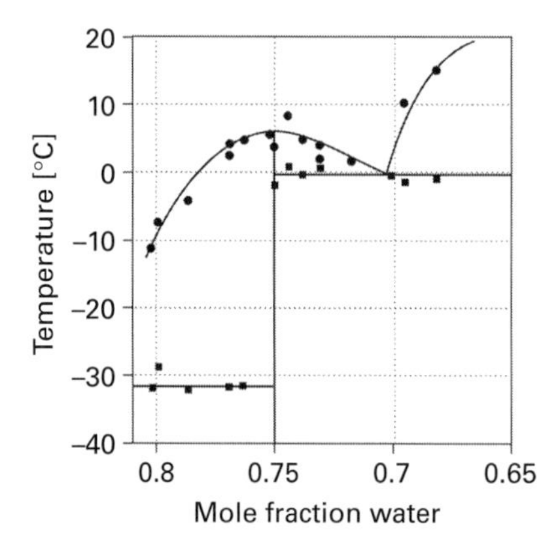

Figure 3.6 Binary phase diagram of zinc chloride and water as determined by Wilcox et al. [125]. Important in the context here is the congruent melting phase at 0.75 mol fractions of water, meaning a complex $ZnCl_2 \cdot 3H_2O$ is formed, which is stable below approximately 6°C. On the right side is a eutectic at 0°C. This diagram also means that in solutions of water and zinc chloride these complexes exist also at somewhat higher and lower water concentrations below the melting point of the trihydrate complex. Reprinted (adapted) with permission from [125]. Copyright (2015) American Chemical Society

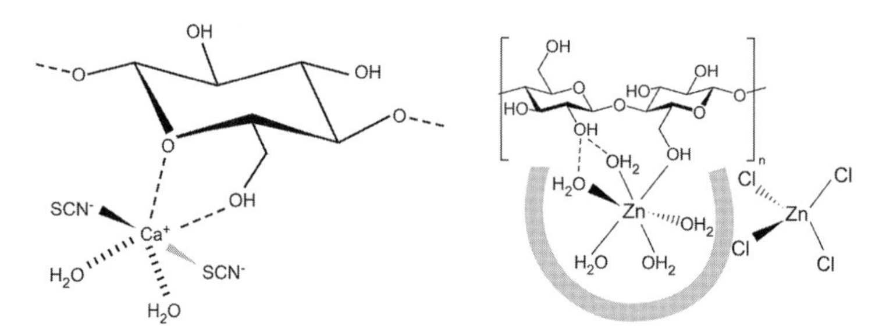

Figure 3.7 A sketch of the action of rhodanide (left) and zinc chloride (right) on a cellulose chain. In the rhodanide case, the calcium cation interacts with the oxygen in the glucose ring and the hydroxymethyl group [131]. In the zinc chloride case, the zinc cation has a hydration shell which is completed by the hydroxyl groups of the cellulose molecule. Attached to it is a tetrahedron of zinc chloride. The right figure is adapted with permission from [126]. Copyright (2016) American Chemical Society

being, however purely inorganic [126, 127]. The phase diagram of zinc chloride and water shows that zinc chloride trihydrate is a congruent melting phase.

The dissolving capabilities of salt hydrate melts were first described and successfully realised by Philip et al [128]. The idea behind these salt hydrates being able to dissolve cellulose is mainly that they are able to break the intermolecular hydrogen bonds [129]. Such a salt hydrate melt system is a solvent for the cellulose system because of its polarity and acidity and its low melting point being far below the crystalline amorphous transition point of cellulose–water mixtures at 320°C [130]. Their interaction with a cellulose chain is schematically shown in Figure 3.7 for rhodanide and zinc chloride.

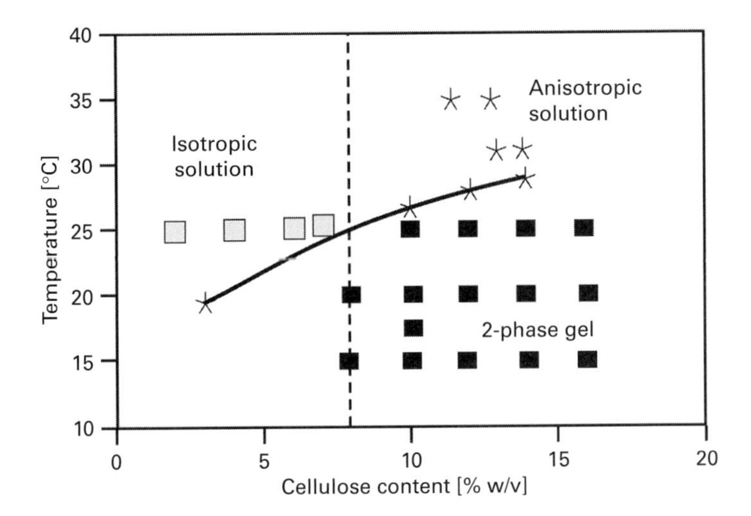

Figure 3.8 Experimental phase diagram of cellulose content (measured in weight per volume) dissolved in a solution of ammonia and ammonium thiocyanate. At low cellulose concentrations, there is an isotropic solution which transforms to an anisotropic one (the cellulose molecules are ordered) but the solid line bounds a miscibility gap in which two phases coexist a gel phase and a solution phase. Redrawn from [132] with permission from Springer-Nature

The main idea put forward by Sen and co-workers for the case of zinc chloride–water is that the Zn-cation is surrounded by a first hydration shell, and the hydroxyl groups of the cellulose molecule complete this hydration shell once they are broken. Attached to such an octahedron of hydrated water is a tetrahedron of Zn-cation on whose vertices chloride ions are located. This special feature makes $ZnCl_2{\cdot}3H_2O$ an excellent dissolving agent for cellulose. In agreement with the phase diagram shown in Figure 3.6 is the experimental observation that a deviation from the trihydrate to a four hydrate also allows it to dissolve cellulose, if the temperature is chosen properly.

The ability of salt melt hydrates to dissolve cellulose also allows us to experimentally determine phase diagrams of cellulose in such salt–water systems. Figure 3.8 shows an experimental phase diagram redrawn after Frey and Theil [132]. The phase diagram first shows that cellulose can be dissolved in a thiocyanate solution and, more importantly, that there exists a binodal line below which two phases coexist, a gel phase and a solution phase. This experimental observation is described by Frey and Theil theoretically with an extended Flory–Huggins model (see Appendix B). One can assume that a similar diagram exists for other salt melt hydrates.

In a light microscopic study, Schestakow [133] investigated the action of zinc chloride tetrahydrate on microcrystalline cellulose. Cellulose samples were taken out of the hot salt hydrate melt after different times and looked at in a light microscope. Figure 3.9 shows the dissolution process. After 5 minutes in the solvent, swelling and ballooning (meaning the amorphous regions swell in the form of balloons constrained by crystalline areas) can be observed. Ten minutes later there are still balloons, but the

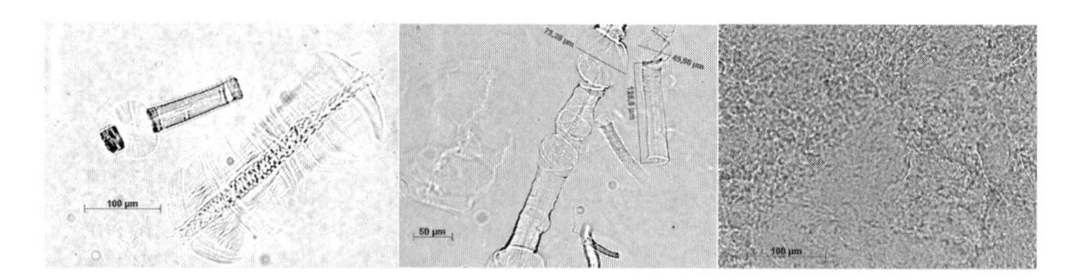

Figure 3.9 Light microscopy of microcrystalline cellulose in a zinc chloride salt hydrate melt. Left picture shows swelling and ballooning (meaning the amorphous regions swell in the form of balloons constrained by crystalline areas) after 5 min, 10 min later there are still balloons but the crystals dissolve and there seems a core left, and after 24 h ageing in the solution no microcrystalline fibres can be seen within the resolution of a light microscope. Courtesy M. Schestakow, DLR [133]

crystals dissolve, always from the periphery to the centre) and there seems a core left. After 24 h, ageing in the solution no microcrystalline fibres can be visualised within the resolution of a light microscope.

3.1.2 Gelation of Cellulose Solutions

Making an aerogel from dissolved cellulose requires as a first step to transform the solution into a wet gel. There are at least three ways to gel a cellulose solution. One is an intrinsic route, one induced by a neutralisation reaction and the third one is possible in systems with a reversible and thermodynamically stable miscibility gap, namely a temperature jump.

Gelation by an Intrinsic Mechanism

The intrinsic way to gelation is observed in sodium hydroxide–based solutions: such a solution tends to form precipitates, which can sediment in a beaker or the solution spontaneously makes a gel. Gavillon and Budtova [134] studied the rheological behaviour of cellulose dissolved in sodium hydroxide water solutions using different cellulose pulps (one with a DP of 180 and one with a DP of 500). They measured the viscosity and the storage and loss modulus (see Section 4.1.1) at different temperatures, and measured the time a solution converted to a gel, which was defined as the point when storage and loss modulus became equal [135]. In this way they extracted the temperature dependence of viscosity, which in their presentation depends exponentially on temperature and also observed that with increasing cellulose content the gelation temperature decreases. There was no clear trend that the DP had an effect on the gel time. The effect of increasing cellulose content can be simply understood: the more molecules are in the solution, the more often they interact and meet and entangle and thus can build a network. The exponential dependence of gel time on temperature in the form $t_{gel} \propto \exp(-\alpha T)$, with α a constant is more than astonishing, since for any thermal-activated process of gelation one would automatically expect an Arrhenius-type behaviour. Plotting their data as a function of inverse temperature

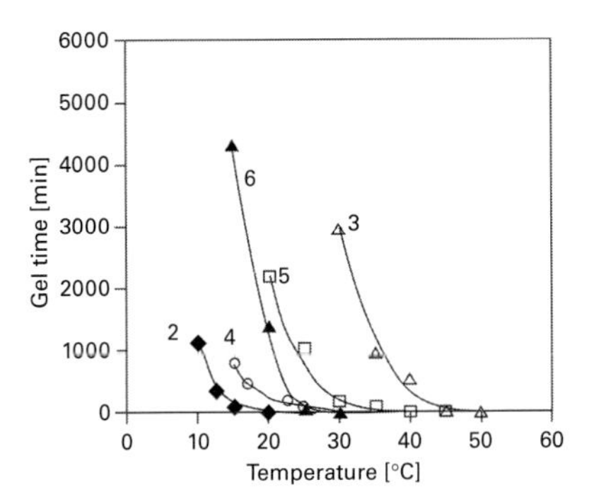

Figure 3.10 Comparison of the experimentally determined gelation time as a function of temperature (data points) and a fit with Eq. (3.1). The filled symbols belong to solutions without ZnO. In all cases, the solutions contained 8 wt% NaOH. (1) is 4 wt% and (2) 6 wt% cellulose. Open points are data with ZnO. (3) 4 wt% cellulose – 1.5 wt% ZnO, (4) 6 wt% cellulose – 0.7% ZnO and (5) 6 wt% cellulose with 1.5 wt% ZnO. Redrawn and reprinted after [136] with permission from Springer-Nature

yields also a good agreement, but yields an activation energy of around 300 kJ/mol. This value is very large compared with the value of the activation energy of the viscosity, which they determined to be around 25 kJ/mol.

In a more recent study, Liu et al. [136] studied the influence of ZnO added to the solution on the gelation using the same sodium hydroxide solution technique. They also found an exponential dependence of the gelation time on temperature and also a dependence on the cellulose concentration as well as the amount of ZnO used to impede gelation. They suggest an empirical model equation to describe their results:

$$t_{gel} = \left(\frac{A}{C^n_{cell} + BC^m_{ZnO}} \right) e^{-\alpha T} \tag{3.1}$$

α, A and B are constants, and they fit their data with kinetic exponents $m = 3.5, n = 9$. A comparison of this model with their experimental results is shown in Figure 3.10.

Gelation by pH Inversion

A classical way to induce gelation of base solutions of cellulose however, is not to wait until gelation happens by itself but to induce gelation by so-called pH-inversion or simply a neutralisation reaction. The wet gel is submerged into an acidic regeneration bath like sulphuric, nitric or hydrochloric acid. For example, using NaOH–water–urea as a solvent for cellulose, the resulting solution can be directly dripped into aqueous salpetric or hydrochloric acid [137, 138]. The originally clear solution immediately after contact with the acid bath turns white, a thin shell builds up and upon further stay in the bath the wet gel becomes more and more stable. After washing with ethanol, they can be directly dried supercritically with carbon dioxide to yield fluffy aerogels.

Let us discuss this a bit further. The gelation process of cellulose dissolved in a base medium (typically at a very high pH value, pH = 14) is different using inorganic acids (HCl, H_2SO_4 etc.) or organic acids (acetic acid, lactic acid etc.), which might be attributed to the neutralisation reaction, which is mostly complete for inorganic acids. They have a high pK value and are completely dissociated in water, compared to most organic acids, which are weak acids, such as acetic or citric acid, and thus the neutralisation reaction can only proceed in the manner dissociated species are available, which can be transformed to a salt.

In acid solutions, the neutralisation of a wet cellulose gel leads immediately to the formation of a solid skin, since the neutralisation reaction is typically very fast. The reaction rate constant of the neutralisation reaction $H^+ + OH^- \rightarrow H_2O$ is of the order of 10^{11} L/(mol s) if inorganic acids are used and they are completely dissociated; see [139, p. 503]. Once this skin is formed (making also the gel opaque), the diffusion of the acid through this porous membrane (skin) determines the process of neutralisation and gel formation. The neutralisation reaction restores the hydrogen bonds between the cellulose molecules and gives them the ability to form fibrils or entangled polymer strands. The process might be diffusion controlled. If so, the gel formation could be estimated, since the characteristic time for a diffusion process of a species into a wet gel depends on the diffusion coefficient and a characteristic size of the gel body L_c. This would yield an estimate for the gelation time as

$$t_{gel} \propto \frac{L_c^2}{D}, \tag{3.2}$$

in which D is the diffusion coefficient of H^+ or OH^- in the solution. Since in liquids the diffusion coefficient is related to the viscosity via the Stokes–Einstein relation (see Section 10.2), the gel time would increase with the viscosity.

A theoretical description could be made by the following more speculative model. Let us assume that the neutralisation reaction is not a simple first-order reaction with the rate constant $k = k(T)$, like

$$\frac{dc_{H^+}}{dt} = -k\, c_{H^+}, \tag{3.3}$$

except that it terminates or is finished if the NaOH concentration, or more specifically the OH^- concentration, is consumed. We then would have a reaction like

$$\frac{dc_{H^+}}{dt} = -k\, c_{H^+}(c^m_{OH^-} - c_{H^+}) \tag{3.4}$$

in which $c^m_{OH^-}$ is the maximum concentration, taken as homogeneous inside the wet gel. Insertion into the diffusion equation (10.7) leads to

$$\frac{\partial c_{H^+}}{\partial t} = D_{H^+}\nabla^2 c_{H^+} - k\, c_{H^+}(c^m_{OH^-} - c_{H^+}). \tag{3.5}$$

We define now several new variables

$$v = \frac{c_{H^+}}{c^m_{OH^-}} \tag{3.6}$$

$$k^* = k\frac{(c^m_{OH^-})^2}{\Omega_m^2} \tag{3.7}$$

$$c = \frac{c_{H^+}}{\Omega_m} \tag{3.8}$$

$$\vec{r'} = \frac{\vec{r}}{L_c} \tag{3.9}$$

$$\tau = k^* t \tag{3.10}$$

with Ω_m as the molar volume of the -OH group. These variables are all dimensionless. Insertion leads to the equation

$$\frac{\partial v}{\partial \tau} = \frac{D_{H^+}}{k^* L_c^2} \nabla^2 v - v(1 - v) \tag{3.11}$$

or

$$\frac{\partial v}{\partial \tau} = D^* \nabla^2 v - v(1 - v) \tag{3.12}$$

with D^* as a dimensionless diffusivity. Note we redefined the concentrations from moles per volume to dimensionless values. This equation is known in the literature as the Fisher equation or more precisely as the Fisher, Kolmogorov, Petrovsky and Piskunoff (FKPP) equation [140, 141]. This equation has fascinating properties. First, in the 1D case, there exist travelling wave solutions or a propagating front, which transform an unstable state into a new stable state. The front speed is determined by

$$v = 2\sqrt{D^*} = \frac{2}{L_c}\sqrt{\frac{D_{H^+}}{k^*}}. \tag{3.13}$$

Network or gel formation as a result of the reaction-diffusion coupling is open for mathematical research. Nevertheless, assuming a travelling wave as a possible solution, the time for gelation can be estimated as

$$t_{gel} \propto \frac{L_c}{v} = L_c^2 \sqrt{\frac{k^*}{D_{H^+}}} \tag{3.14}$$

and thus the smaller the shortest wet gel body size, the shorter the gelation time. If such a model can be applied to the pH inversion in biopolymeric aerogels, and cellulose especially, is an open issue.

Temperature Jump Induced Gelation

Gelation of cellulose dissolved in ionic liquids or salt hydrate melts is different from those treated in the preceding discussion. As reported by Innerlohinger et al. [142] and Liebner's group [143–147] after dissolution of cellulose in NMMO at temperatures above 100°C gelation occurs upon cooling. The same was observed for cellulose dissolved in calcium thiocyanate by Hoepfner et al. [148, 149] and for zinc chloride tetrahydrate by Schestakow et al. [150]. In all these cases, the reversibility of dissolution on increasing the solution temperature and gelation on cooling suggest full agreement with the phase diagram shown in Figure 3.8 that a miscibility gap exists and gelation could be a result of phase separation occurring by different mechanisms, such as a sequence of nucleation, growth and aggregation or spinodal decomposition

[151] or a mixture of both. This will be discussed in Chapter 4 as a mechanism called polymerisation-induced phase separation (PIPS).

3.1.3 Preparation of Bulky Cellulose Aerogels

We already described in Chapter 2, that aerogels can be produced in the form of bulky materials, like plates of several inches thickness and lateral dimensions even in the metre range, but also beads or powders and for biopolymers and especially cellulose also in the form of fibres, such that in principle yarns and woven and non-woven fabrics could be made from them [26, 152]. In this section, we review briefly methods to prepare cellulose aerogel monoliths and present a few properties. Kistler was the first to produce from a cellulose derivative the first cellulose aerogel. His work was long forgotten, and especially in the 1970s and 1980s it was not taken up by the scientists working on aerogels. Biopolymeric aerogels became popular since the work of Tan and co-workers published in 2001. Tan and co-workers published a paper on cellulose aerogels which became well known and popular, and the main results were even cited in newspapers [22]. They did not dissolve cellulose as described earlier, but started from a cellulose-derivative cellulose acetate and de-esterified it. This cellulose ester was cross-linked in an acetone solution using toluene-2,4-di-isocyanate. They could prepare gels if the cellulose concentration was larger than five and less than 30 wt%. They used supercritical drying and were able to get aerogels with a specific surface areas of around 400 m^2/g and densities in the range of 100–350 kg/m^3. They observed that the smaller the cellulose ester concentration and the larger the cross-linker amount, the larger the shrinkage. Although cellulose derivatives as starting materials for aerogel production are still under investigation, as in the work of Pinnow et al. [153] and Gan et al. [154], the focus of cellulose aerogel preparation is along the lines outlined previously, either using sodium hydroxide solutions, ionic liquids or salt melt hydrates to prepare a solution of polymers.

Gavillon and Budtova prepared a material they called aerocellulose dissolving cellulose in non-polluting solvents, such as aqueous NaOH or NMMO, and dried the gels after solvent exchange supercritically with carbon dioxide [134, 155]. Their focus was, however, on the preparation of cellulose aerogels from cellulose–NaOH–water solutions and looking at the influence of the preparation conditions on morphology and properties. As a starting material, they used microcrystalline celluloses with degree of polymerisation ranging from 180–950. They first let the cellulose swell at low temperatures in a mixture of sodium hydroxide and water. The cellulose concentrations were varied between 3–8 wt%. The polymer-water-NaOH solutions were cast into cylindrical molds of 8–10 mm diameter and 30–40 mm length and regenerated after gelation with water. The regeneration step led to a shrinkage of about 10% in volume. The gelation time varied exponentially with temperature as discussed earlier, meaning at 10°C the gelation takes around three days, at 40°C it takes only 6 minutes. Before supercritical drying, the solvent was exchanged with acetone. Gavillon and Budtova observed that the specific surface area slightly decreases as the cellulose concentration increases from 5–7 wt%. Adding a surfactant,

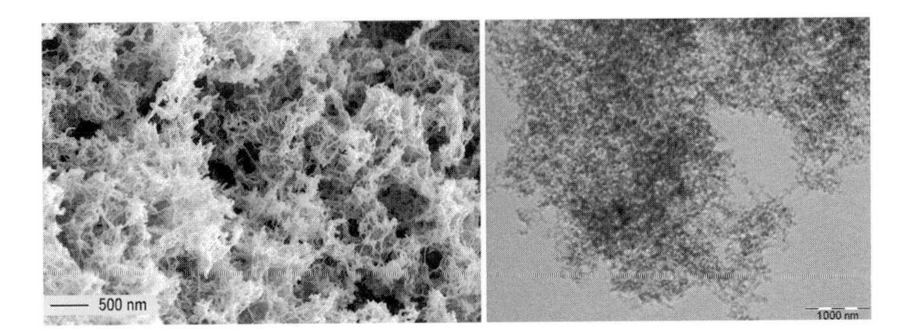

Figure 3.11 SEM picture of a cellulose aerogel (left) made with 5 wt% cellulose dissolved in 7.6 wt% sodium hydroxide, regenerated in water at 50°C and supercritically dried after exchange of the pore fluid with acetone. The TEM picture (right) shows that the fibrils are built by tiny particles, after [155].

alkylpolyglycoside, in low concentrations reduced the envelope density from values of 140 kg/m^3 to 60 kg/m^3 at 1 wt% surfactant. They attribute this to the formation of macropores, or, in other words, the cellulose polymer strands made more compact fibrils. Performing the regeneration in water not at 25°C, but at 50°C changed the microstructure from a fibrillar one with large macropores and thick fibrils to a mixture of fibrils and particles, which they also could view using transmission electron microscopy (TEM). One figure shown in Roxane Gavillon's PhD thesis is reproduced in Figure 3.11 together with the SEM picture of the same material. Budtova and Gavillon attribute the change in pore size and the structure of the aerogels regenerated at 50°C as an effect of temperature, since this increases the mobility of the polymers and thus the chance to rapidly entangle instead of building thicker, compact fibrils at lower temperature, where the mobility is reduced.

Jin and co-workers [156] were the first to produce high-quality cellulose aerogels using salt melt hydrates. They prepared a hot solution of calcium thiocyanate and water with 1:4 mol/mol to dissolve cellulose as described previously. Cellulose dissolved in rhodanide or zinc chloride solutions is a highly viscous liquid, similar to honey. Jin and co-workers therefore spread the hot solution with low cellulose content (0.5–3 wt%) on a glass plate to form layers of 1 mm thickness. After cooling to room temperature, which was accompanied by the sol gel transition, the gels were washed in methanol to extract the salt and regenerate the cellulose. The films were freeze dried after rapid quenching in liquid nitrogen. The densities obtained were in the range of 20 kg/m^3 at 0.5 wt% and 100 kg/m^3 at 3 wt% cellulose. The specific surface area increased with increasing cellulose content from around 105–200 m^2/g from 0.5 wt% cellulose to 3 wt%. The microstructure as shown by SEM displayed an open porous network of cellulose fibrils whose network density increases with the cellulose content.

Hoepfner et al. [148] used the process of Jin [156] to produce monolithic cellulose aerogels. They dissolved microcrystalline cellulose of a DP of 150 in a rhodanide hydrate at around 110–120°C. The hot cellulose solution was poured into polyacrylic

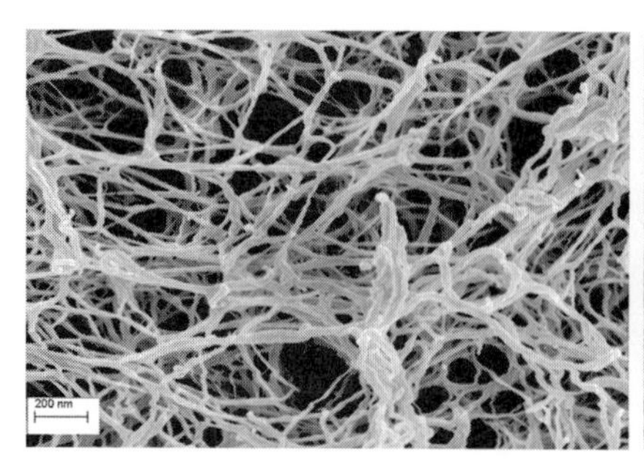

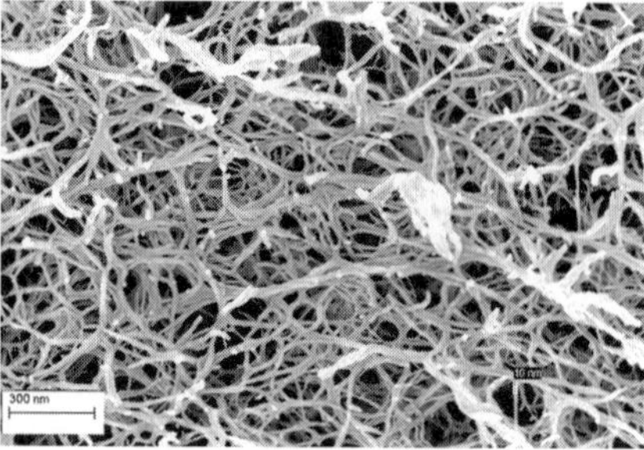

Figure 3.12 SEM pictures of two cellulose aerogels with 2 wt% cellulose (left) and 4 wt% cellulose (right) prepared using rhodanide as a solvent, ethanol as a regeneration and washing agent and supercritically dried with carbon dioxide. Courtesy K. Ganesan, DLR

multi-wall sheets with a cross section of 15 × 15 mm with a length of around 100 mm. The cellulose was regenerated in ethanol with a subsequent step of supercritical drying with carbon dioxide. The monolithic aerogels were white, soft and easily deformable by indentation. The envelope varied between 10 and 60 kg/m^3 with the cellulose concentrations varying between 0.5 and 3 wt% and were thus lower than Jin's aerogels. This can be attributed to the supercritical drying better preserving the gel structure than freeze drying does. The properties, such as envelope density and specific surface area, are described in more detail in Chapters 7 and 8. Figure 3.12 shows two SEM pictures of aerogels prepared according to this route. The picture clearly shows that the nanofleece consist of cellulose fibrils having thicknesses in the range of 20–50 nm.

In a careful study on preparation conditions, Schestakow and co-workers [150] used zinc chloride tetrahydrate as a solvent system and studied especially several regeneration bath fluids and their effect on the microstructure. Cellulose aerogels, concentrations of 1–5 wt% cellulose, were after gelation regenerated in water, ethanol, isopropanol or acetone. The cellulose aerogels regenerated in acetone had a specific surface area of around 340 m^2/g, almost doubling the value of those regenerated in water. A comparison of the microstructures observed by SEM is shown in Figure 3.13. There are remarkable differences induced by the regeneration liquid. Although in principle all four exhibit a fibrillar network, the network compared to those shown in Figure 3.12 is not irregular or random; they instead look always rounded as if during gelation they would have bent around something. The size of the fibrils is in the nanometer range, according to Schestakow around 10–20 nm. The network becomes denser from water to ethanol, isopropanol and acetone. The last two also show broken fibrils, which probably come from preparation for SEM.

The third route to prepare cellulose aerogels utilises ionic liquids. Innerlohinger et al. [142, 157] were the first to produce aerogels from cellulose using NMMO as a solvent. Cellulose was obtained from different pulps having different degrees of polymerisation ranging from 180–4,600. The process of Innerlohinger and co-workers started with a dissolution in an NMMO–water mixture, performed a shaping, regeneration and washing in ethanol and water, solvent exchange and supercritical drying

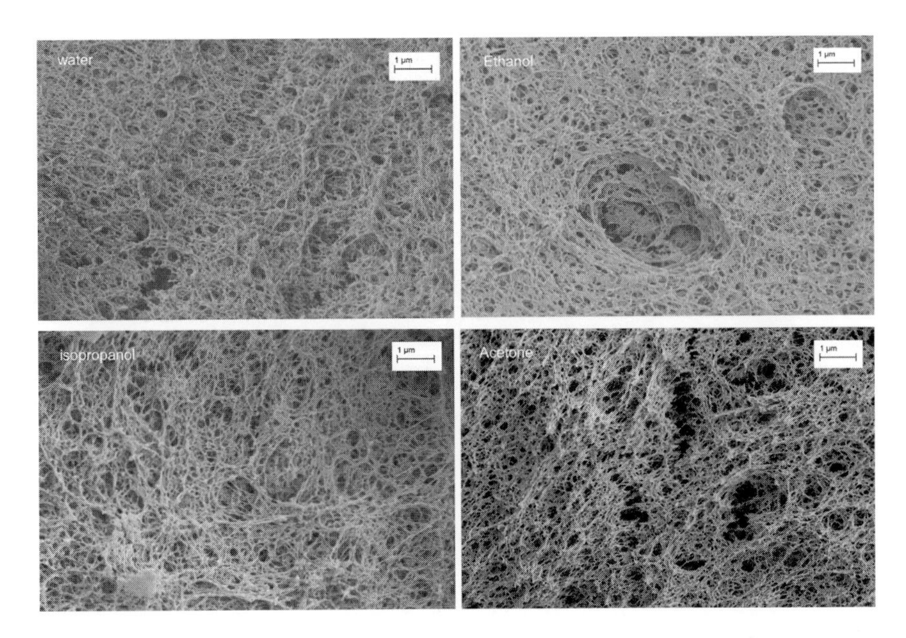

Figure 3.13 Cellulose aerogels prepared with zinc chloride hydrate as a solvent, regeneration in different liquids and supercritical drying with carbon dioxide. The cellulose content in each case is 5 wt%. Reprinted from [150] with permission from Elesvier

with carbon dioxide. Their process allowed to vary the cellulose content between 0.5–13 wt%. The bulk, monolithic aerogels were cylindrical with a 26 mm diameter. From their extensive studies, a few results are reproduced here. The envelope density varied in an almost linear fashion between 50–200 kg/m^3 as the cellulose content changes from 0.5–9 wt%. They reported a total shrinkage of around 40%. The reported internal surface areas range between 100–400 m^2/g. In the line of this work, the group of Liebner in Vienna studied extensively different ionic liquids and other unconventional dissolution routes. Liebner and co-workers [143, 144] also use NMMO as a solvent, but N-benzyl-morpholine-N-oxide (NBnMO) as a stabiliser against oxidation. They were able to produce dimensionally stable cellulose aerogels. The cellulose contents were 3, 5, 10 and 12 wt% and thus in the range of Innerlohinger's work. The regeneration was performed with ethanol or a mixture of ethanol and dimethylsulfoxide (DMSO) in different concentrations. The whole process of dissolution, regeneration, solvent exchange and supercritical drying with carbon dioxide could last several days. Their aerogels exhibit densities in a range from 50–260 kg/m^3 and specific surface areas from 172–284 m^2/g. Liebner and co-workers studied later [145] aerogels produced by their method in more detail and especially looked carefully to shrinkage in every step of processing, since shrinkage especially of biopolymeric aerogels can be rather larger; see Chapter 5. In this chapter, we finally would like to mention a careful study performed by Pircher and co-workers [151, 158] in which they compared different solvents for cellulose and looked at their impact on the porous microstructure. As a cellulose starting material, they used cotton

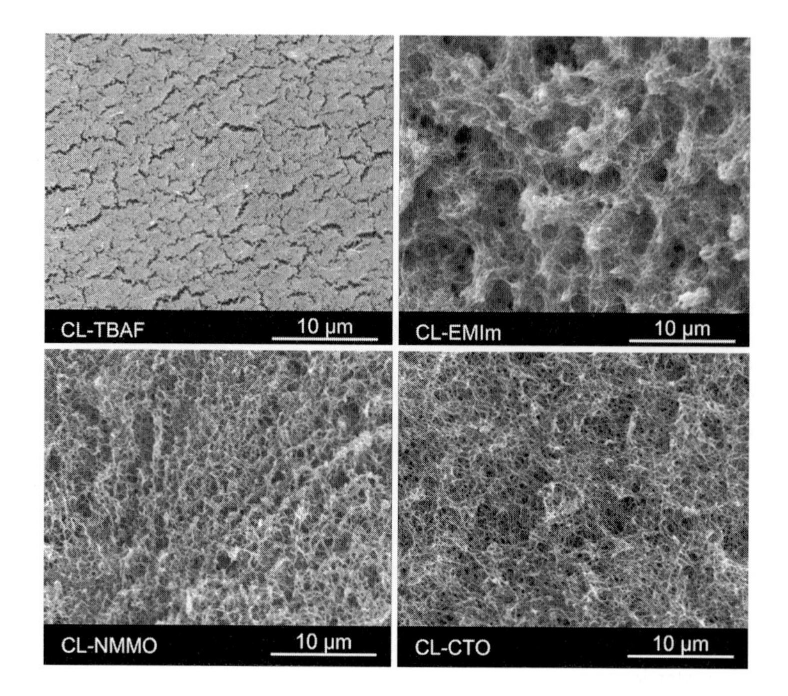

Figure 3.14 Scanning electron micrographs of aerogels derived from solutions of 3.0 w% cotton linters in TBAF/DMSO, EMIm/DMSO and NMMO and 1.5 w% cotton linters in caclium thiocyanate octahydrate/LiCl after Pircher [151]. Reprinted under Creative Commons license CC BY

linters and four different solvent systems, namely mixtures of (1) tetrabutylammonium fluoride and DMSO (TBAF–DMSO); (2) the ionic liquid 1-ethyl-3-methyl-1H-imidazolium acetate and DMSO (EMIm–DMSO); (3) a salt hydrate melt, calcium thiocyanate octahydrate, which in contrast to Jin [156] and Hoepfner [148] was modified with lithium chloride; and (4) NMMO. All gels were regenerated in ethanol and supercritically dried with carbon dioxide. They studied the volumetric shrinkage in all stages of the aerogel production, explored envelope density, performed small and wide angle X-ray scattering (SAXS, WAXS) studies and used thermoporosimetry (see Chapter 9) and mechanical properties in compression tests (see Chapter 13). The SEM analysis of aerogels prepared with NMMO or $Ca(SCN)_2{\cdot}8H_2O$–LiCl revealed random networks of cellulose nanofibrils. Figure 3.14 shows a few examples. All aerogels exhibited large shrinkage from gel preparation to the dry state between 23–52 %. The solvent system used had a pronounced effect. Whereas the ionic liquid EMIm could lead to a shrinkage of 52%, the rhodanide–LiCl solvent led only to a shrinkage of 23%.

3.2 Alginate: From Solution to Gel

The discovery of alginate can be traced back to the 1880s (Edward Stanford), and its commercial production commenced in the early twentieth century. As in the past, the current industrial production exclusively relies on brown seaweed as a raw material. Alginate in a certain sense is complementary to cellulose discussed in Section 3.1 as it plays in seaweeds a role analogous to cellulose in terrestrial plants. Sodium alginate is the most available salt of alginic acid. Potassium and ammonium alginates are also commercially available. Sodium alginate and other monovalent alginate salts are white or yellowish powders, and they are highly soluble in water. Another commercial water-soluble product is propylene glycol alginate, which is produced by esterification of alginic acid with propylene oxide. All alginate salts including insoluble calcium alginate are approved food additives (E400–E405) and widely used in the food industry as emulsion and dispersion stabiliser and gelling agent. Alginate finds its use in the textile industry as well as in pharmaceutical and medical fields for making wound dressings and dental impressions and as a carrier matrix.

3.2.1 Structure of Alginate

Alginate is an umbrella term for a large family of polysaccharides extracted primarily from farmed brown algae. Alginate constitutes up to 40% of the seaweed dry matter providing mechanical strength to the plant as well as acting as a water storing agent to prevent water lost when seaweeds are exposed to air [159].

Alginate is a linear binary copolymer of 1-4-linked residues of mannuronic (M) and guluronic (G) acids (Figure 3.15). The residues are arranged in a blockwise pattern with homopolymeric regions of G residues (called G-blocks) and a homopolymeric region of M sequences (called M-blocks) interspersed by regions in which the two residues alternate [159].

Several bacteria produce alginate as an extracellular polymer with distinctly different structural features. Unlike alginates from seaweed, alginates of bacterial origin have a certain degree of acetylation and may contain no G-blocks or at least no contiguous G residues. Fermentative production of alginate is so far economically infeasible, and materials from bacterial alginates are available at a small scale.

Alginate both from seaweed and bacterial sources have polydisperse molecular weight, i.e., molecules with different molecular weights are present in the sample. If out of total N polymer molecules N_i of molecules have the molecular weight M_i, the number-average molecular weight $\overline{M}_N$ is given as

$$\overline{M}_N = \frac{\sum_i N_i M_i}{\sum_i N_i}. \tag{3.15}$$

The number-average molecular weight is thus the ratio of the first μ_1 to the zeroth μ_0 non-central moments of the distribution. The first non-central moment is the mean value and the μ_0 is equal to one. Another way to quantify the molecular weight is to introduce the weight-average molecular weight ($\overline{M}_W$), where the averaging is performed over the weight fractions $w_i = N_i M_i$:

$$\overline{M}_W = \frac{\sum_i w_i M_i}{\sum_i w_i} = \frac{\sum_i N_i M_i^2}{\sum_i N_i M_i}. \tag{3.16}$$

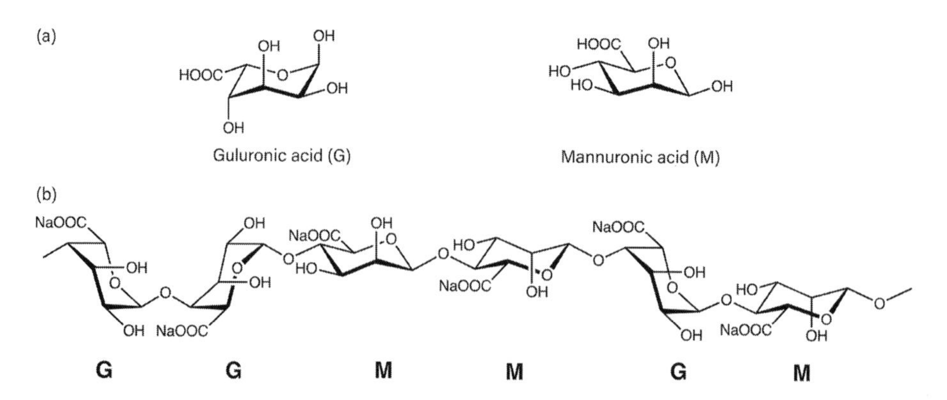

Figure 3.15 (a): Chemical structure of sodium salts of mannuronic (right) and guluronic (left) acids; (b): chain conformation with two G- and M-block and one GM-block.

The weight-average molecular weight as defined by Eq. (3.16) is the ratio of the second μ_2 to the first non-central moments of the distribution. The μ_2 is the mean of the squared molecular weight. Commercial alginates have a weight-average molecular weight ($\overline{M}_W$) of approximately 200 kDa[1] [159]. The molecular weight of alginate as present in plant sources is unknown, but is believed to be much higher than that in commercial samples due to a substantial degradation during the extraction process. From probability theory, it is known that the variance σ^2 of a probability distribution can be expressed through the first and second non-central moments as

$$\sigma^2 = \mu_2 - \mu_1^2. \tag{3.17}$$

To characterise the breadth of the molecular weight distribution, it is customary to introduce the so-called molar-mass dispersity D_{M} (see [160]) as the ratio of the number-average molecular weight to the weight-average molecular weight. Both the quantities can readily be expressed through the non-centred moments, so the molar-mass dispersity reads

$$D_{\mathrm{M}} = \frac{\overline{M}_W}{\overline{M}_N} = \frac{\mu_2}{\mu_1}\frac{\mu_0}{\mu_1} = \frac{\sigma^2}{\overline{M}_N} + 1. \tag{3.18}$$

For a monodisperse polymer σ^2 is zero and the molar-mass dispersity is 1. The molar-mass dispersity for alginates typically ranges between 1.5–3 [159].

The mannuronic and guluronic acids are structurally very similar; the only difference is the configuration at the C-5 atom. Because of this difference, the M and G units adopt two different conformations of the pyranose rings (Figure 3.15). This gives rise to flat ribbon-like and flexible chain conformations when mannuronic acid residues are linked to each other. In contrast, a pair of guluronic acid residues form a buckled and more rigid structure (Figure 3.15). Another common feature of mannuronic and guluronic residues is that they each bear a carboxylic group.

[1] A Dalton (Da) is the unified atomic mass unit defined as the mass of a carbon atom divided by 12.

At a basic level, the bulk composition is characterised by the fraction of M and G units in a single alginate chain, F_M and F_G, respectively. Because $F_M + F_G = 1$, it is sufficient to provide a single value of the so-called mannuronate/guluronate ratio F_M/F_G for characterising the bulk composition. Being the result of complex biosynthetic pathways along with seasonal variations, the guluronate fraction can vary over a wide range ($0.20 < F_G < 0.75$) and may even depend on the part of the plant considered [159].

At a more detailed level, the sequential structure of alginate can be characterised by fractions of diads, i.e., fractions of four possible pairs MM, MG, GM and GG. Fractions of triads, tetrads and higher-order multads would be needed for a complete statistical description of a given alginate chain.

Elucidation of the sequential structure is not a routine task. High-resolution proton and carbon nuclear magnetic resonance (^{1}H-NMR and ^{13}C-NMR), circular dichroism and Fourier-transform infrared spectroscopy (FTIR), sometimes also combined with advanced chromatographic methods, are mainly employed [161, 162]. Up to now, these analytical methods provide access to the the fractions of monads (F_M, F_G), diads (F_{MM}, F_{MG}, F_{GM}, F_{GG}) and eight corresponding triads. The degree of polymerisation of natural alginates is high enough to regard them as having infinitely long chains. In this case, not all the fractions are independent, and conservation relations hold, for example for diads $F_{MG} = F_{GM}$ and $F_{MM} + 2F_{MG} + F_{GG} = 1$. The fractions have been measured for alginates separated from a number of species [163, 164], as shown in Table 3.1, however, such information is only rarely present in the material science publications.

In a random copolymer, the probability of finding a given type of residue at a particular position along the chain is simply equal to the fraction of that residue. As we go along the chain of a random copolymer, the probability of finding a given residue is constant and does not depend on what we just saw in the previous position. Truly random copolymers demonstrate such kind of behaviour and thus obey Bernoullian statistics. The probability of finding a pair of residues is then just a product of the individual probabilities. For alginate with two residues, M and G, the fractions of diads would then be given as follows (again assuming a long chain):

$$F_{MM} = F_M^2 \tag{3.19}$$

$$F_{MG} = F_{GM} = F_M F_G \tag{3.20}$$

$$F_{GG} = F_G^2 \tag{3.21}$$

A simple calculation with Eqs. (3.19)–(3.21) for data in Table 3.1 shows that F_{MM} and F_{GG} fractions significantly exceed the Bernoullian fractions for MM and GG pairs. If Eqs. (3.19)–(3.21) would hold true, the knowledge of the bulk monomeric composition would be sufficient to determine the sequential structure of alginate, but it is evidently not the case. It is out of the scope of this chapter to discuss more advanced models of the sequential structure of alginate. Further information can be found in the specialised literature [161, 165].

Table 3.1 Sequential composition of alginate from different species. Data from [163].

Source	F_{G}	F_{M}	F_{GG}	F_{GM}, F_{MG}	F_{MM}	F_{GGG}	F_{GGM}	F_{GMG}	F_{GMM}	F_{MGG}	F_{MGM}	F_{MMG}	F_{MMM}	$\overline{N}_{G>1}$
A. nodosum	0.43	0.57	0.20	0.23	0.34	0.12	0.08	0.12	0.11	0.08	0.15	0.11	0.23	3.5
M. pyrifera	0.42	0.58	0.22	0.2	0.38	0.17	0.05	0.14	0.06	0.05	0.15	0.06	0.32	5.4
L. digitata	0.40	0.60	0.26	0.14	0.46	0.22	0.04	0.06	0.08	0.04	0.1	0.08	0.38	7.5
L. hyperborea	0.68	0.32	0.56	0.12	0.2	0.51	0.05	0.07	0.05	0.05	0.07	0.05	0.15	12.2

Alginates with a high guluronate fraction $F_G > 0.7$ and an average G-block length of a few tens of residuals yield stable gels with superior mechanical strength and a lower degree of syneresis compared to M-reach alginates. Because gelling abilities of alginate positively correlate with the mannuronate/guluronate ratio F_M/F_G, or the M/G ratio for short, industrial alginates are often broadly classified as low-G and high-G alginates. Thus, it is very desirable to provide at least the M/G ratio when documenting research work with alginate-based materials.

As no stable gels can be obtained for alginates with strictly alternating ...-M-G-... sequences, a certain length of the G blocks is required for formation of junctions. Because every G-block with at least two G residues is terminated by a GGM triad, the frequency F_{GGM} is a measure of the total number of such G-blocks and the average length of G-blocks can be calculated as follows [165]:

$$\overline{N}_{G>1} = \frac{F_G - F_{MGM}}{F_{GGM}} \tag{3.22}$$

The fraction F_{MGM} in Eq. (3.22) excludes all isolated G residues, i.e., those for which the G-block length is equal to one. The required G-centred fractions F_{GGM} and F_{MGM} can be measured by NMR. The average length of G-blocks calculated by Eq. (3.22) is also summarised in Table 3.1. The data show that alginates with similar M/G ratios can have very different sequential structure (cf. the first three species in Table 3.1).

3.2.2 Degradation and Chemical Modifications

The glycosidic linkages between uronate units of alginate are susceptible to cleavage both above and below pH 7. In a routine workflow, it is convenient to track the progress of degradation by measuring viscosity of the alginate solution regularly. To minimise degradation of alginate solutions and gels, they should be kept close to neutral pH of 6–8 and at limited heating conditions, ideally in a refrigerator at +5°C. Alginates are also susceptible to chain degradation in the presence of reducing compounds (ascorbic acid, cysteine, sulphites) at neutral pH values due to formation of free radicals [166].

As we discuss in Section 5.3 on supercritical drying, biopolymer hydrogels cannot be subject to supercritical drying owing to their hydrolytic lability. Alginate quickly depolymerises in both subcritical and supercritical water into monosaccharides and further to a wide range of low-molecular weight products. If hydrolysis is performed in a tubular reactor with a very short residence time, differences in depolymerisation kinetics for M–M, M–G and G–G pairs allow isolation of homooligomers as well as monomeric mannuronic and guluronic acids [167].

Even when sterilised with steam at 120°C, alginate solutions and hydrogels demonstrate a severe shrinkage and irreversible degradation. Instead, alginate solutions can be sterilised by filtration through 0.22 μm filters, but hydrogels should be prepared from chemically modified alginate to become autoclavable [168].

In fact, there is a broad range of derivatisation pathways to enhance existing and introduce new properties in native alginate [166, 169]. Most of the chemical transformations involve two hydroxyl groups (C-2 and C-3 positions) and carboxyl group -COOH (C-6 position). Possible reactions include acetylation, phosphorylation, sulfation, grafting of hydrophobic moieties and cell signalling molecules, among many others. Regioselectivity, both in terms of selective modification of M vs. G residues and 2-OH vs. 3-OH, can be controlled in some cases. Derivatisation can be performed in aqueous solutions, but also in suspensions and gels, i.e., under heterogeneous conditions.

Derivatisation of alginate and also the processing of alginate materials have been limited to aqueous chemistry due to complete insolubility of alginate salts and alginic acid in organic solvents. Only recently, a few aprotic solvents such as DMSO and dimethylformamide (DMF) have been reported be able to dissolve tetrabutylammonium salt of alginate [170, 171]. For other biopolymers, the use of ionic liquids has been utilised, e.g., for dissolution of cellulose [172] and other biopolymers [173–175].

Dissolution of alginate and other biopolymers in solvents which can be directly extracted by supercritical CO_2 or its mixtures with cosolvents would have a great impact on the process scale-up. Already the use of high-boiling nonaqueous solvents as a gelation medium in a combination with a low-boiling solvent for the solvent exchange would greatly simplify the downstream separation purification. Gelation strategies for nonaqueous solutions of biopolymers are still to be developed.

In addition to chemical modifications, composition of natural alginates can also be modified by a set of enzymes (epimerases) which were identified in alginate-producing seaweeds and bacteria. These enzymes are involved in the in vivo inversion of stereochemistry of mannuronan at the C-5 atom yielding guluronan. Such epimerases can also be used in vitro to generate specific nonrandom epimerisation patterns [176], e.g., transform homopolymeric blocks ...M–M–M–M... into strictly alternating ...M–G–M–G ... sequences. This approach opens up a specific tailoring of alginate and creating of new M and G patterns along the polymer chain.

3.2.3 Gelation of Alginate

When aqueous sodium alginate is poured in calcium salt solution, a transparent or opaque gel is immediately formed. If after a few minutes the gel is removed from the Ca^{2+} salt solution and cut open, alginate solution that is not gelled will flow out from the gel (Figure 3.16). This observation demonstrates that gelation, as an ionic reaction, proceeds rapidly leading to the formation of a thin gel layer. At a later stage, gelation becomes diffusion limited as Ca^{2+} cations have to diffuse through a progressively growing gel layer.

Compared with other gelling polysaccharides, the binding of di- and trivalent cations eventually resulting in the gel formation is the most striking feature of alginate as polyanion. The gelation induced by reaction with cations is called ionotropic

Figure 3.16 Pouring sodium alginate solution into gelation bath with aqueous copper nitrate (a) and resulting worm-like Cu-cross-linked alginate hydrogel harvested after a few minutes in the gelation bath (b).

gelation in order to distinguish it from other types of gelation, which are briefly discussed in this section.

In the 1970s, based on a large body of experimental data from circular dichroism, X-ray diffraction, light scattering, viscometric, potentiometric titration and ion-exchange experiments, it was suggested that gelation is a result of metal-mediated parallel association of polymer chains. In this view, calcium, or generally divalent cations M^{2+}, bind to two polyuronate units of alginate chains. Because such structures bear a resemblance to a box for carrying eggs, the model is widely known as the egg-box model.

The detailed quantitative mechanism of metal binding and gelation still remains a subject of scientific debate. At a qualitative level, binding between M^{2+} and two adjacent G units in a single alginate chain occurs at low cation concentrations. The geometrical nest-like configuration of the G–G block (see Figure 3.15) favours interactions of M^{2+} with carboxylate and hydroxyl groups of guluronate. The M–M and M–G blocks adopt much flatter conformations so that they are involved in the complexation only at higher cation concentrations. As the concentration of M^{2+} further rises, dimerisation of G–M^{2+}-G sites takes place yielding zigzag-shaped junction zones. As a result, the chains are zipped into clusters by the autocooperative binding and form what is called "egg-box" junctions (Figure 3.17). Subsequent intra-chain association results in lateral and coil-like multimers which serve as building blocks for fibril-like structures of the emerging gel phase.

From this qualitative picture, it is clear that a certain minimum length of the G-blocks is needed for formation of a junction. Indeed, a body of literature on alginate gels suggests a correlation between the elastic modulus and the average $\overline{N}_{G>1}$; see Eq. (3.22). Therefore, no gelation is possible for alginates comprised of strictly alternating sequences [163].

Depending on the cation concentration, M–G local composition and the length of alginate chains, the intra-chain association can be governed by either hydrogen

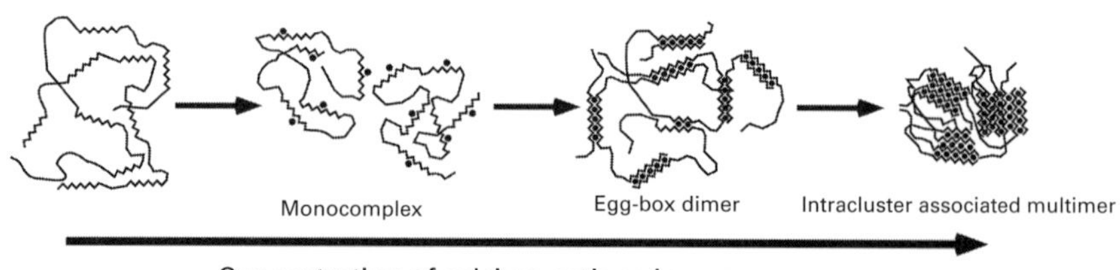

Figure 3.17 Schematic illustration of the multiple-step binding of Ca^{2+} to alginate according to the mechanism described by Fang et al. [177]. The calcium cations are depicted as full black circles.

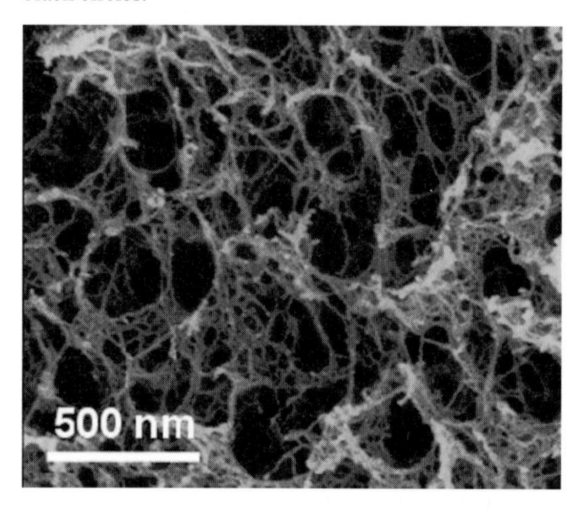

Figure 3.18 Scanning electron micrograph of the Ca-cross-linked alginate aerogel. Reproduced from [179] with permission from the American Chemical Society

bonding or by electrostatic interactions, and leads to multimers with different type of agglomeration: lateral association, chain coiling and entanglement [178]. SAXS and NMR structural studies suggest that multimers composed of a few tens of polymer chains form already with substoichiometric amounts of M^{2+}, i.e., when only a small fraction of G-blocks is occupied. In materials science and in particular in aerogel science, gelation is often performed in an excess of cations by dripping alginate solution into a large batch with metal salt solution (known as diffusion setting method see the next subsection). Available SAXS studies of Ca-alginate gels under such conditions indicate that Ca-saturated alginate gels are made up of nanosized fibrils of 5–15 nm in diameter [179, 180].

Remarkably, rod-like scattering behavior detected by SAXS seems to be almost unaffected by solvent exchange and supercritical drying [179, 180]. Therefore, according to the current view, the fibrillar morphology of alginate hydrogels is built up at the gelation step and to a certain extent preserved in the alcogel and the resulting aerogel. A typical fibrillar microstructure of alginate aerogel processed through stepwise solvent exchange and low-temperature supercritical drying with CO_2 is shown in Figure 3.18. Estimation of the fibril diameter from the cylindrical pore model (see Chapter 9) as $4/S_{BET}\rho_s$ gives values in the range of 5–8 nm (for a typical specific surface area of 300–500 m^2/g and $\rho_s = 1.6$ g/cm^3), in a reasonable agreement with the SAXS results.

In the aerogel literature, it is quite often and unjustifiably broadly stated that delicate structure of wet gels is preserved by supercritical drying. Although limited structural data rather speak for a small effect of the solvent exchange and supercritical drying on alginate hydrogels, the question arises whether it is generally the case. As we discuss in detail in Section 5.4 on solvent exchange, biopolymer hydrogels in general, and alginate in particular, demonstrate a certain degree of shrinkage when exposed to organic solvents prior to supercritical drying. Although detailed structural studies are still pending for all intermediate steps from hydrogel to lyogel and further to aerogels, two boundary cases can be anticipated.

In the first scenario, secondary structures such as fibrils or primary particles are formed at the gelation step and only slightly affected at the subsequent solvent exchange and supercritical CO_2 drying. Silica hydrogel discussed in the previous section is the most common example for such systems. Low shrinkage throughout the entire process is a macroscopic manifestation of moderate rearrangements in the gel microstructure.

In the second scenario, gel undergoes substantial changes induced by the organic solvent or sc-CO_2, or both. With some assumption, resins swollen in organic solvents illustrate this second class of systems. Upon contact with supercritical carbon dioxide, swollen resin loses the volume but often remains partially swollen up to the complete extraction of the solvent. Depressurisation causes a progressive drop of the CO_2 density, resulting in a severe shrinkage down to dense nonporous monoliths. Here at each step the network evolves in response to the changing thermodynamic quality of the surrounding solvent.

Freeze and evaporative drying (Section 5.1 and 5.2) undoubtedly have a significant effect on the gel structure. As for the supercritical drying and preceding solvent exchange, the question to what extent the structure is retained requires special analysis for each individual polymer, type and density of cross-linking as well as regimes of the solvent exchange and supercritical drying.

It seems likely that among biopolymers almost no structural changes occur in gels composed of building blocks with resistance against buckling (cellulose and chitin nanowhiskers, bacterial cellulose, chemically cross-linked lignin and globular protein gels without substantial unfolding). Gels of anionic polysaccharides such as alginate, pectin and carrageenans are likely to be ‘mixtures’ of locally ordered regions (bundles, fibres) interspersed by predominantly disordered polymer networks. The fact that aerogels from all these and similar biopolymers possess fibrillar morphology with a narrow diameter distribution may be a sign of structuring during the solvent exchange and supercritical drying. Very recently, this view has been supported by SAXS for chemically cross-linked chitosan [181]: neither gelation nor solvent exchange to methanol brought about formation of fibrillar morphology: both hydro- and alcogels compose of random walk chains. Surprisingly, the fibrillar network was shown to emerge during low-temperature supercritical drying with CO_2. Whether and to what extent the supercritical drying is the structure forming step for other biopolymers remains an intriguing open question.

To induce gelation, metal cations should be introduced into aqueous solution of alginate salt. The selectivity of the GG-blocks for metal cations depends on the cation nature. Experimentally and later by quantum chemical modelling the following affinity scale was determined for doubly charged cations: Pb > Cu > Cd > Ba > Sr > Ca > Co, Ni, Zn > Mn > Mg [159, 182, 183]. Fragmental data on trivalent cations are also available in the literature [184]. Although such scales are often used when interpreting hydrogel properties (especially mechanical and transport ones), caution should be taken as the scales are only valid at small cation concentrations. Therefore, low affinity does not automatically imply impossibility of gelation: both manganese- and magnesium-cross-linked self-supporting hydrogels can be prepared [164, 185].

To date, there are only scarce studies on properties of aerogels which are cross-linked with metals other than calcium. Most of the results, especially when targeting life science applications, are obtained with Ca^{2+} due to its nontoxic character and well-established gelation protocols. Since magnesium alginate is a water-soluble complex, it is difficult to form stable Mg-coordination biopolymeric hydrogels as was observed with other divalent metal ions.

From a technical point of view, there are two conventional and widely adapted methods to perform cation-induced ionotropic gelation of aqueous alginate salts: the diffusion setting and internal setting methods. Based on the character of interchain interactions, gels can be classified as physical (i.e., non-covalent bonding between chains) and chemical (or covalent, i.e., chemically cross-linked chains). Metal-cross-linked alginate gels belong to chemical gels, whereas alginic acid gel is a physical gel. Alginate gels with more than one type of interactions are also known and often demonstrate unusual properties.

Diffusion Setting Method

In the diffusion setting method, a solution of alginate salt is extruded dropwise into a gelation bath with a soluble calcium salt, e.g., calcium chloride or calcium acetate. Due to rapid gelation in an external layer of the droplet, the spherical shape is usually preserved. After the rapid formation of the outer layer, gelation is diffusion controlled. The gelled droplets are left in the solution for several hours or days to ensure complete gelation. Resting time can be shortened if capsules, i.e., gel beads with a liquid core, are desired, for example as immobilisation matrices. An interesting reverse implementation of the diffusion setting is also possible and especially useful for encapsulation purposes. In this case, calcium salt solution with a substance to be encapsulated and additional components (protecting polymers, buffers, etc.) is dripped into stirred solution of alginate salt; see Figure 3.19.

At a lab scale, a pipette or burette is used for the extrusion, while at a pilot and production scale a number of concepts such as mechanical cutting, vibrating nozzles and electrospraying have been implemented [26] with a throughput up to tens of tons per hour. To reduce time of the subsequent solvent exchange, the diffusion setting can

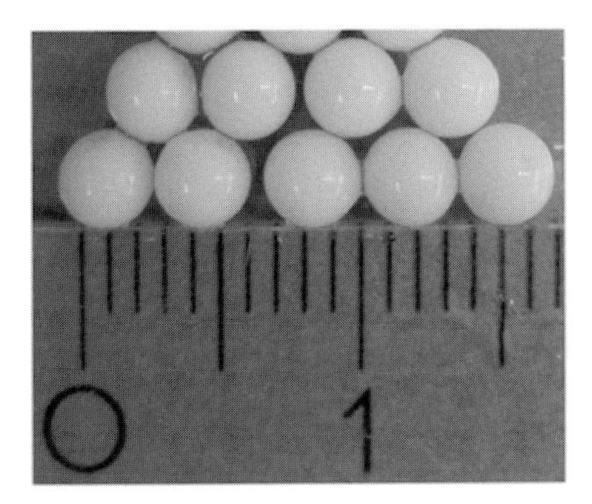

Figure 3.19 Alginate–chitosan gel capsules loaded with yeast cells. The cells in solution of $CaCl_2$ with additives were extruded into aqueous alginate salt and further reinforced in chitosan solution. Major unit of the ruler is in centimetres. Figure is taken from [186] and distributed under the terms of the Creative Commons Attribution 4.0 International License CC BY 4.0

be performed in a mixed solvent, for example, EtOH–H_2O. This approach combines metal-induced and nonsolvent induced phase separation methods (see 'other Gelation Methods of Alginate', later in this section).

The diffusion setting method can also be used for preparing hydrogels with morphologies other than spherical: for instance, fibres are produced by forcing alginate solution through a nozzle into the gelation bath [187]; films are prepared by either spraying onto or infiltration into a plate followed by gelation in metal salt solution [188, 189]; and monoliths are produced with cylindrical symmetry. In the latter case, gelation is performed in moulds connected to the gelation batch through a semi-permeable membrane. In the simplest case, alginate solution in a dialysis tubing is immersed in calcium chloride bath and left until complete gelation.

Heterometallic ionotropic alginate hydro- and aerogels can also be processed by the diffusion setting method using a mixture of two water-soluble salts [164]. In an alternative approach, hydrogel cross-linked with one metal is immersed into a salt solution of another metal with a higher affinity towards alginate. Depending on the contact time, complete or partial substitution may occur, yielding the latter case to heterometallic gradient hydrogels [190, 191]. Ion exchange ability of Ca-cross-linked alginate hydrogels is utilised for removal of heavy metal ions.

Internal Setting Method

A complementary gelation method with a better control of the gelation rate is the internal setting method (see Figure 3.20). In this method, an insoluble precursor is dispersed into aqueous alginate salt. Cross-linking cations are liberated by a pH drop. One of the most studied systems for the internal setting is calcium carbonate with D-glucono-δ-lactone (GDL). Here GDL hydrolyses in water, yielding gluconic acid,

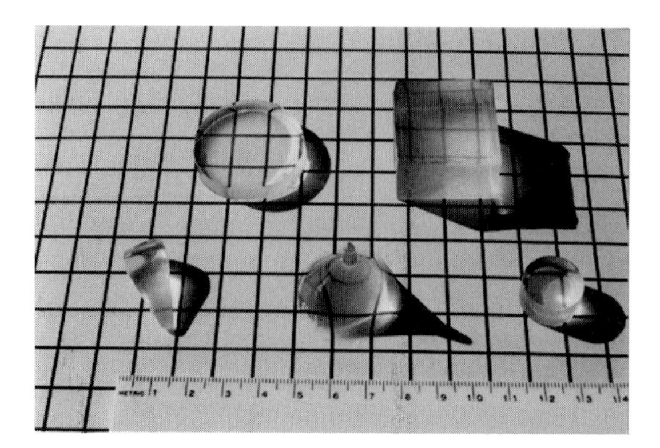

Figure 3.20 Alginate hydrogels of various shapes produced by the internal setting method from 1.5 wt/vol.% sodium alginate aqueous alginate extracted from Laminaria hyperborea at Ca^{2+}/COO^{-} molar ratio of 0.27. Reproduced with permission from [192]

which in turn reacts with water sparingly soluble $CaCO_3$ liberating Ca^{2+} homogeneously within aqueous solution of alginate salt, causing gelation. In this method, alginate solution (pH 6.5–7) and insoluble precursor are mixed first, and only then GDL, either as a solid or freshly prepared solution, is admixed to the suspension.

The internal setting method remedies the problem of the internal setting that the cross-linking density and alginate concentration are not uniform throughout the hydrogel droplet. The choice of GDL is very favourable as it hydrolyses in mild conditions yielding widely occurring in nature gluconic acid. GDL is registered under the number E575 in the list of food additives within the European Union.

To balance gelation rate and gel homogeneity, mixed precursor calcium carbonate with gypsum ($CaSO_4{\cdot}2H_2O$) has been developed [192]. Moderately water-soluble $CaSO_4{\cdot}2H_2O$ ($\simeq$0.015 mol/L ar 25°C), when admixed alone to alginate solution, leads to inhomogeneous gels due to fast gelation, while alginate gelled by $CaCO_3$–GDL needs a few hours to set. The use of $CaCO_3$–$CaSO_4{\cdot}2H_2O$ mixture allows to control gelation rate by varying the molar ratio between these slow and fast dissolving calcium salts.

Pressurised carbon dioxide is an alternative pH reducing method. In so-called carbon dioxide induced gelation, a suspension of metal carbonate or hydroxycarbonate in aqueous sodium alginate is subjected to pressurised carbon dioxide at 30–50 bar and room temperature [193, 194]. Since a significant amount of CO_2 is dissolved in the aqueous phase of the gel at the end of the process, controlled depressurisation introduces macrosized voids into the gel. Further processing steps can also be performed without depressurisation in the same high-pressure autoclave [193].

Introduction of calcium precursors in the form of insoluble salt has the disadvantage that they sediment over time and thus have to be milled finely. To achieve molecular-level homogeneity, water-soluble Ca^{2+} complexes with chelating agents such as egtazic acid (EGTA) can be utilised. In such systems, GDL acts as a pH reducing agent leading to dissociation of the CaEGTA complex with the concomitant effect of making the Ca^{2+} available for cross-linking alginate [195].

When the pH of aqueous alginate salt is lowered below the pK_a values of M and G units (3.4 and 3.6, respectively), so-called acidic gels are formed. Acidification should be done slowly and homogeneously to avoid precipitation of alginic acid. Similar to the internal setting method, hydrolysis of GDL or maleic anhydride can be exploited for pH lowering. Alternatively, alginate salt can be converted into alginic acid gel by dialysis against diluted acid.

It seems that an intermediate gelation mechanism is realised in alginate hydrogels when cross-linked by transition metals. Aqueous solutions of transition metal salts such as $Ni(NO_3)_2$ or $CuSO_4$ have acidic pH due to hydrolysis. When gelled by the diffusion setting method, a significant fraction of free COOH groups remain in the hydrogen form after solvent exchange and supercritical drying [196]. In contrast, the use of $Ba(NO_3)_2$ solution, with pH above the pK_a of alginate, yields aerogels with completely salified carboxyl groups.

Regardless of the chosen gelation approach, freshly prepared alginate hydrogels exudate a certain amount of liquid accompanied with contraction of the gel body during the storage. This process is referred to as syneresis. To indicate the fact of ongoing changes in the physical structure and thus in gel properties, the storage period is sometimes called ageing, analogously to silica and other inorganic gels (see Section 2.1). Molecular mechanisms of the syneresis in alginate gels are not fully elucidated. It is generally agreed that the degree of syneresis increases with the fraction and length of MG-blocks [197, 198], however also depends on the cation nature and cation concentration in the gel. From a practical standpoint, ageing conditions should be monitored and documented to ensure reproducibility.

Other Gelation Methods of Alginate

Aqueous and nonaqueous solutions of some synthetic and natural polymers can undergo so-called cryogelation. In the cryogelation process, the polymer solution is frozen and kept several degrees below the melting point of the system followed by thawing. Gels formed under such conditions are known as cryogels. It is very important to note that in the current literature, the term *cryogel* often has a different meaning, designating a solid material formed by the removal of all swelling agents from a gel by freeze drying (see Section 5.2). In this case, the term does not refer to any particular gelation method. Because of this ambiguity, the use of the term *cryogel* for freeze-dried materials should be avoided when possible.

The mechanism and influencing parameters for such systems are relatively well elucidated [199]. The gelation occurs in unfrozen liquid microphase which is formed in non-deep frozen polymer solutions. The microphase acts as a microreactor, where polymer chains are greatly concentrated forming physical gels upon thawing. If a cross-linker is added, it also accumulates in the unfrozen liquid microphase yielding chemical gels. The pores in cryogels are formed by growing solvent crystals, the size of which is determined by the freezing regime and the solvent composition. An interesting feature of cryogels obtained from aqueous solvents is that they typically possess a large fraction of interconnected macropores and may even have so-called gigapores ($\sim$10–100 μm), whereas conventional gelation methods usually result in hydrogels with extended mesoporosity (2–50 nm).

Studies on cryogelation of alginate and other anionic polysaccharides such as xanthan gum, sodium hyaluronate, carboxymethyl curdlan and carboxymethyl cellulose began only recently [200, 201]. The mechanism of cryogelation is not elucidated but believed to be analogous to cryogelation of extensively studied polyvinyl alcohol, wherein polymer chains are forced to concentrate, align and associate in liquid domains between growing ice crystals. The forced associations survive upon thawing, yielding the cryogel network [202]. What is certainly known is that a successful cryogelation of alginate and other anionic polysaccharides requires a careful selection of the pH. It seems that the formation of junction zones through van der Waals forces and hydrogen bonds between partially protonated (uncharged) alginate chains can be achieved at a lowered pH ($\leqslant$4.0).

Cryogelation of alginate can be realised as an extension of the internal setting method where an insoluble calcium salt is dispersed in aqueous alginate followed by freezing the dispersion [199]. Cryogelation can also be combined with the diffusion setting method. In this approach, alginate solution is extruded into a bath with a cold solution of calcium salt, so that freezing and cross-linking are not spatially separated [203]. After a certain resting time in the cold gelation bath, the frozen beads are filtered and thawed at room temperature. Alternatively, aqueous alginate can be first frozen and then immersed into aqueous ethanol solution of $CaCl_2$ at $-20^{\circ}C$ to induce gelation of alginate [204]. This intermediate case between simultaneous and spatially separated freezing and gelation has the advantage that the final step takes place in ethanol-water mixture, allowing for the partial elimination of the solvent exchange step.

Similar to cellulose, aqueous alginate forms a separate phase when the thermodynamic quality of the solvent decreases due to addition of a nonsolvent. The nonsolvent induced phase separation (NIPS) of alginate is performed by its aqueous solution in alcohols, ketones, dimethyl sulfoxide, dimethylformamide and a few other organic solvents [205, 206]. It seems that an extended aggregation of polymer chains takes place in the NIPS process. Macroscopically it is manifested in an appreciable shrinkage, i.e., the volume of the resulting gel is smaller than that of the starting solution. As we discuss in Section 5.4 about solvent exchange, polymer–solvent interactions play a very important role in the NIPS process. As a rule of thumb, one can say that more porous gels will be obtained when the mutual affinity between the solvent and nonsolvent is high [207]. The fact that water in gels is substituted by an organic solvent directly during the NIPS process opens up attractive opportunities for aerogels processing by supercritical drying without the need of an extra processing step.

3.3 Variables and Symbols

Table 3.2 lists some of the variable and symbols of importance in this chapter.

Table 3.2 List of variables and symbols in this chapter.

Symbol	Meaning	Definition	Units
c_{cell}	Concentration cellulose in solution	$<$	wt%
c_{ZnO}	Concentration ZnO in solution	–	wt%
c_{H^+}	Concentration H^+	–	$\mathrm{kg/m^3}$
c_{OH^-}	Concentration $0\mathrm{H}^-$	–	$\mathrm{kg/m^3}$
$c^m_{\mathrm{OH}^-}$	Maximum concentration OH^-	–	$\mathrm{kg/m^3}$
c	molar fraction of H^+	$c = \frac{c_{\mathrm{H}^+}}{\Omega_m}$	–
D	Diffusion coefficient	–	$\mathrm{m^2/s}$
D_{H^+}	Diffusion coefficient of H^+	–	$\mathrm{m^2/s}$
DP	Degree of polymerisation	–	
k_B	Boltzmann coefficient	1.38×10^{-23}	J/K
k	Reaction rate constant	–	1/s
k^*	Dimensionless reaction rate constant	$k^* = \left(\frac{c_{\mathrm{H}^+}}{c^m_{\mathrm{OH}^-}}\right)^2$	–
L_c	Size of a gel body	–	m
r'	Dimensionless coordinate	$\vec{r}/L_c$	–
t_{gel}	Gel time	–	s
v	Dimensionless concentration	$\frac{c_{\mathrm{H}^+}}{c^m_{\mathrm{OH}^-}}$	$\mathrm{m^3}$
T	Temperature	–	K
α	Fit factor	–	1/K
Ω_m	Atomic or molar volume	–	$\mathrm{m^3}$
τ	Dimensionless time	–	–

4 Gelation

Gelation describes the transformation from a liquid or fluid state to a somehow solid state. Gelation is well known from daily experience, making, for instance, jam or simply a jelly pudding. How does this transformation occur, and what makes a gel special? Gels and gelation are studied quite extensively in chemistry and physics (soft matter), and especially three decades ago theoreticians have discussed the formation of gels and the relation between structure and properties quite extensively (percolation theory, fractals). In this chapter, we will first discuss the viscosity of solutions and how it changes during gelation. Then, for an understanding of modern equipment to analyse gelation, it is important to briefly discuss viscoelasticity, simple models for viscous fluids. We then will describe how gelation is measured, from very simple methods to more elaborate ones. The chapter closes with a survey of theoretical models for gelation. Anyone not really interested in mathematical models of gelation might skip Section 4.2 and restart reading in Section 4.3, where these models are applied to aerogels.

4.1 Viscosity of Gelling Solutions

Viscosity is a material parameter that relates shear stress and a velocity gradient in a liquid. It is related to internal friction in the fluid. To understand this, make a simple experiment of your own: take a cup filled with water and put it on a surface, for instance, a leaf or a thin sheet of a polymer, and then move it. If you do this slowly, you can move it apparently without any force. Increasing the velocity requires more force. The resistance against this movement is the viscosity. Do the same experiment with honey, and you observe that you need a much higher force: it has a higher viscosity. Or take a can filled with water and pour it out. It easily runs into the basin. Do the same with honey or oil; the same amount of liquid takes much longer to empty the can. Therefore, from everyday life, one has a certain idea about viscosity. A clear definition can be made by reducing such experiences to its essence: the movement of a plate on a surface induces in the subsequent liquid layers a fluid flow parallel to the plate. The fluid flow can be measured and even visualised. For the definition, we assume that it varies linearly over the height of the cup (a sketch is shown in

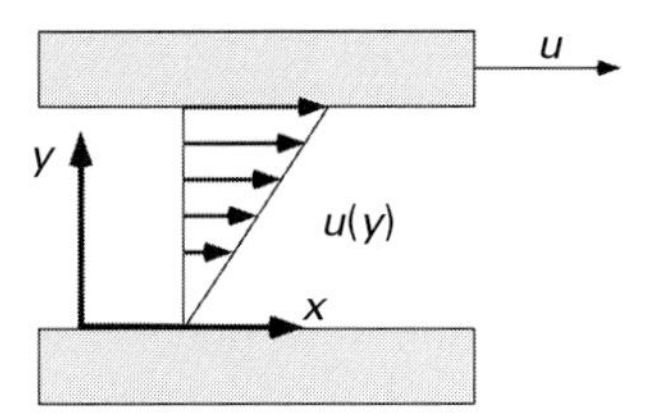

Figure 4.1 Scheme of a so-called Couette flow. One plate is at rest, the other moves with a constant velocity in x-direction. If the movement is at low velocity, a linear velocity profile is established.

Figure 4.1 for this type of so-called Couette flow). Then the shear stress (tangential to the water or liquid surface) is empirically related to the gradient of the velocity

$$\tau = \mu \frac{du}{dy} = \mu \Gamma \quad \Gamma = \dot{\gamma}, \tag{4.1}$$

with Γ the so-called shear rate, most often written as the time derivative of the shear deformation γ as $\dot{\gamma}$. Do not take this as a fundamental law of physics. It is simply an approach to describe certain types of liquids. If one performs a measurement of shear stress and measures the shear rate and both are related linearly, then a fluid obeying this relation is said to be a Newtonian liquid, since Newton used this relation first. In reality, many fluids, especially gelling fluids, do not behave like this. In many fluids, the viscosity is a function of shear rate itself (where opposing things can be observed, such as a decreasing viscosity with increasing shear rate and vice versa). The viscosity might even be history dependent, i.e., it depends on the previous shearing done on the fluid. Think, for instance, about polymers, then the viscosity is a complex function of the length of the polymer molecules, their conformation in space (long rods, ravels, coils, degree of cross-linking etc.).

One more comment on viscosity: from the definition given, one can extract that the coefficient of viscosity μ determines the amount of energy dissipated in fluid motion, since the entropy production in a fluid is directly proportional to it [208–210]. The coefficient μ is also called dynamic viscosity or shear viscosity. In fluid dynamics, the coefficient $\nu = \mu/\rho$ (where ρ is the liquid density) is called the kinematic viscosity and is of eminent importance in scaling analysis of fluid flow. Theoretical models of viscosity relate the coefficient μ to the liquid structure in close analogy to its relation derived in the kinetic theory of gases [211, 212]. Born and Green [213, 214] have derived an expression for the viscosity which relates it to the pair correlation function $g_0(r)$ and the interatomic pair potential U:

$$\mu = \frac{2\pi n_0^2}{15} \sqrt{\frac{m_A}{k_B T}} \int_0^\infty g_0(r) \frac{\partial U}{\partial r} r^4 dr, \tag{4.2}$$

where n_0 is the average number density of molecules, m_A the molecular mass and k_B, T have their usual meanings. Therefore, the viscosity is related to the molecular structure of a fluid, which can be measured by X-ray or neutron diffraction, yielding, for instance, the pair correlation and radial distribution function of molecules or atoms in a liquid. This relation was, however, developed mainly for liquid metals or more generally for hard sphere fluids. Polymeric solution with long molecules that might be straight or folded, clumped or bended, will surely behave differently, and therefore another approach is necessary to describe polymer solutions [54]. Without going into the details, it is sufficient to note, that the dynamic viscosity of polymer solutions (and in a certain sense sols from which aerogels are derived are similar) is related to the molecular weight M_m by a power law (Zimm model)

$$\mu \propto M_m^{\alpha} \tag{4.3}$$

with α a coefficient in the range from 0.77 to 0.8 (see [54], pp. 239 ff). If we think a colloidal particle as a clumped polymer, its molecular weight is related to its diameter and thus the larger the colloidal particles become, the larger is the dynamic viscosity. We therefore expect that with the ongoing formation of particles, their growth and clustering, the dynamic viscosity increases. Einstein derived more than 100 years ago a simple relation between the dynamic viscosity of a pure liquid μ_0 and that liquid in which particles at a volume fraction ϕ are dispersed, namely

$$\mu = \mu_0(1 + 2.5\phi). \tag{4.4}$$

This means, inasmuch as colloidal particles or clusters evolve by a chemical reaction in the liquid, the viscosity increases. Einstein's equation is only valid at low-volume fractions. Extension were made in the literature since then. Most often, the so-called Richardson–Zaki relation is used to describe empirically the dynamic viscosity of a dispersion

$$\mu = \frac{\mu_0}{(1 - \frac{1}{2}\phi)^5}, \tag{4.5}$$

which yields Einstein's result at low-volume fractions ($\phi \ll 1$). Often the exponent 5 in the denominator is also an adjustable constant, and in experiments values between 4 and 5 are fit to data.

In the following, we will give a very brief recapitulation on the theory of viscoelastic fluids and finish it with a short summary of the viscosity of gelling fluids.

4.1.1 Simple Models of Viscoelasticity

For a solid, we have the simple relation, that shear stress is proportional to shear deformation γ as long as the deformation is fully elastic and small. The constant of proportionality is called shear modulus G. This is expressed as

$$\tau = G\gamma \tag{4.6}$$

For a viscous fluid, we take the same simple relation, namely

$$\tau = \mu\dot{\gamma} \tag{4.7}$$

and do not care for the moment about a possible dependence of viscosity on the shear rate itself. There are two ideal models to describe a viscoelastic material, the Maxwell and the Kelvin model.

Maxwell and Kelvin Model

In the Maxwell model of a viscoelastic material, a damping element described by Eq. (4.7) is added to a spring with shear modulus G. The spring, subscript s, should be visualised as representing the elastic or energetic component of the response, while the damping element, most often called a dashpot, subscript d, represents the conformational or entropic component in which energy is dissipated. In such a series connection, the stress on each element is the same and equal to the imposed stress, while the total strain is the sum of the strain in each element:

$$\tau = \tau_s = \tau_d \qquad \gamma = \gamma_s + \gamma_d. \tag{4.8}$$

Then the total shear rate of such a model is given by

$$\dot{\gamma} = \frac{\dot{\tau}}{G} + \frac{\tau}{\mu}. \tag{4.9}$$

In the Kelvin model, it is assumed that a damping element and the spring are arranged parallel such that the stress generated by both elements add to

$$\tau = G\gamma + \mu\dot{\gamma}. \tag{4.10}$$

Both models have their own dependence of shear rate or strain on time and can nicely describe stress or strain relaxation experiments, meaning experiments in which a material is subjected to a constant load and the time evolution of strain is measured, or vice versa. We are not going to describe or discuss this. Here we are going to analyse what happens if strain and stress are not constant for some time period, but vary sinusoidally with time, a situation very often encountered in measurement of viscoelastic fluid behaviour. Let us take, for instance, a Maxwell solid and assume

$$\tau = \tau_0^* \exp(i\omega t) \quad \gamma = \gamma_0^* \exp(i\omega t), \tag{4.11}$$

where we have used a complex notation (note: according to the Euler relation, we have $\exp(i\alpha) = \cos(\alpha) + i\sin(\alpha)$). Insertion into Eq. (4.9) yields after some manipulations and introducing the abbreviation $\varphi = \mu/G$

$$Gi\omega\gamma_0^* \exp(i\omega t) = \left(i\omega + \frac{1}{\varphi}\right)\exp(i\omega t). \tag{4.12}$$

Defining the complex modulus as

$$G^* = \frac{\tau_0^*}{\gamma_0^*} \tag{4.13}$$

gives

$$G^* = \frac{Gi\omega}{i\omega + \frac{1}{\varphi}}. \tag{4.14}$$

This relation can be put into an interesting form, namely

$$G^* = G\frac{\omega^2\varphi^2}{1+\omega^2\varphi^2} + i\frac{G\omega\varphi}{1+\omega^2\varphi^2} = G' + iG''. \tag{4.15}$$

G' increases from zero to the finite value G as the frequency increases, whereas G'' is zero at zero and infinite frequency and therefore has a maximum in between. A similar calculation can be performed for a Kelvin solid.

Especially for polymers, these simple models do not describe their mechanical behaviour accurately. Combinations of a Maxwell solid with another spring in parallel (standard linear solid) can be used to calculate the mechanical response or a Kelvin model with a spring or dashpot in series and even complex models including in principle infinite numbers of dashpots and springs parallel or in series have been suggested to describe the viscoeleastic behaviour of materials not following Hooke's law or Newton's law with a constant viscosity. We are not going to study such complex descriptions of viscoelastic materials. The interested reader is referred to literature on viscoelasticity [215].

Storage and Loss Modulus

In the current context, it is sufficient to note that stress and strain in a viscoelastic material are still related to each other, but that both are not in phase. In an elastic solid, strain and stress are always in phase (see Eq. (4.6)), whereas in a viscous material they are not. To see this, assume $\gamma = \gamma_0 \sin(\omega t)$. Then we have according to Eq. (4.7) $\tau = \mu\gamma_0\omega\cos(\omega t) = \mu\gamma_0\omega\sin(\omega t + \pi/2)$. The stress is out of phase by 90°. In a viscoelastic solid, being a mixture of viscous and elastic components, the phase lag between stress and strain will be in between 0 and $\pi/2$. If we generalise these considerations, we can write, using a complex notation for convenience,

$$\gamma = \gamma_0 \exp(i\omega t) \tag{4.16}$$

$$\tau = \tau_0 \exp(i\delta)\exp(i\omega t), \tag{4.17}$$

in which now δ describes the phase lag between stress and strain. If we divide stress by strain, we get a complex shear modulus as

$$G^* = \frac{\tau}{\gamma} = \frac{\tau_0}{\gamma_0}\exp(i\delta), \tag{4.18}$$

and using Eulers expansion of complex numbers, we have

$$G^* = \frac{\tau_0}{\gamma_0}(\cos(\delta) + i\sin(\delta)). \tag{4.19}$$

The first term is called the *storage modulus*,

$$G' = \frac{\tau_0 \cos\delta}{\gamma_0}, \tag{4.20}$$

because it describes the energy stored in the elastic deformation and the second term describes the damping or energy loss and is therefore called *loss modulus*:

$$G'' = \frac{\tau_0 \sin\delta}{\gamma_0}. \tag{4.21}$$

If we calculate the ratio of both, we arrive at a term called the *loss tangent*:

$$\tan\delta = \frac{G''}{G'}. \tag{4.22}$$

In a typical rheometer, all three quantities can be measured as a function of temperature, time and rotation frequency. We can also use Eq. (4.19) to obtain the value of the complex shear modulus as

$$|G^*| = \sqrt{G^* \bar{G^*}} = \sqrt{(G')^2 + (G'')^2} \tag{4.23}$$

to see that the apparent modulus of a viscoelastic material is a weighted sum of elastic and viscous contributions.

4.1.2 Evolution of Viscosity during Gelation

Having measured the storage and loss modulus, one can calculate the dynamic viscosity below the gel point:

$$\mu = \frac{\sqrt{(G')^2 + (G'')^2}}{\omega}, \tag{4.24}$$

where ω is the frequency used in a rheometer setup. The schematic dependence on time is shown in Figure 4.2.

An example taken from a real measurement using an Ubbelohde viscosimeter is shown in Figure 4.3. The basics of an Ubbelohde viscosimeter is as follows. A long capillary is dropped into a bath of the liquid and the time is measured that the liquid needs to rise in the capillary as well as the pressure difference driving the capillary rise. Then the viscosity can be calculated by applying Hagen–Poisseuille's law with some corrections reflecting the necessity to accelerate the fluid. Thiel [216]

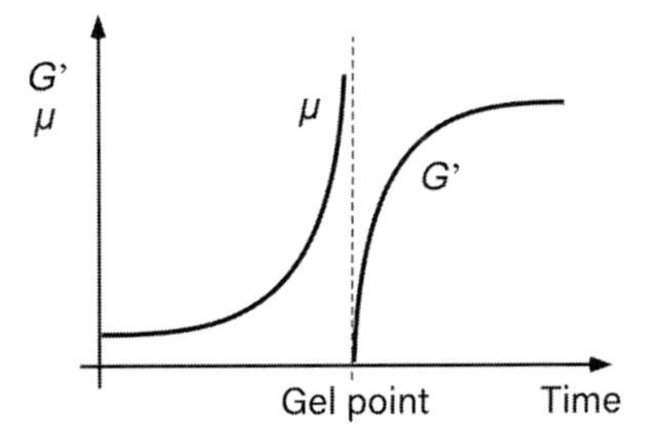

Figure 4.2 Schematic dependence of the viscosity and the storage modulus as a function of time in a gelling system. Note not only the increase in viscosity as the system starts to gel, but also that the storage modulus is zero in a liquid and attains a constant value after gelation has set in.

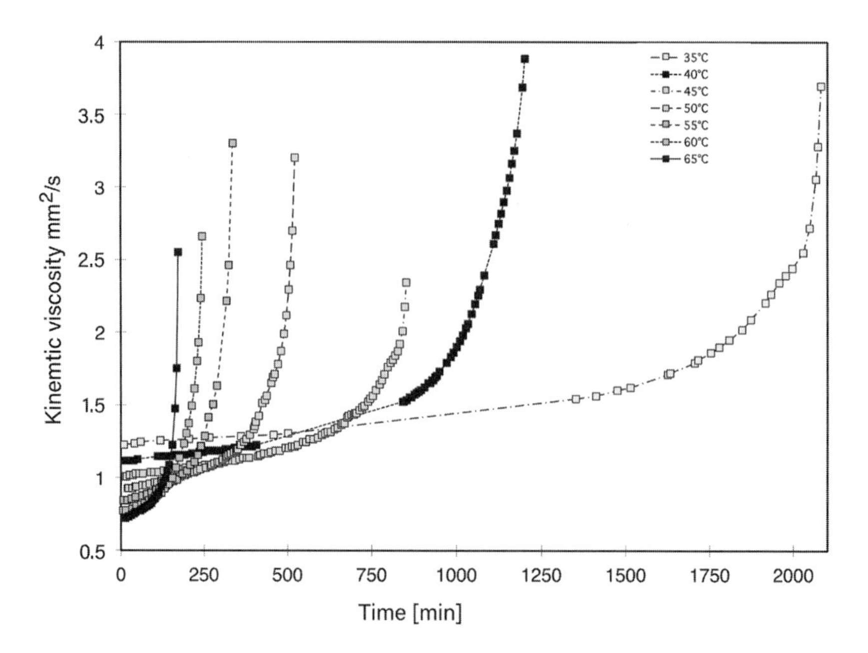

Figure 4.3 Kinematic viscosity as a function of temperature in solutions that make RF aerogels (R/F = 1/2, R/C = 1500) [216].

measured the viscosity of RF–gel solutions at various temperatures as time passed by and determined from such curves the point of gelation and its dependence on temperature. Figure 4.3 clearly demonstrates that at constant temperature the viscosity of the solution does not change appreciably with time (there is a slight increase, which could be explained by the formation of dispersed particles in the solution; see Eq. (4.4)), but within a short time interval it suddenly rises enormously. Before this change, the liquid viscosity was in the range of water; after that, it rose by orders of magnitude. The gel point still is hard to define on the curves of this diagram. Where does one read it off? One could define it, for instance, at a certain percentage of deviation from the baseline or a certain value of the dynamic viscosity, meaning once it is reached, the gel formation has set in. In doing so, one can determine, that the gel time depends in Arrhenius fashion on temperature. Thiel [216] determined for the RF–gels, an activation energy of 72 kJ/mol. Such a dependence on temperature is quite often observed. The gel time determined in this way depends also on other parameters of the sol preparation. In RF aerogels, it depends on the resorcinol catalyst (R/C) ratio; the larger R/C, the larger the gel time.

4.1.3 Tilting as a Simple Measure of Gelation

In any type of lab experiment, the simplest approach to determine the time to gelation is to take a test tube filled with a gelling liquid and tilt it. If the meniscus moves during tilting, no gelation has occurred. If the meniscus follows completely the tilting

Figure 4.4 Gelation of an RF sol. In this sequence (to be read from the upper-left corner to the bottom-right corner as a time sequence), the catalyst was HCl. Initially there is a clear solution, which upon shaking the beaker slightly became turbid until it gelled to first a white gel, which became after a short time piggy pink (here a light grey colour).

operation or even if one can turn it upside down without any motion of the material inside the test tube, gelation has occurred. If one touches such a gel, it feels wet and soft. One can typically deform it more or less slightly (it depends on the gel; there are gels that can be squeezed, whereas other gels, such as wet silica gels from TEOS, cannot be squeezed, but only pressed slightly and gently a bit). If one measures the time from preparation of the starting solution to the point of gelation (tilting operation), a gel time can be defined and recorded. It might not be precise, but in many cases it is for aerogels sufficient, since the gelation time easily varies between minutes and days. An example of such an observation method is shown in Figure 4.4.

4.1.4 Rotating Bob Viscosimeter

Another type of viscosimeter is the so-called *rotating bob viscosimeter*, which was invented by Couette utilising that the flow field between to coaxial cylinders in which the inner cylinder rotates at a given angular velocity is a simple shear field (or vice versa, the outer cylinder rotates and the inner one is at rest). The usual construction employs two concentric cylinders, one on the bob of a torsion pendulum, the other rotating at constant speed; see Figure 4.5. The viscous drag of the liquid between the cylinders induces a torque which is measured. From the torque, the rotation frequency and the geometric layout of the cylinders, the viscosity can easily be calculated. Such a viscosimeter was used by Vogelsberger and co-workers to measure the time dependent viscosity of silica sols [217]. One of their results is shown in Figure 4.6. Evaluating the data by fixing the gel point at a fixed reduced viscosity of 50 and making an Arrhenius plot yields a value for the activation energy of around 53 kJ/mol.

Most often today, so-called *rheometers* are available and used, in which the shear rate and the temperature can be varied in wide intervals. A sketch of the central element of modern rheometers is shown in Figure 4.5. On a plate, a drop of the liquid to be investigated is placed and the upper plate is moved downwards onto the liquid until a predefined slit distance between both plates is achieved, such that a thin liquid film establishes between the plates. The movable plate is then set into oscillating movements with a preset oscillation frequency and the angular torque to set the

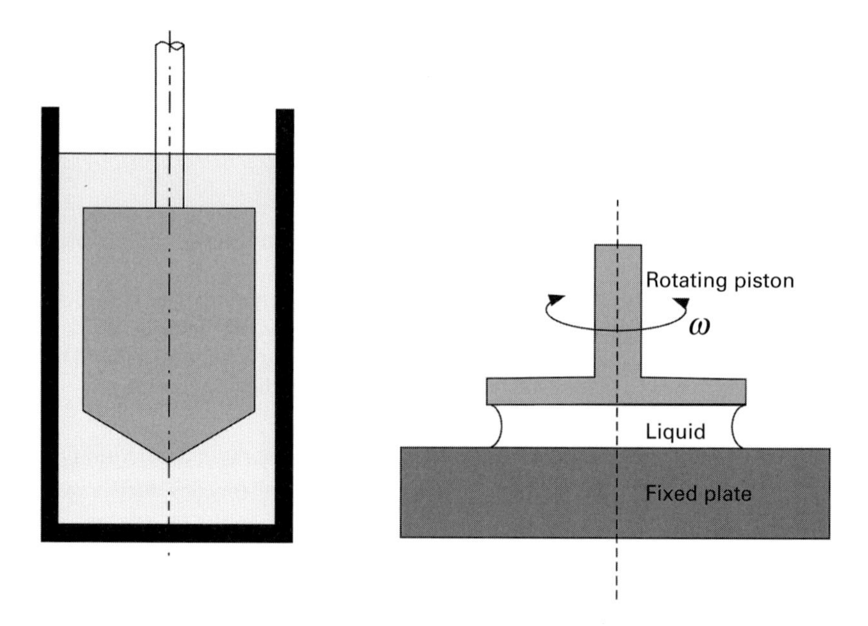

Figure 4.5 The left picture shows a scheme of a Couette or Searle rheometer. In the Couette case, the black container moves, while the bob is fixed and vice versa in the Searle case. The right picture shows a scheme of a rotating disc to measure the viscoelastic properties of gelling liquids.

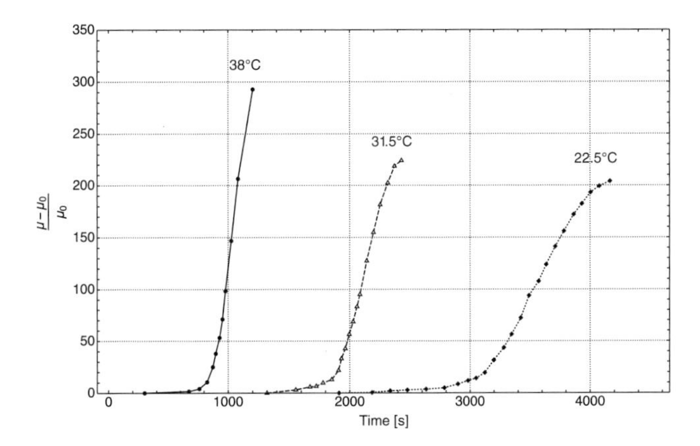

Figure 4.6 Dynamic viscosity of silica sols prepared from sodium silicate. On the ordinate, a normalised viscosity is plotted, defined as $\mu/\mu_0 - 1$ with μ_0 the viscosity of the sol initially. Reprinted from [217] with permission from Elsevier

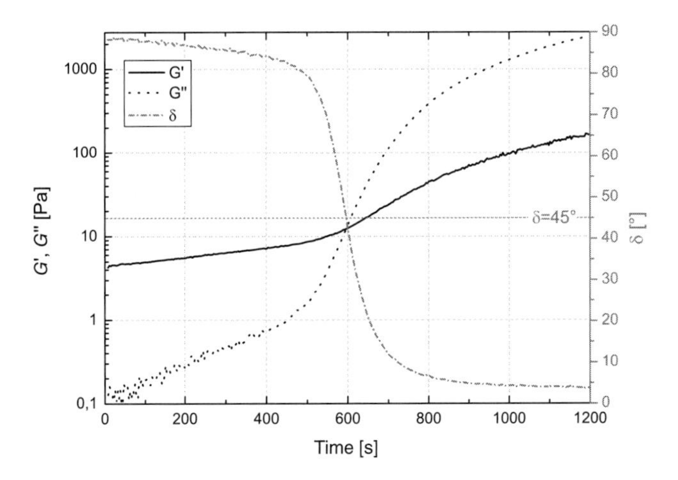

Figure 4.7 Measurement of storage and loss modulus and the phase lag δ in a rotating disc viscosimeter. The solution is a 3 wt% cellulose dissolved in zinc chloride tetrahydrate. The measurement was performed at 68°C and shows that after 600 s both moduli intersect. Courtesy of M. Schestakow [133]

fluid into a rotating motion is measured. The shape of the plates can vary. From such measurements, one obtains the storage modulus G', the loss modulus G'' and the loss tangent. From both moduli, the viscosity could be calculated, but this calculation is usually not performed. Especially the gel point can be determined from both shear moduli. At the gelation point, G' and G'' intersect in a plot of their time dependence at isothermal conditions and at constant rotating frequency [135]. This technique was used by Schwan [89] to define gelation in RF aerogels and by Buchtová and co-workers [218] and Schestakow [133] to measure the gelation of cellulose aerogels. An experimental result is shown in Figure 4.7. The viscoelastic properties of a solution prepared from microcrystalline cellulose dissolved in zinc chloride tetrahydrate were recorded in a rotating disc viscosimeter. The curves for the storage and loss modulus intersect, and at the point of intersection the loss tangent is almost 45°. The time at which both moduli attain the same value can be interpreted as the gelation time [135].

4.1.5 Flow Damping

Another experimental approach was used by Krause et al. [219]. The essence of the technique is to put into a gelling solution tracer particles, set the solution into motion and observe the tracer particles with particle image velocimetry (PIV), which is easily able to handle thousands of tracer particles and calculate their velocity vector in a solution [220]. This is done by a laser sheet illuminating a plane in the fluid, and the scattered light from the tracer particles is measured with a digital camera. Comparing frame by frame the light scattered from particles and identifying the location of the particles, one can calculate the distance they have moved and their local velocity (as long as they do not move out of the thin laser sheet). Since the process of transformation of a solution to a gel is irreversible (at least in all types of gels starting from monomers), they prepared two solutions. One with $BaTiO_3$ particles with a diameter of 1 μm in a concentration of only $\approx 2.7 \cdot 10^{-2}$ g/l and one solution of tetra-n-butylorthotitanate, ethanol, water and hydrochloric acid (this gives a TiO_2 aerogel). Just before the measurement started, both solutions were mixed in a beaker, shaken and filled into an optically transparent cuvette. High flow velocities in the range of 1–2 mm/s were observed in the centre of the cuvette. The flow velocity decreases in an almost linear fashion with time, as shown in Figure 4.8, indicating that the viscosity during gelation also increases continuously. Drawn in this figure is a linear plot whose extrapolation to zero velocity would define the gel time. Krause and co-workers used for the theoretical description of their experiments a model that is based on percolation theory (to be dealt with later in this chapter).

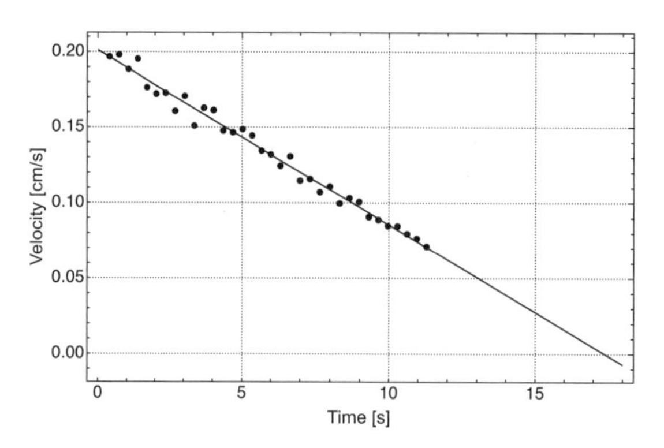

Figure 4.8 Damping of the tracer particle velocity in a titanate solution as a function of time. Drawn in is a simple linear fit. The extrapolation to zero velocity could define the gel time, which then would be 17.4 s.

4.1.6 Light Transmission and Scattering

In many cases, the transition of a solution to a gel is coupled to a transition from transparent to translucent or opaque. For instance, RF solutions are clear and transparent after mixing of the components, but as time passes by they become milky, white, piggy pink or even light brown, depending on the pH value used to catalyse gel formation. For cellulose, alginate and silica gels, such a transition cannot be observed, as they stay transparent even after gelation. They become translucent or white after supercritical drying. For gels showing a transition from transparent to opaque, the change in light transmission can be observed by the naked eye and therefore used to determine the time for gelation either electronically with a photodiode or manually with a stopwatch. An example is shown in Figure 4.9 taken from the work of Hajduk and Ratke [221]. They took RF–aerogel solutions, which are optically clear after mixing, put them into a paraffin bath and measured the gelation time using a stopwatch and carefully looking at the immersed drop. On gelation, the initially clear solution becomes first turbid and then opaque, grey-white. This is shown in Figure 4.9, which denotes two different gel times. The change from clear to turbid was called the starting of gelation. This time was recorded as t_g^{start}. The change to opaque was recorded as the finishing time of gelation t_g^{fin}. This also is a simple way to characterise the onset of gelation. A plot of the gelation time as a function of drop volume is shown in Figure 4.10.

We will later discuss these results in more depth in Section 4.3.3. Another way of measuring gelation, can be done by light scattering. Prior to gelation, many colloidal particles nucleate and grow being centres of scattering for light (of the right wavelength). Illuminating a cuvette with a solution using a laser and measuring the scattered light intensity as a function of scattering angle yields curves, which initially focus around the direction of the laser beam but spread out inasmuch as particles nucleate and grow in size and number. The scattering curve contains much information about the particle growth, number density and aggregation. This technique can be used at least for all solutions not turning turbid during gelation but at least translucent. In the case of solutions turning opaque, one could measure the scattering in the

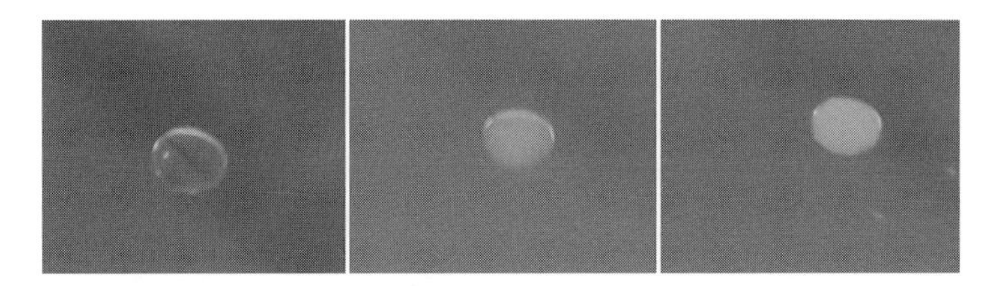

Figure 4.9 The left figure shows the drop freshly immersed in oil, and in the middle the onset of gelation can be seen, denoted as t_g^{start}. At this moment, the drop is turbid or translucent but still wobbling around and thus deformable like a highly viscous drop. In the right figure, gelation led to a drop with a stable size and the material now has become white. This time is denoted as t_g^{fin}. The picture is from [221] and here reproduced under the terms and conditions of the Creative Commons Attribution license 4.0

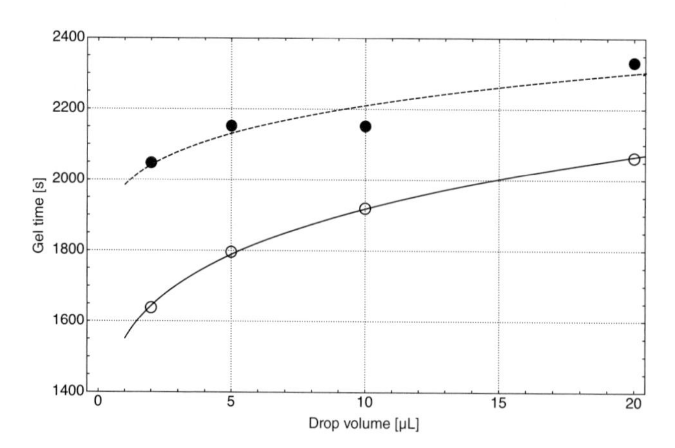

Figure 4.10 The two gel times defined in the text as a function of drop volume. The lines drawn in curves are fits to a power law dependence on drop volume. The open circles denote the onset of gelation, t_g^{start}, and the full circles the end of gelation t_g^{fin}. This figure is redrawn after [221] and here reproduced under the terms and conditions of the Creative Commons Attribution license 4.0

backscatter mode. The technique has been applied by many researchers for silica, RF solutions and cellulose. For more details, see Appendix C.

4.2 Theoretical Descriptions of Gelation

As mentioned several times, a gelling solution always starts with mixing the chemicals or dissolution of polymers (as in the case of biopolymeric aerogels) yielding always a solution in which chemical reactions start. In the case of gels starting with monomers, these react such that oligomers are formed and eventually particles. Once such particles appear in the solution, they do not, however, rapidly form a contiguous network: the system still is in a colloidal state. Network formation requires that these particles aggregate (in a solution in which polycondensation still takes place and the particles still grow by addition of monomers and oligomers).

In the literature on gelation, various descriptions or models have been developed to describe the transition from a solution or a colloidal sol to a gel. Most famous are percolation models, since they are able to show that once a certain threshold is reached, for instance the volume fraction of particles, a spanning cluster of particles exists such that the physical properties of the system change. (For instance, in a electrically non-conducting matrix, one can introduce electrically conducting spheres and ask at what volume fraction a continuous cluster or network of particles is established, such that the whole sample becomes electrically conducting.) Percolation models will be briefly discussed in Section 4.2.1. Also famous and being in the centre of a few decades of scientific research in the 1980s and 1990s were so-called models of diffusion-limited aggregation (DLA) of particles or diffusion-limited aggregation of clusters (DLCA) combined with, for instance, reaction controlled growth or ballistic aggregation. These models were interesting, and most of them done with numerical studies, since the aggregated clusters could be evaluated with respect to their fractality and the investigation of fractals and fractal patterns, after the term was coined by Mandelbrot [55, 56, 222]. We discuss this also briefly in Section 4.2.2. Percolation as well as DLA and DCLA models were always also studied with a very classical mean field approach, developed 100 years ago by Smoluchowski [223] for aggregation of particles by Brownian motion and used since then in thousands of papers to describe aggregation in aerosols, clouds, immiscible liquids, dispersions, colloids etc. [224–226]. We will briefly recapitulate such mean field aggregation models, then apply them to gelling systems and discuss possible mechanisms of network formation in aerogels. As also mentioned already several times, the aerogel literature discusses from time to time if gelation is not a kind of phase separation. This idea is taken up at in Section 4.2.5, discussing in some detail possible mechanisms of polymerisation-induced phase separation (PIPS) in aerogel forming solutions.

4.2.1 Percolation and Fractals

The principle of percolation is best illustrated by Figure 4.11. Take a square lattice and put on a square a quadratic black particle. The location on the lattice is chosen randomly. Take another one and throw it onto the lattice. If the site is already occupied, make another choice. With the increasing number of particles on the lattice, there will be a situation that some black squares make a connection from one side to the other. The number of black squares needed in relation to the total number of squares

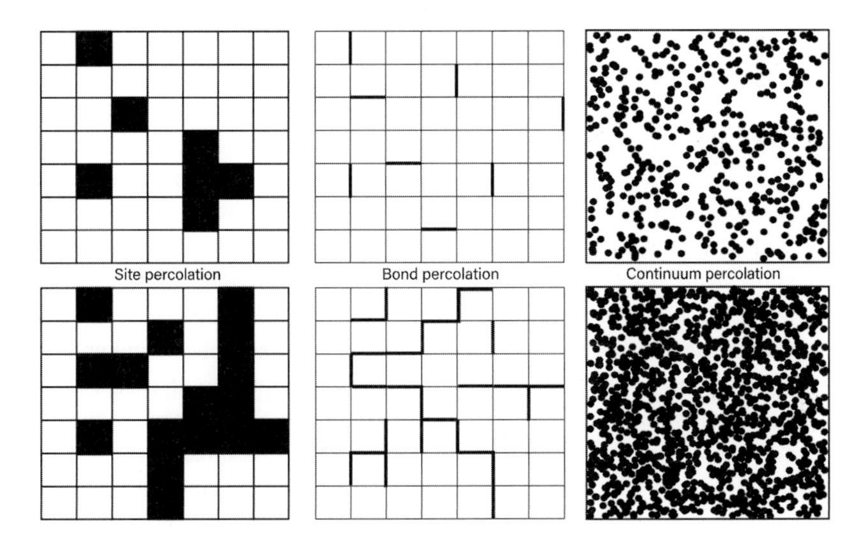

Figure 4.11 Illustration of site, bond and continuum percolation below the percolation threshold and beyond the threshold, which is characterised by the appearance of a spanning cluster of black boxes or bonds, reaching from one side of the enclosing box to the other.

in the lattice defines an occupation probability, p. The probability once a cluster of squares is able to connect the borders is called the percolation threshold, and this probability is named p_c and the cluster is said to be a percolating one. Instead of squares on a lattice, one can also look for the edges of the squares and randomly put sticks on them. This is then termed 'bond percolation'. Once the sticks on the edges connect opposite sides, the so-called bond-percolation threshold is reached. In both cases, long before the percolation threshold is reached, clusters of either black squares or bonds are touching each other but not the border. In contrast to lattice or bond percolation, one can also perform so-called off-lattice or continuum studies, by just throwing circles or spheres randomly into square or a box, allowing them to overlap or not and asking for the area or volume fraction of circles or spheres necessary to make a path connecting opposite sides. Some of the tasks of percolation theory are to analyse the threshold; the critical occupation probability p_c; the probability that a site belongs to the infinite cluster $P_\infty(p)$; the average cluster size and its distribution before the threshold is reached and after; and the coherence or correlation length of clusters, as both depend on the site occupation probability p and how they change in the neighbourhood of the threshold. There are excellent textbooks on percolation theory, and the interested reader is recommended to read them. See the classical textbook of Stauffer and Aharony [111], the book of Viscek [49] on fractals and the book going into all details by Christensen and Moloney [227]. There also is a Wiki webpage available which summarises results on all types of percolation on any type of lattice or continuum [228]. The idea to describe with percolation the sol–gel transition is intriguing, since below the percolation threshold or the appearance of a spanning cluster, the system would behave like a fluid and above somewhat like a solid. The simplest model to discuss percolation and perform some calculations is a one-dimensional model. Take an infinite long line and divide it into an infinite number of units. Then randomly place sticks onto these units. Depending on the number of sticks one has placed, some sticks will contact and build a one-dimensional cluster. It is easy to imagine that only if sticks are placed on all units a cluster exits that connects both ends of the line. Therefore, in this case the percolation threshold is $p_c = 1$. We are not calculating the size distribution of clusters before this threshold is reached, but instead look to another analytically solvable case of percolation on a Bethe lattice (or Cayley tree). Such a lattice is defined by the following rule. Fix a point in space and allow two, three, four or five and so on bonds to extrude from this point. Place on the end of such a bond another point and repeat the procedure with exactly the same number of bonds per point (site) as for the starting point. Such a tree of bonds can be imagined as a mathematical description of the formation of branched polymers, and was so in the 1940s by Flory and Stockmeyer. In our case, imagine a site to be an activated resorcinol molecule that has three active bond possibilities, or silicic acid, which would have four active bonds. Figure 4.12 shows two examples with three bonds and four bonds per site. For such lattices, one can show that the percolation threshold is dictated by the functionality z, i.e., the number of bonds extending from each site, illustrated in Figure 4.12 as full circles. The percolation threshold is $p_c = 1/(z-1)$. This for $z = 3$, $p_c(3) = 1/2$ and for $z = 4$, $p_c = 1/3$.

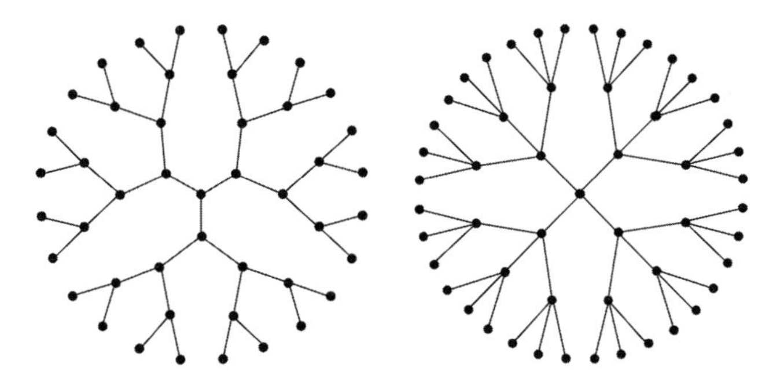

Figure 4.12 Illustration of a Bethe lattice with a functionality of $z = 3$ (left) and $z = 4$. From each site (dots), three or four bonds extend to the next site.

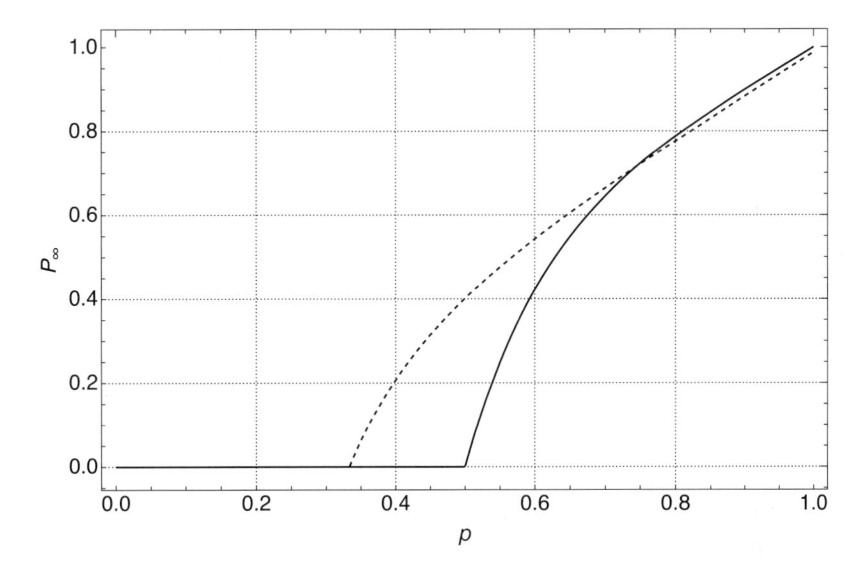

Figure 4.13 Probability of a site to belong to the infinite cluster beyond the percolation threshold. The solid line corresponds to the lattice with a functionality of $z = 3$, the dashed line to one with $z = 4$.

The calculations of cluster size, cluster size distribution etc., especially on a Bethe lattice with $z = 3$, is standard in the literature. We therefore do not repeat such calculations, but give the results. The probability that a site or bond belongs to the infinite cluster or the fraction of atoms which belong to the infinite network is then

$$P_\infty = p - p\left(1 - \frac{p(z-1)-1}{1/2p(z-1)(z-2)}\right)^z \cong \frac{2z}{z-2}(p - p_c). \tag{4.25}$$

In both cases, p is the occupation probability or concentration. This relation is valid above $p = p_c$. Figure 4.13 shows its dependence on p for the two interesting cases $z = 3, z = 4$. The average cluster size on the Bethe lattice with $z = 3$ below the percolation threshold can be calculated as

$$S(p) = \frac{p_c(1+p)}{p_c - p} \propto (p_c - p)^{-1} \quad \text{for} \quad p \to p_c^- . \tag{4.26}$$

The average cluster size diverges at the percolation threshold, since that simply is the occupation probability where an infinite cluster appears. Unfortunately, besides one-dimensional and Bethe lattice percolation, no other system for percolation, such as site or bond percolation on a square or triangular lattice, or continuum percolation with spheres or boxes, can be treated analytically, but needs computer simulations and analysis of the structures created thereby. From such simulations, however, one can extract scaling laws for the properties of interest. First, the probability that a site or bond belongs to an infinite (spanning) cluster P_∞ is given by

$$P_\infty \propto (p - p_c)^\beta, \tag{4.27}$$

in which β is a universal exponent. For lattice site percolation in 3D, $\beta_{3D} = 0.417$. The correlation length ξ being the mean distance between two sites on the same finite cluster also has a universal behaviour:

$$\xi \propto |p - p_c|^{-\nu}, \tag{4.28}$$

with $\nu = 0.875$ on a 3D lattice. It diverges at the percolation threshold. The size of a finite cluster also diverges and scales as

$$S(p) \propto |p - p_c|^{-\gamma}, \tag{4.29}$$

with $\gamma = 1.795$. The size of a characteristic cluster or the typical size of the largest cluster is

$$s_\xi(p) \propto |p - p_c|^{-1/\sigma}. \tag{4.30}$$

The size distribution of clusters being composed of s sites, $n_s(p)$, has the form

$$n_s(p) \propto s^{-\tau} f((p - p_c)s^{-\sigma}), \tag{4.31}$$

in which $\tau = 2 + \beta/(\beta + \gamma)$ and $\sigma = 1/(\beta + \gamma)$. Here $f(z)$ is a lattice independent scaling function, which cannot be derived analytically but has to be derived from computer simulations. The only point that can be said is that the function has a maximum and, looking to simulation results, it looks in many cases like a skewed Gaussian distribution. The percolation threshold values vary largely for 2D or 3D, lattice, bond or continuum percolation; see [228, 229]. Here we would like to summarise a few results being important for aerogels: if spheres are cast randomly into space, the critical volume fraction is $\phi_c = 0.2895$; if cylinders with spherical caps are cast into a box, the percolation threshold depends strongly on the aspect ratio. If height to diameter has a ratio of 1, $\phi_c = 0.2439$, it decreases to 0.06418, and if the aspect ratio is 200, it decreases to 0.007156. For particulate aerogels, it looks as if large volume fractions are needed to make a percolating network, whereas for cellulose aerogels, in which typically the aspect ratio of fibril length to diameter is around 10, comparatively low volume fractions are sufficient. Looking a bit deeper into percolation theory, one also finds that the transition at $p = p_c$ is not a sharp one, especially in continuum percolation, when spheres, circles or cylinders also have a radius or aspect ratio distribution. It also is well known that in finite systems the percolation threshold is continuous around the value of the infinite systems (as discussed later). Such a behaviour would support the observations that around the sol–gel transition, the viscosity does not change abruptly but exhibits a smooth but strong increase.

Another important aspect of percolation models is that the structures of the clusters below and above the threshold are ramified, meaning they are not smooth, compact objects, but so-called fractals. In all classical, compact geometrical objects, their volume is always related to the cube of a characteristic length, $V \propto L^3$. This is not the case for fractal objects. Their volume is related to a fractional power d_f, and this power always is smaller than the embedding Euclidian dimension (in 3D, it is just 3). We then have $V_f \propto L^{d_f}$. Let us explain the concept of fractals coined by Mandelbrot [222] with two mathematical examples in two and three dimensions. The first is a so-called Sierpinski carpet: take a black square and divide it into nine equal-sized squares. Take out the central one. Then divide the remaining eight squares again into nine subsquares and take out the central one. Repeat the procedure ad infinitum. One is then first left with a carpet full of holes. Secondly, one can enlarge the carpet to any magnification and it looks the same. This is one aspect of mathematical fractals: self-similarity. The other aspect is: how does the area fraction of black squares change with the magnification or the length scale L of a square (the area fraction also could be termed mass)? For the Sierpinski carpet and many other mathematical fractals, one can give an exact answer; see [55, 56]. The Sierpinski carpet is a two-dimensional fractal object. The same procedure can be done in 3D, and is then called a Menger sponge. Figure 4.14 shows the carpet and the sponge to visualise a mathematical fractal. In both cases, the area or the volume converges to zero after infinite subdivisions, whereas the surface area goes to infinity. For real objects, one has to state that they are, first, not exactly self-similar (only in a statistical sense) and they typically exhibit fractal behaviour only in an interval of their length scales. How can one measure that an object like a percolation cluster is a fractal? Several methods are available. For instance, in a computer simulation on a square lattice, one can vary the lattice size and

Figure 4.14 Two mathematical fractals. The left panel shows a Sierpinski carpet and the right one its 3D analogue, the Menger sponge.

then plot the size of the largest cluster as a function of lattice size L. If the cluster would behave like a compact object, its mass would vary like L^2. If it is, however, a fractal object, a double logarithmic plot will not have a slope of 2, but a different one expressed as

$$M(L) \propto L^{d_f} \quad \text{or} \quad d_f = \frac{\log M(L)}{\log L}, \tag{4.32}$$

with $M(L)$ the mass (or volume or area) of the cluster. Having generated in a computer a cluster (or in reality, like in aerogels or in colloids after a 2D random aggregation process), one can take a 2D picture and cover the structure with boxes (squares), starting from large-size boxes and step by step decreasing the size and count in each step the number of boxes needed to cover the cluster structure. Plotting (always double-logarithmically) the number of the covering boxes as a function of the box size yields a power law relation. The slope is the fractal dimension. Another method would be to measure the density–density autocorrelation function defined as

$$C(\vec{r}) = \frac{1}{V} \sum_{\vec{r}'} \rho(\vec{r}')\rho(\vec{r}' + \vec{r}), \tag{4.33}$$

in which the density $\rho(\vec{r}')$ is 1 if at $\vec{r}'$ there is an occupied site of the object and otherwise zero. The total volume is $V = \sum \rho(\vec{r}')$. One can imagine $C(\vec{r})$ is the average density at a distance $\vec{r}$ from a site on a fractal. If the fractal would be isotropic, one would expect that the correlation function behaves in a power law fashion like $C(r) = Ar^{-\alpha}$. The mass in a radius $\vec{r}$ from a site at $\vec{r}'$ is

$$M(R) = \int_0^R C(r) d^d r = R^{-\alpha+d} = R^{d_f}, \tag{4.34}$$

which allows us to determine the fractal dimension from measuring α and knowing the Euclidian dimension ($d = 2$ for a two-dimensional object, $d = 3$ for a 3D object). Experimentally there is another nice way to determine the fractal dimension: scattering. In Appendix C, we explain that the scattering of X-rays, neutrons or light is determined by the shape factor and the structure factor $S(q)$. The latter is important in non-dilute systems with a continuous phase, which is surely the case in aerogels. We also show in the appendix that the structure factor is the Fourier transform of the density–density autocorrelation function. For fractals stemming from a percolation process, we therefore would expect that the structure factor can be expressed as a power law of the wave vector q like

$$S(q) \propto q^{-d_f} \propto I_s. \tag{4.35}$$

The measurement of scattering intensity I_s as a function of wave vector would then show at sufficiently large q-values a deviation from Porod's law (for which a power law decrease of I_s with an exponent 4 is predicted). Many experimental techniques to determine the fractal dimension of ramified objects are nicely described in the book of

Kaye [57]. On percolating clusters, one can show that the preceding given universal exponents can be related to the fractal dimension

$$d_f = \frac{1}{\nu\sigma} \quad \text{or} \quad d_f = d - \frac{\beta}{\sigma}. \tag{4.36}$$

For a 3D site percolation on a lattice, one therefore obtains a fractal dimension of $d_f = 2.53$, which is thus a bit smaller than the embedding Euclidian dimension of $d = 3$.

We have to ask, how does percolation and gelation fit together? The simple answer is, at the percolation threshold of a real system, gelation has started by the appearance of a spanning cluster and all measurable properties change then continuously inasmuch as the still existing finite clusters (down to monomers) exist in the system and may add to the spanning cluster. While this answer generally will be correct, one is left with the problem: what type of percolation model (Bethe lattice, site or bond percolation on what kind of lattice, continuum percolation with spheres or rods of a size distribution) should one apply, and how should one deal with the real chemistry, meaning, for instance, that even after the occurrence of a spanning cluster there are still subcritical clusters down to monomers, which will further modify the structure and the properties? The answer is, unfortunately, not simple. One could start with a simple translation of the concept of percolation to aerogels, by identification of the occupation probability with the volume fraction of particles or fibrils. The percolation threshold would then be the critical volume fraction above which a spanning cluster exists. In almost all aerogels, the fraction solid is between 1–10%. For all lattice site or lattice bond percolation models, the percolation threshold is always above 15% volume fraction; only if one takes into account next nearest neighbour interactions or even second next nearest neighbour ones, the threshold might be just below 10%. In continuum models, the situation is not much different, besides for rods, when their aspect ratio is above 10. Then a percolation threshold would be below 10% down to even 0.3% (for numbers, see the webpage cited [228]). This simple idea obviously does not work.

Unfortunately, the theoretical description of sol–gel transition also does not really look appropriate for aerogels. In the wonderful review by Stauffer and co-workers [230], the sol–gel transition is focussed on monomers building polymers and eventually leading to one big macromolecule spanning the whole volume. In contrast to this idea, very useful for polymers in general, in gels leading to aerogels, monomers react indeed to form oligomers, which build particles, and these form a spanning network in all particulate aerogels. These particles are the building entities of a percolation process and not the monomers. In aerogels starting from dissolved polymers, they are the starting point and their aggregation leads to gelation. Therefore, the classical ideas of a sol–gel transition from monomers to spanning polymers look not appropriate for aerogels. How do particles then appear? As mentioned in Section 2.2 on RF aerogels and in the discussion on cellulose aerogel synthesis in Chapter 3, there might always be first a phase separation leading to liquid polymer-rich droplets in a liquid polymer diluted matrix (the solvent). The polymer or oligomer rich droplets could undergo a percolation transition, and these solid-like particles then aggregate to make a spanning cluster. Thus, instead of having an infinite system in which a percolation threshold is reached, we would have a finite one, given by the drop size emerging from phase separation. Although therefore the gelation kinetics in polymer building system will be different from those leading to aerogels, the ideas of Stauffer's approach relating percolation theory to gelation might be applicable to the formation of particles starting from monomers.

The average degree of polymerisation, DP, should increase in as much as the percolation threshold is reached. From Eq. (4.29), one could postulate

$$DP = C(p_c - p)^{\gamma} \quad p \to p_c^-, \tag{4.37}$$

and above the percolation threshold we have a similar behaviour

$$DP = C'(p_c - p)^{\gamma'} \quad p \to p_c^+. \tag{4.38}$$

The radius of the macromolecules R_s (radius of gyration for example) should scale with the number of monomers s like

$$R_s \propto s^{h(p)}. \tag{4.39}$$

Numerous models have been developed in polymer science to relate radius of a macromolecule to the number of monomers in it. In a simple random walk model $h(p) = 0.5$ [54], which agrees completely with the classical model of diffusive growth, see for an in-depth discussion [50]. The gel fraction defined as the number of monomers belonging to the infinite cluster behaves as shown in Eq. (4.27), meaning the gel fraction vanishes at the critical point p_c. The viscosity also should diverge at the critical point

$$\mu \propto (p_c - p)^k. \tag{4.40}$$

Identifying p with the volume fraction of clusters of size s, Stauffer and co-workers argue that $k = 0.7$ in 3D percolation, but also show that other models yield different values. Application of these ideas to polymer-rich droplets emerging from phase separation means that these due to their finite and small size will make a finite but droplet spanning cluster almost immediately after their appearance in the liquid matrix. Their viscosity will change from liquid to solid-like. Whatever model of percolation one would like to apply, they all predict that such droplets change their nature to solid-like with a percolation transition in a finite-sizes system. In addition, the structure of the solid-like droplets would be a fractal, and thus one has a substructure in the meso-sized particles or fibrils. Therefore, we have to look at percolation in finite-sized samples in contrast to everything described earlier, which holds only if the sample is infinite.

Finite-sized effects change percolation drastically. The threshold is no longer a very precisely defined value, but the transition is smooth. Even at occupation probabilities below p_c, spanning clusters are possible. This is called in the percolation literature *finite-sized scaling*. We do not repeat the derivations but write down some results which we think are important in the context of gelation leading to aerogels. We had already defined the radius of gyration in Chapter 2 on silica aerogels, but will repeat it here for convenience and express it a bit differently. The radius of gyration is the average square distance of an occupied site positioned at $\vec{r}_i$ from the centre of mass, located in a cluster at $\vec{r}_{cm}$. We call it here R_s instead of R_G to denote the number of sites s occupied in the cluster,

$$R_s^2 = \frac{1}{s}\sum_{i=1}^{s} |\vec{r}_i - \vec{r}_{cm}|^2 = \frac{1}{2s^2}\sum_{i,j} |\vec{r}_i - \vec{r}_j|^2. \tag{4.41}$$

This radius of gyration is needed to calculate the correlation length ξ. The square of the correlation length is defined with the size distribution of clusters as

$$\xi^2 = \frac{\sum_s 2R_s^2 s^2 n_s(p)}{\sum_s s^2 n_s(p)} \tag{4.42}$$

and thus depends on the occupation probability and the number of occupied sites in the cluster. We already have shown, that in an infinite system, the correlation length scale with the distance from the percolation threshold like

$$\xi \propto |p - p_c|^{-\nu} \tag{4.43}$$

with $\nu = 0.88$ in a 3D lattice. If the system has a finite size L, the mass of a cluster would scale like $M(L) \propto L^d$ if the object were compact. As already used in the discussion of fractals, this relation changes if the object is fractal in nature. We then have $M(L) \propto L^{d_f}$ with the fractal dimension d_f. We now assume that the number of occupied sites in a cluster varies in the same way, meaning

$$s \propto R_s^{d_f} \quad \text{or} \quad R_s \propto s^{1/d_f} \qquad p \to p_c. \tag{4.44}$$

We now have to make an important distinction. If the size of the system $L \gg \xi$, the system behaves like a compact, smooth object, otherwise it behaves like a fractal. This difference is obvious looking to daily examples. For instance, looking from above, any desert looks like a smooth object, but going closer and closer to the sand surface the desert becomes rough and ramified with a structure defined by the sand grains. For the mass of a cluster, we then write

$$M(L) = L^d P(p) = L^d (p - p_c)^{\beta} = L^d \xi^{-\beta/\nu} \tag{4.45}$$

using expression (4.43). If $L \approx \xi$, we can assume that there might be a transition between a fractal to compact behaviour:

$$M(L) = L^{d_f} = \xi^{d_f}, \tag{4.46}$$

and then, using Eq. (4.45), we get

$$M(L) = \xi^{d-\beta/\nu} \quad \text{and thus} \quad d_f = d - \frac{\beta}{\nu}. \tag{4.47}$$

We therefore generally write a new relation for the mass of a cluster

$$M(L, \xi) = \left(\frac{L}{\xi}\right)^d \xi^{d_f}. \tag{4.48}$$

We can also rewrite this as

$$M(L,\xi)=\begin{cases} L^{d_f} & \text{if } L \ll \xi \\ \xi^{d_f}\left(\frac{L}{\xi}\right)^d & \text{if } L \gg \xi \end{cases}. \tag{4.49}$$

Let us now apply the ideas of finite-sized scaling to other properties of interest. The average cluster size was defined previously in a infinite system as $S(p) = |p - p_c|^{-\gamma} \propto \xi^{\gamma/\nu}$. In finite-sized systems, there also is a crossover dictated by the relation between size and correlation length. One can derive the following relation:

$$S(L,\xi)=\begin{cases} L^{\gamma/\nu} & \text{if } L \ll \xi \\ \xi^{\gamma/\nu} & \text{if } L \gg \xi \end{cases}. \tag{4.50}$$

Recall the values of the exponents. We have in 3D $\beta = 0.41, \quad \gamma = 1.8, \quad \nu = 0.88$. Therefore, the size of the system has an influence on the cluster mass and the cluster size (as well as on the percolation threshold) [231].

We are left with two problems: how do solid-like particles aggregate, and how does phase separation occur? These issues are treated now.

4.2.2 Diffusion-Limited Aggregation – DLA and DLCA

The standard approach in the aerogel literature is that Brownian motion of particles, which might be pre-nucleated entities or particle-like oligomers, lead to a percolating network [3]. Brownian motion is a random motion of particles. The trajectories of moving particles do not follow straight, distinct or predictable paths, but a zig-zag one. The motion is random in direction and in jump size, such that it can be described at best as a diffusive motion, in which the mean square displacement $\overline{x^2}$ of a particle from a given point in space can be described as $\overline{x^2} = 2D_B t$ with D_B the Brownian diffusion coefficient. The Brownian diffusion coefficient was calculated already by Einstein, and he derived an expression as $D_B = k_B T/(6\pi\mu R)$ with μ the viscosity of the fluid and R the particle of radius. The idea of Brownian motion as the building mechanism of aggregates, clusters and fractal structures was taken up especially 40 years ago by Witten and Sander [232], Meakin [233], Kolb et al. [234] and many others in the following decades. The technique they and others developed was called diffusion-limited aggregation (DLA) or, when clusters also could move, diffusion-limited cluster aggregation (DLCA).

The DLA or DLCA technique starts in its simplest version with a lattice and puts somewhere in the lattice a so-called seed, a particle fixed at a point in space. Then particles are added to the system from the borders and move randomly from lattice point to lattice point, meaning a particle being added has on a square lattice four possible directions to move, which are chosen randomly. The step size is typically one lattice point. Once such a particle hits the seed particle, it is bonded to it and fixed in space and a new one is released randomly from the border. Instead of adding particles in series from a border, one can also add into the lattice after setting a seed (or multiple seeds) a given number of particles and let them move randomly, follow how they attach to the seed(s) and look at the resulting aggregate. It is interesting that by this simple method, fractal structures appear, meaning highly branched clusters, which are self-similar. An advanced method of this technique does not use a lattice, but puts seed(s) arbitrarily in a box and adds an arbitrary number of particles in the box which make a Brownian motion in all possible directions (six in 3D) with an arbitrary step size. Such an approach was used by Halperin and co-workers to simulate the nanostructure of highly porous silica aerogels [235, 236]. There are meanwhile many programs for free available on the internet, especially Java codes which simulate DLA or DLCA, even biased by some force field etc. The interested reader should look at the wonderful website [237], where he also can find the Java library and run the simulations by himself, and of course look into the excellent book of Tamás Viscek on fractal growth phenomena [49]. Before we present some real results, let us illustrate DLA and the structures that can be simulated with it.

Let us start with the simple 2D model provided by Hiroki Sayama as a Wolfram Demonstrations Project [238]. Figure 4.15 shows two snapshots from this simulation in which 500 particles of fixed size move randomly in a square, and once they either hit the seed or already arrested particles, they are fixed and stop moving. One can clearly see that a branched structure evolves. The clusters can reach the boundaries of the square and thus make a spanning cluster, if the number of particles or, in other words, the volume fraction, is increased. Kolb et al. [234] showed that there is scaling between radius of gyration and the number of particles in a clusters, like $R_G \propto N^{\nu}$ in the limit of large N; the authors also state that eventually the clusters cover the whole space, independent of how low the initial number of particles is. The exponent ν was found to be $\nu = 0.73 \pm 0.04$. Therefore, random motion of particles in a finite box can lead to spanning clusters, and this would be the onset of gelation.

An example of a more advanced simulation is the results obtained by Halperin and co-workers [236], where they intended to model aerogels. They chose from a truncated Gaussian distribution of particle sizes a fixed number of particles, put them into a 3D box and let them perform a random motion in all six possible directions

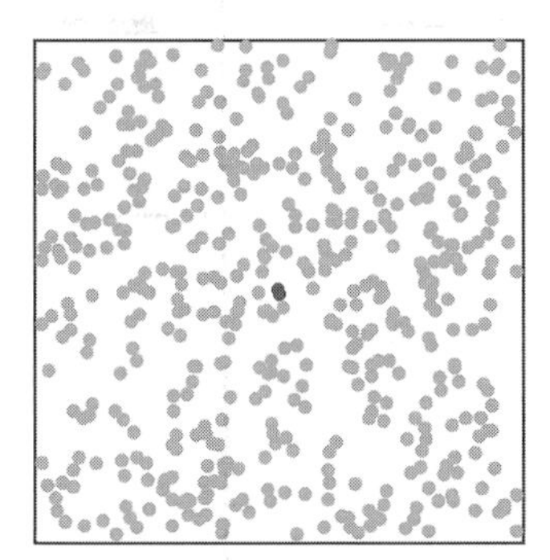
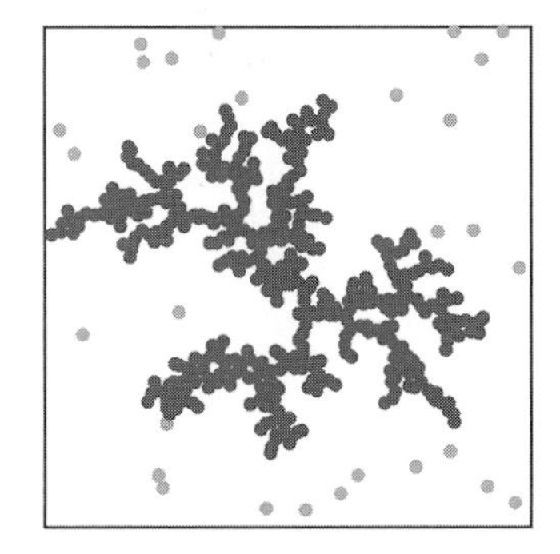

Figure 4.15 DLA of 500 particles moving randomly with one seed particle in the centre. They are arrested if they meet the seed or particles already attached to the seed. The arrested particles are coloured as dark grey.

with a fixed step-size. The probability of motion was calculated according to a scaled probability $P = (m_i/m_0)^\alpha$, with m_i the mass of the ith particle, m_0 the mass of the lightest particle and α an index chosen as smaller than zero, typically around -0.5, which reflects on the one hand diffusive Brownian motion and on the other hand allows already aggregated particles also to perform a random motion. Therefore, their simulation is a DLCA. Once particles touch, they stick together. They obtained highly branched clusters of particles. They were able to vary the volume fraction solid particles and measured in their images the 3D mean free path, which is an excellent description of the pore diameter (for a definition of the mean free path, see Chapter 9).

Their simulations exhibited interesting results. First, it seems to be most important, that aggregation by random motion of particles and clusters always led in the simulated box to the occurrence of a spanning cluster and thus gelation. The second interesting result they obtained is that the path length distribution does not depend on the diffusion characteristic (step-size) nor the α-parameter in the range of -0.45 to -0.75. Third, the path length distribution has two branches; at small path lengths it is characterised as if the particles are fractally aggregated, and at larger path lengths it agrees with a randomly distributed correlated arrangement of particles. From their interesting results, we show the inverse of the mean free path as a a function of envelope density (see Figure 4.16). Their result suggests that this is directly proportional to the simulated aerogels density. This observation could have a meaning for the Knudsen effect of gas diffusion and thermal conductivity in aerogels; see Chapters 10 and 12. Although this simulation has the same deficiencies as all DLA and DLCA simulation with a fixed distribution of particle sizes, since it neglects concurrent changes in particle size due to curvature induced coarsening, growth and polycondensation, it leads to valuable results. Unfortunately, current microstructural analysis is not able to do a similar analysis of nanostructured aerogels as is done with the 3D structures generated in a computer. This might be possible in the future with increased resolution

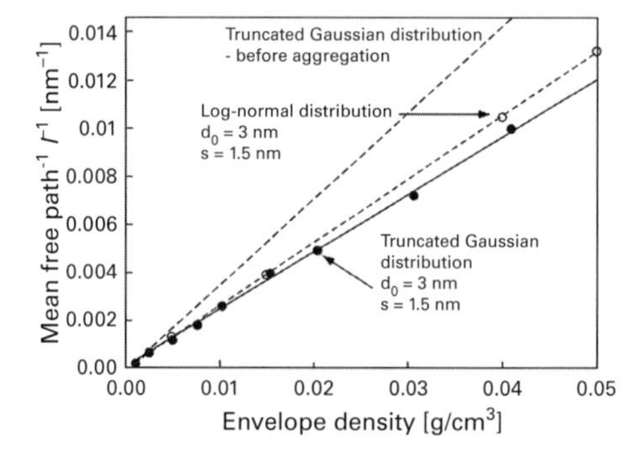

Figure 4.16 Mean free path in the simulated aerogels as a function of simulated aerogel envelope density for particles with an initial mean particle size of $d_0 = 3$ nm and a standard deviation s of the distributions of 1.5 nm. Note the linear relation between envelope density and the inverse mean free path length. Reprinted with permission of Elsevier after [236]

of nanotomography. We can conclude that DLA and DLCA algorithms are able to simulate a microstructure looking in a certain sense similar to what is observed with SEM and TEM; see Chapter 6. However, the origin of a string-of-pearls or necklace-like structures cannot be simulated by this kind of numerical analysis.

4.2.3 Mean Field Model of Smoluchowski

We start with the famous Smoluchowski equation for aggregation. Instead of particle mass, we use particle volume. Both are related via the particle density, provided this is the same for all particles (the particles should not have a porous substructure). If $n_V(v,t)dv$ denotes the particle number density per unit volume in the interval $v, v+dv$, the rate of change of $n_V(v,t)$, with time is given by

$$\frac{\partial n_V}{\partial t} = -\frac{1}{2}\int_0^v W(v-\tilde{v},\tilde{v})n_V(v-\tilde{v},t)n_V(\tilde{v},t)d\tilde{v} + n_V(v,t)\int_0^\infty W(\tilde{v},v)n_V(\tilde{v},t)d\tilde{v}. \tag{4.51}$$

The first term on the right side describes the gain of particles due to aggregation or collisions, characterised by the scattering cross section of particles moving on a straight line multiplied with their relative velocity, $W(\tilde{v},v)$. It essentially describes the probability of collisions to occur. For more details, see [224, 239]. Without going into details of this equation, two things can be obtained directly without knowledge of the collision kernel $W(\tilde{v},v)$. First, integrating Eq. (4.51) over the volume yields the time rate of change of the total number of particles

$$\frac{dn}{dt} = \int_0^\infty \frac{\partial n_V(v,t)}{\partial t}dv = -\frac{1}{2}\int_0^\infty\int_0^\infty W(\tilde{v},v)n_V(v,t)n_V(\tilde{v},t)d\tilde{v}dv \tag{4.52}$$

with

$$n(t) = \int_0^\infty n_V(v,t)dv \tag{4.53}$$

as the number of aggregates per unit sample volume, and second, multiply Eq. (4.51) with v and integrate again of the volume yields the time rate of change of particle volume fraction

$$\frac{d\phi}{dt} = 0 \tag{4.54}$$

with

$$\phi = \int_0^\infty v n_V(v,t)dv, \tag{4.55}$$

This shows that the volume fraction is a conserved quantity. Any solution of the Smoluchowski equation depends strongly on the exact expression for the collision kernel. Numerous expressions are discussed and treated in the literature cited, and it also was shown decades ago that the equation admits for certain mathematical conditions of the kernel a unique solution. A general solution, however, is not possible.

Since we are here only concerned with aggregation leading to gelation, we only consider a few expressions of the collision kernel (we omit mostly details of the physics, but just are looking to the dependency on the particle volume).

The most simple case treated in the literature is a constant collision volume, which is according to Fuchs [224] a good approximation for Brownian motion of small particles with a size distribution having a small width of span:

$$W = W_0 = \frac{8k_B T}{3\mu}, \tag{4.56}$$

with μ the dynamic viscosity of the solvent. In this case, Eq. (4.52) leads directly to

$$\frac{dn}{dt} = -\frac{1}{2}W_0 n^2, \tag{4.57}$$

which upon integration yields a hyperbolic decrease of the total particle number

$$n(t) = \frac{n_0}{1 + \frac{1}{2}W_0 n_0 t}. \tag{4.58}$$

The size distribution $n_V(v,t)$ can be derived analytically (and this was done 70 years ago by Schumann), but we omit it here. It is worth mentioning that the average particle (or aggregate) volume obtained from $\phi_0 = n(t)\overline{v}$ yields a simple expression

$$\overline{v} = v_0 \left(1 + \frac{1}{2}W_0 n_0 t\right). \tag{4.59}$$

This means that the average particle radius grows with the cube root of time. Such a situation is discussed in the literature on gelation and it is quite often stated that it never will lead to a transformation of a sol to a gel. We come back to this point later.

If aggregation of particles occurs in a shear flow (mild stirring of a solution), one can use the simplified expression given by Golovin [240] to

$$W_{sf}(v,\tilde{v}) \approx \Gamma(v^{1/3} + \tilde{v}^{1/3})^3 \approx \Gamma(v + \tilde{v}), \tag{4.60}$$

with Γ containing the shear rate and some numerical constants (see [239]). The simple additive kernel is an approximation often used in the literature.

With this simple additive kernel a solution is possible and was derived by Golovin [240] for an initial distribution of a simple exponential type. At large particle volumes and large times, his complex solution can be simplified to the expression

$$n_V(v,t) \approx \frac{n(t)}{2\sqrt{\pi}}\sqrt{\frac{\phi}{n_0}\frac{1}{v^{3/2}}}. \tag{4.61}$$

The number density can be obtained directly from Eq. (4.52) again by inserting Eq. (4.60), leading to

$$n(t) = n_0 \exp(-\Gamma\phi_0 t), \tag{4.62}$$

where ϕ_0 denotes the constant volume fraction of particles. The average particle volume can also be calculated directly, since $\phi_0 = n(t)\overline{v}$, leading to

$$\overline{v(t)} = \frac{\phi_0}{n_0} \exp(\Gamma \phi_0 t) = v_0 \exp(\Gamma \phi_0 t). \tag{4.63}$$

It is important to notice that the average particle volume (aggregate volume) is unbounded. This is an essential feature of gelling systems [111, 241], but typically in gelling systems this should happen in a finite time and not at $t \to \infty$.

Another important collision kernel is constructed for particles performing sedimentation. Assuming that the sedimentation velocity follows Stokes' law, i.e., proportional to the square of the particle radius, the collision kernel can be written as

$$W_S(v, \tilde{v}) = C_s \frac{(\rho_p - \rho_f)g}{\mu} |v^{2/3} - \tilde{v}^{2/3}|(v^{1/3} + \tilde{v}^{1/3})^2. \tag{4.64}$$

In this equation ρ_p, ρ_f is the particle density and the solvent density respectively, and g denotes gravity acceleration. C_s collects all constants stemming from Stokes' law for the sedimentation velocity of solid particles in a fluid. The symbols $|\cdots|$ denote the absolute value. Unfortunately for this collision kernel, not even an approximate analytical solution of the aggregation equation is available. Nevertheless, it is important in the context of RF aerogels, since it was observed that in very dilute solutions the particles forming through polycondensation sediment and form a gel-like layer at the bottom of a beaker [89]. Therefore, we make an approximation to this kernel, which is mathematically more pleasant:

$$W_S^{app}(v, \tilde{v}) = C_s \frac{(\rho_p - \rho_f)g}{\mu} (v^{1/3} - \tilde{v}^{1/3})^2(v^{1/3} + \tilde{v}^{1/3})^2. \tag{4.65}$$

This approximation has the same degree of homogeneity (for a definition see Eq. (4.72)) and it is zero if both particles have the same volume, but it underestimates the collision rate.

Inserting this collision kernel into Eq. (4.52) allows calculation of the time evolution of the particle number density. If we define the moments of the aggregate distribution as

$$M_q(t) = \int_0^\infty v^q n_V(v, t) dv \tag{4.66}$$

and assuming that the moments

$$M_{2/3}(t) = \int_0^\infty v^{2/3} n_V(v, t) dv \qquad M_{4/3}(t) = \int_0^\infty v^{4/3} n_V(v, t) dv \tag{4.67}$$

are approximately constant (although they will be time dependent), one can derive the time dependence of the number density:

$$n(t) = \frac{M_{2/3}^2}{M_{4/3}} + \frac{n_0 M_{4/3} - M_{2/3}^2}{M_{4/3}} \exp(-C_{s0} M_{4/3} t) \tag{4.68}$$

with

$$C_{s0} = C_s \frac{(\rho_p - \rho_f)g}{\mu}. \tag{4.69}$$

We will use this expression later to calculate the gelation time of a sedimenting colloidal dispersion.

Another important collision kernel is the full expression used for particles in a liquid with dynamic viscosity μ performing Brownian motion. The kernel was originally derived by Smoluchowski himself [223] as

$$W_B(v, \tilde{v}) = \frac{2k_B T}{3\mu}(v^{1/3} + \tilde{v}^{1/3})\left(\frac{1}{v^{1/3}} + \frac{1}{\tilde{v}^{1/3}}\right). \tag{4.70}$$

Solutions of the Smoluchowski equation for this type of kernel cannot be derived analytically, but approximate solutions are possible and will be discussed later. This kernel is especially important, first because one approximation is the constant kernel discussed earlier, and secondly, it is the underlying basis of all diffusion limited cluster aggregation models, i.e., their mean-field description. One can generalise the sum kernel of Eq. (4.60) to an expression, intensively discussed by Lushnikov to describe gelation [242, 243]

$$W_g(v, \tilde{v}) \propto v^p \tilde{v}^q + v^q \tilde{v}^p, \tag{4.71}$$

with p, q being arbitrary exponents. To summarise the statements on collision volumes or kernels, let us make an important comment. All the preceding kernels are homogeneous functions of their arguments, meaning that if we scale v with a constant a, they behave like

$$W(av, a\tilde{v}) = a^{\lambda} W_B(v, \tilde{v}), \tag{4.72}$$

with λ the degree of homogeneity. One can immediately see that the Brownian and the shear flow kernel have a degree of homogeneity $\lambda = 1$, the Stokes case has a $\lambda = 4/3$ and the kernel of Eq. (4.71) a $\lambda = p + q$. We will make use of this property in deriving scaled solutions of the aggregation equation.

4.2.4 Scaling Solutions and Gelation

In this section, we are concerned with so-called scaling solutions, which gives a hint at the size evolution of aggregates and conditions for gelation for a given mechanism of particle motion in a solvent. A lot of complex mathematical equations in physics allow for so-called scaling and even self-similar solutions [244]. The aggregation equation also has such solutions under certain conditions. To our knowledge, the first authors having derived such a self-preserving size distribution as a solution of the aggregation equation for Brownian collisions were Friedlander and Wang in 1966 [245], later taken up by Pulvermacher and Ruckenstein for many coagulation kernels of interest, especially in the context of aerogels [246]. Many papers have discussed and extended these first papers on self-similar solutions so that nowadays it quite often is simply stated that such solutions exist if the kernel is a homogeneous function, and they are typically written in the following form:

$$n(v,t) \propto \frac{1}{s(t)^2} F(z) \quad \text{with} \quad z = \frac{v}{s(t)} \tag{4.73}$$

and $s(t)$ is the typical cluster volume, e.g., the average volume. Note: such scaling solutions do not really solve the Smoluchowski equation, but they are valid only for extended times, meaning all full solutions can be approximated by such a scaling solution if $t \to \infty$. We will repeat here the derivation of Friedlander and Wang to get the conditions for scaling and self-similarity and then give some results for kernels of interest for aerogels. In order to obtain a scaling solution, we first look for suitable scaled variables. Friedlander and Wang used the following new variables:

$$\eta = \frac{Nv}{\phi_0} \qquad \tau = \left(\frac{N}{N_0}\right)^2. \tag{4.74}$$

Note that $\phi_0 = N(t)\overline{v}$, and thus we could write $\eta = v/\overline{v}$.[1] We then try to write the size distribution in a new way as

$$n(v,t) = \frac{N(t)^2}{\phi_0} \psi(\eta, \tau). \tag{4.75}$$

With the definitions in Eqs. (4.74, 4.75) and using Eq. (4.52) one can derive the following new form of the aggregation equation, which looks only at first glance more ugly then the original one, namely

$$\left[\int_0^\infty \int_0^\infty W(\tilde{v}, v) n(v,t) n(\tilde{v},t) d\tilde{v} dv\right] \cdot \left[\eta \frac{\partial \psi}{\partial \eta} + 2\psi + 2\tau \frac{\partial \psi}{\partial \tau}\right]$$
$$+ \int_0^\eta W(\tilde{\eta}, \eta - \tilde{\eta}) \psi(\tilde{\eta}, \tau) \psi(\eta - \tilde{\eta}, \tau) d\tilde{\eta} = 2 \int_0^\infty W(\tilde{\eta}, \eta) \psi(\tilde{\eta}, \tau) \psi(\eta, \tau) d\tilde{\eta}. \tag{4.76}$$

This result is valid for homogeneous functions of arbitrary degree λ. The derivation of this equation is cumbersome and tedious, but straightforward.[2] Important are the relations

$$\int_0^\infty \psi(\eta, \tau) d\eta = 1 \tag{4.77}$$

$$\int_0^\infty \eta \psi(\eta, \tau) d\eta = 1, \tag{4.78}$$

[1] Remember that ϕ_0 is the volume fraction of pre-nucleated particles as discussed previously. In principle, this is not really a constant, but it is slightly time dependent.

[2] For the derivation use $\frac{\partial n}{\partial t} = \frac{2N}{\phi_0}\frac{dN}{dt}\psi + \frac{N^2}{\phi_0}\frac{\partial \psi}{\partial t}$ and calculate $\frac{\partial \psi}{\partial t} = \frac{\partial \psi}{\partial \tau}\frac{\partial \tau}{\partial t} + \frac{\partial \psi}{\partial \eta}\frac{\partial \eta}{\partial t}$. One also uses the identity $\int_0^\infty W(v,\tilde{v}) n(v,t) n(\tilde{v},t) d\tilde{v} = \int_0^\infty W(\phi_0/N\eta, \phi_0/N\tilde{\eta}) N^2/\phi_0 \psi(\eta,\tau) N^2/\phi_0 \psi(\tilde{\eta},\tau) \phi_0/N d\tilde{\eta}$, and using the homogeneity of degree λ in the kernel, this turns out to become $(\phi_0/N)^\lambda N^3/\phi_0 \int_0^\infty W(\eta,\tilde{\eta}) \psi(\eta,\tau) \psi(\tilde{\eta},\tau) d\tilde{\eta}$. The rest is tedious work.

which prove the good choice of the new variables. This, however, is not the end of the story. Look at Eq. (4.76). The only position in which the scaled time τ explicitly enters is the term $2\tau\frac{\partial\psi}{\partial\tau}$. If this term vanishes at large times, the function ψ depends only on η and thus it is self-preserving, or, self-similar or in other words, we have split the distribution function of aggregates into a pure time-dependent part and a volume-dependent one ($\psi(\eta)$). Let us also look at how the scaling of Friedlander and Wang fits to the scaling in Eq. (4.73). Rewrite using $\phi_0 = N(t)\overline{v}$:

$$\frac{N^2}{\phi_0} = \frac{\phi_0^2}{\overline{v}^2\phi_0} = \frac{\phi_0}{\overline{v}^2} \propto \frac{1}{\overline{v}^2}, \tag{4.79}$$

as it should be. Let us assume that the term $\tau\frac{\partial\psi}{\partial\tau} \to 0$ as $t \to \infty$. Then Eq. (4.76) reduces to an ordinary non-linear integro-differential equation for ψ. We are not discussing the mathematical conditions for a unique solution of this equation; for some considerations, see [246]. The general scaling solution for ψ, valid in the long time limit can be written as

$$\psi(\eta) = A\eta^{\lambda}\exp(-\alpha\eta), \tag{4.80}$$

with λ the homogeneity of the kernel and A, α constants to be determined via normalisation.

We now may ask if we can give an answer to gelling systems quite generally without solving the preceding equations in all depth. Gelation in the mean-field theory of Smoluchowski requires that the number density of single particles goes to zero in a finite time while the average particle volume goes to infinity. We have already seen, that for the constant collision kernel as well as the simple sum kernel, this does not happen. In Stokes sedimentation, it can happen in a finite time. One is therefore guided to assume that if a collision kernel has a degree of homogeneity larger than 1, gelation, meaning the appearance of an infinite large cluster, might occur in a finite time. Family and Viscek [247] and Kolb and co-workers [234] suggest that the size distribution has a long-time behaviour describable as the following (known as Family–Viscek dynamic scaling):

$$n_V(v,t) \propto t^{-w} v^{-\tau} f(v/t^z), \tag{4.81}$$

with three free exponents, w, τ and z. Note that this approach is different from that given in Eqs. (4.73) and (4.75) but accounts better for the fractal nature of aggregates; see [248]. These three exponents are not independent but are related through the constancy of volume fraction:

$$\phi_0 = \int_0^\infty v n_V(v,t)dv \propto t^{-w}\int_0^\infty v^{-\tau} f(v/t^z)dv \propto t^{2z-w-z\tau}\int_0^\infty \xi^{1-\tau} f(\xi)d\xi, \tag{4.82}$$

with

$$\xi = v/t^z. \tag{4.83}$$

The volume fraction is only a constant if

$$w = z(2 - \tau) \tag{4.84}$$

and provided the integral has a finite value. If, as is typically assumed, $f(\xi) = \text{const.}$ at $t \to 0$ and $f(\xi) = 0$ at $t \to \infty$, this restricts possible values of τ[3] to $\tau < 2$. We can rewrite Eq. (4.81) using the definition of ξ as

$$n_V(v,t) \propto t^{-2z} \xi^{-\tau} f(\xi). \tag{4.85}$$

Let us first check if this scaling solves the Smoluchowski equation (4.52). We insert it and after some manipulations we get the following result if we assume that the kernel has a degree of homogeneity λ:

$$-2zt^{z(1-\lambda)-1} \xi^{-1} f(\xi) = 1/2 C_1(\xi) - \xi^{-1} f(\xi) C_0, \tag{4.86}$$

with $C_0(\xi), C_1(\xi)$ being abbreviations for

$$C_0(\xi) = \int_0^\infty W(\xi, \tilde{\xi}) \tilde{\xi}^{-\tau} f(\tilde{\xi}) d\tilde{\xi} \tag{4.87}$$

$$C_1(\xi) = \int_0^\xi W(\xi - \tilde{\xi}, \tilde{\xi})(\xi - \tilde{\xi})^{-\tau} \tilde{\xi}^{-\tau} f(\xi - \tilde{\xi}) f(\tilde{\xi}) d\tilde{\xi}. \tag{4.88}$$

In Eq. (4.86), the left side is time independent if

$$z(1 - \lambda) - 1 = 0 \quad \text{or} \quad z = \frac{1}{1 - \lambda}. \tag{4.89}$$

The scaling exponent z depends thus critically on the homogeneity of the kernel. It is interesting to note that as $\lambda \to 1$, this exponent goes to infinity and thus immediate growth of a cluster would occur, which means gelation is to happen if the degree of homogeneity is larger or equal than 1. This confirms our previous considerations. If z fulfills condition Eq. (4.80), then Eq. (4.86) can be rewritten as

$$f(\xi) = \frac{C_1(\xi)}{2(C_0(\xi) - 2z)} \xi^\tau. \tag{4.90}$$

If the dependence of $C_0(\xi), C_1(\xi)$ on ξ would only be weak, this result means that $f(\xi)$ scales with a power of ξ and the exponent τ is important. But looking closer at Eq. (4.90) in connection with Eq. (4.87), this defines an integral equation for $f(\xi)$.

Let us use this approach to calculate the moments of the distribution. Any moment is defined as stated in Eq. (4.66). Using Eq. (4.81), we obtain after some manipulations

$$\begin{aligned} M_q &= \int_0^\infty v^q n_V(v,t) dv \propto \int_0^\infty v^q t^{-w} v^{-\tau} f(v(t^z) dv \\ &= t^{z(q-\tau+1)-w} \int_0^\infty \xi^{q-\tau} f(\xi) d\xi. \end{aligned} \tag{4.91}$$

[3] Let us assume, that $f(\xi) \propto \xi^{-p}$ for large times. i.e., it follows a power law decay at large aggregate volumes. Then the integral converges only at the upper boundary if $p > 2 - \tau$. The integral then diverges at the lower boundary, but since the power law decay would only be valid for larger volumes and times, one might not care about it.

Using Eq. (4.84) yields

$$M_q \propto t^{z(q-1)} \int_0^\infty \xi^{q-\tau} f(\xi) d\xi. \tag{4.92}$$

If the integral converges, the time dependence of any moment is fixed by the value of z and thus by the degree of homogeneity of the kernel. If we look at the second moment of the distribution $z = 2$, we see that it diverges as λ approaches unity:

$$M_2(t) \propto t^z \propto t^{\frac{1}{1-\lambda}} \quad \text{and thus} \quad \lim_{\lambda \to 1} M_2 \to \infty \quad \text{for all t.} \tag{4.93}$$

This again shows that inasmuch as the homogeneity of the kernel approaches unity (or is above), a huge cluster develops, meaning gelation occurs. The application of scaling analysis and methods for an exact solution of the Smoluchowski equation to understand the sol–gel transition have been extended in the last decades and numerous papers appeared. Especially with respect to the kernel of Eq. (4.71), Lushnikov performed detailed analyses [242, 243] showing under what conditions gelation occurs. Although these and other studies, such as [249–251], easily found in the literature are very valuable and fascinating, the interesting point in this context is, that they apply to infinite systems. In a finite system, and all real experiments are finite systems, gelation can occur by Brownian motion, although the preceding analysis shows that for the constant kernel or the full Brownian kernel with a degree of homogeneity of zero, this should not occur. Computer simulations based on the Brownian diffusion clearly show that this happens for single-particle aggregation or clusters moving by Brownian motion.

4.2.5 Polymerisation-Induced Phase Separation (PIPS)

All aerogels are formed by the transformation of a solution to a wet gel followed by a drying process being adopted to the envisaged application or the properties to be achieved. The conventional approach to gel formation is that the chemical reactions between the monomers in the solution lead to polymers with a variable degree of polymerisation, DP, and these polymers coil or arrange somehow themselves to particles, which by suitable motion in the solvent, for instance Brownian motion (DLA, DLCA), shear or gravity induced buoyant flows or residual flows from stirring, aggregate and eventually form a spanning cluster, the gel, once the percolation threshold is reached [3, 45]. This spanning cluster reacts further with unreacted monomers and oligomers being still in the solution, strengthening the network and changing the morphology and of course the properties.

It was essentially Gommes [99, 100], who questioned this approach and raised the question of whether gel formation, especially in RF aerogels, could not be a kind of phase separation, especially spinodal decomposition. Gommes adopted the classical Flory–Huggins (FH) theory of polymer solutions to RF aerogels. Once the solution enters the miscibility gap, either a phase separation proceeds by formation of a polymer-rich phase and their growth and aggregation or by spinodal decomposition.

It should be mentioned that the idea of phase separation was originally cast into the discussion by Pekala and co-workers, who studied small angle X-ray scattering of RF aerogels and came to the conclusion that 'the structure of the aerogels results from nanoscale phase separation in the solution precursor' [252]. This idea was not followed since then, but newer investigations using SAXS and DLS seemed to show that colloidal aggregation determines the gelation process [85, 94, 253].

We already have mentioned in Section 2.2 on RF aerogels that the miscibility gap in some aerogels, if it exists, would be irreversible and not an equilibrium one. We also saw that in some cellulose solutions, discussed in Chapter 3, a stable miscibility gap exists and one might suppose from the dependence of the shape of the miscibility gap on the degree of polymerisation (see the Appendix B on the Flory–Huggins model) that the gap for different sources of cellulose will differ due to their different degree of polymerisation. This brings up the point that the process of phase separation and gelation might be describable as a polymerisation-induced one, as discussed in the literature for many polymer systems, usually, however, with a large volume fraction of polymer in a solvent, and this idea has never been taken up for aerogels until recently. The models, known in the literature for PIPS, are typically based on the Cahn–Hilliard equation. For a survey on PIPS, see the articles by Hajime Tanaka and many others [254–268]. We therefore start with a short repetition of this equation and then try to apply it to aerogels.

Cahn–Hilliard Equation

The Cahn–Hilliard (CH) model starts with the idea that the free energy of a system is not constant everywhere in a material but depends on the local volume fraction of polymers in solution and in a heterogeneous system on the gradient of the polymer fraction. The free energy density of the whole system, being in a volume V, is then written as a functional:

$$F(\phi, \nabla\phi) = \int_V (f(\phi) + \kappa|\nabla\phi|^2) dV. \tag{4.94}$$

The term $f(\phi)$ denotes the isotropic volume free energy without taking into account that local gradients of the polymer concentration might exist. The term $\kappa|\nabla\phi|^2$ accounts for gradients in the concentration, or in other words, for the existence of interfaces between phases. The square root of κ is related to the interface tension. Models have been developed to calculate κ from, for instance, a regular solution model or a nearest neighbour model and bond energies and distances. A derivation of Eq. (4.94) can be found in, for instance, [269]. Here we take this equation for granted.

The volume free energy $f(\phi)$ can be taken from the Flory–Huggins model; see Appendix B or any other model describing polymer solutions [270, 271]. The next step in the derivation of a kinetic equation describing the spatial and temporal evolution of ϕ is to define a generalised chemical potential as the functional derivative of the free energy:

$$\mu = \frac{\delta F}{\delta\phi}, \tag{4.95}$$

and recalling the definition of a functional derivative[4] leads to

$$\mu = \frac{\partial f}{\partial \phi} - 2\kappa \nabla^2 \phi. \tag{4.96}$$

The diffusion current density is in a material always related linearly to the gradient in chemical potential. The proportionality factor is the mobility $M(\phi)$, which generally depends on ϕ:

$$\vec{j} = -M(\phi)\nabla\mu = -M(\phi)\nabla\left(\frac{\partial f}{\partial \phi} - 2\kappa\nabla^2\phi\right). \tag{4.97}$$

Together with a continuity equation

$$\frac{\partial \phi}{\partial t} = -\nabla \cdot \vec{j} \tag{4.98}$$

we finally get a non-linear equation for the evolution of the polymer concentration:

$$\frac{\partial \phi}{\partial t} = \nabla \cdot \left[M(\phi)\nabla\left(\frac{\partial f}{\partial \phi} - 2\kappa\nabla^2\phi\right) + \xi\right]. \tag{4.99}$$

We added here artificially a noise term ξ, which gives rise to fluctuations having a time average of zero. If the mobility and the interface parameter κ would be constant, say M_0, κ_0, one would obtain the equation

$$\frac{\partial \phi}{\partial t} = M_0\left[\frac{\partial^2 f}{\partial \phi^2}\nabla^2\phi - 2\kappa_0\nabla^4\phi + \xi\right]. \tag{4.100}$$

This equation has been discussed in the literature quite often especially for simple versions of the free energy functional $f(\phi)$. It is not necessary to repeat anything from the classical literature. We instead are going to further develop the model equation, since in the polymer solution things are more complicated and a constant mobility and interface term are not a good approximation.

Cahn–Hilliard Equation with Fluid Flow

Before we discuss the mobility and interface term, let us first further generalise the CH equation by introducing fluid flow, since one can be sure that in solutions fluid flow is possible. Fluid flow can have various origins: local density variations due to concentration gradients, residual flows from stirring, flow from sedimenting droplets or solid-like particles and interface tension-driven flows of droplets in a liquid matrix (Marangoni flows) [274, 275]. Any flow transports kinetic energy $\frac{1}{2}\rho|\vec{u}|^2$. Here ρ is the density of the solvent and $\vec{u}$ is the mean velocity of the fluid. If there is fluid flow, the mass flux contains not only diffusive fluxes but also convective ones, which are

[4] Functional derivatives are calculated using the variational calculus. Minimisation of the free energy expression in Eq. (4.94) with respect to the function ϕ leads to the Euler–Lagrange equation. Eq. (4.96) is exactly this equation using Eq. (4.94). For more details, see the classical textbooks on mathematical physics by Courant and Hilbert [272] or Morse and Feshbach [273].

described by an additional term $\nabla \cdot (\vec{u}\phi)$ (see the standard textbook on physicochemical hydrodynamics by Levich [276]). This changes Eq. (4.99) to

$$\frac{\partial \phi}{\partial t} + \nabla \cdot (\vec{u})\phi = \nabla \cdot \left[M(\phi)\nabla \left(\frac{\partial f}{\partial \phi} - 2\kappa \nabla^2 \phi \right) + \xi \right] \tag{4.101}$$

but also introduces another issue, namely how to calculate the fluid flow velocity. In general, this is clear: fluid flow is described by the Navier–Stokes equation. In a recent review paper by Fei Wang an co-workers, a formulation of a generalised Navier–Stokes equation compatible with the CH equation was derived [277] and used to analyse phase separation in polymer solutions, leading also to (aero)gels. This variant of the Navier–Stokes equation reads

$$\rho(\phi)\left(\frac{\partial \vec{u}}{\partial t} + \vec{u} \cdot \nabla \vec{u} \right) = -\nabla p + \nabla \cdot \left[(\kappa(\nabla \phi)^2 + f(\phi))\mathbf{I} - 2\kappa \nabla \phi \otimes \nabla \phi \right] + \nabla \cdot \mu_D(\phi)\mathbf{e}, \tag{4.102}$$

with $p(\phi), \mu_D, \mathbf{I}$ as the pressure, the dynamic viscosity and the identity tensor. $\mathbf{e} = (\nabla \vec{u} + \nabla \vec{u}^T)$ is the shear rate tensor. This equation implicitly assumes that the densities of the polymer-rich phase and the solvent-rich phase are similar, otherwise an additional term would arise (Boussineq approximation). One can immediately recognise that the addition of fluid flow makes the whole process describing phase separation in polymer solutions much more complicated: one has to solve the CH equation and the Navier–Stokes equation simultaneously. If one would like to include Brownian motion or any other type of stochastic motion, one can add on the right side of the Navier–Stokes equation a random noise term (Langevin approach) as done in Eq. (4.99), whose time average is zero.

Expressions for Mobility and Interface Gradient Term

To proceed further, one needs expressions for the mobility $M(\phi)$ and the interface term κ. We simply stated earlier that the mobility relates the diffusion flux to the gradient in chemical potential. This reads

$$J_\mu = -M\nabla \mu. \tag{4.103}$$

Any diffusion flux, however, can also be written as a gradient of concentration, which here is the polymer concentration ϕ:

$$J_D = -D\nabla \phi. \tag{4.104}$$

Since μ is a function of ϕ, the chemical potential gradient $\nabla \mu$ can be further written as

$$(\partial \mu / \partial \phi)\nabla \phi. \tag{4.105}$$

By this reformulation, the second flux is rewritten as

$$J_D = -M(\partial\mu/\partial\phi)\nabla\phi. \tag{4.106}$$

Both fluxes J_μ and J_D must be identical and thus we have

$$M\frac{\partial\mu}{\partial\phi} = D \quad \text{or} \quad M = \frac{D}{\frac{\partial\mu}{\partial\phi}}. \tag{4.107}$$

If $N = 1$, the partial derivative $(\partial\mu/\partial\phi)$ using the chemical potential of the Flory–Huggins theory (see Appendix B) reads as follows:

$$\frac{\partial\mu}{\partial\phi} = k_B T\left(\frac{1}{\phi(1-\phi)} - 2\chi\right). \tag{4.108}$$

If N is greater than 1, the partial derivative $(\partial\mu/\partial\phi)$ reads

$$\frac{\partial\mu}{\partial\phi} = k_B T\left(\frac{1}{N\phi} + \frac{1}{1-\phi} - 2\chi\right). \tag{4.109}$$

The χ-term leads to negative mobilities, which sounds strange and thus one might consider neglecting this term in any numerical solution of the Cahn–Hilliard equation. We then have in the case of $N = 1$ the mobility as

$$M(N=1) = \frac{D}{k_B T}\phi(1-\phi) \cong \frac{D\phi}{k_B T}. \tag{4.110}$$

The last approximations holds in aerogels, if the fraction of polymers is less than a few percent. In the case of N greater than 1, the mobility is expressed as

$$M(N>1) = \frac{D}{k_B T}N\frac{\phi(1-\phi)}{1-\phi+N\phi} \cong \frac{D}{k_B T}\frac{N\phi}{1+N\phi}, \tag{4.111}$$

which seems interesting, since now the mobility depends on the polymer concentration and the degree of polymerisation. When N is extremely large, the mobility $M(N > 1)$ can be simplified as $D/(k_B T)$. One can also state that in aerogels, if the degree of polymerisation is sufficiently large (for example, $N > 50$) and the volume fraction polymers is small, the mobility is constant and simply related to the diffusion coefficient $Mk_B T = D$. We then would end with the afore mentioned simple CH equation. A remark at this point is important: the degree of polymerisation N is a function of time, therefore the mobility is a function of the polymer fraction and time.

The possible concentration dependence of κ needs special attention. In dilute dispersions of a polymer, or an emerging polymer, one can state, as done by Chan and Rey [278], that κ must change in as much as a polymer-solvent interface builds up. Chan and Rey use an approach of

$$\kappa = \kappa_0 N(t), \tag{4.112}$$

arguing that the length of the polymer chain increases linearly with N. The surface of a polymer varies proportional to N only if it is a straight rod. Generally the radius of gyration of a polymer varies like $\sqrt{N}$ and therefore the surface, exposed to the solvent, varies like $N^{3/2}$. Therefore, a more general approach would be

$$\kappa = \kappa_0 N(t)^{3/2}. \tag{4.113}$$

This relation mimics that the surface of a polymer rod is exposed to the solvent. With these relations, the problem is fixed and the CH equation, including fluid flow, can be solved numerically. A first approach of this was performed recently by Fei Wang and co-workers [110, 277]. We are not discussing their first results on PIPS to describe aerogels structures via phase separation with either a binodal-spinodal or a pure spinodal decomposition. The interested reader is advised to take a look at these recent papers.

4.3 Predictions of Gel Time

Using the results derived in the preceding section we construct different models of gelation. First we derive simple models based on aggregation, then we use the Family–Viscek scaling and finally we describe a model based on PIPS.

4.3.1 Gelation by Aggregation

Gelation occurs once a spanning cluster of solid-like particles appears. We do not assume an infinite sample volume, but take a finite volume as usual in a real experiment. The sample volume will be V_T and it will simply be a cube. A sketch of this situation is shown in Figure 4.17. We can define that for a cluster of length $\ell \approx V_T^{1/3}$ the number of particles at the gelation time t_g is given by

$$N_c = \left(\frac{V_T}{v(t_g)} \right)^{1/3}. \tag{4.114}$$

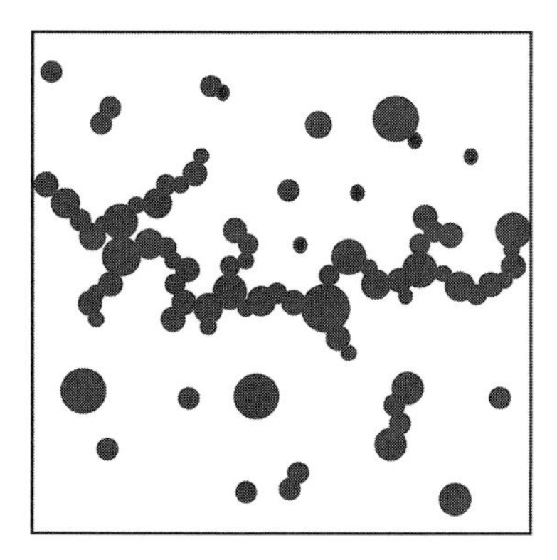

Figure 4.17 Gelation is assumed to occur if a spanning cluster of particles having all different sizes expands from one side of the box to the other.

The number can be calculated from the number density $n(t)$ and the box volume V_T to be $N_c = n(t_g)V_T$. Putting these equations together leads to an expression for the gel time

$$V_T^{-2} = n(t_g)^3\overline{v(t_g)}. \tag{4.115}$$

Using the relation between number density of particles and their average volume $\phi = n(t_g)\overline{v(t_g)}$ at the gelation time allows simplification of this expression, since ϕ is constant:

$$V_T^{-2} = n(t_g)^2/\phi. \tag{4.116}$$

We just have to solve Eq. (4.52) for a given collision kernel. For three kernels, we have previously given a solution, the constant kernel, mild shear flow and the Stokes sedimentation kernel. We first use the result for the constant collision kernel, given in Eq. (4.57), to yield after some manipulations and inserting also Eq. (4.56)

$$t_g \approx \frac{3V_T}{4k_B T\sqrt{\phi}}\mu. \tag{4.117}$$

First this result would mean that the larger the total volume fraction, the shorter the gelation time is. The second issue is the larger the sample volume, the greater gelation times. Both relations are in general agreement with experimental observations. The dependence on the viscosity μ of the solution also is interesting: the higher the viscosity, the larger the gelation times. This is to be expected: in a water-like solution, Brownian motion and thus aggregation are fast, whereas in a honey-like solution, both are impeded. There is, however, a problem with this relation: putting in some typical values for a water-rich solution such as an RF or a silica aerogel means that viscosity is around 1 mPa·s, and taking typical temperatures for gelation between room temperature and 80°C leads to gelation times in a sample volume of a few cubic centimetres of the order of 10^{15} seconds plus or minus a decade. These numbers are orders away from reality. Although the physics seems to be covered with this model, it is much to far from reality.

Let us take the expression for mild shear given in Eq. (4.62). This leads after some manipulations to

$$t_g = \frac{1}{2\Gamma\phi_0}\ln\frac{n_0^2V_T^2}{\phi_0}. \tag{4.118}$$

Again, the larger the initial volume fraction, the shorter the gelation time. The sample volume enters only logarithmically but with the same tendency as in the Brownian case. Interesting enough, a higher shear rate would reduce the gelation time. In a certain sense, this is in contradiction to observations; but not completely; see [221]. The problem here with observations is that at higher shear rates one would have to take into account that the fluid flow might lead to a breakup of aggregates. Such effects are not covered in the original model but would need extra terms [239]. Using the preceding derived expression for the number density in Stokes sedimentation, Eq. (4.68), one obtains for the gelation time the expression

$$t_g = \frac{1}{C_{s0} M_{4/3}} \ln \left(\frac{(n_0 M_{4/3} - M_{2/3}^2) V_T}{\sqrt{\phi_0} M_{4/3} - M_{2/3}^2 V_T} \right). \tag{4.119}$$

Using the expression of Eq. (4.69) shows that the gelation time is proportional to the viscosity of the solution, and this defines its temperature dependence. The sample volume and volume fraction effects are hidden in the logarithmic term, such that there is only a slight dependence on it. Nevertheless, the logarithmic term has some peculiarities: both the numerator and denominator can have zeros, and then either gelation is instantaneous (because the gelation time would be negative, numerator zero) or it is infinite (denominator zero). Only in between a finite gelation time is observed. Making some estimates, however, one can show that these peculiarities for sedimentation are formally nice, but realistically the logarithmic term can well be approximated by $\ln n_0 V_T$, and thus gelation would be determined by the viscosity of the solution and the density difference between particles and the solution. Since without the size distribution the broken moment $M_{4/3}(t)$ cannot be further handled, its impact on the gelation time is hard to characterise.

There is another possibility to calculate the time for gelation with the simple aggregation equations that we have used so far. We have derived for different collision volumes the time dependence of the average cluster volume. Let us assume that gelation occurs once this average volume is equal to a fraction α of the sample volume V_T. We then have for Brownian motion of particles the relation $V_T = \alpha \bar{v}$. Using the expression derived earlier, it is easy to show that the gelation time should follow the following relation:

$$t_g \propto \frac{3}{4} \frac{V_T \mu}{\phi_0 k_b T}. \tag{4.120}$$

The principle physical features of this equation are the same as in Eq. (4.117). The only difference is that the volume fraction enters directly and not in a square root. Therefore, the estimates made earlier concerning the gelation times are of the same order. They are too large to make any physical sense, although, as stated previously, the dependence on sample volume V_T, temperature and volume fraction look physically meaningful. This points to the fact that single particles do not lead to gelation, but once a few particles have aggregated to clusters, these move and cluster–cluster aggregation determines the onset of gelation.

Let us check if these simple considerations also fail in the case of gelation under a mild shear flow. Using the same idea of the preceding paragraph, we arrive at an expression

$$t_g \propto \frac{1}{\Gamma \phi_0} \ln \left(\frac{V_T n_0}{\phi_0} \right), \tag{4.121}$$

which is essentially similar to Eq. (4.118); in particular, the dependence on the shear rate is equal as well as the logarithmic dependence on sample volume. A numerical

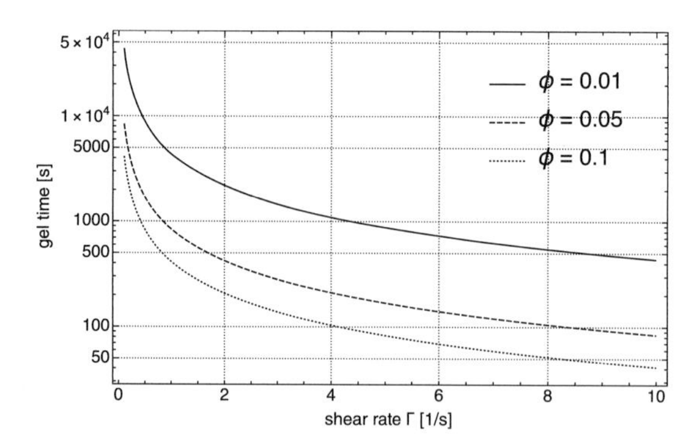

Figure 4.18 Gelation of particles with an initial number density of 10^{23} per m^3 for mild shear flows with shear rates of 0.1–10 s^{-1} after Eq. (4.121) using a proportionality factor of 1. The range of shear rates is in accordance with values used by Hajduk and Ratke for RF aerogels [221]

comparison of both equations shows that the differences between them are negligible. The gelation time is essentially determined by the volume fraction and the shear rate. In Figure 4.18, we show some calculated gelation times. In contrast to the calculations made for Brownian motion, the calculated gel times are reasonable. For shear rates around 1, they are between 500 and 5000 seconds, meaning something like 10 minutes to 1.5 hours. With an increasing volume fraction, the gel times decrease as they should. Although these considerations look at a first glance convincing, one should not overestimate them.

In all three cases, this simple model of gelation shows that the temperature dependence is dictated by the viscosity of the solution: the higher the viscosity, the larger the gelation times, or the higher the temperature, the shorter the gelation times (note that the viscosity also depends on the composition of the solution). In all cases, the sample volume is important: the smaller it is, the shorter the gelation time.

4.3.2 Family–Viscek Scaling for Gel Times

The scaling analysis of the Smoluchowski equation allows us to at least make an estimate of the time for gelation. Viscek [49] uses the prevlous derived Family–Viscek scaling to calculate the effective volume $V(t)$ of clusters using the relation

$$V(t) = \sum_s V_s n_V(t) \propto \phi_0 \int_1^\infty v^{d/d_f - 2} f(v/t^z) dv \propto v_1 t^{(d/d_f - 1)z} \tag{4.122}$$

with ϕ_0 as initial the volume the diffusing particle. If at the time of gelation t_g this volume is 1, one obtains an estimate for the gelation time as

$$t_g \approx \phi_0^{\frac{1}{z(1-d/d_f)}}. \tag{4.123}$$

In this equation, it first is interesting that the exponent is always negative, since $d > d_f$. Therefore, the gel time decreases with increasing volume fraction. At small

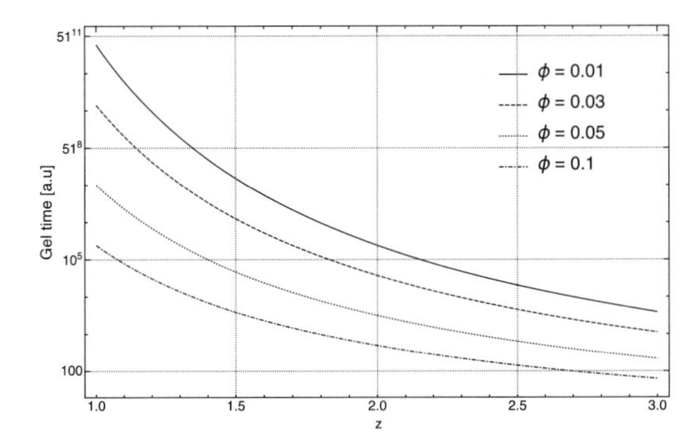

Figure 4.19 Gel time as predicted by the Family–Viszek scaling for a 3D fractal structure with $d_f = 2.53$. The parameter on the abscissa is the scaling parameter z. Note that the ordinate is logarithmic, and thus a small change in volume fraction at a fixed z can lead to drastically different gelation times.

values of z, the gel time rapidly goes to large values. If z is larger than 1, the gel time is finite and typically short. Figure 4.19 shows some calculations. Although all mean-field theories are fascinating and give important general rules for gelation processes, they lack of a decisive relation to the chemistry and physics of aerogel forming solutions as they were outlined in the chapters on synthesis. So their predictive character is limited.

4.3.3 Gel Time Prediction by PIPS

As an illustrative example for the application of the previously developed model on polymerisation-induced phase separation, PIPS, we apply it to RF aerogels. Let us review the processes going on in RF solutions until gelation. First the polycondensation reaction takes place and leads to oligomers with various degrees of polymerisation until the evolving miscibility gap with a lower critical point touches the operating point of the system (temperature and fraction of polymers). This takes time, and it can be calculated approximately as if spinodal decomposition has occurred. (In reality, first the binodal line is passed and the system will phase separate by nucleation of droplets, their growth followed by spinodal decomposition. Here we neglect it, because for a spinodal decomposition we can perform analytical calculations.) After phase separation, aggregation of the solid-like mixture of polymer-rich particles starts, and this leads after some time to gelation, which can be calculated with the simple equations given previously, at least in a first approximation. Let us recapitulate the relevant equations for the spinodal derived in Appendix B. The spinodal is a function of the degree of polymerisation and of course the volume fraction polymer ϕ, and we write the following as if we denote N as the degree of polymerisation:

$$T(\phi) = \frac{1}{2\chi_1} \frac{1 - \phi + N\phi}{N\phi(1 - \phi)}. \tag{4.124}$$

Let us now assume that one has put a solution of resorcinol, formaldehyde and a catalyst in a furnace being at a temperature T_g, which also will lead to a concentration of polymer ϕ_0 after all resorcinol molecules have reacted. We then ask for the time it needs that the critical point is down to this temperature. The time dependence stems from the fact that the degree of polymerisation changes with time according to $N(t) = 1 + k_r t$, as derived in Appendix D. We then have to solve the equation

$$2\chi_1 T_g = \frac{1}{1+\sqrt{1+k_r t}}\left(\frac{1+k_r t}{1-\phi_0} + \frac{1}{\phi_0}\right) \tag{4.125}$$

and define the time of gelation exactly at that time this happens. This leads to the following expression for the gel time:

$$t_{gel} = \frac{1 - 2T_{gel}(1-\phi_0)\phi_0\chi_1}{k_r\phi_0(2T_{gel}(1-\phi_0)\chi_1)} \approx \frac{1}{2k_r T_{gel}\phi_0\chi_1}. \tag{4.126}$$

Figure 4.20 shows a result for one furnace temperature and different values of the maximum polymer concentration. The rate constant for the polymerisation reaction was taken from the values given by Yamamoto and co-workers [90] transforming the values given there in mol/(L·s) using the molar volume of resorcinol to a frequency of reaction with the unit 1/s. The only open parameter is χ_1, which was adopted to reach values which are experimentally reasonable, meaning in between a few minutes to hours or even a day. With the parameters set, one can first see that with increasing volume fraction polymer the time to gelation decreases, in agreement with all experimental data, and second, that the gel times are below a day. Since no thermodynamic investigations of RF solutions are available and thus no better description of the system with a lower critical point, this seems to be acceptable. The temperature dependence is included in the reaction rate constant $k_r(T)$ of the polycondensation, which in the case of an Arrhenius behaviour would yield a classical temperature dependence.

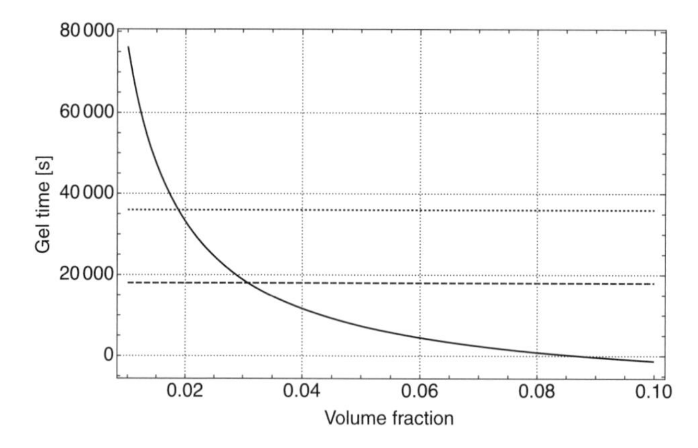

Figure 4.20 Gel time defined as that time, when the spinodal temperature at a given volume fraction of polymer touches the temperature of annealing the solution in a furnace. $\chi_1 = 0.018$, $k_r = 0.001$, $T_{gel} = 350$ K. The two horizontal lines indicate the time fo 5 and 10 hours.

As mentioned in Section 2.2 on RF aerogels, the reaction between resorcinol and formaldehyde has an activation energy of around 78 kJ/mol [83]. The observations made by Thiel and co-workers [216] yield an activation energy or gelation of around 72 kJ/mol. This is close by, and one might think that the approach outlined here is not that bad, and it would suggest that the viscosity of the solution does not play any decisive role. This picture of gelation to be essentially a process of spinodal decomposition might be an oversimplification, since the phase separation process typically would start with a nucleation of liquid RF-oligomer-rich droplets in a liquid solution once the binodal is passed, and only further progress of the condensation reaction leads to a situation that the spinodal is reached. If the process of particle formation starts with a nucleation event, the droplets would be able to move inside the solution by Brownian motion, Marangoni motion, sedimentation etc. and then aggregate thereby, while also making a percolation transition as outlined earlier. In such a case, the viscosity of the solution must have an influence on the gelation kinetics. To clarify the process of gelation in RF aerogels, further experimental studies are necessary, accompanied by theoretical investigations such as those outlined in Section 4.2.5.

Size Dependence of Gel Time

Scaling up of aerogel production from granular materials of a few micrometres in diameter to tiles of a few ten millimetres or larger would profit from a knowledge, how gel formation depends on container size. Astonishingly, this aspect has not sparked much interest in the literature of aerogels.

Anglaret et al. [279] studied experimentally and theoretically the gelation time as a function of container size for neutral and weakly base catalysed TMOS gels. They filled cylinders with the aerogel forming solution. Varying the height of fill and thus the aspect ratio height to diameter allowed the authors to extract a size dependence. They found a size-independent gel time for neutrally catalysed gels. For base catalysed gels, they observed that the gel time increased with container size. In order to interpret their observations, they developed a fluctuating bond aggregation (FBA) model, which is similar to a DLCA algorithm but needs first a critical cluster concentration below which gelation is impossible. and allowed a flexible bonding between particles in a cluster. If the clusters are bonded weakly or flexible, there is almost no size dependence of gelation time. These ideas are in agreement with observations of silica gelation as discussed in Chapter 2.

In a recent study, Ratke and Hajduk [221] analysed with a new experimental method the size effect in the gelation of RF gels. For their study, they used resorcinol-formaldehyde gels catalysed under base condition with Na_2CO_3 as a catalyst (C) with an R/C ratio of 1515. In contrast to Anglaret et al. [279], they did not use cylinders of different aspect ratios to study the size effect but decided to use a spherical shape. The spherical RF solutions (and gels) were prepared by inserting RF solution drops into a oil bath at 80°C. They could vary the volume of the drops between 2–20 μL. The bulb container with the oil bath was set into rotation to avoid settling of the drops before they gel. The rotation speed was adopted to the volume of the drop. Defining a shear rate as described in Eq. (4.1), the experimentally used shear rates varied with the drop volume like

$$\Gamma = \xi V^{-4/5}, \tag{4.127}$$

with ξ a numerical constant. They measured the gelation time simply with a stopwatch looking how the immersed drops changed their transparency. A description of their results was already given; see Figure 4.9. The gelation time depends in a non-trivial way on the drop volume as shown in Figure 4.10. Hajduk and Ratke developed a new model yielding an analytically tractable relation reflecting specialties of the experimental setup based on first the hydrodynamics of the drops inside the oil reservoir and second on the Smoluchowski equation.

First they took into account that the paraffin oil, which rotates in the bulb at low frequencies and fulfils a simple shear flow with an azimuthal velocity of $u_\phi = r\omega \sin\theta$ [274], with r as the radial coordinate in the bulb. The linear shear flow has a shear rate $\Gamma = \omega \sin\theta$. They used a simplified expression for the shear rate, $\Gamma_T = \pi f/\sqrt{2}$, with f as the rotating frequency of the bulb. They showed that the aggregation and concurrent growth of particles by polycondensation reactions is essential and extended the Smoluchowski equation used a with a growth term.

Taking growth and aggregation into account leads to a modified Smoluchowski equation:

$$\begin{aligned}\frac{\partial f}{\partial t} + \frac{\partial}{\partial v}\left(\frac{dv}{dt} f\right) = &\frac{1}{2}\int_0^v W(v, v - v')f(v', t)f(v - v', t)dv' \\ &- f(v, t)\int_0^\infty f(v', t)W(v, v')dv'.\end{aligned} \tag{4.128}$$

They did not solve this equation but calculated the number density by intergrating both sides of the equation with respect to the particle volume, multiplying both sides by the particle volume v and integrating again over the particle volume. Performing the integration leads to two ordinary differential equations:

$$\frac{dn}{dt} = -\Gamma_T n(t)\phi(t). \tag{4.129}$$

Instead of a constant volume fraction of particles, this is now varying with time:

$$\frac{d\phi}{dt} = \int_0^\infty \frac{dv}{dt} f(v, t)dv. \tag{4.130}$$

To evaluate these equations, one has to make an assumption on the growth rate dv/dt. For simplicity, the authors assumed a constant volumetric growth of the particles with a chemical reaction of first order [51] through its interface area A at a rate ν_0, meaning

$$\frac{dv}{dt} = \nu_0 A. \tag{4.131}$$

Assuming that the particles remain spherical, their surface area A can be expressed in terms of particle volume and Eq. (4.130) can be evaluated using $\phi(t) = n(t)\overline{v(t)}$:

$$\frac{d\phi}{dt} = v_0 \alpha \phi^{2/3} n^{1/3}, \tag{4.132}$$

with $\alpha \approx 4.84$. Solving Eqs. (4.129) and (4.132) leads first to an expression for the number density as a function of volume fraction:

$$n(\phi) = n_0 \left(1 - \frac{\Gamma_T}{4\alpha v_0 n_0^{1/3}}\right)^3 \tag{4.133}$$

Abbreviating $u = \phi/\phi_s$, the volume fraction ϕ can be expressed as

$$\phi_s = \left(\frac{4\alpha v_0 n_0^{1/3}}{\Gamma_T}\right)^{3/4}, \tag{4.134}$$

leading to

$$n(u) = n_0(1 - u^{4/3})^3. \tag{4.135}$$

Inserting Eq. (4.133) into Eq. (4.132) and using again $\phi = n(t)\overline{v}$ gives a differential equation for the volume fraction alone:

$$\theta = \frac{v_0 \alpha n_0^{1/3}}{\phi_s^{1/3}} t. \tag{4.136}$$

In their analysis, they did not use the full expression, but a series expansion to first-order expansion with respect to the normalised volume fraction:

$$u \cong \frac{\theta^3}{27}. \tag{4.137}$$

Integrating Eq. (4.129) directly and using the normalised time leads to an expression of the number density

$$n(\theta) = n_0 \exp\left(-4 \int_0^\theta u(\theta) d\theta\right), \tag{4.138}$$

and for the average particle volume:

$$\overline{v(\theta)} = \frac{\phi_s}{n_0} u(\theta) \exp\left(4 \int_0^\theta u(\theta) d\theta\right). \tag{4.139}$$

Using the definition of the gel time as given in Eq. (4.116) and inserting the expressions for the number density and the volume fraction leads to

$$V_T^2 n_0^2 \phi_s = \frac{27}{\theta_g^3} \exp\left(\frac{2}{27}\theta_g^4\right), \tag{4.140}$$

with θ_g as the obvious definition of a scaled gelation time. Solving for the gelation time in non-dimensional coordinates led the authors to express the gel time as

$$t_g = \frac{(an_0^{1/3}\alpha v_0)^{1/4} V_T^{1/5}}{\alpha v_0 n_0^{1/3}} [2 + \ln(bn_0^2 V_T^2 (n_0^{1/3} V_T^{4/5} \alpha v_0)^{3/4})]^{1/4}, \tag{4.141}$$

with a, b being numerical constants. For fitting this expression to their experimental data, they added an incubation time, defined as a time in which monomers react to oligomers and stable particles are established in the solution. Neglecting the logarithmic term leads to

$$t_g = t_{incub} + \frac{(1.5an_0^{1/3}\alpha v_0)^{1/4} V_T^{1/5}}{\alpha v_0 n_0^{1/3}} = t_{incub} + cV_T^{1/5}. \tag{4.142}$$

A fit of their experimental data to this expression is shown above in Figure 4.10. They showed that the constants of fit are physically reasonable. This model predicts gel times as a function of size and their growth rate in a solution if the aggregation of growing particles is due to shear flows.

4.4 Variables and Symbols

Table 4.1 lists some of the variable and symbols of importance in this chapter.

Table 4.1 List of variables and symbols in this chapter.

Symbol	Meaning	Definition	Units
$C(r)$	Density–density correlation function	–	–
d_f	Fractal dimension	–	–
DP, N	Degree of polymerisation	–	–
$F(\phi)$	Free energy functional	–	–
$g_0(r)$	Pair correlation function	–	–
G	Shear modulus	–	N/m^2
G'	Storage modulus	–	N/m^2
G''	Loss modulus	–	N/m^2
$\vec{j}$	Diffusion current density	–	$moles/m^2s$
k_B	Boltzmann coefficient	1.38×10^{-23}	J/K
k_r	Reaction rate constant for polycondensation	–	1/s
L	Lattice size	–	–
M_m	Molecular weight	–	kg
$M_q(t)$	qth moment of the particle size distribution	–	–
$M(\phi)$	Mobility	–	–
N	Particle number	–	
n_0	Initial number of particles	–	–

Table 4.1 Continued

Symbol	Meaning	Definition	Units
$-n_s(p)$	Cluster size distribution	–	–
$n_V(v,t)dv$	Particle number density in the volume class $v, v+dv$	–	$1/m^6$
$n(t)$	Particle (aggregate) number density	–	$1/m^3$
p	Occupation probability	–	–
p_c	Percolation threshold	–	–
p_c^- and p_c^+	Percolation threshold probabilities when approaching p_c from below and above, respectively	–	–
q	Absolute value of the wave vector $\vec{q}$	–	1/m
P_∞	Probability that a site belongs to an infinite cluster	–	–
R_G	Radius of gyration	–	m
R_s	Radius of a macromolecule	–	m
$S(p)$	Average cluster size	–	–
$S(q)$	Structure factor	–	–
$s_\xi(p)$	Cluster size	–	–
$s(t)$	Cluster size	–	–
t_g	Gel time	–	s
T_{gel}	Temperature at which gelation takes place	–	K
T	Temperature	–	K
$U(r)$	Pair potential	–	–
$\bar{v}$	Average particle or aggregate volume	–	m^3
$W(v,v')$	Collision volume	–	m^3/s
x, y, z	Cartesian coordinates	–	–
z	Functionality of a site on a Bethe lattice	–	–
z, τ	Scaling exponents	–	–
$\beta, \nu, \beta, \sigma, \gamma$	Universal exponents	–	–
δ	Loss tangent	–	–
γ	Shear deformation	–	–
Γ	Shear rate	–	1/s
κ	Parameter in the CH equation	–	–
λ	Degree of homogeneity	–	–
μ	Dynamic viscosity	–	Ns/m^2
μ	Chemical potential	–	–
η	Scaled aggregate volume	–	–
ϕ	Volume fraction	–	–
ϕ_s	Volume fraction solid	–	–
ϕ_p	Pore volume fraction	–	–
$\psi(\eta)$	Scaling solution of the aggregation equation	–	–
τ	Shear stress	–	N/m^2
τ	Scaled time	–	–
ξ	Correlation length	–	m

5 Drying of Wet Gels

Aerogel technology provides lightweight materials with an outstanding combination of properties as mentioned in Chapter 1. One major problem for the preparation of aerogels is how to eliminate the liquid solvent from the wet gel while avoiding shrinkage, cracking and collapse of the gel structure. Several drying techniques have been used and are still under development.

5.1 Ambient Drying

The most obvious method to replace the liquid solvent from a wet gel is evaporation, maybe using an oven at any temperature above room temperature but below the boiling point of the solvent to accelerate the process. In the literature about aerogels, the process of drying at normal pressure is most often called ambient or evaporative drying. As simple as this method is, it typically does not lead to an aerogel. Performing such an operation with a wet silica gel leads to crumbles of dry silica, which even have a low porosity, typically less than 50%, a high density and thermal conductivity. Therefore, these types of dry gels are not called aerogels but xerogels, especially in the older literature. The same is true for RF aerogels, which after such a simple treatment look like bakelite or cellulose gels look like a match. Therefore, it was noted by Kistler that other procedures are necessary to preserve the wet gel structure. Nevertheless, many papers are published annually on drying of gels under ambient conditions, and indeed in the past there have been developed methods, first for silica aerogels, to perform ambient drying and keeping the wet gel structure to a large extent intact. Quite often it is stated in the literature that the production of aerogels is cost intensive, not only in terms of precursors but especially due to the drying method often applied, namely supercritical drying. Therefore, it also is quite often stated that in order to make aerogels interesting for large-scale commercial applications, one must avoid this seemingly expensive part of preparation. We will not discuss such statements presented by the advocates of ambient drying and contradicted vehemently by the specialist for supercritical drying, but deal in this section with the basics of ambient drying, try to explain the problems and show a few solutions. A comprehensive treatment of ambient drying can be found in the book of Brinker and Scherer [3].

5.1.1 Some Observations during Drying

Take a wet gel, put it into a furnace and ask, what can be observed? First of all, if one does such an experiment with having the wet gel on a balance, one definitively will observe a mass loss over time such that at the end the dry gel mass is measured. One then can plot the mass loss as a function of time and ask if there are models available describing the measurement. Second, if the wet gel body keeps its form during drying, with no fractures or fissures, one can simply measure its size and compare it to that of the wet gel and will definitely observe a more or less large shrinkage, meaning the dimensions will have changed. Third, during drying one might observe, as in the case of most oxidic gels, that they lose their transparency. Instead of looking like a pane of glass, they might become turbid, quite often only for a short period of time, and the turbid region might only invade into the material such that after drying the gel is transparent or at least translucent again. Fourth, one might observe that during evaporative drying, the piece of wet gel deforms, as a rectangular bar bends if only dried from one side. Preparing sheets or tiles of aerogels or xerogels could then lead to an irreversible bent sheet, which might not be useful, and special techniques are sometimes involved to keep the sheets planar.

Mass Loss

Let us go to the first observation, the mass loss. Unfortunately, there are only a few studies dealing with wet gels during evaporative drying that explored the mass loss and tried to understand it. Whereas the drying of porous materials or beds of grains, crumbles, sand, food, seeds, salt, sugar, cement mixtures etc., is an engineering discipline both experimentally and theoretically [280–284], and thousands of studies have been published, the drying of wet gels leading to xero- or aerogels, although performed quite often, is generally only studied with respect to the final result: the microstructure and properties. A few studies, however, looked in more detail at the mass loss during drying.

Although silica aerogels are the most often studied ones, there are only a few papers which publish drying data. Bisson and co-workers [285] studied the drying of three differently prepared silica gels. The authors used pre-polymerised tetraethoxysilane and employed a two-step acid-base route with sulfuric acid and ammonia dissolved in isopropanol. The gels were dried supercritically with carbon dioxide. Two series of samples were treated with trimethylclorosilane before drying to modify the surface, making it hydrophobic (we come to this point in more detail later), and then dried for 24 h or 48 h at 60°C in an oven. They published a drying curve of the sample without surface modification, which is reproduced in Figure 5.1. The figure reveals a standard drying curve of a material: first there is a linear decrease in mass, which levels off later. The linear decrease is described in the literature on drying as a constant evaporation period (CRP), since the time derivative, meaning the mass loss per unit of area and time, is constant. This period is followed by a so-called first falling period (FRP1), meaning the rate of evaporation (recall it is the time derivative of the mass loss curve) or the mass loss rate decreases with time. Sometimes this period is followed by a

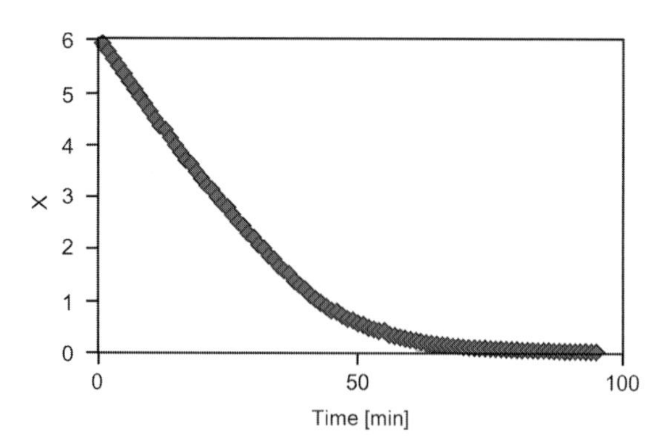

Figure 5.1 Loss of liquid mass, expressed as a fraction $X = m_{liq}/m_{dry}$ with time at 25°C. The liquid was isopropanol and the drying performed under flowing nitrogen. Reprinted from [285] with permission from Elsevier

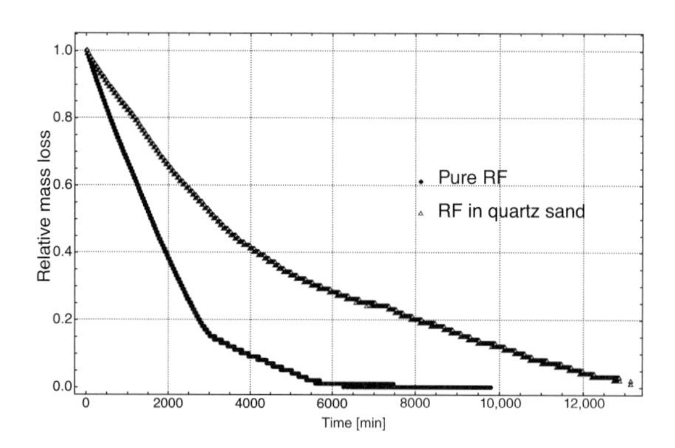

Figure 5.2 Loss of mass with time of an RF aerogel and a quartz sand bed filled with RF solution. Evaporation occurred at 25°C, after [288]

second falling period. The transition between the CRP and the FRP1 period marks the so-called critical point, characterised as that point when the liquid menisci move into the porous body (see the following subsection). A similar behaviour was observed by Smith and co-workers [286] when they prepared silica aerogels with modification of the surface using also trimethylclorosilane (we come later back to their observation, since it was a milestone in silica aerogel synthesis). The transition from CRP to FRP1 cannot clearly be seen in Figure 5.1 but is more pronounced in data published on RF aerogels.

Although RF aerogels can be prepared without supercritical drying at ambient conditions if the synthesis parameters are chosen properly (see Section 2.2), there is only one paper in which mass loss curves are published. Reuss and Ratke [287] measured the mass loss during drying of sand beds, in which the pore space between the sand grains was filled with a RF solution. From their original data, we reproduce the drying curve of a pure RF aerogel and one with a quartz sand bed in Figure 5.2. In both cases, one observes first a linear decrease in mass loss with time. The pure RF gel dries faster than the gel in the interstices of the sand. After a certain time, there is a change in drying rate or mass loss: in the pure RF gel, the obvious change is around 3000 minutes, marking the transition from CRP to FRP1, and for the RF gel inside the sand bed there is only a smooth transition of around 3000–4000 minutes.

Shrinkage

The second observation concerns the change in sample dimensions during or after drying. Even after supercritical drying, it is observed that the dimensions of the wet gel differ from that of the dry one. The shrinkage is usually more pronounced during ambient drying and needs special measures to reduce it. The shrinkage during evaporative drying might be intended, for instance, if one would like to prepare from a gel a dense ceramic, but in order to prepare aerogels it is a negative effect, because accompanied with it is a reduction in pore size and an increase in density. Both effects alter the thermal and mechanical properties. Shrinkage was measured in gels, for instance, by placing a rod of a wet gel in a dilatometer and measuring the length change upon drying, or simply, if the drying takes enough time, by measuring the size of a specimen with a calliper from time to time. One example of such a measurement is shown in Figure 5.3, showing the effect of evaporative drying on a silica gel without and with surface modification by trimethylclorosilane (TMCS). This figure clearly demonstrates that shrinkage upon drying of a silica gel is irreversible if the surface is not modified. If it is, the sample retrieves its original size almost exactly (there is a bit of shrinkage left, a few percent). This was called 'springback' effect, and since its discovery by Smith and co-workers it has been widely used, also industrially.

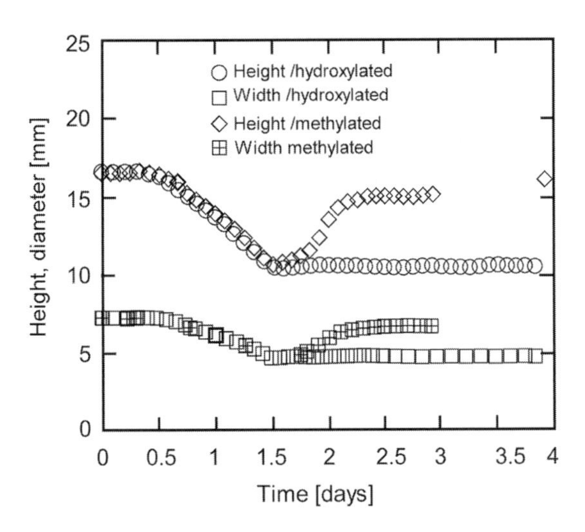

Figure 5.3 Shrinkage of silica gels (height and diameter) having simple -OH groups at their inner surface (hydroxylated) or treated with trimethylclorosilane replacing an -OH group with -Si–$(CH_3)_3$ leading to a hydrophobic surface. Reprinted from [286] with permission from Elsevier

A variation of this approach to surface modification was made a few years later by Schwerdtfeger and co-workers [289] to make it more feasible for industrial production. The springback effect was nicely photographed by Bisson and co-workers; see [285]. The origin of this springback is in the chemistry of the modified surface. Each silanol group at the surface is replaced by an -O–Si–$(CH_3)_3$ group. The methyl groups are non-polar, and although the wet gel shrinks upon drying, these groups repel each other. There are other sources of shrinkage, one is solvent exchange, meaning exchanging the pore fluid after gelation to a liquid miscible with carbon dioxide for supercritical drying. Such effects are discussed in Section 5.4 and Chapter 7. In order to explain the observations made during drying, we have to think about several aspects: first of all, the evaporation itself; second, capillary stresses on a liquid in nanosized pores; third, transport of liquid and gas through the pores; and fourth, the stresses induced during drying on a porous body and the deformation accompanying them. We are not going to discuss full aspects in full detail. First, the book of Brinker and Scherer [3] is a wonderful source for the physics behind drying, and second, treating all details would mean writing another book. Therefore, we give only a brief survey of evaporative drying.

5.1.2 Evaporation Rate

Evaporation is a thermally activated process of transferring a molecule from a liquid into the gas phase. At its boiling temperature, any liquid is in equilibrium with the gas (vapor) phase. The pressure in the gas phase is equal to the external pressure, and the vapor contains molecules of the liquid only. When the liquid is exposed to an intert gas (e.g., air) below the boiling temperature, an equilibrium is also attained, but the gas phase is a mixture now: molecules of both the liquid and the inert gas are present. The fraction of the total pressure due the molecules of the liquid is called partial pressure. The partial pressures are well tabulated for many liquids as a function of temperature. The simplest way to look at drying was already made by Hertz more than 100 years ago. He connected kinetic theory of gases to evaporation using that in kinetic theory of gases the number of moles evaporating from a surface per unit area and time $Z(T)$ is

$$Z(T) = \frac{p_V(T) - p_A}{\sqrt{2\pi M_m R_g T}}, \tag{5.1}$$

in which $p_V(T)$ is the vapour pressure above the liquid, the pressure according to the evaporation line in the phase diagram, p_A the partial pressure of the gas, R_g the universal gas constant and M_m the molar mass of the gas. This would yield a linear decrease of mass with time, since the rate $Z(T)$ is constant for a given temperature. (Note that the pressure and temperature are not independent along the evaporation line. This is a univariant line in the phase diagram and thus $p_V(T)$ is fixed, once the temperature is fixed.) This simple model, however, does not describe reality. First of all, even above a surface of water or whisky in a glass, the transport of the evaporating molecules away from the surface is not only given by the equilibrium evaporation rate but determined by mass transport in the gas phase, which can be purely diffusive or in most cases a mixture of diffusion and convection away from the interface [284].

A simple treatment of purely diffusive transport of gas molecules through a porous body can be a first step. This approach was, for instance, performed by many researchers and for silica gels also by Bisson [285]. A mathematical description of diffusive evaporation can be found in the textbook of Crank on diffusion [290]. Second, evaporation is not only a matter of gas transport, but evaporation means that there is also a transport of heat, in two ways. The liquid cools during evaporation (heat of evaporation), and the evaporated molecules transport heat away and might even in a non-isothermal sample condense at other locations inside the pore network. Third, in a porous solid the pores might have a range of a few tens of micrometre down to a few nanometres. Any liquid in such a pore space builds a meniscus of a rather large curvature, and thus there will be two effects: first capillary pressure and then the curvature will change the evaporation rate.

5.1.3 Capillary Stress and Vapour Pressure

A liquid in a pore has a shape that depends on its wetting capabilities with the pore walls, described by the wetting angle θ. The basis of its description is the Young–Laplace equation relating the pressure p_c exerted on a liquid in a cylindrical pore of radius R:

$$p_c = \frac{2\sigma_{LV} \cos\theta}{R}, \tag{5.2}$$

in which σ_{LV} is the liquid-gas interfacial tension and θ the wetting angle between the pore wall and the liquid. A sketch of it is shown in Figure 5.4. A wetting angle below 90° means the liquid is pulled into the capillary (positive pressure or tension) and the meniscus has a concave shape: at the walls, the liquid is more advanced than at its centre. This also means the liquid is under stress and the solid walls are under compression. If the wetting angle is larger, the pressure becomes negative and the liquid is pressed out of the pores (e.g., mercury in a glass capillary). The capillary pressure can reach very high values in aerogels. Surface tensions of some liquids against air are tabulated and data can easily be found on the World Wide Web. Let us take three examples: water, ethanol and 1-butanol. Water has at 20°C an interfacial tension of 0.07286 N/m, ethanol 0.02239 N/m and 1-butanol 0.0018 N/m against air. For convenience, we assume perfect wetting in all cases, meaning $\theta = 0$.[1] Then the capillary stress in a pore with 1 nm is for water around 100 MPa, for ethanol around 50 MPa and 1-butanol 3 MPa. Changing the radius to, let us say, 1 μm reduces the capillary stress by a factor of 1000. Since for small pores, below 10 nm, the capillary pressure reaches values in the MPa regime, huge pressure differentials arise during drying. Assume two pores of 10 nm diameter and 100 nm lie close to each other, then during evaporation, once the meniscus enters the pore space, a pressure differential over a wall of 10 nm thickness would lead to a pressure gradient of around 10^9 MPa/m. Such gradients lead to a pore collapse, since the solid backbone of the gel built by nanosized particles is not able to withstand such pressure gradients.

[1] The assumption of the wetting angle of most alcohols and water against for instance silica or polymers like in RF aerogels as close to zero is justified [291–294].

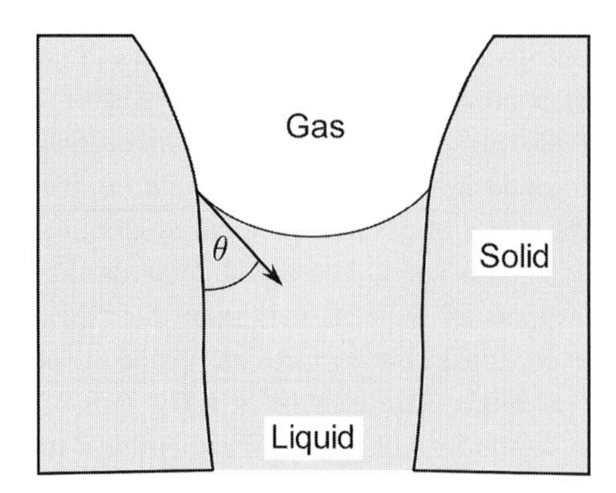

Figure 5.4 Schematic of partially filled capillary in a solid body.

There are several ways to avoid or reduce pore collapse: first, exchange of the solvent with a liquid having a low surface tension or adding surfactants; secondly, strengthening the solid backbone by ageing; and thirdly, modifying the surface of the solid by suitable surface groups, such that they repel each other during shrinkage. Adding surfactants to the liquid solution is quite often used in the literature. Surfactants generally are surface active molecules having a polar head group and a non-polar hydrocarbon chain. The head groups might be negatively charged (anionic), positively charged (cationic), zwitterionic or even non-ionic. Numerous surfactants are on the market and can be used, both in organic and inorganic aerogels. They are used especially in cosmetics, shampoos and household cleaning liquids. The idea using surfactants is to reduce as much as possible the surface tension of the liquid and thus capillary stresses. Especially in silica aerogels using trifunctional (MTMS) or difunctional (DMDMS) silanes in addition to or replacing TEOS, surfactants are used to modify or suppress phase separation [69, 70, 73].

The second effect of capillarity is on vapour pressure. The vapour pressure above a curved interface differs from that above a flat one. One easily can imagine that at a convex surface (e.g., sphere) the vapour pressure is higher than above a concave surface, since in the first case the bonds at the surface are already stretched and it needs at the same temperature less energy for a molecule to evade from the surface. We derive later in Chapter 9 the so-called Kelvin equation, which expresses this relation and place it here for a spherical surface with radius R:

$$p(T) = p_0 \exp\left(\frac{\sigma_{LV}\Omega_m \mathcal{H}}{k_B T}\right), \tag{5.3}$$

in which Ω_m is the atomic volume and $\mathcal{H}$ is the mean curvature of the vapour–liquid interface. If the interface has a spherical shape, $\mathcal{H} = 2/R$. The curvature $\mathcal{H}$ is positive for convex bodies and negative for concave ones. Note: $\sigma_{LV} \cdot \Omega_m/(k_B T) = \zeta$ has the unit of a length and is called the capillary length. Typical values are in the range of a few nanometres. Having aerogels or wet gels with pore sizes below 50 nm (mesoporous) means that the ratio ζ/R is of the order 0.1 to 1 and thus the vapour pressure is either increased or reduced by a large amount. Since in aerogels the meniscus usually is negatively curved, the vapour pressure is reduced such that large pores evaporate faster than small ones.

5.1.4 Flow of Fluid and Gas in the Network during Drying

It was already mentioned that the capillary pressure always pulls the pore liquid to the surface, since the wetting angle of most liquids used in aerogel production is small. If the liquid evaporates at the surface with the rate given in Eq. (5.1), two things could happen: first the loss of liquid at the surface is compensated by fluid flow inside the pore network; or second, the menisci could enter the pores, meaning the evaporating surface moves into the pore network.

If the pore network is very rigid and not deformable, the fluid flow can only supply the surface of the body until a critical pressure inside the liquid is reached and then the evaporation surface will move into the porous body. If, however, the pore network is deformable, which is the case in many wet gels from inorganic or organic precursors, the compression of the solid network by the capillary stresses and the flow to the sample surface is accompanied by a deformation: the pores become thinner or smaller, the sample shrinks and this shrinkage supports fluid flow.

In any case, the situation will arise that the menisci enter the pore space. This does not mean that the pores above the liquid–gas interface are completely dry, since the liquid wets the solid backbone. There always will be a very thin layer of liquid in which fluid flow assists evaporation. The evaporating gas will be removed from the pore space by diffusion, probably Knudsen diffusion (see Chapter 10). The evaporating front or liquid–gas interface will not move everywhere into the pore space to the same degree, since the pores have different sizes. The typically broad size spectrum means that the larger pores will empty first (provided all pores are connected).

What happens was simulated in the literature as an invasion percolation problem [49]. The gas entering the originally fully saturated gel invades the pore space. This leads to interesting situations: there can be a percolating network of a gas path through the body and islands of fully wet gel patches are still there. This has been studied in many papers for porous bodies with macropores and a non-deformable solid backbone; see, for example, [282, 283, 295, 296]. Yiotis et al. [295] simulated a porous medium with a random distribution of pores which all were connected by throats. Pore size and throat size were selected randomly and distributed in space (throats were always much thinner than the pores). Initially the pores and throats were completely saturated with a liquid, and evaporation could take place at the top surface only with a mass transfer rate dictated essentially by the vapour pressure difference between the liquid surface and the ambient pressure. The gas transport was simulated as a steady-state diffusion in which the throat size determined the transport current. Inasmuch as evaporation reduces the liquid phase volume, the liquid–gas interface inside the porous medium recedes locally following invasion percolation. Always the larger pore adjacent to a throat is invaded. An example of the simulation is shown Figure 5.5.

Figure 5.5 Simulated drying of a cube. The porous medium is simulated as pores which are connected by throats. The top-left figure shows a situation after 20% of liquid has been replaced by gas, the top-right figure after 40%, the bottom-left after 60% and the bottom-right after 80% gas has invaded the porous body. Shown are only the liquid clusters. Reprinted from [295] with permission from Elsevier

Their simulation led them to conclude that drying has four steps: (1) The first period is a surface evaporation regime with a rapidly falling evaporation rate. (2) The second regime is one of almost constant evaporation rate once a spanning liquid cluster exists only (or a spanning gas cluster). Diffusion through the pore space determines in this step the loss of liquid. (3) The third period exists when liquid clusters are isolated as islands or 3D patches and diffusion from their surface determines the mass loss. (4) The fourth regime is the receding front regime: the surface is dry and diffusion takes place through the dry zone.

In a simulation performed by Prat [283], viscous flow of the liquid phase was taken into account as well as the wetting angle between liquid and solid. He came to the conclusion, which we think is important in connection with wet gels leading to aerogels, that at very small wetting angles and small pore sizes the liquid films extend up to the surface of the porous medium during the whole drying process. This always leads to the fastest drying rate determined by Eq. (5.1). In his simulation, the pore shape and contact angle have no effect on drying as long as the films do not recede within the porous material. They do this, however, at the end of drying. One can summarise the considerations on drying using a picture adopted from Brinker and Scherer's book [3]. This is shown in Figure 5.6. The picture illustrates that initially liquid evaporates always from the surface and the surface is kept wet by capillary forces pulling the liquid to the surface. The flow of liquid to the surface could be modelled as a Darcy flow, meaning the mass transport is proportional to the permeability of the network and inversely to the viscosity of the liquid and proportional to the pressure gradient driving the flow (see also Chapter 11). The permeability is proportional to the square of the pore size, which means in gels with mesopores that it is rather low and thus huge

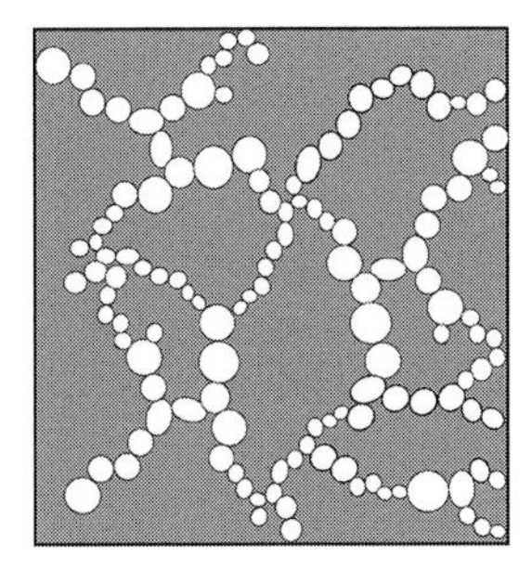
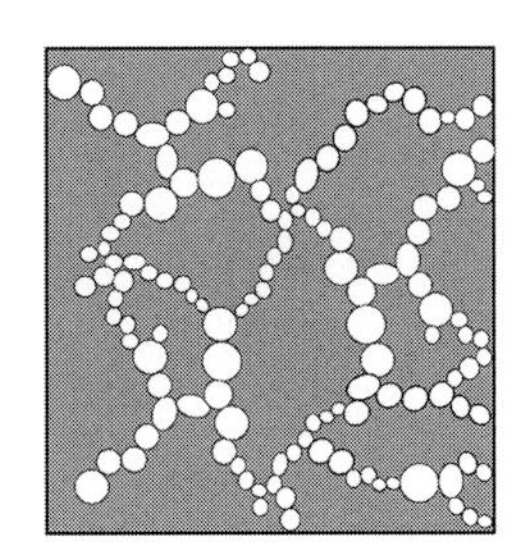
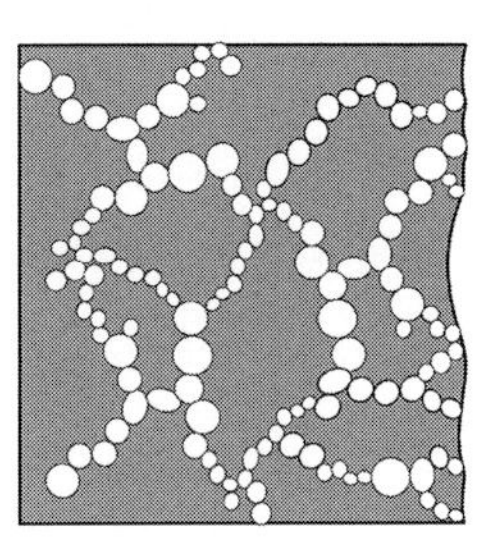

Figure 5.6 Schematic of a wet particulate gel filled with a liquid. Evaporation at the surface leads to mass loss, and flow of liquid to the surface occurs by capillarity and is compensated by shrinkage of the network (if this is compliant). Once a volumetric compression limit is reached, the vapour–liquid interface moves into the body and further drying is determined by gas transport through the pore space.

pressure gradients are necessary to drive the flow towards the surface. The capillary forces do not only drive the liquid to the surface but also set the solid backbone under compression. If the solid backbone is compliant, it will shrink and thereby support the transport of liquid to the surface. The compliance of the solid network and its deformation (shrinkage) determines how much and how long the tension in the liquid is able to maintain a fluid flow to the surface. The critical point is reached, when the network cannot compress any further. Then the evaporation surface recedes into the solid backbone structure and further drying is due to evaporation at the liquid menisci and gas transport through the mesopores.

5.2 Freeze Drying

A rather simple and elegant method to remove the solvent from a gel is to freeze the gel fluid and evaporate the solid along the sublimation line of the phase diagram shown schematically in Figure 5.7. This technique, industrially widely used, is called freeze drying or lyophilisation, and commercial freeze dryers are available in any size or volume. In contrast to a liquid–solid interface, where there is always capillary pressure, there is none at a solid gas interface. Thus evaporation from the frozen state cannot affect the pores or better the solid nanonetwork. At a state below the triple point of a solvent, the molecules directly sublimate from the solid phase into the gas phase (see Figure 5.7). In principle, one has to set the temperature T and pressure p of the drying chamber to a point on the phase boundary below the triple point. (Note that the sublimation line has only one degree of freedom. If one sets the temperature, the pressure is fixed on the line, or vice versa.) There are some good reviews on freeze drying, especially concerned with dispersion or suspensions of ceramic particles or biopolymers [297, 298].

Let us elaborate a bit more the processes involved in freeze drying. First of all, the gel fluid has to be frozen. Second, the solid shall sublimate along the sublimation line. Therefore, we first discuss freezing a bit more and then sublimation.

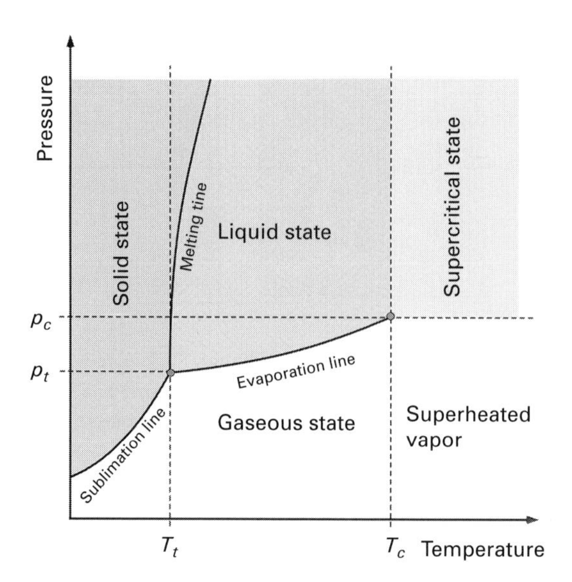

Figure 5.7 Schematic pressure temperature phase diagram. p_c, T_c denotes the critical point, and p_t, T_t denotes the triple point. Sublimation best occurs along the sublimation line in the phase diagram. There is a direct transformation of the solid to the gaseous phase.

5.2.1 Freezing of Single-Component Liquid

The freezing process of any type of fluid or liquid is a first-order phase transition and intensively treated in textbooks on solidification or crystallisation [299–302]. Such a first-order transition is characterised by a finite change in heat content at the melting/solidification temperature and a discontinuous change in Gibbs free energy as shown schematically in Figure 5.8. Below the melting point T_m, the Gibbs free energy of the solid is lower than that of the liquid and vice versa above it. On solidification or freezing, the latent heat of melting ΔH_m is released at the solid–liquid interface. This heat is transported by heat diffusion into the solid and often also by convection into the liquid. At the melting point, the solid and liquid phases are in equilibrium. To initiate the freezing or solidification, a nucleus (or many nuclei) of the solid phase has to be established and must be able to grow inside the liquid. In order to provoke nucleation, a certain undercooling ΔT below the melting temperature is necessary, since an interface with its interface energy has to be established. This interfacial energy has to be provided by the undercooling. Once a critical nucleus size is reached, the solid can grow, since its volume free energy is lower than that of the liquid and the gain of volume free energy during growth is much bigger than the energy that has to be spent to build the interface (the interface energy is proportional to the square of the solid particle size, and the gain in volume free energy is proportional to the cube of the size; this term wins above a critical particle size). The nucleation can be triggered substantially by foreign impurities in the liquid, or by container or pore walls, provided they reduce the interfacial energy needed to nucleate the solid. If these were perfectly wetted by the solid, the growth would proceed without supercooling

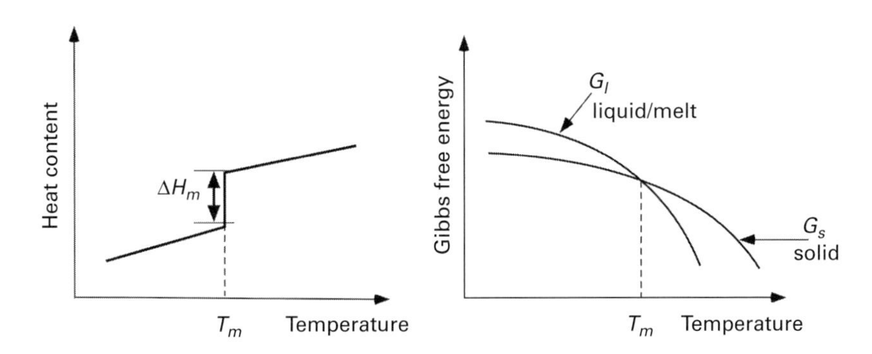

Figure 5.8 Schematic dependence of the heat content above and below the melting point T_m and the Gibbs free energy as a function of temperature of a one component material.

below T_m. Typically in liquids cooled from the surface, by which the heat is extracted, the solid is formed at the container walls first and grows into the liquid perpendicular to the isotherms of the thermal field. Also typically in crystallisation, not one crystal is nucleated, but along the container wall many nuclei appear and grow almost independently into the liquid (there is a competition between these crystals and only those survive whose crystallographic orientation with respect to the normal of the thermal field provides the largest growth speed). Inasmuch as the freezing front runs into the liquid, the residual liquid heats up by the aforementioned release of latent heat of melting until, depending on the size of the container, its interior approaches the melting temperature and crystals nucleate everywhere. If the container size is small or the liquid volume is small, like in droplets, the cooling can be fast enough and crystals grow only from the outer surface to the interior like pillars. The cooling rate, meaning the temperature decrease with time, $\dot{T} = dT/dt$, determines the freezing rate and the solidified microstructure to a large extent. In porous materials, the pore walls are preferential sites for nucleation of the solid and the solidification or freezing will be determined by the pore size, the type of solid pore walls (described by the interfacial tension between the forming solid and the pore wall) and of course the freezing rate. A rough estimate of the temperature profile in a piece of material during cooling can be made as follows.

Suppose we have an arbitrarily formed body of volume V having a surface area A. The sensible heat content of any body with density ρ_e and specific heat capacity c at a temperature T and volume V is

$$H = \rho_e \cdot c \cdot T \cdot V. \tag{5.4}$$

The time change of heat content is

$$\frac{dH}{dt} = \rho_e \cdot c \cdot V \frac{dT}{dt}. \tag{5.5}$$

The sensible heat content only changes if heat is extracted via the surface to the surrounding. Let the temperature of the surrounding environment be T_S. The heat transport at the surface is proportional to the difference between the actual temperature of the body and the environmental temperature (Newtonian cooling). Then the surface heat flux is

$$j_A = -h_T \cdot (T_S - T(t)). \tag{5.6}$$

j_A is a heat current density, measured in W/(m^2). Here we have introduced a so-called heat transfer coefficient h_T, which has to be determined experimentally (or theoretical models have to be developed). A few examples relevant in this context are given in Table 5.1. The total heat transport through the surface is

$$I_O = -h_T(T_S - T(t))A. \tag{5.7}$$

The change in heat content is balanced by the heat extraction at the surface, and we therefore have

$$\begin{aligned} \frac{dH}{dt} &= I_O \\ \rho \cdot c \cdot V \frac{dT}{dt} &= -h_T(T_S - T(t))A. \end{aligned} \tag{5.8}$$

This differential equation of first order can be integrated directly yielding the temperature of the body. Assuming that initially at $t = 0$ the body had a temperature T_0, we get

$$T(t) = T_S - (T_S - T_0) \exp\left(-\frac{h_T A}{\rho_e c \cdot V} t\right). \tag{5.9}$$

The body cools exponentially with time from T_0 to T_S, the temperature of the environment. Essential here are the ratio of the surface to its volume A/V, the heat transfer at the surface h_T and the temperature difference between the material and the cooling environment. We have neglected here the latent heat release and assumed that the material is in its interior always at a constant temperature equal to its surface temperature. This is definitively not true, but the simple derivation made here captures essential issues. If the material to be cooled would be a droplet or bead, A/V is essentially the inverse bead diameter. This means that smaller beads cool faster than larger ones. Much better and realistic models are available, especially for beads or droplets cooling in air or a liquid including the latent heat of melting/solidification. They show that the cooling rate is a complex function of droplet radius and of course the heat transport away from its the surface [303–305]. Especially the heat transfer coefficient h_T is not a constant, but depends on the fluid flow around the sample, possible evaporation (boiling) and film formation and can be influenced by stirring the cooling liquid, etc.

5.2.2 Freezing of a Multicomponent Liquid

In systems with more than a single component, the situation is more complicated. If the pressure is not an important variable, one can express the areas of stability in a so-called phase diagram, with temperature and concentration of a second species as coordinates [107]. In certain regions, a single phase is stable, but in other regions two phases coexist in equilibrium. In addition, these diagrams always show that generally

Table 5.1 Some heat transfer coefficients.

Cooling media	Heat transfer coefficient h_T in W/(m^2 K)
Still air	30
Slightly convecting air	40
Water	100–600
Liquid nitrogen	5–300

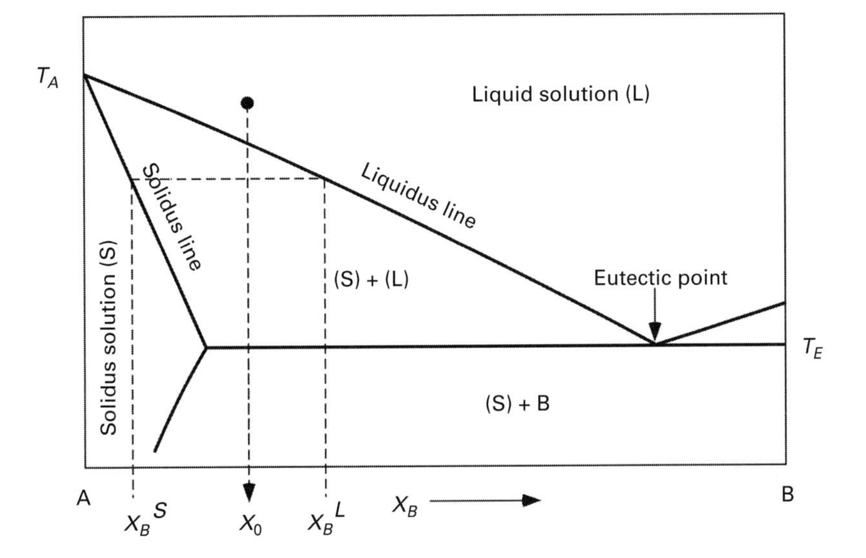

Figure 5.9 Sketch of a two-component phase diagram of a solvent (A) in which another component (B) is dissolved, e.g., A = water, B = ethanol, or A = water and B = NaCl. The straight line from the melting point of A, T_A, down to the eutectic point is called the liquidus line. The straight line from T_A to the end of the eutectic line is called the solidus line, because below is the solid solution phase field denoted as (S). Between both lines, a two-phase regime liquid and solid exists, (S) + (L). In some cases, the solidus line is degenerated and the solid solubility of B in A is zero.

the solid state dissolves a smaller amount of second species than does the liquid state. The existence of two phase regions and the difference in solubility between solid and liquid state is of utmost importance in solidification, since this is the basis of the large variety of microstructures that evolve from the liquid state in passing such an area on cooling. Figure 5.9 sketches a two-component phase diagram (like it is found, for instance, in water–ethanol or water–methanol). There are two important lines: above the so-called liquidus, line both components A and B mix completely. Component A can dissolve a small concentration of B even in the solid state (this is, for instance, not the case in water–ethanol). This defines the region of solid solution (S), which is bounded by the solidus line towards the two-phase solid–liquid region, denoted as (S) + (L). Below the eutectic temperature T_E the solid solution (S) and the crystals of pure B form a two phase solid region. Take a solution with a concentration x_0 of B molecules in A and follow what happens if the temperature of the solution would be lowered. Once the temperature reaches the liquidus line a small amount of solid with a concentration given by the solidus line appears. On further cooling the amount of solid increases and the amount of liquid decreases fixed by the lever rule. The whole system is in the two-phase solid–liquid region. Due to the different solubility of B in the liquid and the solid the liquid enriches in B, following the liquidus line, in as much as the temperature decreases. At the eutectic temperature T_E the residual liquid freezes as two solid phases, B and (S). Drawn in is for a certain temperature a so-called tie line, defined by two compositions: the solid phase has the concentration x_B^S and the liquid x_B^L of B molecules. The fraction of solid and liquid phases can be calculated as $f_S = (x_B^L - X_0)/(x_B^L - x_B^S)$ and $f_L = (X_0 - x_B^S)/(x_B^L - x_B^S)$. Both are temperature dependent.

Whereas in the case of a single component liquid essentially the heat transport determines the crystallisation, in multicomponent liquids mass transport is generally more important. The ratio of heat and mass diffusivity is typically of the order 10 to 1000. The externally imposed temperature gradient during cooling from the surface of a gel inwards of the gel body leads to a solidification of the multicomponent solution, mainly composed of alcohols, water and of course monomers and oligomers of different degrees of polymerisation. There might be additional trace impurities such as salts, especially in biopolymeric gels. At the growing interface between the liquid and the solid, all components in the solvent which cannot be incorporated into the liquid (as dictated by the phase diagram) are rejected into the liquid. They have to diffuse into the liquid. In order to explain the fundamental processes in solidification of multicomponent melts (solutions), we have to make a few definitions and clarify some technical terms.

In the most simple case, the interface between the liquid and the growing solid is flat. This interface is, however, unstable to perturbations. In solidification research, it is well known that planar interfaces develop first cells, mostly of a hexagonal arrangement, expanding from the liquidus temperature in the phase diagram to the solidus temperature. The origin is the so-called constitutional supercooling. To understand the basic idea of constitutional supercooling, consider a binary liquid (A with a minority component B) and let it solidify at a constant velocity (see the vertical line in Figure 5.9 at composition x_0). The difference in solubility between solid and liquid induces a pile up of solute at the solid–liquid interface. Whereas the advancing solidification front rejects continuously species (B) into the liquid, the liquid tries to equilibrate the concentration by diffusion of molecules. This gives rise to an exponential decay of the solute concentration ahead of the interface (the decay length is roughly given by D/v, with D the diffusion coefficient and v the solidification velocity). The concentration profile can be converted into a virtual temperature profile indicating a position ahead of the interface as if the concentration there would be in equilibrium. The real temperature is, however, determined by the heat transport, which is orders of magnitude faster than species transport. The difference between the real temperature and the virtual, equilibrium phase diagram temperature defines a region in which two conditions can prevail: the real temperature always is much larger than the virtual temperature or vice versa.

The first case means whenever a perturbation grows in advance of the interface, it sees a region which is hotter and thus melts back. If the virtual temperature is larger than the real one, a perturbation sees a region which is in reality colder, and thus it can grow [299, 301], leading to cells. These cells, looking like cigars [306, 307], are also unstable to perturbations; they develop further branches and are then called dendrites. The concept of constitutional supercooling only states when a front becomes unstable; it does not predict the most stable growth mode, its shape and branching.

This problem was solved later by Mullins and Sekerka [308], who introduced another important aspect: the curvature of a perturbation. The melting point of a planar front differs from that of a curved interface. Convex interfaces have a lower melting point than concave-shaped ones. This means that if a perturbation is highly curved, it melts back easier than would a shallow, curved interface. Capillarity therefore tries to avoid sharp and highly curved interfaces. This is counteracted by diffusion, or more generally the transport, of heat or solute ahead of a curved, wavy interface. A perturbation induces a perturbed solute or thermal field and either heat or solute (or both) have to be transported away from the tips by diffusion. The longer the wavelength of the perturbation, the larger the diffusion distance to equilibrate the concentration profile. The competition between transport and curvature determines the wavelength that is selected.

The stability analyses performed in the last decades show that at a given temperature gradient, a low solidification velocity will yield a planar interface, which becomes cellular at higher velocities. These cells become unstable and develop into dendrites. At even larger velocities, a planar solidification front can be achieved again. The region between the fully liquid state and the fully solid state is called a mushy zone, schematically shown with a section of a simplified phase diagram in Figure 5.10.

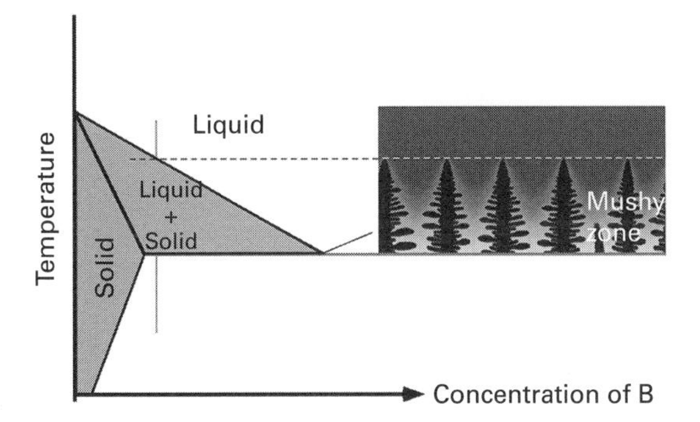

Figure 5.10 Schematic of a mushy zone as existing during solidification of a binary solution. The solid pattern evolves at a liquidus temperature with a solid, whose concentration is given by that of the solidus line at that temperature. The amount of solid phase increases inasmuch as the temperature decreases towards, for instance, the eutectic temperature.

The cells and dendrites have shapes and spacing determined by the temperature gradient and the solidification velocity or the cooling rate. The larger the cooling rate, the smaller the size of the dendrites and their spacing. The typical size range achieved in solidification of a multicomponent crystal from a multi-component liquid is 1 μm–1 mm. This is important to know for the usage of freezing as the first step to produce porous bodies from gels. Many models have been developed to predict the spacing of the primary dendrites and their secondary arms (see [299–301]). Without going into detail, which does not make sense, since for most solutions used in sol–gel processing the phase diagrams are not known, one can state that all models yield the following two relations for the primary dendrite stem or the cell spacing λ_1:

$$\lambda_1 \propto v^{-1/4} G^{-1/2} \quad \text{or more generally} \quad \lambda_1 \propto v^{-a} G^{-b} \tag{5.10}$$

with G the temperature gradient ahead of the solid–liquid interface and v the solidification velocity. a, b are adjustable parameters which are ideally the ones given in the left equation, but experimentally one finds also other values. The side arm or secondary dendrite arm spacing λ_2 always follows a relation

$$\lambda_2 \propto t_f^{1/3} \propto \dot{T}^{-1/3}, \tag{5.11}$$

in which $t_f = \Delta T_s / vG$ is the time to complete solidification, which is related to the solidification interval, $\Delta T_s = T_L(x_0) - T_s(x_0)$ and the cooling rate $\dot{T}$. The exponent 1/3 is characteristic if fluid flow in the liquid is absent, which definitively will be the case in solidification of the liquid in a nanostructured gel phase.

Figure 5.11 sketches relations between cooling rate $dT/dt = G \times v$ and the ratio G/v and observable microstructures. Such diagrams go back to Kurz and Fisher [301]. The cooling rate determines the fineness or the scale of the microstructure whereas the

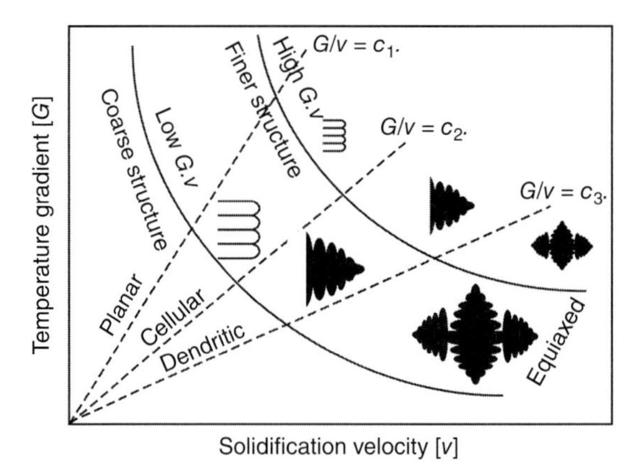

Figure 5.11 Sketch of solidification/crystallisation microstructures of solutions. The cooling rate $dT/dt = G \times v$ determines the fineness or the scale of the microstructure whereas the ratio G/v determines the type of structure. At high G/v, the solidification front can be planar, at lower G/v cells develop, at even lower ratios dendrites appear and at very low G/v the dendrites are equaixed, filling the whole volume of a sample.

ratio G/v determines the type of structure. At high G/v, the solidification front can be planar, at lower G/v cells develop, at even lower ratios dendrites appear and at very low G/v the dendrites are equaixed, filling the whole volume of a sample.

5.2.3 Freezing of a Gel Liquid

In gels, the solid network of nanoparticles modifies the solidification or crystallisation. How does freezing, crystallisation or solidification occur in thin pores of a few to a few hundred nanometre diameters? Does freezing change the solid nanonetwork? Gross et al. [309] studied the melting and solidification of succinonitrile (SCN) in silica aerogels prepared by Airglass^{TM}. They found that substantial supercooling below the freezing point of SCN up to 10 K is necessary to initiate solidification. Their analysis is given in detail in Chapter 9. For the current context, it is sufficient to understand that the supercooling ΔT below the equilibrium melting point needed to initiate crystallisation in pores depends on the pore size r_p like

$$\Delta T = \frac{-2\sigma_{SL} \cos\theta}{r_p \Delta S_m}, \tag{5.12}$$

with $\Delta S_m = \Delta H_M / T_m$ as the entropy of melting, θ the wetting angle between the liquid and the pore wall and σ_{SL} the interface tension between solid and liquid. This relation means that at a certain undercooling ΔT below the equilibrium freezing point, the liquid is solidified in all pores larger than a certain critical radius r_c. In all pores which are smaller, the liquid still prevails.

Adopting the preceding sketched ideas to freeze drying simply means that a huge temperature difference between the wet gel and the cooling liquid should be used. Small particles of the wet gel (beads) are preferable and will lead to a faster cooling. Slow cooling in a bath which is only slightly cooler than the wet gel (room temperature to water-ice temperature) is probably a bad idea. One must, however, take into account further issues with lyophilisation.

The first-order phase transition from liquid to solid is usually also accompanied by a change in specific volume. Water has at standard temperature and pressure (STP) conditions at 0°C at specific volume of 0.001 m^3/kg whereas ice has a specific volume of 0.00109 m^3/kg. They differ by around 10%. This jump in specific volume will induce a pressure on the solid network. Since solidification does not start everywhere in a gel sample homogeneously but starts in the largest pores first, while the smaller ones are still filled with liquid, a huge deformation of the pore network and thus the solid network is likely to occur. Therefore, freeze drying is prone to a modification of the solid nanonetwork. Another issue, quite often overlooked in the literature on freeze dried aerogels, is the complex issue of a process, called in the solidification literature *pushing and engulfment* of foreign particles [310]. This means that depending on the solidification front speed, foreign particles (almost independent of their chemical nature) are either incorporated into the growing solid front or are pushed ahead of it and are then enriched in the residual liquid, as shown schematically in Figure 5.12. This process was studied experimentally and theoretically many times in the last six

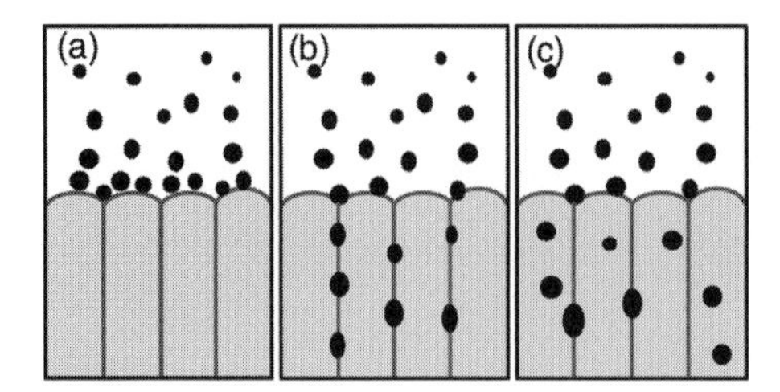

Figure 5.12 Three possible modes in which a solidification or crystallisation front can interact with particles in the solvent. These particles can be, for instance, oligomers, polymeric particles not attached to a cluster or loose fibrils: Panel (a) shows that the solidification front moves them into the liquid, (b) shows that they are entrapped at cell boundaries and (c) shows that they can be engulfed into the growing solid.

decades, since for metals it is essential, if one would like to make metal ceramic-particle composites, and of course it is essential if one would like to conserve, for instance, living cells in a water-based solution by freezing the solution.

Although there exist many models to describe this effect, the general rule is that there exists a critical solidification velocity v_{cr} below which particles are pushed and above which they are entrapped into the moving solidification front. The critical velocity is generally inversely proportional to a function of the particle diameter. In many models, this is a hyperbolic dependence. The smaller the particles are, the larger the solidification front velocity must be to engulf them. Since in wet gels these particles might be oligomers or polymeric particles or polymeric fibrils and their size is in the nanometre range, the solidification front velocity must be very high. The solidification speed v is proportional to the cooling rate dT/dt and inversely proportional to the temperature gradient G. A high speed to entrap particles in the solvent therefore asks for a high cooling rate. For wet gels, this means that the freezing of a solvent mixture can push the solid network ahead of the growing ice front and considerably deform the solid network. Such a behaviour is quite often observed in freeze drying of biopolymeric gels.

An example was recently given by Saelices and co-workers [311]. They found as one interesting result that freeze drying of monolithic gels leads to 2D structures (in full agreement with the aforementioned statements) compared to freeze drying of droplets, which revealed a fibrillar 3D structure in the cellulose nanofibre gels. A wonderful study on the 3D structure of aerogels prepared from a polysaccharide, namely guar galactomannan and the tamarind seed xyloglucan cross-linked with galactose oxidase forming hydrogels and transformed to aerogels by conventional and unidirectional freeze drying, was performed by Ghafar and co-workers using the synchroton radiation facility at Grenoble (ESRF) to make X-ray microtomographs [312]. The materials obtained were macroporous with pore sizes up to 500 μm having a maximum around 100 μm as to be expected from solidification of water. The freezing was performed either at $-70°$C after the hydrogels were injected into capillaries of 0.8 mm inner diameter. For directional solidification, the hydrogels were cast into petri dishes and cooled from one side with liquid nitrogen. Freeze drying was then

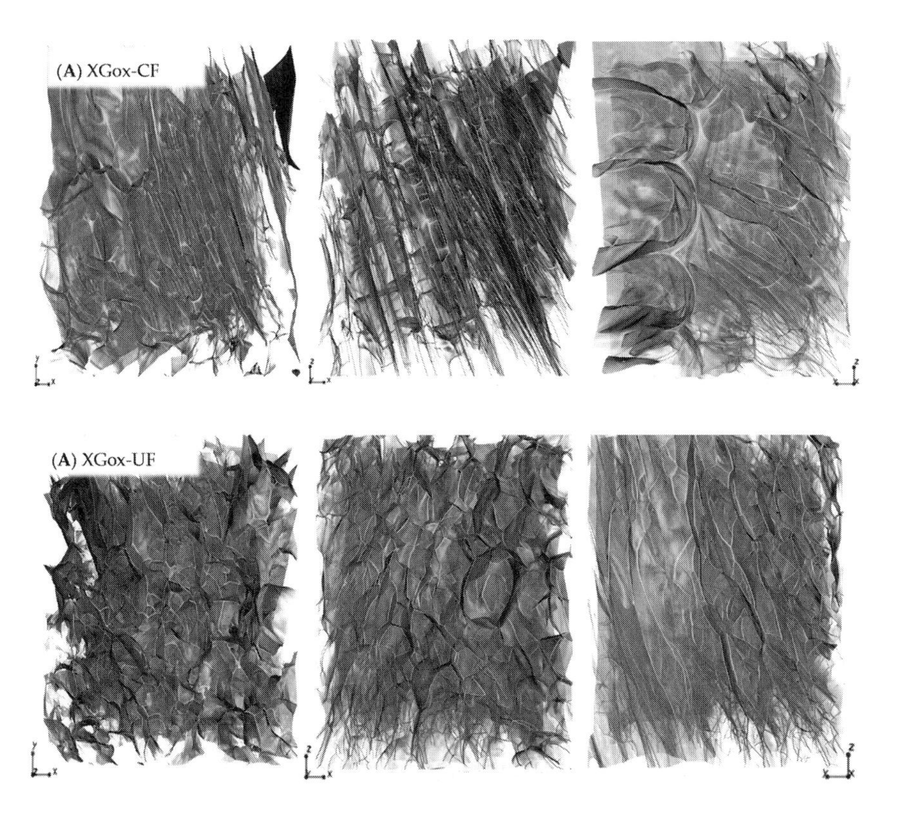

Figure 5.13 Three-dimensional reconstructed images from synchrotron phase contrast microtomography of cross-linked galactomannan (XGox) aerogel. The upper row shows the microstructure of a conventional freeze dried aerogel and the lower one a hydrogel solidified from one side. The size of the box is around 500 μm^3. Reprinted under Creative Commons licence from [312].

performed after freezing at a pressure of 100 Pa and at $-20°C$. From the interesting pictures as well as the videos that their study produced, we present here one example in Figure 5.13 illustrating in our point of view the issues discussed concerning the microstructures formed during solidification and the interaction of the solidifying water with the gel network. Figure 5.13 shows that the polysaccharides were obviously entrapped into the cellular network of the growing ice crystals. In that sense, the structure is templated by the microstructure of the ice. The size of the structures also confirm the previous statement that cells and dendrites will typically have a diameter in the few ten to hundred micrometre range and mesoporous materials are hard to prepare by this method.

Summarising, if freeze drying is taken as a method to produce from wet gels dry ones (if they can be called aerogels is a matter of taste; see Chapter 1), it definitively is essential to control the freezing process. Simply putting a wet gel into a freeze dryer and working as if the wet gel would be a conventional material, like foods or seeds, is not a good idea. A better idea is to perform a directional freezing with a well-defined shape of the wet gel, as schematically shown in Figure 5.14, and well-defined

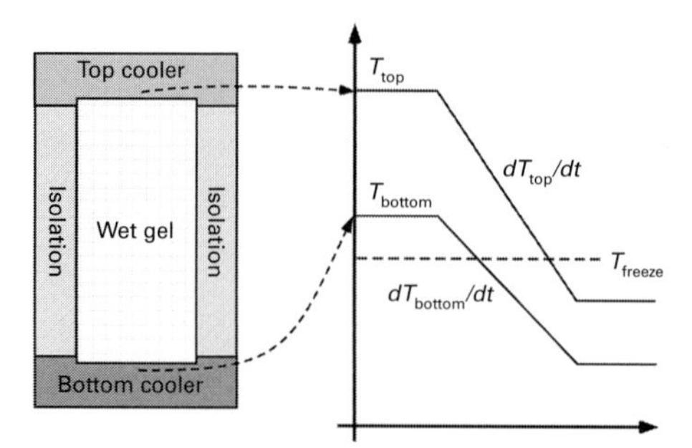

Figure 5.14 Uniaxial directional freezing of a wet gel. Heat is extracted at the bottom and lateral or radial heat transport should be reduced as much as possible with a good isolation. At best, the top and bottom are cooled with a slightly different cooling rate to guarantee a constant solidification velocity and a constant temperature gradient at the freezing front.

thermal boundary conditions. In the schematic arrangement shown in Figure 5.14, heat is extracted at the bottom, and lateral or radial heat transport should be reduced as much as possible with a good isolation (for instance, a silica aerogel [313]). At best, the top and bottom are cooled with a slightly different cooling rate to guarantee a constant solidification velocity and a constant temperature gradient at the freezing front. Assume, for instance, that gelation was done at a temperature above room temperature, say 50°C, and then the wet gel is cooled to room temperature and put onto a shelf in a freeze dryer. In order to induce freezing, first a temperature gradient must be established and therefore the bottom temperature must be decreased by a suitable amount. After establishing the temperature difference between the top and bottom, both temperatures can be reduced with different cooling rates $dT_{\text{top}}/dt, dT_{\text{bottom}}/dt$.

The freezing front will then move from bottom to top with a constant velocity, which can be understood with the following considerations. At the moving solid–liquid interface, the Stefan condition

$$\Delta H_V \frac{dy}{dt} = \lambda_S \left.\frac{\partial T^S}{\partial x}\right|_y - \lambda_L \left.\frac{\partial T^L}{\partial x}\right|_y \tag{5.13}$$

must hold. This condition simply states that in any movement of the solidification front, the heat liberated at the solid–liquid interface must by transported away by the heat flux defined by the temperature gradient in the solid and liquid. Here λ_L, λ_S are the thermal conductivities in the wet gel and the frozen gel, ΔH_V is the enthalpy of melting per volume and $T^{L,S}$ are the temperature fields in the fluid and the frozen part of the gel respectively. For the ease of calculation, we will assume that both gradients in the solid and liquid can be approximated as

$$\left.\frac{\partial T^L}{\partial x}\right|_y = \frac{T_{top} - T_m}{L - y} \tag{5.14}$$

$$\left.\frac{\partial T^S}{\partial x}\right|_y = \frac{T_m - T_{bot}}{y}. \tag{5.15}$$

In these equations, the temperatures T_{top}, T_m, T_{bot} are well defined (see also Figure 5.14); the length L is somewhat arbitrary and must not necessarily identified with the length of the wet gel, since in principle the temperature profile during freezing will be approximately linear in the solid part but in the liquid part only linear close to the front and will level off to a constant value farther off from the solid–liquid interface. In order to solve Eqs. (5.13–5.15), we make the following assumption:

$$y(t) = y_0 + vt, \tag{5.16}$$

in which the freezing front moves at a constant speed v starting from a position y_0. For the cooling rates of the top and bottom coolers, we assume that the following relations hold:

$$T_{top}(t) = T_{top0} - \dot{T}_{top}t \tag{5.17}$$

$$T_{bot}(t) = T_{bot0} - \dot{T}_{bot}t, \tag{5.18}$$

with $\dot{T}_{top} \neq \dot{T}_{bot}$ but both being constant. If we require the gradient in the liquid to be a constant, we would have

$$G_L = \frac{T_{top}(t) - T_m}{L - y(t)} = G_{L0}. \tag{5.19}$$

Eqs. (5.16) and (5.19) lead to

$$T_{top0} + G_{L0}(y_0 - L) - T_m = (\dot{T}_{top} - G_{L0}v)t. \tag{5.20}$$

The left side is time independent. One solution ensuring the right side being always time independent is

$$\dot{T}_{top} = G_{L0}v. \tag{5.21}$$

This fixes the cooling rate of the top cooler. The cooling rate of the bottom cooler is obtained by inserting Eqs. (5.16–5.19, 5.21) into the Stefan condition, Eq. (5.13). After some algebraic manipulations, we obtain for the cooling rate of the bottom cooler

$$\dot{T}_{bot} = \frac{\Delta H_V}{\lambda_S}v^2 + \alpha G_{L0}v. \tag{5.22}$$

Here α is defined as

$$\alpha = \frac{\lambda_L}{\lambda_S}. \tag{5.23}$$

The necessity to also extract at the bottom the heat of melting requires that both cooling rates are different. If one only cools at the bottom, the freezing front moves initially with a certain speed into the sample and slows down continuously, since the latent heat evolving at the interface heats up the wet gel in front of it, and it might be possible that at the top freezing might occur before the front from the bottom has reached the top.

5.2.4 Sublimation

Once the solvent in wet gels is fully frozen, the process of sublimation can start. The sublimation line in the phase diagram can be described as [109]

$$p_{sub}(T) = p_s \exp\left(-\frac{\Delta H_{sub}}{R}\left(\frac{1}{T} - \frac{1}{T_s}\right)\right) = p^* \exp\left(-\frac{\Delta H_{sub}}{RT}\right) \tag{5.24}$$

with the obvious abbreviation

$$p^* = p_s \exp\left(\frac{\Delta H_{sub}}{RT_s}\right), \tag{5.25}$$

with ΔH_{sub} as the heat of sublimation and p_s a reference vapour pressure at the reference temperature T_s (for instance, the triple point). The heat of sublimation can be calculated from tabulated values for the heat of melting ΔH_m and the heat of evaporation ΔH_{vap}, since $\Delta H_{sub} = \Delta H_m + \Delta H_{vap}$. For many substances used in aerogel processing, the p, T phase diagrams and the functional dependence of the sublimation line on temperature are known (water, ethanol, methanol and mixtures of them). The sublimation pressure defines the rate of atoms or molecules leaving the solid, frozen surface. Kinetic theory of gases derives for the sublimation rate the expression

$$Z(T) = \frac{p_{sub}(T) - p_A}{\sqrt{2\pi M_m R_g T}}, \tag{5.26}$$

with M_m the molar weight and p_A the partial pressure of the subliming species in the space above the sample. The sublimation rate is measured in moles per second and area. The total mass of molecules leaving the surface of the porous body per unit area and time having a porosity of ϕ_p is then

$$\frac{dm_{sub}}{dt} = \frac{p_{sub}(T) - p_A}{\sqrt{2\pi M_m RT}} \phi_p M_m, \tag{5.27}$$

This equation shows two things: first, if the right side is constant throughout the sublimation process, the mass loss is proportional to time; and second, that it is essential for a fast drying process to work with a high temperature, best close to the triple point. This equation, however, overestimates the sublimation from ice inside a frozen gel body. Inasmuch as ice at the surface of the body has sublimated into the low-pressure environment of a freeze dryer, the diffusion through the pore space becomes more and more important and determines the time needed to dry a sample. One has to take into account that the ice–gas interface moves into the body. Therefore, the problem of sublimation is that of gas transport through the mesopores of an aerogel with a moving interface. The characteristic time is inversely proportional to the Knudsen diffusion constant (see Chapter 10 for more details). Let us make a comment: sublimation of ice from a porous ice–mineral body into vacuum is the essential process that makes the wonderful million kilometres long tail of comets visible, because mineral aggregates are taken away from comet surfaces from the sublimating gas with supersonic speed [314–317].

5.3 Supercritical Drying

5.3.1 Supercritical Fluids

As we ascertained in Sections 5.1 and 5.2, the coexistence of the liquid and vapour phase inevitably causes capillary forces, which collapse the gel matrix. To eliminate the capillary forces, let us first examine the phase diagram of the solvent present in the gel again with the example of the $p - T$ diagram of carbon dioxide. In the $p - T$ diagram for a single component – the solvent in our case – the liquid–vapour equilibrium ends at a point called the critical point; see Figure 5.15. The coordinates of the critical point (p_c, T_c) are tabulated for various compounds. To grasp qualitatively how the supercritical state is related to the gas and liquid states, let us perform a thought experiment [318]. Starting in the gas phase with no liquid (point A in Figure 5.15), we rise temperature to reach the point A' and then compress the sample to the final state F. On the way from A to F, no liquid phase appeared because we did not cross any line in the diagram. The final state would appear to be all gas. Alternatively, starting from the liquid phase (point B) and first compressing the sample to reach point B', we then warm up to the same state F; no phase change has occurred, and the sample would appear to be all liquid.

We see that there is no sharp boundary between the liquid and gas states: supercritical fluids share properties of both gases and liquids. The density of the supercritical fluid phase spans a wide range with values typical at ambient conditions for a gas or conventional organic solvents (>0.6 g/cm^3). As an example, Figure 5.16 shows lines of constant density (isopycnic lines) for carbon dioxide. Other properties of supercritical fluids may tend to be close to one of the states or take intermediate values. For instance, dynamic viscosity of supercritical CO_2 is of the same order of magnitude of the gaseous CO_2 ($\sim 10^{-5}$ Ns/m^2), whereas the diffusion coefficient (10^{-7} m^2/s) is

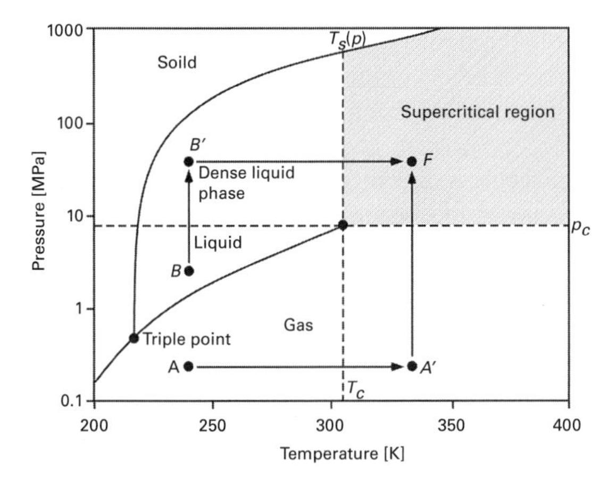

Figure 5.15 Section of the $p - T$-phase diagram of carbon dioxide.

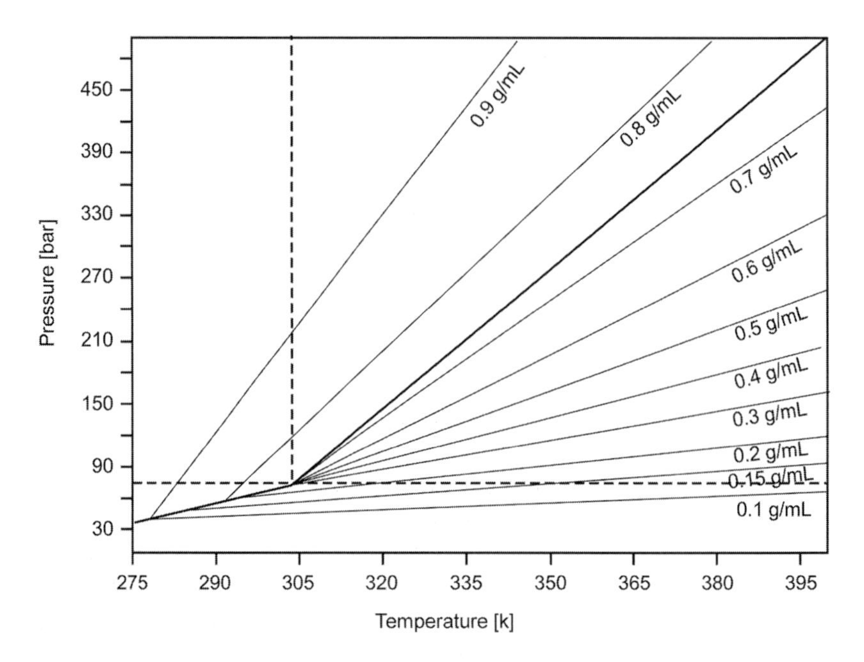

Figure 5.16 Lines of constant density for carbon dioxide. Density of CO_2 at the critical point is 0.467 g cm^{-3}. Reprinted from [319] with permission from Elsevier

intermediate between that of typical values for gases and liquids, 10^{-5} and 10^{-9} m^2/s, respectively.

The existence of the critical point is a manifestation of intermolecular forces. In fact, any gas of interacting molecules with finite volume should exhibit a critical point. In the gaseous state, the kinetic energy of molecules ε_k dominates over the potential energy of intermolecular interactions u, whereas in the liquid state the attractive forces keep molecules close to each other, constraining their molecular movement. The physical state of a substance can be characterised by the balance between the kinetic and potential energy [320] and can be expressed as the ratio of both, as shown in Eq. (5.28):

$$\xi = \frac{\varepsilon_k}{|u|}. \tag{5.28}$$

The kinetic energy of a molecule is directly proportional to the temperature, as shown in Eq. (5.29):

$$\varepsilon_k = \frac{3}{2} k_B T. \tag{5.29}$$

The analytical expression for the potential energy u may take quite complex forms depending on the molecular structure of the gas molecules. For the present discussion, we make a general assumption that the potential energy is a function of the distance between a pair of molecules, r_{ij}. The function u goes to infinity as r_{ij} approaches zero (repulsion at small distances) and tends to zero as r_{ij} infinity (no interaction at large distances). At an intermediate value of r_{ij}, the potential energy goes through a minimum.

Now let us relate the average distance between any two molecules, $\langle r \rangle$, to the gas pressure p. Using the virial expansion for a gas with N molecules in the volume V leads to Eq. (5.30):

$$\frac{p}{k_B T} = \frac{N}{V} + B_2(T)\left(\frac{N}{V}\right)^2, \tag{5.30}$$

with $B_2(T)$ a factor being only temperature dependent (in principle, this is a treatment of a non-ideal gas going back to Van der Waals). In an ideal gas, $B_2(T)$ would be zero and Eq. (5.30) would be the classical ideal gas law relating pressure temperature and the number density of molecules. The fact that the average intermolecular distance $\langle r \rangle$ is proportional to $\left(\frac{N}{V}\right)^{-1/3}$ and neglecting all higher-order terms, we obtain at a fixed temperature

$$p \sim \frac{1}{\langle r \rangle^3}. \tag{5.31}$$

Assume that the interaction between the molecules in a non-ideal gas can be described by a simple (6,12) Lennard–Jones potential $u(r) = A/r^{12} - B/r^6$. Then we can use Eq. (5.31) to express the potential energy u not as a function of intermolecular distance as it is usually done, but as a function of pressure. This yields Figure 5.17, in which the intermolecular potential is shown as a function of pressure and compared to the kinetic energy ε_k, which depends linearly on temperature. The figure can be read as follows: at a temperature T_1, the balance between the kinetic and potential energy is achieved at a certain pressure p_1, i.e., $\xi = 1$. At this temperature, we can always condense the gas by increasing the pressure such that ξ becomes less than unity. There is, however, a

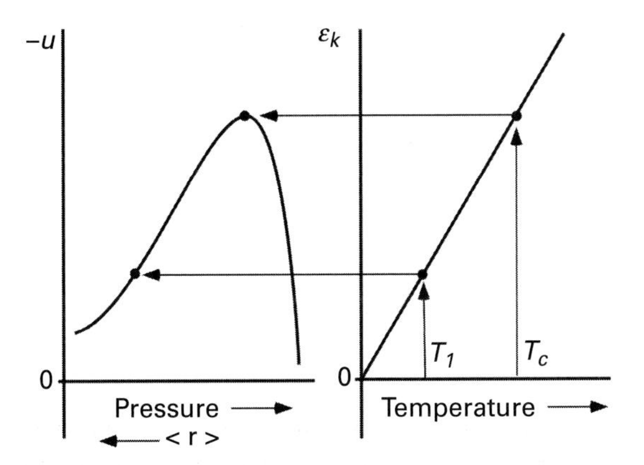

Figure 5.17 Relationship of the critical temperature T_c to the intermolecular potential energy u expressed as a function of pressure (left panel) and in the right panel the kinetic energy as a function of temperature. The maximum of the potential defines the critical temperature T_c.

Table 5.2 Critical points for several organic solvents.

Solvent	T_c, °C	p_c, MPa
Diethyl ether	194	3.6
Methyl tert-butyl ether	224	3.4
Methanol	239	8.1
Ethanol	241	6.4
Isopropanol	235	4.8
Acetone	235	4.8
Methyl ethyl ketone	262	4.2
Water	374	22.1
Nitrous oxide (N_2O)	36.4	7.2
Carbon dioxide (CO_2)	31.1	7.4

temperature T_c above which it is no longer possible to keep $\xi < 1$, because at $T > T_c$ the absolute value of the potential energy u can only decrease (which corresponds to repulsion). This is how the critical temperature T_c emerges in such a simple qualitative model for a gas of interacting molecules.

This model further suggests a relationship between the molar potential energy of a liquid and its critical temperature. Indeed, taking the vaporisation enthalpy ΔH_{vap} (far below T_c) as an estimate for the molar potential energy, for one mole of a gas at T_c we have

$$\frac{3/2 R_g T_c}{\Delta H_{vap}} \lesssim 1. \quad (5.32)$$

The condition given by Eq. (5.32) is satisfied for all known fluids, but numerical values of the quotient on the left-hand side differ from unity being around 0.3 for simple fluids (with a close-to-ideal behaviour), 0.20 ± 0.02 for common aprotic organic solvents and 0.15 ± 0.03 for protic ones (Kopp–Lang rule). Further empirical correlations between ΔH_{vap} and boiling temperature [321] allow for a rough estimate of the critical temperature. Reliable experimental data for critical temperature and critical pressure for those solvents that usually come into question in the aerogel processing are reported in the literature. A brief excerpt is given in Table 5.2. A number of group-, bond- and atom-contribution methods are available for rare situations when T_c and p_c are unknown. In particular, the method of Joback is quite reliable for estimating critical properties if the normal boiling point of the solvent is known [322]. The estimation methods are also useful for thermodynamic calculations of phase behaviour of multicomponent mixtures.

Coming back to the elimination of capillary forces that arise in microscopic pores filled with a solvent, it is now evident that the liquid and vapour phases no longer coexist at pressures above p_c and temperatures above T_c. A natural idea emerging from the phase diagram in Figure 5.15 is to make use of this fact and to transform the intra-gel solvent into a supercritical fluid without letting the solvent evaporate from the gel pores. This process is often referred to as high-temperature supercritical drying and will be discussed in the next section. Another approach goes beyond the phase behaviour of a pure solvent and relies on extraction of the solvent with supercritical carbon dioxide (so-called low-temperature supercritical drying; see Section 5.4). In the latter case, phase behaviour and critical properties for solvent/CO_2 mixtures are to be considered.

In the following sections, we discuss both approaches. Regardless of the drying method, the core idea remains the same: solvent in the gel body has to be transferred into the state where no phase boundary exists and thus the capillary forces cannot originate.

5.3.2 High-Temperature Supercritical Drying

Samuel Kistler was the first who in the beginning of 1930s realised the essential importance of the supercritical state for the elimination of capillary forces [12, 13]. His brilliant idea was to heat up the gel with an organic solvent in its pores, covered with an excess of the same organic solvent above the solvent T_c in a robust stainless steel vessel (Figure 5.18). In this process, the liquid and vapour phases coexist all the way up to the critical point, but the pores remain filled with the solvent so that there is no evaporation from the pores. During this heating process, system pressure in the vessel changes along the liquid–vapour line of the phase diagram, shown in Figure 5.15. Once temperature exceeds T_c, the system pressure exceeds p_c and the phase boundary disappears. At the same time, the surface tension drops and becomes zero at the critical temperature [323–325]. Finally, the vessel is depressurised by opening gently the outlet valve (Figure 5.18) so that cooling due to the Joule expansion can be compensated (path $F \rightarrow B'$ in Figure 5.15). In this case, expansion from the supercritical fluid to the gas is close to isothermal and does not lead to the formation of a liquid phase.

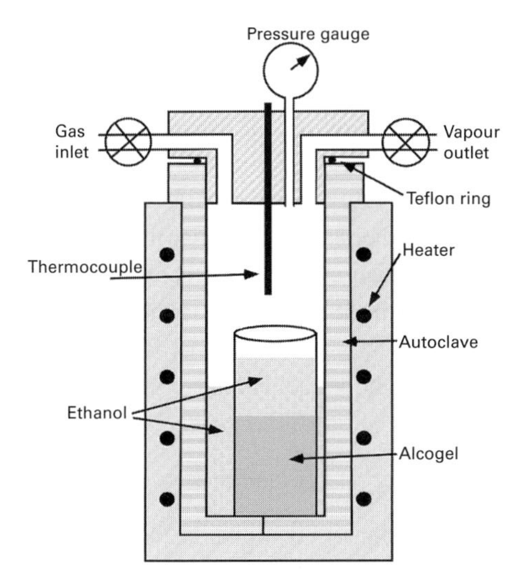

Figure 5.18 Schematic of an autoclave for high-temperature supercritical drying [326].

The fluid is therefore released from the gel pores only at the very last step, at isothermal expansion, and leaves behind the dry skeleton of the initially wet gel – the aerogel. Because the critical temperature is relatively high for most of the organic solvents (Table 5.2), this process is called high-temperature supercritical drying (HTSCD, or hot supercritical drying for short). In a typical experimental protocol, a pre-formed and aged gel is placed in the high-pressure autoclave in a glass vial. The remaining volume of the autoclave is filled with solvent. Upon heating at a rate of 20–100 K/h, the solvent excess evaporates into the gaseous phase and builds up the pressure, thereby precluding evaporative drying from the gel. The autoclave is kept sealed until the temperature 10–20 K above the T_c is reached. The heating rate is sometimes reduced when approaching the critical point to prevent cracking. The fluid is relieved through a valve (0.03–0.1 MPa/min) while the temperature is held constant at $(1.1–1.2)T_c$. As soon as atmospheric pressure is reached, the heating is turned off. The autoclave can be evacuated or purged with a dry gas to remove remaining solvent. The overall process duration is between 3–30 h depending on the gel size and its density: monoliths of low target density should be processed slowly to minimise the risk of cracking.

In the publication appeared in the *Journal of Physical Chemistry* in 1932 [13], Kistler observed that the critical temperature of water is too high for gels to withstand: the gel is peptised, losing the monolithic structure as temperature approaches T_c. The key of Kistler's idea was to replace the original solvent, in which the gel was synthesised, by a possibly inert solvent with a sufficiently low T_c. For example, water (T_c = 647 K) can be exchanged with ethanol (T_c = 514 K) and then further with diethyl ether (T_c = 467 K) and even further with liquid propane (T_c = 370 K). Importantly, each preceding solvent in this row is completely miscible with the following one allowing for complete solvent exchange. Since then, the idea of the solvent exchange is widely used in aerogel processing (see Section 5.4).

HTSCD is mostly utilised for drying of oxide gels. When applied to silica alcogels, the high-temperature drying leads to esterification reaction with silanol groups on the gel surface according to Eq. (5.33):

$$\equiv \mathrm{Si-OH} + \mathrm{R-OH} \rightleftarrows \equiv \mathrm{Si-OR} + \mathrm{H_2O} \tag{5.33}$$

Due to esterification, hydrophobic silica aerogels are obtained in HTSCD with alcohols. Surface hydroxyl groups of other oxides such as alumina, titania and zirconia are less prone to a reaction with supercritical alcohols. To recover silanol groups from alkoxy groups, the latter should be oxidised by heating or plasma treatment upon drying.

Supercritical alcohols can also depolymerise siloxane bonds ≡Si–O–Si≡ via Eq. (5.34):

$$\equiv \mathrm{Si-O-Si} \equiv + \mathrm{R-OH} \rightleftarrows \equiv \mathrm{Si-OR} + \equiv \mathrm{Si-OH} \tag{5.34}$$

Excess of the solvent favours both reactions (5.33) and (5.34) that lead to a partial dissolution of the silica backbone with the formation of tetraalkyl silicate $Si(OR)_4$ as the ultimate result of reaction 5.34. Therefore, during drying silica in monomeric form

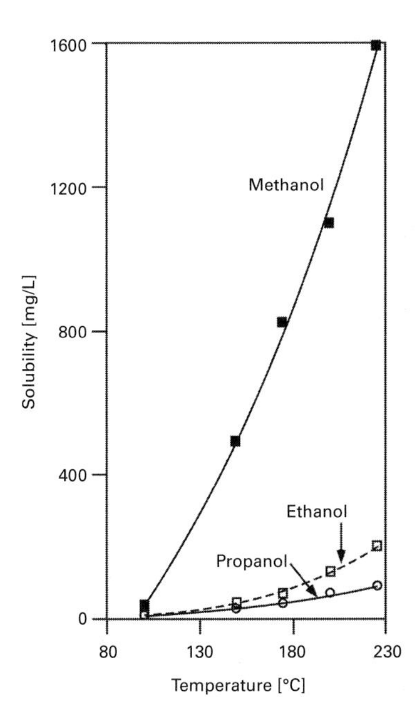

Figure 5.19 Solubility of calcined amorphous silica in methanol, ethanol and propanol. Reprinted from [328] with permission from Elsevier

can be transported through the gel and even washed away from the autoclave. In other words, the HTSCD is a solvothermal treatment.

For given pressure and temperature, the equilibrium solubility of silica shows to what extent the cleavage of siloxane bonds proceeds. The solubility of SiO_2 decreases in solvents with a low dielectric constant ε; see [327]. This is illustrated in Figure 5.19 for three alcohols: methanol ($\varepsilon = 32.7$), ethanol ($\varepsilon = 24.5$) and propanol ($\varepsilon = 20.1$). Wherever solvent for HTSCD can be chosen, aprotic solvents with possibly low dielectric constant should be preferred.

We see that a supercritical fluid might play an active role in the drying process, especially when performed in protic solvents with high dielectric constants. Dissolved species can be carried over from the autoclave or re-deposited within the gel. The latter may affect the primary particles size, pore size distribution and pore connectivity. Furthermore, the solvothermal-induced re-dissolution and re-precipitation may convert gels with amorphous backbone into crystalline aerogels.

Therefore, drying conditions have multiple effects on the structural features at both microscopic (connectivity and size of primary particles, pore size and shape) and macroscopic (shrinkage, cracks, density gradients) scales. Effects of the HTSCD on routinely measured aerogel properties such as skeletal and envelope densities, specific surface and pore volume, pore size distribution, crystallinity as well as mechanical and optical properties have been reported [329–331].

As we noted previously, additional solvent is needed to preclude evaporative drying and to ensure that the liquid phase is present until the critical point is reached. At the same time, the reactions Eq. (5.33) and Eq. (5.34) imply that a solvent excess promotes the gel re-dissolution. To minimise this effect, the autoclave can be pre-pressurised with nitrogen (e.g., to 10 MPa) before heating. During the heating, some amount of the solvent will still evaporate into the gaseous phase and the pressure will further increase. The presence of the inert nitrogen does not significantly affect the critical temperature of the solvent, and the drying can begin once the actual temperature exceeds $(1.1–1.2)T_c$. Additionally to suppressing esterification and alcoholysis, the required amount of extra solvent is also reduced.

In the preceding discussion, we implicitly assumed that all by-products formed at the gelation step are washed out and the intra-gel solvent is an individual compound. Relative rates of the esterification and alcoholysis can be reduced by addition of small quantities of water, aqueous acids or bases. Although such additives increase critical parameters of the mixture and thus force running the drying at high temperature and pressure, they considerably reduce the amount of silicon extracted from the high-pressure autoclave and have an impact on textural properties of the resulting aerogels [332]. An elegant way to produce small quantities of water in situ is by methanol dehydration when zeolite pellets are placed at the bottom of the drying autoclave [333].

Presence of water in the solvent may, however, facilitate unwanted chemical transformations in the gel (e.g., formation of oxyhydroxides, [334]). In this case, an extra solvent exchange is required to extract water from the gel pores which remains after hydrolysis and condensation. When selecting solvent for the solvent exchange, preference should be given to those solvents that are miscible with water.

Time during the heating process can be used to reinforce or functionalise gels. For this, functionalisation agents such as partially hydrolysed or hydrophobic precursors are dissolved in the solvent added to the autoclave. Due to higher rates of hydrolysis and condensation under such conditions, gel surface reacts with the precursor chemically modifying the primary particles [335, 336].

Lightweight alcohols and ketones with the critical pressure in the range 5–8 MPa are usually employed in the HTSCD. In an attempt to reduce working pressure and temperature, diethyl and methyl tert-buthyl ethers were recently employed in the processing of oxide aerogels at milder drying conditions (at 20°C and 1.5 MPa above T_c and p_c for the corresponding ether; see Table 5.2) than typically used in drying with alcohols (260–300°C, 5.0–9.0 MPa). Ethers are aprotic, less reactive solvents with lower critical point and dielectric constant, and thus are expected to have less effect on the gel morphology and surface chemistry. It has indeed been demonstrated that silica, alumina and zirconia aerogels possess on average two times larger specific surface when dried in ethers compared to drying in supercritical ethanol [337]. Furthermore, ether-dried alumina and zirconia aerogels remain amorphous, while drying in ethanol yields crystalline aerogels [337, 338]. These examples again emphasise a rather active role that a supercritical fluid may play in the drying process: it is not just a fluid to be vented out from the gel pores, and can be utilised to preserve existing or impart a new functionality to the emerging aerogel.

We discussed so far application of the HTSCD process for oxidic gels. For these gels, HTSCD is an established practice due to their relatively high chemical stability against solvents in the supercritical state. Very little is reported on HTSCD of organic gels. Instead of hot, supercritical-drying, low-temperature supercritical drying with carbon dioxide is commonly employed for organic polymer (including biopolymer) gels due to high chemical inertness of CO_2 and low process temperature ($<60°C$). Among very few types of organic gels that can withstand the HTSCD process, resorcinol-formaldehyde (RF) gels is the most studied system so far.

In principle, RF aerogels dried via HTSCD possess very similar structural and textural properties as compared to nowadays routine drying with supercritical carbon dioxide. Expectedly, the active role of the supercritical fluid is even more pronounced in this case: depolymerisation products as well as products of the solvent decomposition have been detected in the outflowing stream when dried in acetone [339–341], methanol [342] and ethanol [340].

Decomposition of the solvent makes its downstream purification and recirculation problematic. Another limitation of the HTSCD of organic gels is that the autoclave has to be cooled and only then opened; otherwise, the resulting aerogel may begin to smoulder.

Although the accumulated experience on the topic is only moderate, HTSCD with non-polar aprotic solvents such as petroleum ether [343] seems to be a feasible alternative to conventional fluids such as acetone or ethanol. Discussed earlier, pre-pressurisation with nitrogen seems to be an efficient strategy [344] for controlling the shrinkage degree. Other chemistries such as phenolic-furfural gels [345] may also be more stable in the harsh conditions of HTSCD. It seems, however, that direct drying should be ruled out for delicate biopolymer gels due to their low thermal and hydrolytic stability. Low-temperature drying with carbon dioxide is generally applicable instead (see Section 5.3.4).

5.3.3 Rapid Supercritical Extraction

In attempts to overcome limitations of the previously described, classical HTSCD, namely the long processing time, other approaches have been suggested. They are known under the generic term rapid supercritical extraction (RSCE) that dates back to the work by John Poco, Paul Coronado, Richard Pekala and Lawrence Hrubesh from the Lawrence Livermore National Laboratory (LLNL) [346]. For this reason, their approach is often abbreviated as LLNL RSCE, to distinguish it from later implementations. The LLNL RSCE process can be run in two modes: in a vacuum chamber or in an external high-pressure autoclave. In both modes, gelation and drying are combined in one single step.

In the first mode, a two-piece sealed mould with a small opening in one of them (Figure 5.20) is first fastened with bolts and wrapped with a heating coil. Sol is injected into the mould to fill it completely. The assembled and filled mould is then located in a vacuum chamber for thermal insulation and safety reasons, and ready for drying.

The process begins with a fast heating of the mould (tens of kelvins per minute). While the temperature rises, the sol reacts until it gels, and hydrostatic pressure in the mould builds up very quickly due to expansion of the liquid phase. To avoid

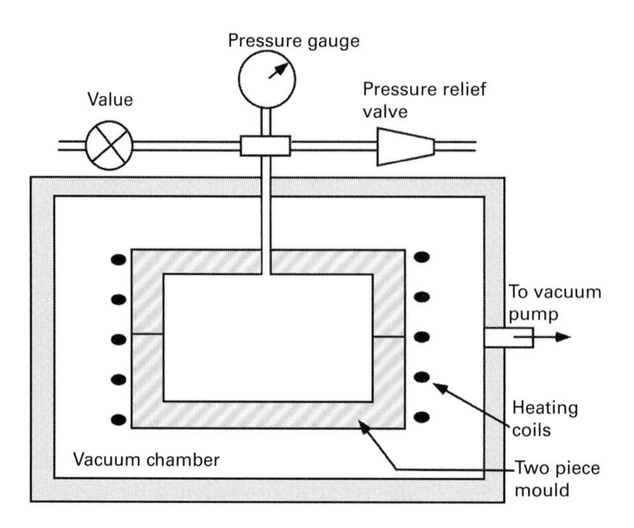

Figure 5.20 Scheme of the equipment for Rapid Supercritical Extraction (RSCE). From [346]. Redrawn from the original figure with permission of Cambridge University Press

overpressure, the mould is connected to a relief valve with a set value a few tens of bar higher than the solvent critical pressure p_c. As soon as pressure in the mould exceeds p_c, the solvent is released through the valve down to atmosphere pressure.

In the second mode, a similar metallic mould is employed (Figure 5.20), but without connecting tube and heating coils. The mould is first filled with the sol, sealed and then placed into a bigger high-pressure autoclave. The remaining volume of the autoclave is filled with solvent, sealed and quickly heated up. This causes a rise of the hydrostatic pressure in the autoclave due to the solvent expansion. During the pressure rise, the gel is formed in the mould. At some point, a significant pressure builds up inside the mould, forcing the fluid to leak through a gap between two pieces of the mould (or through a fritted lid). The heating of the external autoclave is continued until the solvent critical temperature is exceeded. Finally, the autoclave is depressurised isothermally to ambient pressure, followed by cooling.

Time from start (liquid sol in the mould) to finish (complete depressurisation) in any of these modes is usually 1–4 h for a lab-scale equipment, whereas the conventional HTSCD process may take tens of hours. This speed up is achieved due to very low compressibility of the liquid phase: fast heating causes hydrostatic pressure to build up in the mould so quickly that it soon exceeds the solvent critical pressure. Once temperature is also above T_c, isothermal depressurisation can begin immediately. Other implementations of the previously described RSCE high-temperature supercritical drying have been developed. One of the most widely explored, called the Union RSCE method, was developed by Ann Anderson and Mary Carroll at the Union College (USA).

The method utilises a hydraulic hot press to seal and heat a mould. In this method, sol is poured into a metal mould. For sealing the mould, it is covered with a thick sheet (stainless steel) sandwiched between two pieces of a temperature-resistant

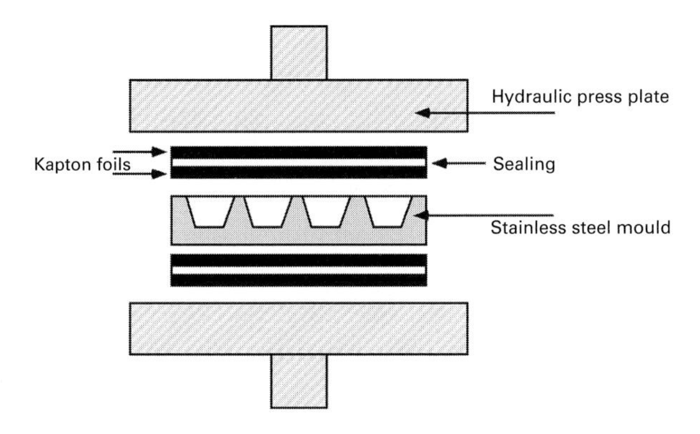

Figure 5.21 Schematic of the mould and press configuration in the Union RSCE process. From [346]. Redrawn from the original figure with permission of Cambridge University Press

gasket material (graphite). The entire assembly is then placed between the platens of a hydraulic hot press and sealed with a predetermined restraining force (Figure 5.21).

To promote gelation of the sol, the press is first maintained at a slightly elevated temperature. Gelation and drying are not spatially separated in the Union RSCE, and gelation takes place directly in the mould. Upon gelation, the mould with the wet gel is heated above the critical temperature. The restraining force is maintained during the heating to prevent untimely evaporation of the solvent. After a short dwell time for equilibration, the restrained force is gradually reduced while maintaining the mould temperature above T_c. During this phase, the supercritical fluid is released from the gel pores. Finally, the platens are cooled.

A number of chemistries are compatible with the Union RSCE: TMOS-derived silica aerogels [347], titania aerogels from titanium butoxide [348] and alumina aerogels from aluminum isopropoxide [349] have been reported, among a few others.

5.3.4 Low-Temperature Supercritical Drying

In the previous section, we concluded that solvents above the critical point are far from being inert: they can dissolve and chemically modify gels to be dried. In the case of hydrogels, water becomes a powerful solvent above 374°C and 220 bar and will likely disintegrate the gel. Furthermore, these conditions are not easy to achieve with standard laboratory equipment.

We already envisaged and explored one possible solution to overcome this issue: replacement of water in the gel with another solvent with lower critical pressure and temperature as well as with lower chemical activity in the supercritical state. That is how we arrived at the idea of the solvent exchange process.

The art of employing solvent exchange for aerogel production was first documented by Samuel Kistler [13]: he succeeded by drying a few organic hydrogels through a series of successive solvent exchanges from water to ethanol (T_c = 514 K) or further to liquid propane (T_c = 370 K). Even the use of inert liquid propane limits the scope of the process due to tedious multiple-step solvent exchanges (water to ethanol, ethanol to diethyl ether, diethyl ether to petroleum ether and finally from petroleum ether to liquid propane!) causing long processing times, high solvent consumption and the need of a downstream purification. The drying temperature of $\cong 1.05T_c$ (389 K) poses high flammability risks in case of organic solvents such as propane.

Static Low-Temperature SCD

The emergence of a much more benign methodology for processing aerogels has waiting for more than 50 years. The work by Tewari et al. [350] from 1985 marked the advent of what we call, after further developments, low-temperature supercritical drying. Their ground breaking idea was to exchange ethanol in the gel pores with an inert solvent which meets two requirements. First, it should have a readily accessible critical point, especially a low critical temperature, as it would minimise the risk of chemical reactions between solvent and gel ('low temperature' is not a well-defined term, but one might define it as a temperature slightly above room temperature). Second, the new solvent should be miscible with ethanol. The latter condition is very useful, as it allows immediate low-temperature drying of silica gels synthesised from silicon alkoxides in ethanol (see Section 2.1).

A few compounds meet these two criteria: a number of fluorine-containing freons, nitrous oxide and carbon dioxide. Only carbon dioxide with $T_c = 31°C$ and $p_c = 74$ bar came into use due to its availability, relative non-toxicity and nonflammability. Other fluids mentioned had occasional been applied for drying but were sidelined by CO_2, which became overwhelmingly the fluid of choice now. In fact, Kistler attempted to dry swollen rubber in CO_2 but did not succeed.

The symmetric molecule carbon dioxide does not have permanent dipole moment and is chemically quite inert at low temperatures. It nevertheless can carbonate materials like cement. In such cases, other, more inert fluids with a low critical point can be employed, for example trifluoromethane CHF_3 with $T_c = 25.7°C$ and $p_c = 4.82$ MPa [351].

The process suggested by Tewari et al. is essentially a solvent exchange process wherein liquid ethanol is replaced by liquid carbon dioxide. To visualise the drying process, let us locate a piece of alcogel in a high-pressure vessel. As a source of carbon dioxide, we use a standard gas cylinder at room temperature. The peculiarity of carbon dioxide and also a few other gases such as pentane is that they liquefy in the gas cylinder forming two distinct phases: dense liquid phase at the bottom and light gas phases at the top. The pressure in the gas cylinder changes with temperature along the liquid–vapour coexistence curve shown in Figure 5.7.

We now use this gas cylinder at room temperature, say at 25°C, to feed a high-pressure vessel with the ethanolic gel. At this temperature, the pressure in the cylinder is around 64 bar. To form liquid carbon dioxide in the vessel, we cool it down to, say, 10°C. According to the phase diagram, the liquid phase will be formed in the autoclave at a lower pressure of 45 bar. Due to the pressure difference between the cylinder and the vessel, carbon dioxide accumulates in the vessel, covering the gel.

Once the gel is covered with liquid CO_2, the inlet valve is closed and the system is left for equilibration. Carbon dioxide and ethanol form a homogeneous mixture as they are miscible in all proportions. This phase behaviour opens up a way to displace ethanol by repetitive purging over a time period of several hours. For that, the outlet valve is opened to push out the liquid phase while the inlet valve is also kept open to make up to the initial volume with fresh liquid CO_2.

After a few flushes with CO_2 (allowing for equilibration), all ethanol will be extracted from the gel. Now the gel pores are all filled with liquid carbon dioxide. The following steps are identical to the HTSCD process with organic solvents detailed in Section 5.3.2: both inlet and outlet valves are closed and the vessel is heated above the critical temperature causing simultaneous rise in pressure. Once the supercritical region is reached, the drying ends with isothermal depressurisation to ambient pressure and cooling to ambient temperature (path $F \rightarrow A'$ in Figure 5.7).

One remarkable feature of this implementation is that operating temperature and pressure do not significantly exceed the critical pressure and temperature of pure CO_2. Indeed, the drying usually runs at 35–40°C and 90 bar creating mild conditions for liable gels and giving the process its name – low-temperature supercritical drying (LTSCD). Because carbon dioxide does not flow through the high-pressure vessel during the process, this implementation is called static LTSCD. Neither a compressor nor a pump is required akin to the HTSCD: the pressure is built up by heating the liquid solvent in the closed vessel.

In fact, this way of drying was long known in biology and biomedical sciences before the work by Tewari et al. [350]. In the 1940s, soon after scanning electron microscopy (SEM) had become an extremely powerful technique to explore what Richard Feynman called 'plenty of room at the bottom', it was realised that wet biological samples are not suited for the high-vacuum conditions in SEM instruments. Such samples must first be freed of all liquids, which constituted a great handicap to the usefulness of the SEM technique for probing biological matter. Attempts to use classical evaporative drying had met with little success. That is how biologists, one decade after Samuel Kistler, had encountered essentially the same problem.

Thomas Anderson in 1951, referring to Kistler, described a drying process in which fixed and stained samples were exposed to what we would now call solvent exchange with ethanol followed by another exchange with amyl acetate. Finally, liquid carbon dioxide was used to replace amyl acetate followed by heating above the critical temperature and gentle pressure release. Anderson called this drying protocol the critical point method and demonstrated that it preserves many aspects of biological structures. Figure 5.22 shows SEM pictures of a critical point dried microorganism under a condition where capillary forces have partially collapsed the object due to incomplete immersion of the sample in liquid CO_2 (Figure 5.22a). Supercritical drying with full submersion in liquid CO_2 yields a well-preserved probe (Figure 5.22b).

No need of a pump or compressor for transport of pressurised fluids makes this process, nowadays called critical point drying, one of the standard sample preparation method for SEM. A wide range of semi- and fully automated critical point dryers is available.

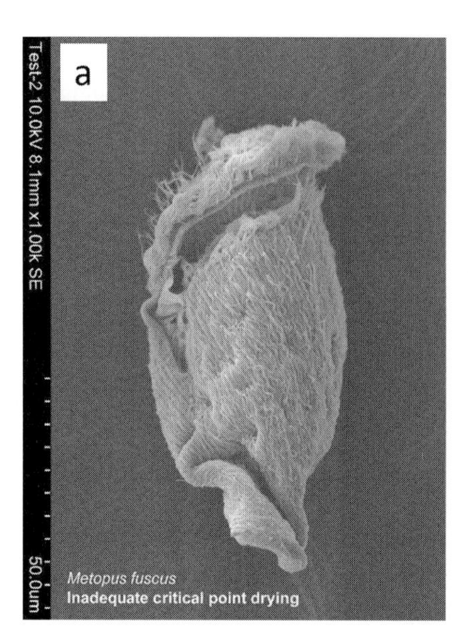

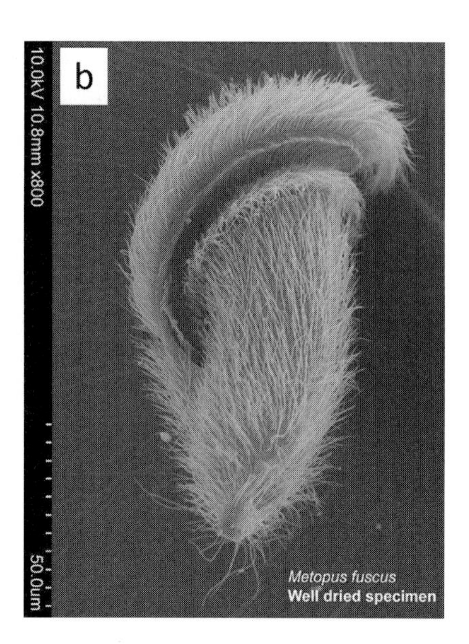

Figure 5.22 SEM image of chemically fixed water-borne protozoa anaerobic organism (Metopus fuscus) after critical point drying: (a) the sample was not fully submerged in liquid carbon dioxide, causing shrinkage and collapse of tiny details such as cilia; (b) well-dried sample for comparison. Courtesy of Dr William Bourland

The (p, T)-process trajectory in critical point dryers is essentially identical to that in the previously described static LTSCD. Commercial and custom-made critical point dryers are often used at lab scale to preserve solvent-containing objects with nanosized architecture.

Dynamic Low-Temperature SCD

The replacement of ethanol or another organic solvent in the gel pores with liquid carbon dioxide is largely a diffusion-limited process. In the condensed phases, i.e., in liquids and solids, the binary diffusion coefficient D_{12} is mainly influenced by temperature, while in gases and supercritical fluids both temperature and pressure have an influence on diffusivity.

The characteristic time for setting diffusional equilibrium for a sample with the characteristic size l is inversely proportional to the diffusion coefficient l^2/D_{12}. As we mentioned in Section 5.3.1, diffusion coefficients in supercritical fluids are typically many times greater than in the liquid phase. Can we take advantage of this fact and run the drying process directly in the supercritical region?

Furthermore, it would be beneficial to avoid any heat transfer during the drying. The reason for this is the following. In the static LTSCD and also in the HTSCD processes, the liquid (e.g., CO_2 or ethanol) in the gel and that covers the gel is heated to bring it above its critical point. Imagine scaling up the process. The amount of liquid in a high-pressure vessel scales with its linear size L as L^3 and so does the overall amount of heat Q required for reaching the working temperature above T_c. The heat

Q has to be transferred through the cross-sectional surface area A of the vessel which scales like L^2. Therefore, the larger the high-pressure vessel becomes, the longer time t is needed to transfer the heat for reaching the operating temperature. Because $Q \sim tA$, this time scales as $t \sim L^3/L^2 \sim L$, i.e., linearly increases with the size of the equipment.

The static LTSCD was greatly improved by van Bommel and de Haan in 1995 [352]. They presented a method that circumvents both the problems: slow replacement of ethanol with liquid carbon dioxide and non-isothermal operation before reaching the supercritical region. Their brilliant idea was to keep the high-pressure vessel at a constant temperature above the critical point for the binary mixture of ethanol and CO_2 and extract ethanol dynamically from the gel. To appreciate this improvement, we first consider the phase behaviour of binary systems with carbon dioxide as one component.

The phase behaviour in binary systems with low-molecular-weight organic solvents such as ethanol or acetone exhibits a positive deviation from the Raoult's law. A pressure concentration diagram, (p, x), for the system ethanol/CO_2 is shown in Figure 5.23 at a constant temperature of 40°C. The very left point (A) of the diagram corresponds to 100% liquid ethanol under its saturated vapour pressure (~0.18 bar at 40°C). Now carbon dioxide is introduced, keeping the system temperature fixed. This causes pressure to rise. Carbon dioxide is present now in both liquid and gas

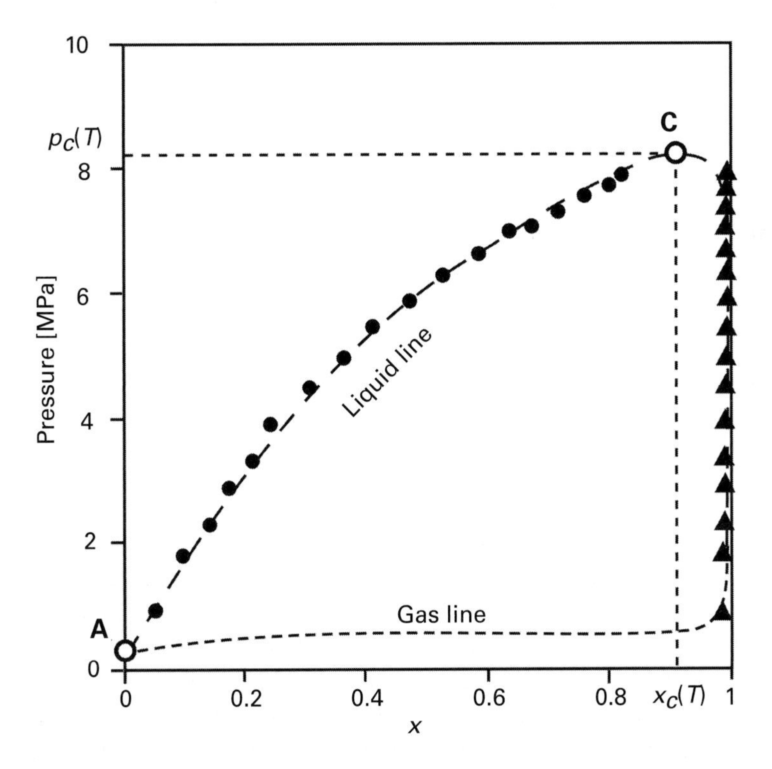

Figure 5.23 Vapour–liquid equilibrium (p, x)-diagram for ethanol–CO_2 system at 40°C.

phases. Concentration of the dissolved CO_2 (in mol.fr.) is given by the liquid line, concentration in the gas phase – by the gas line (Figure 5.23). The liquid line is nothing but a solubility line for carbon dioxide as a function of pressure: at 10 bar, the liquid phase contains 0.05 mol.fr. CO_2, and 20 bar 0.11 mol.fr. and so on.

The liquid and gaseous phases coexist at intermediate pressures. Whereas the amount of CO_2 in the liquid phase changes along the liquid line, the gaseous phase soon becomes almost pure CO_2 with less than 0.01 mol.fr. of EtOH. As more carbon dioxide comes into the vessel, the system pressure rises further and the liquid phase enriches in CO_2 while the gaseous phase solvates more ethanol. At a certain pressure, the composition of the liquid and gaseous phases become identical (point C in Figure 5.23). This pressure is called the critical pressure $p_c(T)$, while the corresponding composition is the critical composition $x_c(T)$. Similar to one-component systems, the density of both phases is also equal at the critical point. Note that the position of the critical point varies with temperature, contrary to what we have for a one-component system (Figure 5.7). We indicate this fact explicitly when writing $p_c(T)$ and $x_c(T)$. For 40°C in our example, $p_c(T) = 80$ bar and $x_c(T) = 0.89$ mol.fr. The region above the critical pressure at a given temperature can arbitrarily be called 'supercritical fluid'.

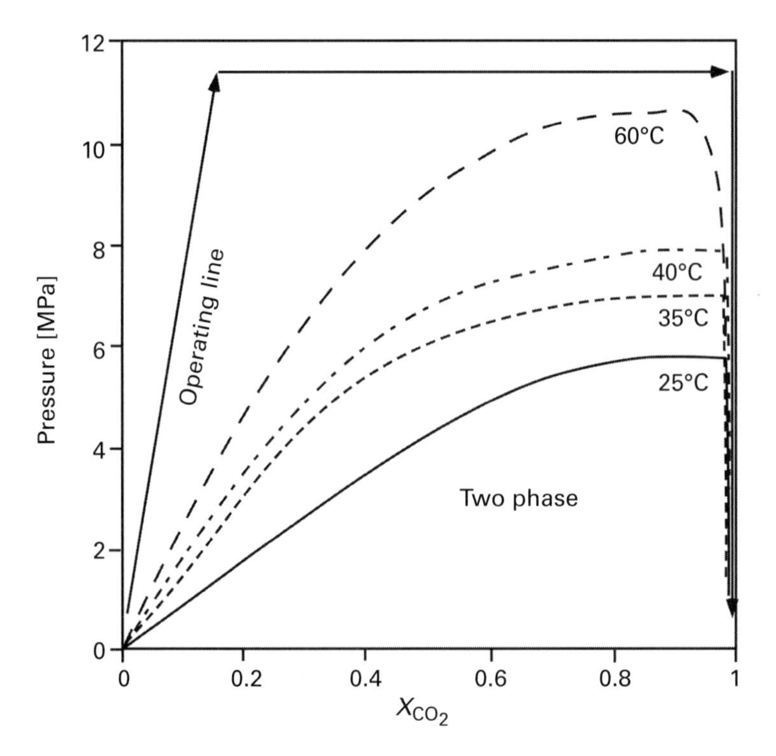

Figure 5.24 Vapour–liquid equilibrium (p, x)-diagrams for ethanol–CO_2 system at 25 and 50°C.

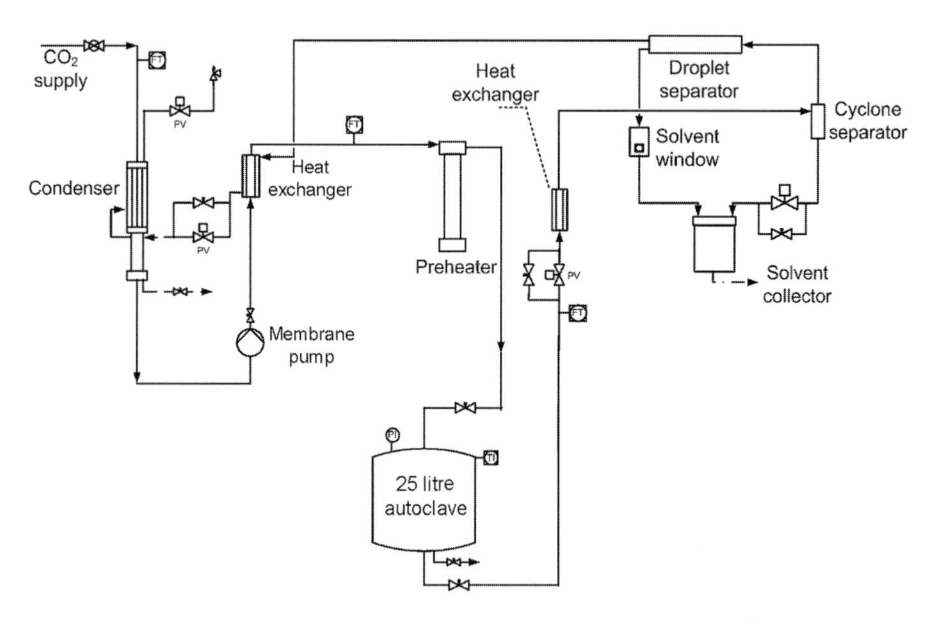

Figure 5.25 Schematic diagram of the drying setup for dynamic LTSCD at the Hamburg University of Technology. Courtesy of Dr Raman S.P.

Now let us consider the influence of temperature on the phase behaviour. Figure 5.24 demonstrates the phase diagram at several temperatures. Above $T_c(CO_2)$, the two-phase loop leaves the right-hand axis and becomes smaller. If we now depressurise the system isothermally from a point above the liquid line and close to the right axis (pure carbon dioxide), no second phase will be formed at 60°C (Figure 5.24), whereas a liquid–gas split will occur at 25°C.

The (p, x)-phase diagrams in Figures 5.23 and 5.24 define a basis for the most advanced drying technique, called dynamic low-temperature supercritical drying. In this technique, a lyogel is placed in a preheated (40–60°C) high-pressure vessel (Figure 5.25) with a small portion of the corresponding organic solvent. This extra solvent evaporates within the vessel, creating a saturated environment to prevent solvent evaporation from the gel. For heating the vessel, an electrical or oil jacket is typically used. Depending on the source of carbon dioxide, a liquid pump or a compressor is used to supply CO_2 into the vessel through an inlet valve. Before entering the autoclave, the CO_2 stream can be preheated in a heat exchanger to maintain roughly isothermal conditions. At a lab scale, the amount of entering CO_2 is low, so it does not significantly affect the temperature inside the vessel. At larger scales, however, temperature instabilities during pressurisation are normal. Isothermal conditions are a sufficient but not a necessary condition. During the pressurisation step, the outlet valve is kept closed and the pressure can build up.

The drying itself begins once the pressure reaches 10–20 bar above the critical pressure for the corresponding binary mixture CO_2–organic solvent. The outlet valve is then slowly opened. The pressure drop that may then occur does not pose a risk of forming two phases because the working pressure is set 10–20 bar higher than the critical pressure of the mixture. The outlet valve is heated to prevent freezing due to the Joule–Thomson effect. Various regimes have been suggested for the drying step ranging from slow continuous washing with CO_2 to short fast flushing interspersed with equilibration periods. Choice of a particular regime highly depends on the gel size, geometry of the high-pressure vessel, arrangement of the drying periphery (e.g., buffer tanks for liquid CO_2) and overall process optimisation strategy (e.g., use of recirculation circuits).

Once the solvent is extracted, the inlet valve is closed and system is depressurised followed by cooling down to room temperature. From the engineering perspective, it would be beneficial to depressurise possibly fast to reduce the residence time an aerogel stays in the high-pressure vessel. This time is measured from sealing to a complete unsealing of the autoclave and largely contributes to operational expenses of the overall process. Too fast depressurisation can either burst the aerogel porous structure due to a large pressure gradient or collapse due to CO_2 condensation within the aerogel (the Joule–Thomson effect). A few studies and personal experience with depressurisation of mesoporous materials speak for a second scenario, controlled, slow depressurisation rates below 5 bar/min for almost all inorganic and biopolymer aerogels for up to 1 cm characteristic length. Larger lengths probably require slower depressurisation rates and should be determined on a case-by-case basis. Faster depressurisation rates are not typically used, but seem to be tolerable as well [353].

Thermodynamic behaviour of the solvent–CO_2 mixture determines the operating conditions of the main drying step. Importantly, the same thermodynamics gives an idea when the CO_2 flow should be stopped. From a practical standpoint, one would prefer not to extract all solvent completely – this would correspond to reaching a point at the right-hand CO_2 axis of the phase diagram. Instead, one may stop supplying carbon dioxide at any composition which exceeds the very right border of the gas line (Figure 5.23). Only above the CO_2 mole fraction of x_{end} there is no risk of the formation of a liquid phase at the depressurisation step. On the contrary, if a fluid with the composition below x_{end} is depressurised, a certain amount of the liquid phase will necessarily be formed within the gel. This may be ruinous for the gel integrity or at least transparency: even a small amount of the liquid phase condensed in the middle of the gel may cause an avalanche-like distortion of the aerogel structure.

We see now how binary phase diagrams provide the minimal working pressure, but also a natural criterion for the extent to which the drying should proceed. Although the real depressurisation step is not always strictly isothermal, analysis of the phase behaviour lays a basis for the estimation of the drying time and energy requirements.

Can water be directly extracted by carbon dioxide in the dynamic LTSCD? A water–CO_2 system demonstrates essentially the same phase behaviour with a critical pressure at a given temperature. The main problem with a water–CO_2 binary mixture is that reasonably low critical pressures occur at very high temperatures (shown in Figure 5.26 for a few selected temperatures). We face here with water the same limitation as in the HTSCD approach: water possesses too high critical temperature to be converted into supercritical fluid without causing considerable gel

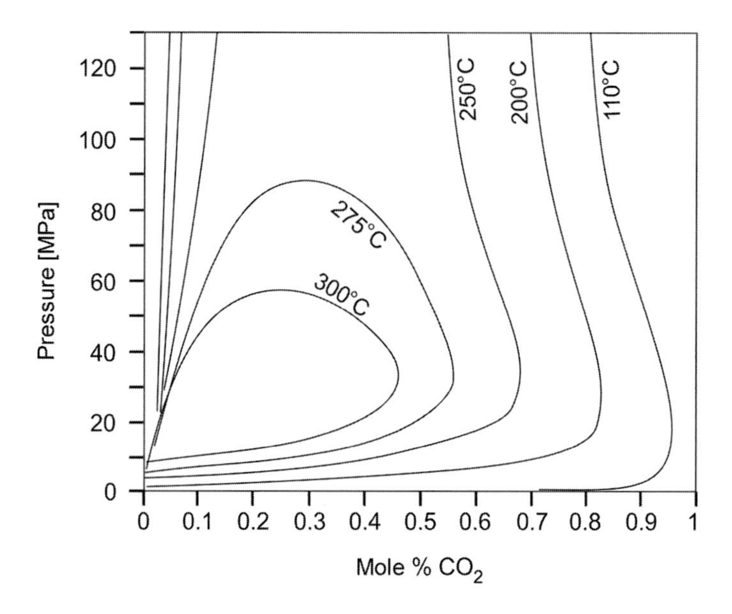

Figure 5.26 Some isothermal lines in the vapour–liquid equilibrium diagram of the water–CO_2 system. The curves were replotted from data extracted from [354].

degradation. It is in a striking contrast to ethanol: while working temperature in the HTSCD would be 270–290°C, the dynamic LTSCD is routinely run at 40°C.

Which solvents can be extracted with carbon dioxide in the dynamic LTSCD at reasonably low working pressures and temperatures? To address this question, let us re-plot the p,x-diagrams from Figures 5.23 and 5.24. We noticed that for this type of phase behaviour at any given temperature T, there is a single value of critical pressure $p_c(T)$. Above $p_c(T)$, the system is composed of only one fluid phase regardless of the composition. We plot now the mixture critical pressure a function of temperature, shown in Figure 5.27. Here temperature can vary only between the critical temperatures of pure components, e.g., for CO_2–EtOH, between 31.1 and 243.1°C. The line $p_c(T)$ forms a so-called critical locus. Any p,T combination above the critical locus line corresponds to one fluid phase and thus provides a suitable operating condition for drying. For example, at 50°C the drying pressure should be above 80 bar, at 100°C above 140 bar and so on.

For binary systems with an organic solvent, the critical locus line begins at the critical point of CO_2 and ends at the critical point of the solvent. Figure 5.28 demonstrates critical loci for a number of solvents. At working temperatures below 50°C, it is safe to choose the drying pressure at 110–120 bar for any solvent used in lab and industrial practice. If a solvent being chemically dissimilar to carbon dioxide should be extracted at $T > 50$°C, the corresponding (p,x)-phase behaviour at the desired temperature is required to select the working conditions.

Lastly, let us turn to a question of the mass transfer during the dynamic LTSCD. Recent research has shown it to be a combination of diffusive and convective mass

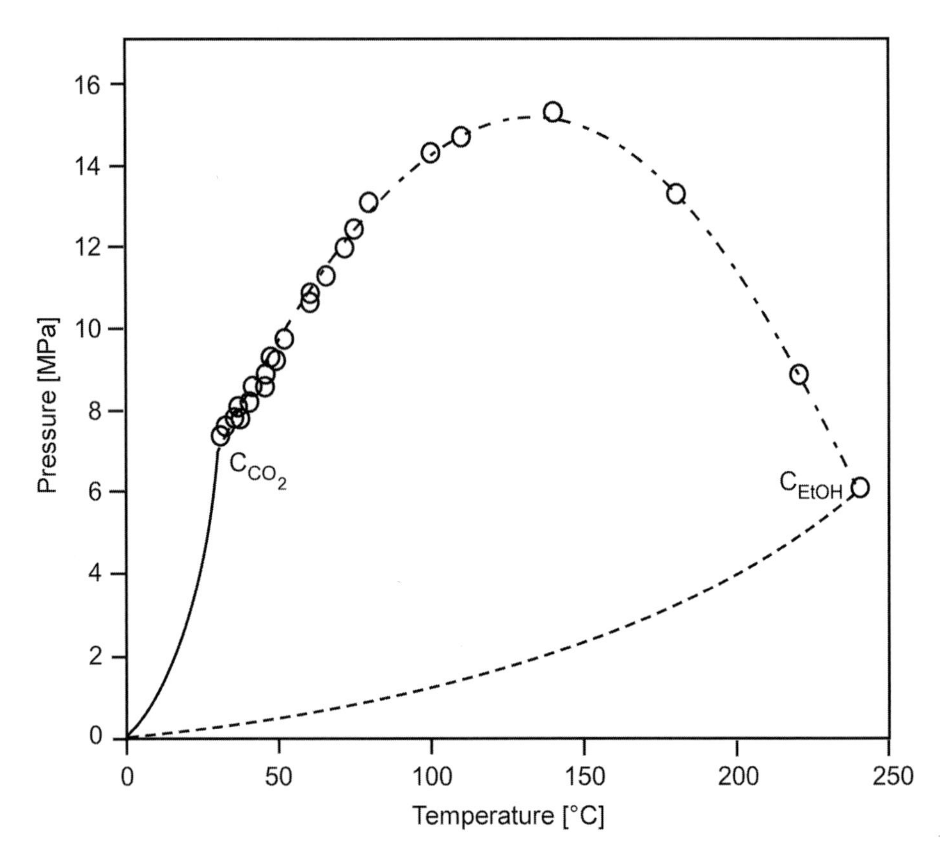

Figure 5.27 (p, T)-equilibrium phase diagram (critical locus). The region above the line corresponds to the supercritical region of the binary mixture.

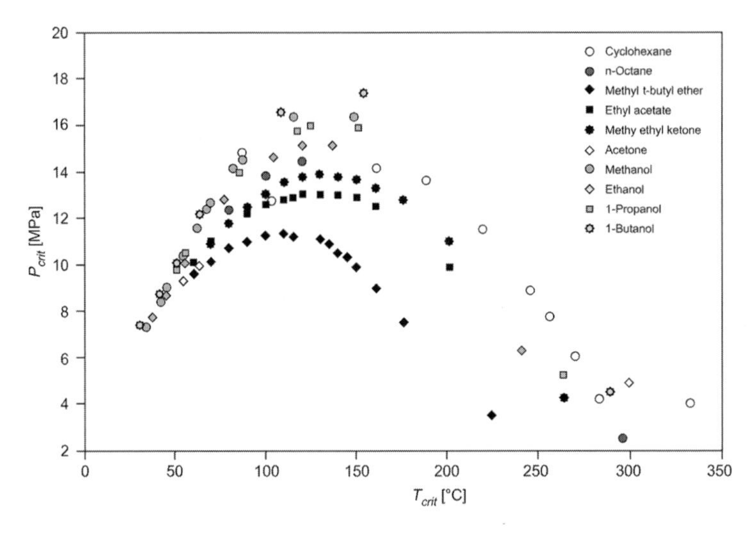

Figure 5.28 (p_{crit}, T_{crit}) equilibrium critical points for several solvent mixtures with CO_2. Inasmuch as the solvents are mixed to CO_2, the critical point first increases in pressure and temperature, but decreases at higher temperatures, where it ends at the critical points of the pure solvents.

transfer [355–360]. It is common to subdivide the drying system into three domains: gel body, bulk fluid around the gel (autoclave domain) and boundary layer. The latter is a thin film where the velocity of the supercritical fluid changes from zero to the velocity of the surrounding flow. During the supercritical drying at a fixed pressure and temperature, the mass transfer in the gel domain is dominated by diffusion. Free convection is the major mechanism of mass transfer in the autoclave domain. For lab-scale drying autoclaves, the solvent concentration in the bulk fluid can often be assumed equal to zero due to small gel-to-autoclave volume ratio; however, it is not the case for industrial autoclaves. Mass transfer in the boundary layer is described using a mass transfer coefficient, which can be expressed as a function of the Reynolds and Schmidt numbers (the Schmidt number is the ratio of the kinematic viscosity to the diffusion coefficient) and ultimately is a function of the bulk fluid velocity, pressure, temperature and local solvent concentration [361]. Calculation of the mass transfer coefficient relies on empirical correlations and requires knowledge of binary diffusion coefficient, viscosity and density as functions of pressure, temperature and solvent concentration. Such data are not readily available for most of solvents, limiting further progress of the predictive modelling of drying kinetics. Theory of mass transfer in dynamic LTSCD is also far from being completed. Due to the gel shrinkage during drying the gel–fluid interface is not fixed. This gives rise to inhomogeneous distribution of porosity and tortuosity within the gel and convective flow due to squeezing a fraction of the solvent from the gel into the bulk phase, among other phenomena. Furthermore, analytical techniques had until recently been allowed only for measurements of the integral mass loss during drying. Spatially resolved concentration profiles in both gel and bulk phases [362] are expected to shed new light on the unresolved issues.

The drying time depends on numerous factors, both gel and equipment specific. It is thus not surprising that earlier attempts to quantify the drying kinetics using the Fick's law with a fitted effective diffusion coefficient were only partially successful. From a practical standpoint, it is still useful to operate with an effective diffusion coefficient as it allows for a quick estimation of drying time. Let us first assume that the mass transfer coefficient is infinitely large so that there are no mass transfer limitations in the boundary layer. Second, the solvent from the gel surface is not only immediately transported to the bulk phase, but the bulk phase itself is assumed to have infinitely large volume. The drying time is called in this case the theoretical minimal drying time t_{min}, and is defined for a gel with a characteristic size R.

The minimal drying time t_{min} depends on the ratio between porosity ϕ_p and tortuosity τ_g of the gel [361], which lies in the range 0.05–0.25. To relate the minimal drying time and the gel size, we formally define a parameter K_{eff} with dimension of the diffusion coefficient, Eq. (5.35):

$$t_{min} = \frac{1}{K_{eff}} \frac{\tau_g}{\phi_p} R^2. \tag{5.35}$$

While K_{eff} is defined as an effective diffusion coefficient, it is concentration independent, contrary to the binary diffusion coefficient D_{12}. Numerical values at different pressures and temperatures are reported in literature [361], being approximately in the range of 3×10^{-8} m^2/s. The theoretical minimal drying time calculated by Eq. (5.35) decreases for lower pressures and higher temperatures due to an enhanced diffusional transport at these conditions. For a gel with a size R of 1 cm, the minimal drying time amounts to 1–3 h depending on porosity and tortuosity. Particles of micrometre to millimetre size require much lower drying time in the range of a few seconds to a few minutes. Such short drying times for microparticles have indeed been observed experimentally, giving hope of developing a continuous drying process in the near future.

It is an entrenched practice to describe the dynamic LTSCD as a diffusion-controlled process. Very recent studies show that this view is only correct for the gel domain during isothermal washing with CO_2 above the mixture critical pressure. When some organic solvents (alcohols, ketones and others) with an initial volume of V_{solv} are pressurised with CO_2, the resulting volume of the liquid phase V_{mix} increases by many times. Such pressurised solvents are known as gas-expanded liquids. The extent to which the solvent volume increases at given pressure and temperature is given by relative volume expansion η_V, Eq. (5.36):

$$\eta_V = \frac{V_{mix}(p,T) - V_{solv}(p,T)}{V_{solv}(p,T)} \tag{5.36}$$

Solvent expansion can also occur in the gel. In this case, an excess of the solvent is squeezed from the gel body and collects at the bottom of the autoclave. Figure 5.29 shows a series of snapshots for an ethanolic alginate gel exposed to carbon dioxide at 99 bar and 60°C, i.e., in the two-phase region of the phase diagram (see Figure 5.29). Although the capillary forces do exist under these conditions, alginate and other gels demonstrate remarkably no visible shrinkage. The relative volume expansion is an exponential function of the mole fraction of the solubilised carbon dioxide. This function seems to be a universal function across many solvents and allows us to estimate the amount of the solvent squeezed from the gel.

Although the phenomenon of volume expansion has been known for a long time, it has not yet employed to intensify supercritical drying. In particular, compression requirements can be optimised because a considerable amount of solvent can be

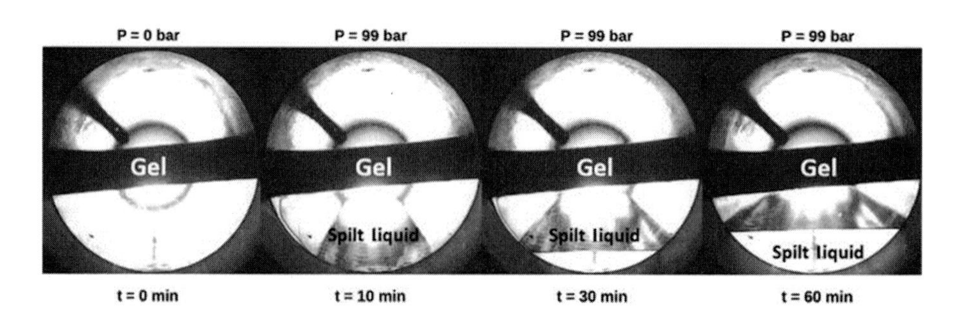

Figure 5.29 A liquid mixture of ethanol and carbon dioxide is spilt out of the gel and accumulated at the bottom of the high-pressure autoclave. Gel: calcium alginate 3.0 wt%. Conditions: 99 bar and 60°C. Adopted from [363] with permission from the American Chemical Society (ACS)

removed from the gel during the pressurisation step. We hope that the reader is now assured that with current understanding and future improvements, dynamic LTSCD drying with carbon dioxide is a mild approach and so far the most universal approach towards aerogels. Thanks to dynamic LTSCD, significant progress in biopolymer aerogels has occurred recently at lab, pilot and industrial scales.

5.4 Solvent Exchange and Gel Shrinkage

5.4.1 Solvent Selection

The supercritical drying process with supercritical carbon dioxide as we briefly discussed in Section 5.3 is only applicable to lyogels containing what we called light organic solvents. Whenever the gelation chemistry can be realised directly in such a solvent, no extra step is required: the lyogel can be dried directly with sc-CO_2. Table 5.3 lists several classes of aerogel precursors along with corresponding solvents employed at the gelation step.

Table 5.3 summarises in the first column the class of aerogels and their chemistry, gives in the second column examples and in the third column lists the typical solvents used. Studying the table reveals that polymerisation, polycondensation and gelation of many precursors can only be performed in non-polar organic solvents (e.g., toluene or 1,2-dichloroethane) because the precursors are only poorly soluble in other solvents. Almost all oxide gels can be prepared in light polar organic solvents such as C_1–C_4 alcohols and C_3–C_4 ketones. Therefore, lyogels obtained from these precursors can be immediately dried with sc-CO_2. In many cases, however, the solvent used for gelation has to be exchanged by non-polar organic solvents.

Table 5.3 shows, that a large class of precursors can exclusively be gelled in aqueous systems or ionic liquids. As we briefly mentioned in the previous chapters and

Table 5.3 Precursors for aerogel processing with corresponding solvents.

Aerogel chemistry	Examples of precursors	Solvent inside the lyogel after gelation
Inorganic oxide	Sodium silicate	Water
	Sodium aluminate	Water
	Tetraalkyl orthosilicates and derivatives	$C_1 - C_4$ alcohols and $C_3 - C_4$ ketones
	Aluminum alkoxides, titanium alkoxides	$C_1 - C_4$ alcohols
	Salts (e.g., $AlCl_3$) with propylene oxide	Mixture of $C_1 - C_4$ alcohol with H_2O
Chalcogenide	colloidal sulphides, selenides and tellurides	Water; N-methylformamide
Polyester	Polylactic acid	Chloroform; ethyl lactate

Table 5.3 Continued

Aerogel chemistry	Examples of precursors	Solvent inside the lyogel after gelation
Polyether	Polyphenyleneoxide	1,2-Dichloroethane
	Resorcinol and formaldehyde	Water
	Tannin and formaldehyde	Water
	Cresols and formaldehyde	Water
Polyurea	4,4'-Methylene diisocyanate with aromatic diamines and triamines	N-Methyl-2-pyrrolidone
	Tris(4-isocyanatophenyl)methane with mineral acids	N,N-Dimethylformamide with ethyl acetate
Polyimide	Pyromellitic dianhydride, 2,2'-dimethylbenzidine and tris(2-aminoethyl)amine	N,N-Dimethylformamide
	Pyromellitic dianhydride with 4,4'-oxydianiline	N-Methyl-2-pyrrolidone
Polyamide	Melamine with isophthaloyl chloride	Dimethyl sulfoxide with N-methyl-2-pyrrolidinone
Aliphatic and aromatic polymers	Syndiotactic polystyrene	Chloroform with acrylonitrile; 1,2-dichloroethane; o-dichlorobenzene
	Monomers from ring opening metathesis polymerisation	Toluene
	Poly(4-methyl-pentene-1)	1,3,5-Trimethylbenzene; decalin; cyclopentane
Polybenzoxazine	Bisphenol A, aniline and formaldehyde	Dimethyl sulfoxide
Biopolymers	Alginate	Water
	Cellulose	Water (with NaOH, urea, ZnO); ionic liquids; salt hydrates; N-methylmorpholine-N-oxide
	Chitin	Ionic liquids; water (with NaOH and urea); N,N-dimethylacetamide (with LiCl)
	Chitosan	Water (with an acid)
	Humic acids	Water
	Lignin	Water with a base
	Pectin	Water
	Proteins (gelatine, silk fibroin, whey)	Water (with an acid or base)
	Starch	Pressurised water (>100°C)

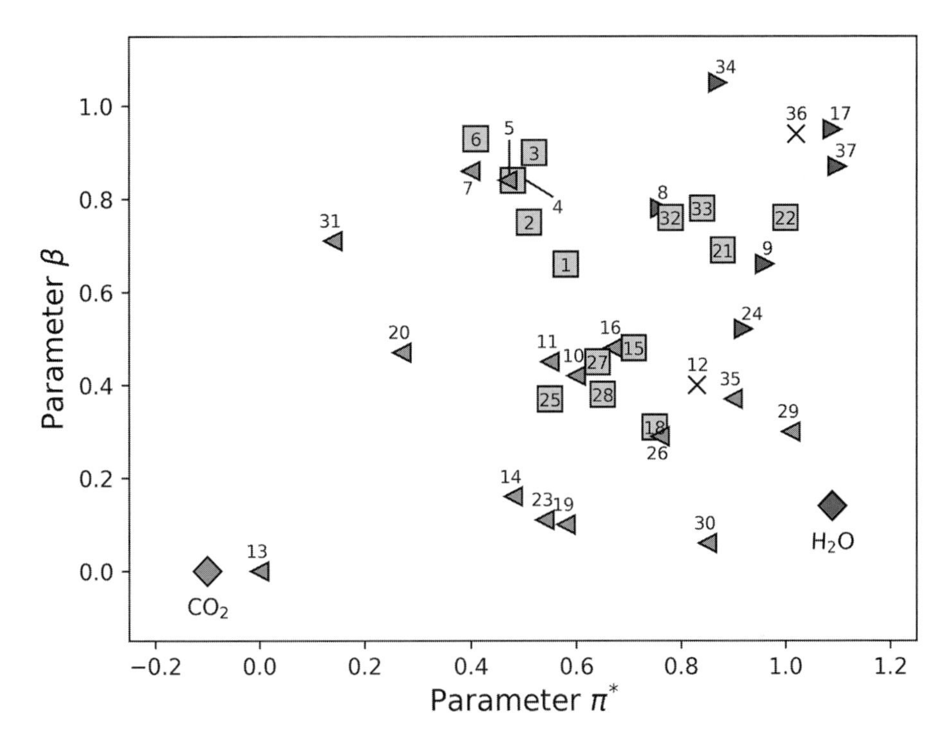

Figure 5.30 Kamlet–Taft plot for several solvents. Supercritical carbon dioxide and water are marked with diamonds (◇), respectively. Symbols ◁, ▷ and × indicate solvents that are miscible only with sc-CO_2 (at 35–50° C, below 15 MPa), only with water (at room temperature) and with neither solvent. Solvents miscible with both sc-CO_2 and water are shown as squares (□). (1) methanol; (2) ethanol; (3) 1-propanol; (4) 2-propanol; (5) 1-butanol; (6) t-butanol; (7) 1-pentanol; (8) 1,2-propanediol; (9) glycerol; (10) methyl acetate; (11) ethyl acetate; (12) propylene carbonate; (13) cyclohexane; (14) o-xylene; (15) acetone; (16) 2-butanon; (17) [Emim][OAc]; (18) acetonitrile; (19) chloroform; (20) diethyl ether; (21) dimethyl formamide; (22) dimethyl sulfoxide; (23) toluene; (24) ethylene glycol; (25) 1.4-dioxane; (26) acetic anhydride; (27) acetic acid; (28) formic acid; (29) nitrobenzene; (30) nitromethane; (31) triethylamine; (32) N,N-dimethylacetamide; (33) N,N-diethylacetamide; (34) hexamethylphosphoramide; (35) benzonitrile; (36) [Hmim][Cl]; (37) [Bmim][Cl].

discuss in detail later, water–CO_2 mixture demonstrates a miscibility gap at moderate pressures and temperatures. Similarly, other polar solvents show only a limited miscibility with carbon dioxide. Such miscibility gaps do not allow for direct supercritical drying of hydrogels and lyogels containing polar solvents.

Thus, the polar solvent (L1) has to be replaced with another solvent (L2) which is miscible in all proportions with both L1 and carbon dioxide above the upper critical solution pressure (see Figure 5.30). The exchange process of L1 with L2 is termed the solvent exchange.

Figure 5.30 illustrates in a compact way which solvents are miscible only with water (at room temperature), only with CO_2 (below 15 MPa at 35–50°C) and with both of them using two solvatochromic parameters β and π^*. The parameters β and π^*, also called Kamlet–Taft parameters, quantify basicity (i.e., ability to accept hydrogen bond) and polarisability, dipole–dipole and dipole-induced dipole interactions, respectively [364].

It is evident that the solvent exchange procedure is required if the intra-gel solvent L1 has simultaneously a high polarity/polarisability ($\pi^*_{L1} > 0.8$) and high basicity ($\beta_L > 0.6$). Such a combination of basicity and polarity makes the solvent L1 only partially miscible with CO_2 (which at drying conditions has $\beta \cong 0.0$ and $\pi^* \cong -0.1$). For the solvent L2 to be miscible with both L1 and sc-CO_2, an intermediate polarity/polarisability ($\pi^*_{L2} \cong 0.5\pi^*_{L1}$) and certain basicity ($\beta > 0.4$) are required.

Gelation of polysaccharides such as alginate, pectin, carrageenans, chitosan and starch takes place in water so that L1 = H_2O with $\beta = 0.14$ and $\pi^* = 1.09$ (assuming no influence of additives, cross-linkers etc.). Therefore, solvents such as C_1–C_3 alcohols, *t*-butanol, acetone, acetonitrile, dioxane, dimethyl formamide and dimethyl sulfoxide can potentially be employed for the solvent exchange. In the majority of cases, ethanol is used due to its availability and relatively low toxicity.

As we discussed in Chapter 3, molten salts and ionic liquids are often used as dissolution media for cellulose [172], chitin [365] and few other biopolymers. Such ionic solvents (points (17), (36) and (37) in Figure 5.30) are somewhat far from the groups of the solvents miscible with sc-CO_2 in terms of the parameter π^* and thus often only partially miscible with them. The solvent exchange can nevertheless be performed with a large excess of the organic solvent or even in the two-phase region. In the latter case, the ionic liquid-rich phase is separated at the bottom due to a density difference [366]. However, especially for molten salts, washing with water is first employed to remove ionic species followed by a standard solvent exchange, e.g., with ethanol. As discussed in Chapter 3, water and organic solvents act in this case as non-solvents for biopolymers dissolved in 'ionic' solvents.

Miscibility between L1 and L2 and between L2 and CO_2 is a necessary but not sufficient criterion for the solvent selection. The gel matrix should not be soluble in a new solvent, nor be appreciably swellable in it. The latter may bring about distortion of the matrix when the solvent L2 is finally extracted with CO_2. Further constraints such as solvent safety (e.g., flammability, toxicity and corrosivity), availability of the solvent with a desired purity and recyclability determine the final choice. Apart from recyclability, all these aspects are determined by individual properties of the solvent L2 which are quantified and known in engineering practice. Recyclability is an overall efficacy of the downstream processing of L1–L2 and L2–CO_2 binary mixtures. All together this information provides a comprehensive basis for rational process design (see typical preparation procedures in Chapter 14).

However, these miscibility and safety constraints are still insufficient when selecting a solvent for the solvent exchange. On the way from as-synthesised gel to lyogel, the gel matrix demonstrates a certain volume reduction, so-called shrinkage (Figure 5.31). Clearly, the processing should be designed to preserve as much of the initial volume of the gel as possible.

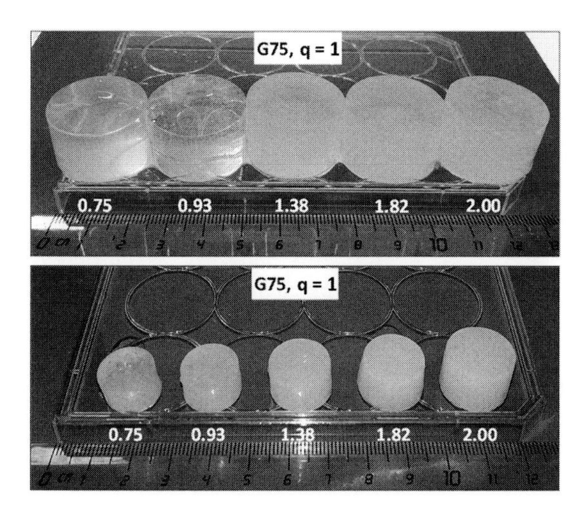

Figure 5.31 Shrinkage of alginate hydrogels (G/M ratio 75:25) with different initial polymer concentration (in wt%, indicated below each sample). Upper panel: as-synthesised hydrogels; lower panel: gels upon a multistep solvent exchange with acetonitrile. Cross-linking degree (q) was fixed at 0.36 g $CaCO_3$/g Na-alginate. Adopted from [367] with permission from ACS

5.4.2 Gel Shrinkage: General Considerations

Already Samuel Kistler in 1932 noted that he 'found it practically impossible to make the transfer from hydrogel to aerogel without large shrinkage' [13]. He also noted that the volume loss from hydrogel to aerogel was due to exchange of water to organic solvent and due to subsequent supercritical drying. Long before preparation of first biopolymer aerogels by Kistler, the desire to preserve the original structure of biological objects motivated a search for delicate methods of solvent exchange for fixation, staining and microscopic imaging [368].

It was quickly realised that the shrinkage is reduced in a stepwise solvent exchange process: the gel is first subjected to diluted solutions of an organic solvent in water, with progressively increasing concentration of the solvent. The stepwise solvent exchange with ethanol was applied in first studies on alginate aerogels by Françoise Quignard and co-workers [369, 370] and has been widely adopted since this direction expanded in the beginning of the 2000s.

Since Kistler's time, the solvent exchange process has not endured significant changes: rigid gels are soaked in an excess of the new solvent, whereas soft and delicate gels are exposed to solvent–water mixtures with increasing concentration of the solvent L2. They are conventionally called one-step and multistep solvent exchange.

To describe quantitatively the shrinkage phenomenon, the volumetric shrinkage Sh is defined as the difference in volumes of the original gel V_{L1} (in solvent L1) and lyogel V_{L2} (in solvent L2) relative to the original volume (see also Chapter 7):

$$Sh = \frac{V_{L1} - V_{L2}}{V_{L1}}. \tag{5.37}$$

Similarly, the linear shrinkage S_l can be defined using a linear dimension l instead. Writing generally the volume as $V = bL^{\alpha}$ allows us to write it also as $\ln V = \ln b + \alpha \ln l$. Thus we can say the volume scales as $\alpha \ln l + \beta$, with $\alpha = 3$ and β as a shape factor. Then the volume shrinkage can be easily related to the S_l for gel of *any* shape. From Eq. (5.37), it readily follows that

$$\ln(1 - Sh) = \ln V_{L2} - \ln V_{L1}. \tag{5.38}$$

The right-hand side of Eq. (5.38) is then expressed in terms of linear dimensions before and after the solvent exchange:

$$\ln V_{L2} - \ln V_{L1} = \alpha \ln l_{L2} + \beta - \alpha \ln l_{L1} - \beta = \ln \left(\frac{l_{L2}}{l_{L1}} \right)^{\alpha}. \tag{5.39}$$

Substituting the right-hand side of Eq. (5.39) in Eq. (5.38) and rearranging the terms, we obtain

$$1 - Sh = \ln \left(\frac{l_{L2}}{l_{L1}} \right)^{\alpha} \tag{5.40}$$

or equivalently

$$Sh = 1 - (1 - S_l)^{\alpha}. \tag{5.41}$$

Eq. (5.41) holds for isotropic shrinkage, i.e., when the gel shrinks to the same extent along all directions. Although the assumption of isotropic shrinkage is not generally the case, Eq. (5.41) allows us to estimate Sh from easily measurable linear shrinkage. The volumetric or linear shrinkage is a measure for the efficiency of the volume preservation during the solvent exchange. Supercritical drying (see Section 5.3), also may cause a certain shrinkage that is quantified analogously.

The volumetric shrinkage as defined by Eq. (5.37) ranges from zero ($V_{L1} = V_{L2}$, i.e., no shrinkage) to one ($V_{L2} \rightarrow 0$, complete shrinkage). Sometimes it is more practical to express the fraction of the preserved volume in the reverse range: zero for complete shrinkage and one when no shrinkage occurs. The volumetric yield Y_v is defined for this purpose, as shown in Eq. (5.42):

$$Y_v = \frac{V_{L2}}{V_{L1}}. \tag{5.42}$$

By comparing Eqs. (5.37) and (5.42), the volumetric yield can readily be related to the volumetric shrinkage as $Y_v = 1 - Sh$. The total volumetric shrinkage Sh_t in the overall process from hydrogel to aerogel is directly related to the final aerogel envelope density, ρ_e^f, a key property of the resulting aerogel. Indeed, assuming that the total mass of the gel backbone m_0 remains constant upon the solvent exchange and supercritical drying (no leaching), from the conservation of mass we get

$$m_0 = \rho_e^0 V_{L1} = \rho_e^f V_f, \tag{5.43}$$

where V_f stands for the final volume of the aerogel that can be measured by Archimedes' principle (see Chapter 7). The density ρ_e^0, sometimes called the target density, can be estimated from concentrations of all constituents (see Chapter 7 on the relation between monomer content and final density). From Eq. (5.43), we readily obtain

$$\rho_e^f = \rho_e^0 \frac{V_{L1}}{V_f} \tag{5.44}$$

or substituting the total shrinkage

$$\rho_e^f = \rho_e^0 \frac{1}{1 - Sh_t}. \tag{5.45}$$

Eq. (5.45) has a clear meaning: the larger the shrinkage, the higher the final envelope density. Therefore, if one would like to prepare a low aerogel density, one has to control the shrinkage and keep it possibly low. As we discuss later in this chapter, the shrinkage depends on the compressibility of the gel network. For rigid silica and RF gels, a very low shrinkage of a few percent is typically observed upon solvent exchange. By contrast, for cellulose, alginate and hydrogels from almost any other biopolymers, a severe shrinkage usually occurs that may reduce about half of their volume during solvent exchange. What solvent should be chosen for the solvent exchange to minimise shrinkage? What are molecular mechanisms of the gel contraction during the solvent exchange? What governs the solvent exchange kinetics? Many more related questions may be posed in this regard.

To date, such questions remain mainly unanswered for polymer systems which have been processed into aerogels. One reason for this is that the polymers often come from natural sources and are insufficiently characterised (composition for copolymers, molecular mass distribution and impurities are often unknown). In addition, the actual cross-linking degree is not verified in the final gel, but only set at the beginning with a certain polymer-to-cross-linker ratio. Consequently, at the moment it is practically impossible to interpret gel shrinkage in terms of existing thermodynamic models. The aim of this section is to discuss a few general qualitative observations that might serve as a basis for our further theoretical comprehension.

Our starting point is Figure 5.32, which depicts published data on the volumetric yield for various biopolymer hydrogels subjected to one- or multiple-step solvent exchange with a variety of solvents. It is natural to expect that such raw data demonstrate a high variability due to polymer nature, cross-linking degree and gel geometry (monoliths, microparticles). Nevertheless, we see that hydrogels with low solid content of a few weight percent are prone to a significant shrinkage: the volumetric yield of around 40–60% is usually observed for hydrogels with polymer content $\lesssim$2 wt%. This result supports an unstated consensus among aerogel scientists: hydrogels with the biopolymer content above 3–5 wt% are generally more robust against shrinkage. To observe more clearly the effect of the biopolymer concentration c_p, let us plot the median volumetric yield within concentration intervals of as shown in 1 wt%, Figure 5.33. The data are much smoother now and can be approximated by an exponential function, Eq. (5.46), suggesting that on average it is exponentially more difficult to control the shrinkage with decreasing polymer concentration:

$$Y_v = 1 - (1 - k_1)\exp(-k_2 c_p). \tag{5.46}$$

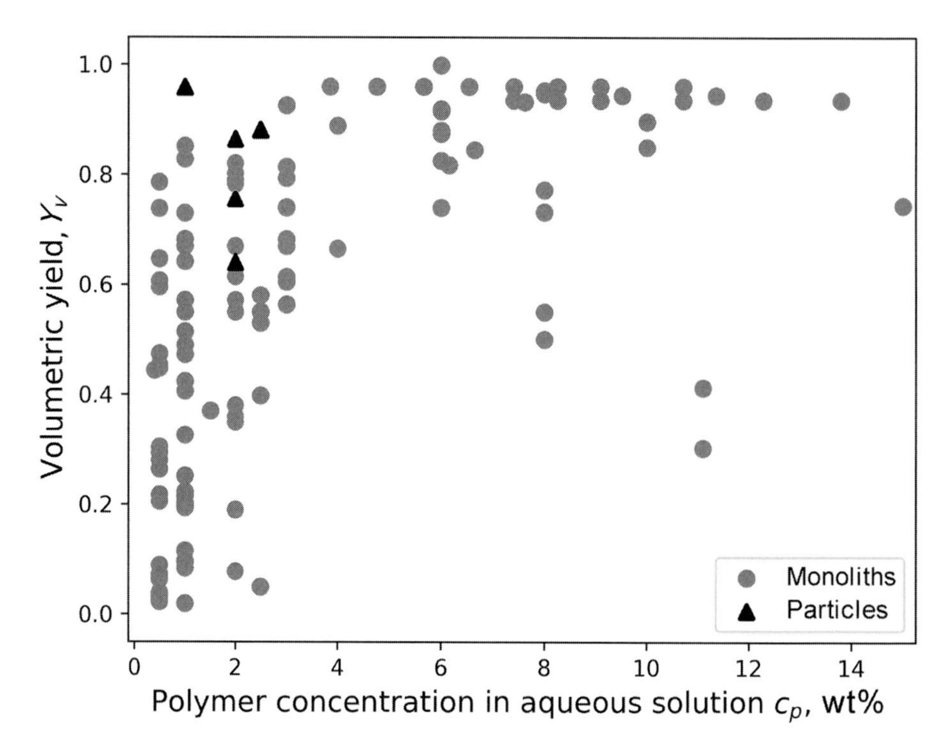

Figure 5.32 Literature data on volumetric yield for hydrogels subjected to one- and multistep solvent exchange with various solvents. Data presented separately for monoliths (grey •) and particles (full black triangles).

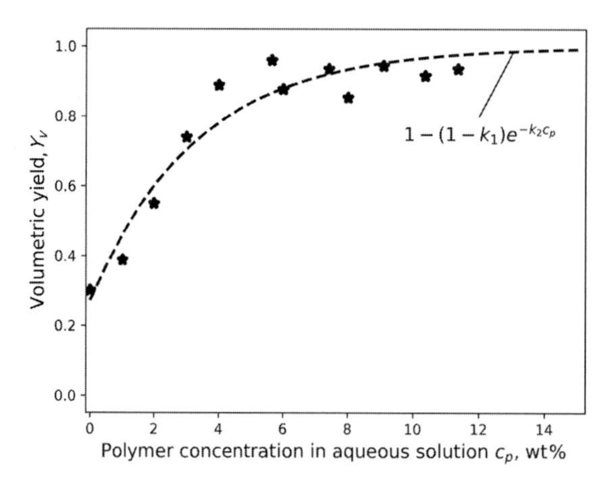

Figure 5.33 Smoothed data (★) on volumetric yield from Figure 5.32. Best fit is shown as a dashed line [371].

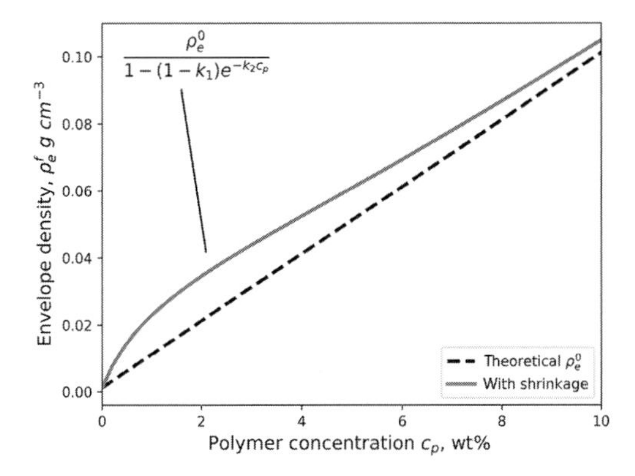

Figure 5.34 Density of a fictitious aerogel (g/cm^3.) as a function of biopolymer concentration (wt%) if shrinkage at the solvent exchange step would be the only reason for the density increase (solid line, values of k_1 and k_2 in Eq. (5.46) are fitted to the median shrinkage in Figure 5.33). The dashed line represents the minimal aerogel density that can be achieved if no shrinkage occurs throughout the process. Adopted from [371] with permission from ACS

Parameters k_1 and k_2 describe the finite volumetric yield at zero polymer concentration and sensitivity of the gel network towards shrinkage, respectively. It is of interest to estimate the final envelope density from Eq. (5.45) if solvent exchange would be the only reason for the density increase ($Sh_t = Sh$). Combining Eq. (5.45) with the definition of the volumetric yield, Eq. (5.42) and Eq. (5.46), we get

$$\rho_e^f = \frac{\rho_e^0}{1-(1-k_1)\exp(-k_2c_p)}. \tag{5.47}$$

In a shrinkage-free process the envelope density ρ_e^0 of the imaginary aerogel (in g/cm^3) prepared from an aqueous solution with biopolymer concentration c_p wt% would be numerically equal to $c_p/100$ (solution density assumed to be concentration independent and equal 1 g/cm^3; see Chapter 7 for the general case). The envelope density calculated from Eq. (5.47) is shown in Figure 5.34 along with the theoretical density (dashed line). The effect of the gel shrinkage on the envelope density is now clearly seen and pronounced for hydrogels with the polymer content $\lesssim$5 wt%.

Numerical values of k_1 and k_2 for the envelope density shown in Figure 5.34 are estimated by fitting to the set of diverse data. Polymer-specific values of the parameters can be obtained when volumetric yield is measured for a series of hydrogels with various polymer content. Essentially similar approach is also applicable to the total volumetric shrinkage upon solvent exchange and supercritical drying.

5.4.3 Gel–Solvent Interactions

Even when the solvent exchange is performed as close to equilibrium as possible, e.g., by adding the solvent drop by drop to a large bath with the gel inside, the equilibrium

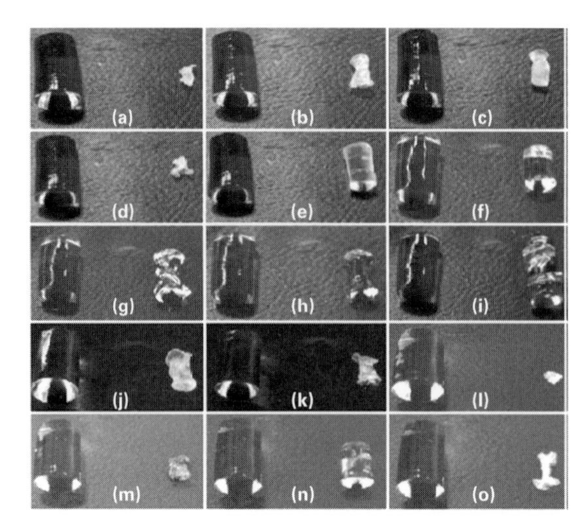

Figure 5.35 Shrinkage occurred after immersion of alginate hydrogel (0.5 wt%, left) into an excess 16 different solvents (right). Solvents used: (a) methyl ethyl ketone; (b) isopropanol; (c) acetone; (d) 1-butanol; (e) methanol; (f) dimethyl sulfoxide; (g) glycerol; (h) propylene glycol; (i) ethylene glycol; (j) ethanol; (k) 1,4-dioxane; (l) propylene carbonate; (m) furfuryl alcohol; (n) N,N-dimethylformamide and (o) acetonitrile. Reproduced from reference [24] under the Creative Commons Attribution License (CC BY 4.0)

volumetric yield remains finite. This finite shrinkage is solvent specific and can be influenced by the solvent nature.

Many researchers noticed, especially when working with biopolymer hydrogels, that the shrinkage highly depends on the solvent nature. Figure 5.35 visualises this fact for soft alginate hydrogels which were intentionally subjected to one-step solvent exchange with 16 different solvents in order to simulate the worst-case scenario and markedly see the difference across the solvents [24]. Curiously, the alginate gels retained more than 10% of their original volume only in a few solvents. Even ethanol, the solvent of choice in the majority of studies, causes a drastic volume decrease when used in the one-step procedure.

In the original Flory–Huggins model of polymer solutions and its numerous extensions, affinity between a solvent (component 1) and a polymer segment (component 2) is quantified by interaction parameter χ_{12} (see Appendix B). It appears in the expression for the enthalpy of mixing and is equal to zero for athermal mixing (ideal mixture). For exothermic and endothermic mixing, the parameter χ_{12} is smaller or larger than zero, respectively. The binary parameter χ_{12} can be related to individual parameters, separately for the solvent and for the polymer, as demonstrated in Eq. (5.48) [372].

$$\chi_{12} = \frac{v_1}{RT}[(\delta_{d,1} - \delta_{d,2})^2 + 0.25(\delta_{p,1} - \delta_{p,2})^2 + 0.25(\delta_{h,1} - \delta_{h,2})^2], \quad (5.48)$$

where v_1 is the molar volume of the pure solvent. The parameters δ_d, δ_p and δ_h are called Hansen solubility parameters. They represent contributions to the cohesion energy density E_{coh} of a pure compound, which is a measure of the totality of intermolecular forces acting between the molecules of the compound with unit volume [373]. The overall cohesion energy is split into contributions due to dispersive (δ_d) and dipolar (δ_p) forces and due to hydrogen bonding (δ_h):

$$E_{coh} = \delta_t^2 = \delta_d^2 + \delta_p^2 + \delta_h^2, \tag{5.49}$$

where δ_t is the total solubility parameter.

A solvent and polymer with similar solubility parameters δ_d, δ_p and δ_h are likely to interact with each other, resulting in solvation, swelling or dissolution of the polymer chains. Conversely, distinctly different Hansen solubility parameters are an indication of incompatibility between the solvent and the polymer.

An approach based on the total solubility parameter δ_t was first proposed by Hildebrand and Scott in 1950, further extended by Hansen in form of Eq. (5.49) and since then has proven its utility for screening and semiquantitative predictions of polymer swelling, wettability and compatibility of various materials [374]. From Eq. (5.48), it immediately follows that the most favourable solvent–polymer interactions correspond to the interaction parameter χ_{12} close to zero. For a cross-linked gel, we anticipate gel swelling in this case. On the other hand large χ_{12} values point to weak polymer–solvent interactions. The latter means that polymer–polymer interactions dominate, causing gel shrinkage and collapse in the worst case.

Hansen solubility parameters δ_d, δ_p and δ_h for a large variety of solvents are tabulated in several sources [374, 375]; a short excerpt is given in Table 5.4. As for polymers, the Hansen parameters are available for many synthetic polymers (see selected examples in Table 5.4), but there are only very limited data for biopolymers. The Hansen parameters of linear polysaccharides can be estimated from structural formula by group contribution methods [376, 377]. Such estimations are given in Table 5.4 for chitin and chitosan.

The approach based on Hansen solubility parameters is illustrated in Figure 5.36 for alginate hydrogels: the volumetric yield of the gels in different solvents from Figure 5.35 is rationalised when plotted as a function of χ_{12}. The original state, i.e., hydrogel, has the highest volumetric yield (100%), all other solvents, as expected, cause a certain shrinkage proportionally to the increase of the interaction parameter.

The extent to which the shrinkage can progress depends on how the gel network resists compression. As a first approximation, it is reasonable to assume that the bulk modulus K of the gel network scales with a power law of the polymer volume fraction $K \sim \phi_{pol}^m$ (with m as a polymer-specific component; such a relation holds also for dry aerogels, as discussed in Chapter 13). This implies that above a certain polymer concentration, the shrinkage may be countervailed by the mechanical response of the gel network. Gels that are prepared from starting solutions with $\lesssim$2 wt% polymer are soft and usually cannot resist changes in thermodynamic quality of the surrounding solvent, demonstrating a significant shrinkage ($\chi_{1,2} > 3$ in Figure 5.36). In this case, the solvent–polymer interactions play the key role in keeping the gel network from collapse. In contrast, the bulk modulus of the network increases with increasing biopolymer concentration and thus the volumetric yield is on average higher as

Table 5.4 Hansen solubility parameters for common solvents, selected polymers and monomers.

Compound	Hansen parameters, $MPa^{1/2}$			Reference
	δ_d	δ_p	δ_h	
Solvents				
MEK	16.0	9.0	5.1	–
2-propanol	15.8	6.1	16.4	–
Acetone	15.5	10.4	7.0	–
1-butanol	16.0	5.7	15.8	–
Methanol	15.1	12.3	22.3	–
DMSO	18.4	16.4	10.2	–
Glycerol	17.4	12.1	29.3	–
Propylene glycol	16.8	9.4	23.3	–
Ethylene glycol	17.0	11.0	26.0	–
Ethanol	15.8	8.8	19.4	–
Propylene carbonate	20.0	18.0	4.1	–
DMF	17.4	13.7	11.2	–
Water	15.6	16.0	42.3	–
Furfuryl alcohol	17.4	7.6	15.1	–
Acetonitrile	15.3	18.0	6.1	–
1,4-Dioxane	16.8	5.7	8.0	–
Polymers and monomers				
Chitin	23.3	15.0	22.5	[377]
Chitosan	22.9	16.7	26.6	[377]
Cellulose, microcrystalline	19.4	12.7	31.3	[378]
Lactose monohydrate	17.6	28.7	19.0	[379]
Lactose anhydrous	19.6	26.2	23.1	[378]
Mannitol	16.2	24.5	14.6	[379]

manifested by data in Figure 5.32 and Figure 5.33. In other words, for dense gels the solvent choice has less substantial effect on the resulting shrinkage.

5.4.4 Minimising Gel Shrinkage

Primary particles in systems such as silica and RF hydrogels are stiff. The network formed by such primary particles is also quite resistant against buckling. This is, qualitatively speaking, the molecular reason why such systems demonstrate a low shrinkage in the solvent exchange. On the contrary, biopolymer gels are often composed of primary fibrils which are also stiff along their axis, but bend easily transversely. That is why polymer strands can be drawn together when the thermodynamic quality of the liquid phase decreases, leading to domination of polymer–polymer interactions over polymer–solvent ones.

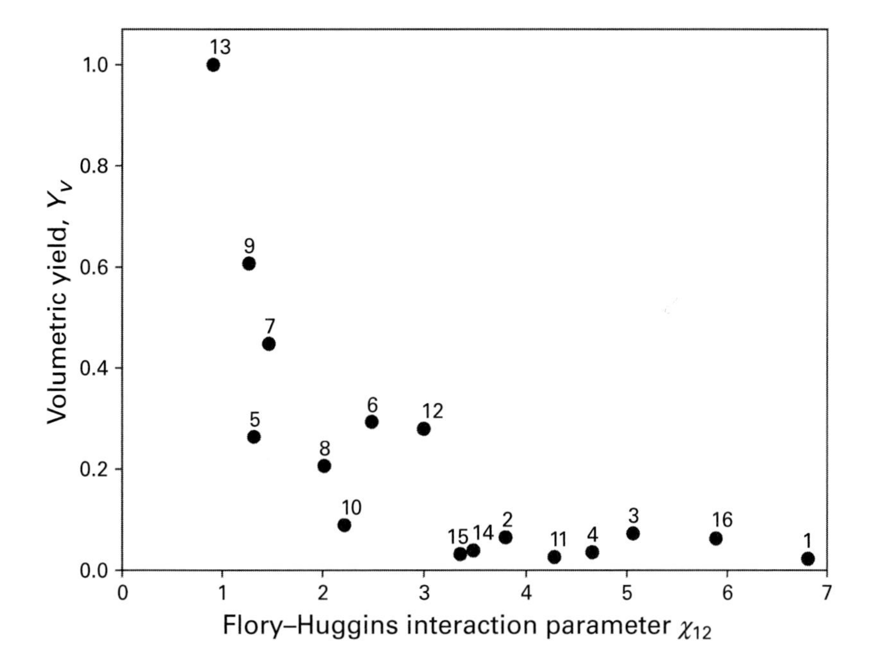

Figure 5.36 Hydrogel volume is partially preserved upon solvent exchange in solvents with similar Hansen solubility parameters as expressed by the Flory–Huggins interaction parameter χ_{12} calculated from Eq. (5.48) at 298 K. The solvent molar volume v_1 is obtained from the molar mass and solvent density. (1) MEK; (2) 2-propanol; (3) acetone; (4) 1-butanol; (5) methanol; (6) DMSO; (7) glycerol; (8) propylene glycol; (9) ethylene glycol; (10) ethanol; (11) propylene carbonate; (12) DMF; (13) water; (14) furfuryl alcohol; (15) acetonitrile; (16) 1,4-dioxane.

When building blocks of a gel are resistant to buckling, very moderate shrinkage is to be expected, even in a one-step solvent exchange: the volumetric shrinkage of a few percent was experimentally observed for hydrogels from bacterial cellulose [380] as well as from cellulose and chitin nanowhiskers [381, 382]. Such fibrils and nanowhiskers are crystalline with a stiffness in the range of a few tens of gigapascals. During gelation and solvent exchange, individual fibrils and nanowhiskers aggregate in bundles due to extensive hydrogen bonding between them and can hardly slip against each other. As a result, hydrogels can be converted into corresponding lyogels (and further to aerogels) at a volumetric shrinkage of 1–6.5% [380–382].

As we mentioned earlier, the multistep solvent exchange is the most widely adopted way to reduce shrinkage: hydrogel is immersed in a series of water–solvent mixtures with progressively increasing concentration of the organic solvent. With more intermediate steps in the solvent exchange protocol, the concentration gradient that acts on the gel at each step is diminished, resulting in higher volumetric yields. Figure 5.37 demonstrates this effect for alginate hydrogels with three different biopolymer concentrations. Being performed at finite concentration gradients, the final volumetric yield is path dependent: even if we wait long enough at each solvent exchange step and let the

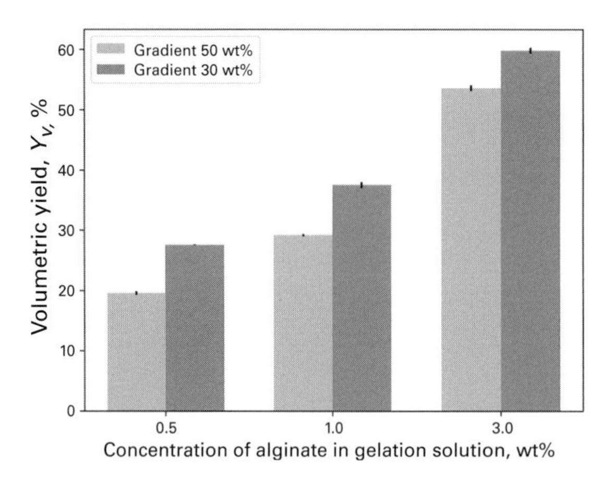

Figure 5.37 Higher concentration gradients of ethanol during a multistep solvent exchange result in reduction of the volumetric yield. Data for alginate hydrogels with 0.5, 1.0 and 3.0 wt% polymer concentration in the starting solution [24]. Reprinted under Creative Commons license CC BY

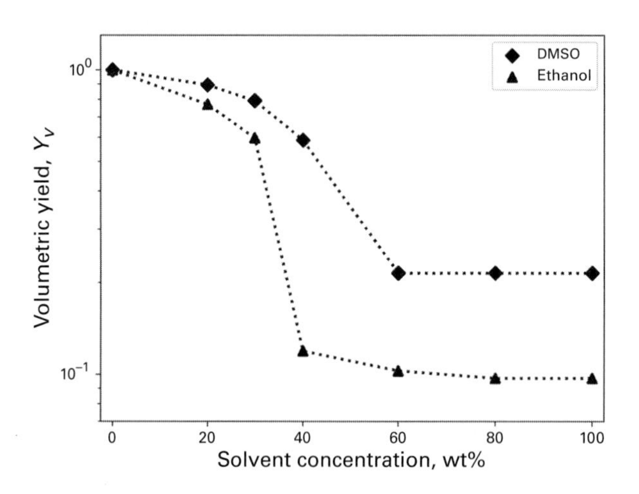

Figure 5.38 Most shrinkage of enzymatically cross-linked guar galactomannan gels occurs between 20 and 40 wt% for ethanol and 40 and 60 wt% for DMSO. Final concentration at each solvent exchange step is plotted along the solvent concentration axis. Standard deviation of each point does not exceed twice the point size. Adopted from [371] with permission from ACS

gel reach a constant volume, the resulting volumetric yield will nevertheless depend on the concentration gradient at each step. With lower gradients, i.e., close-to-equilibrium conditions, we may anticipate the lowest possible shrinkage. In practice, however, maximum of five or six solvent exchange steps are employed, as the provision of more steps increases the overall process duration. The gel volume is not equally sensitive to changes in the solvent concentration during multistep solvent exchange [24]: extensive shrinkage usually takes place below solvent concentration of ~50 wt% (Figure 5.38).

For this reason, solvent exchange protocols usually advise to immerse gels first into 10 and 30% solvent–water mixtures. As the last steps often lead only to a marginal shrinkage, they can be performed at larger concentration gradients, e.g., with 60 and 90% mixtures and finally with pure solvent. Strengthening of hydrogels by adding a reinforcing agent is another common strategy for reducing the hydrogel shrinkage in the course of the solvent exchange. Various reinforcing agents have been reported such as powder fillers and macro-, micro- and nanofibres as well as nanocrystals. So far, this area is driven by empirical results and has not been accompanied by a theoretical rationale.

Although addition of reinforcing agents is aimed to reduce the volumetric shrinkage and thus to help preparing aerogels with lower densities, at the same time such agents increase the overall mass of the solid backbone, causing an increase in the envelope density. To formalise this optimisation problem, let us consider an aerogel, free of reinforcing agents, of volume V_f that encompasses a solid backbone with the mass m. We assume that the backbone mass is preserved during all processing steps and consequently was present in the as-synthesised gel. Let us denote V_0 the volume of the starting solution that was gelled. The cumulative effect of the syneresis and shrinkage at the solvent exchange and supercritical drying is thus given by the total volumetric yield $Y_{v,t} = V_f/V_0$. For the envelope density ρ_e, we readily obtain

$$\rho_e(c=0) = \frac{m}{V_f} = \frac{1}{Y_{v,t}(c=0)}\frac{m}{V_0}. \tag{5.50}$$

Now we add a reinforcing agent at a concentration c into the starting solution keeping its volume V_0 constant. To simplify the further treatment, we select for the concentration c the same unit as for the envelope density, e.g., g/cm^3. The final volume of the aerogel and its envelope density now depend on the concentration of the reinforcing agent:

$$\rho_e(c) = \frac{m + cV_0}{V_f(c)} = \frac{1}{Y_{v,t}(c)}\frac{m + cV_0}{V_0} = \frac{1}{Y_{v,t}(c)}\left(\frac{m}{V_0} + c\right). \tag{5.51}$$

Here $Y_{v,t}(c)$ is the total volumetric yield upon addition of cV_0 units of mass of the reinforcement agent, $Y_{v,t}(c) > Y_{v,t}(c = 0)$. Eliminating the term m/V_0 from Eq. (5.51) using Eq. (5.50), the envelope density of the reinforced aerogel reads

$$\rho_e(c) = \frac{1}{Y_{v,t}(c)}\left(Y_{v,t}(0)\rho_e(0) + c\right). \tag{5.52}$$

The function $Y_{v,t}(c)$ is a measure for the efficacy of the reinforcing agent. Although its analytical form is unknown, a modelled function can be assumed, for example analogously to Eq. (5.46):

$$Y_{v,t}(c) = 1 - (1 - Y_{v,t}(0))\exp(-\alpha c), \tag{5.53}$$

where the coefficient α is a concentration-independent measure of how efficient the reinforcing agent is in preserving the gel volume throughout the entire process ($\alpha > 0$). It can be seen qualitatively from Eq. (5.52) that the first term monotonically

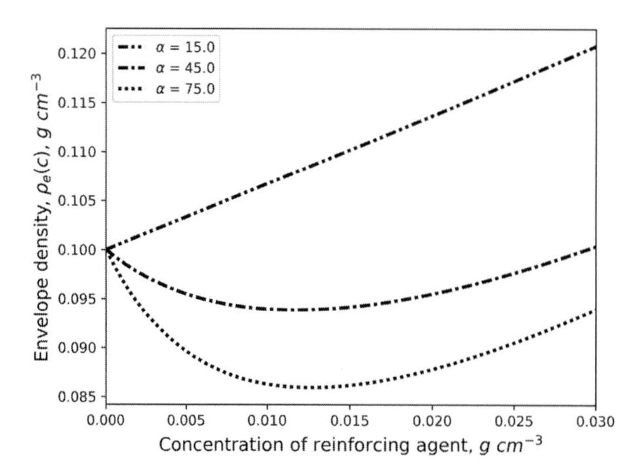

Figure 5.39 Effect of reinforcing agent with different efficiency α on envelope density of the resulting aerogel (envelope density of agent-free aerogel $\rho_e(0) = 0.1$ g/cm^3 and total volumetric yield $Y_{v,t}(0) = 0.6$).

decreases whereas the second term linearly increases. Therefore, there should be a single concentration of the reinforcing agent at which the lowest aerogel density is achieved.

A formal analysis of the model can be performed by equating to zero the first derivative of the envelope density from Eq. (5.52) with respect to c, taking the model volumetric yield function, Eq. (5.53). A detailed analysis shows that the density minimum exists not for all physically meaningful combinations of the parameters α, $Y_{v,t}(0)$ and $\rho_e(0)$: depending on the potency of the reinforcing agent to strengthen a given gel network α, the resistance of the network $Y_{v,t}(0)$ and its density $\rho_e(0)$, the addition of the agent may or may not result in the desired density lowering. A few representative cases for an aerogel with the envelope density of 0.1 g/cm^3 and the total volumetric yield of 0.6 are shown in Figure 5.39.

Leaving further analysis of the model to the reader and omitting cumbersome algebra, we give here a relation between α, $Y_{v,t}(0)$ and $\rho_e(0)$ that should be fulfilled for the density minimum to appear. For the minimum concentration, we have

$$\frac{d\rho_e(c)}{dc} = 0, \tag{5.54}$$

which for Eq. (5.52) with Eq. (5.53) gives

$$(\gamma + \alpha c + 1)\exp(-\alpha c) = \frac{1}{1 - Y_{v,t}(0)} \tag{5.55}$$

with $\gamma = \alpha Y_{v,t}(0)\rho_e(0)$. For small concentrations of the reinforcing agent ($\alpha c \ll 1$), the exponent can be approximated as $\exp(-\alpha c) \approx 1 - \alpha c$ and Eq. (5.55) simplifies to

$$\alpha^2 c^2 + \alpha\gamma c - \gamma - 1 + \frac{1}{1 - Y_{v,t}(0)}. \tag{5.56}$$

For a density minimum to appear, the quadratic equation (5.56) must have at least one real root from whence we finally get

$$\alpha \geqslant 2\frac{\sqrt{\frac{1}{1-Y_{v,t}(0)}} - 1}{Y_{v,t}(0)\rho_e(0)}. \tag{5.57}$$

From Eq. (5.57), we see that strengthening becomes more and more difficult as the envelope density of the reinforcing-agent-free aerogels goes lower. Note that due to approximate analysis, the condition given by Eq. (5.57) is only an estimate: to observe an appreciable density decrease the potency coefficient α has to be approximately twice larger than the right-hand side of Eq. (5.57). For the curves depicted in Figure 5.39, the right-hand side of Eq. (5.57) is approximately equal to 19.4 for $\gamma = 15, 45$ and 75.

5.5 Variables and Symbols

Table 5.5 lists some of the variable and symbols of importance in this chapter.

Table 5.5 List of variables and symbols in this chapter.

Symbol	Meaning	Definition	Units
A	Surface area of a body	–	m^2
c	Specific heat	–	–
c_p	Biopolymer concentration	–	wt%
D	Solute diffusion coefficient	–	m^2/s
D_{12}	Interdiffusion diffusion coefficient	–	m^2/s
dT/dt	Cooling rate	$= G \cdot v$	K/s
ϵ_k	Kinetic energy of a gas	–	J
E_{coh}	Cohesion energy	–	–
$\mathcal{H}$	Mean curvature	–	1/m
H	Enthalpy	–	J
ΔH_V	Volumetric enthalpy of melting/solidification	–	J
ΔH_m	Enthalpy of melting/solidification	–	J
ΔH_{sub}	Sublimation enthalpy	–	J
ΔH_{vap}	Vapourisation enthalpy	–	J
G	Temperature gradient ahead a solid–liquid interface	–	K/m
k_B	Boltzmann coefficient	1.38×10^{-23}	J/K
h_T	Heat transfer coefficient	–	W/m^2K
I_O	Heat transfer through an interface	–	W
M_m	Molar mass	–	kg/mol
N	Number of molecules	–	–
$p(T)$	Pressure	–	N/m^2
p_c	Capillary pressure	–	N/m^2
p_a	Ambient pressure	–	N/m^2

Table 5.6 Continued

Symbol	Meaning	Definition	Units
$p_{sub}(T)$	Sublimation pressure	–	N/m^2
$p_V(T)$	Pressure along the evaporation line	–	N/m^2
$\langle r \rangle$	Intermolecular distance	–	m
t_f	Solidification time	–	s
R	Pore radius	–	–
R_g	Universal gas constant	8.314	J/(mol K)
S_f	Entropy of melting	–	–
Sh	Shrinkage	–	–
S_l	Linear shrinkage	–	–
Sh_t	Total shrinkage	–	–
T	Temperature	–	K
T_s	Surface temperature	–	K
T_0	Initial temperature	–	K
T_E	Eutectic temperature	–	K
ΔT	Solidification interval	–	K
$\|u\|$	Potential energy of a gas	–	J
v	Solidification velocity	–	m/s
V_{L1}	Volume wet gel in liquid L1	–	m^3
V_{L2}	Volume wet gel in liquid L2	–	m^3
X	Mass loss	$\frac{m_{liq}}{m_{dry}}$	–
$y(t)$	Position of the solid–liquid interface	–	–
Y_v	Volumetric yield	–	–
$Z(T)$	Evaporation rate	–	kg/m^2s
α, β	Geometrical exponents	–	–
χ_{12}	Flory–Huggins parameter	–	–
δ_k	Hansen solubility parameters, k = d,p,h	–	–
λ_1	Primary dendrite stem or cell spacing	–	m
λ_2	Secondary dendrite arm spacing	–	m
λ_S	Thermal conductivity solid	–	W/(mK)
λ_L	Thermal conductivity liquid	–	W/(mK)
μ_L	Dynamic viscosity of a liquid	–	Ns/m^2
ϕ_p	Pore volume fraction	–	–
ρ_e	Envelope density	–	kg/m^3
Ω_m	Atomic or molar volume	–	m^3/atom or m^3/mol
σ_{LV}	Gas–liquid interfacial energy	–	Nm/m^2
σ_{SL}	Solid–liquid interfacial energy	–	Nm/m^2
θ	Wetting angle between solid and liquid	–	–
ζ	Capillary length	$\sigma_{LV} \cdot \Omega_m/(k_B T)$	m

6 Morphology of Aerogels

In this chapter, we present some microscopic pictures to illustrate their variety observed in aerogels. The selection is not a complete survey of possible microstructures. Before looking at these, we have to briefly review the methods by which these picture were prepared. More details on electron microscopic techniques in particular can be found in books such as [383] and of course searching in wikipedia for appropriate keywords.

6.1 Imaging Techniques

Two techniques are commonly used to get an impression of the nanostructure of aerogels: scanning electron microscopy (SEM) and transmission electron microscopy (TEM). The first one is most often used because its application looks simple, the machines are readily available and the pictures suggest an understanding. On SEM pictures, which allow to resolve structural entities on nanometre scale, quite often particles, clusters of particles, interconnected particles or fibrils, networks of fibrils can be seen and therefrom diameters can be measured. But SEM pictures give an impression of the pores, including their shape and size. TEM imaging is less often used because the interpretation does not look as simple as in the SEM case, and TEM operation is more complex and needs special knowledge to interpret the structures.

6.1.1 Scanning Electron Microscopy

SEM is used to investigate surfaces and near-surface regions of materials. A sample, having an arbitrarily shaped surface, is scanned with a high-energy electron beam. Acceleration voltages may vary between a few hundred V up to 30 kV and beam currents in the range of a few pA to a few hundred nA. Different kinds of signals result from interactions with the sample atoms and are detected as shown schematically in Figure 6.1. In most modern SEMs, an electron source is based on the field emission principle allowing a nanometre imaging resolution. The electro-optical column accommodates several lens systems made of electromagnetic coils. They allow focussing the electron beam and changing its diameter. A deflection coil is used to scan

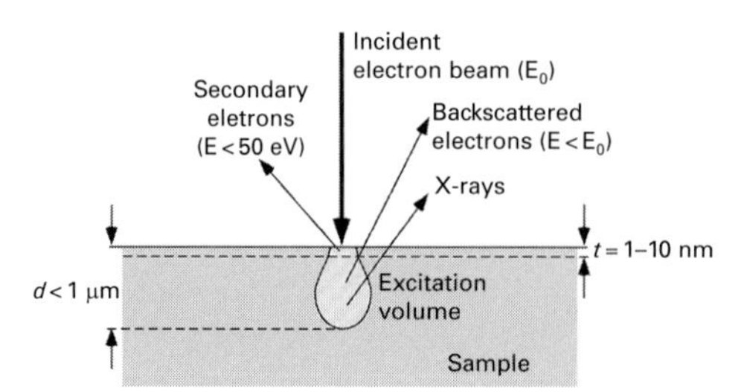

Figure 6.1 The picture shows schematically that a primary electron beam impinging on the surface yields essentially three types of electrons leaving the surface: secondary electrons from a thin layer of thickness t, backscattered electrons from elastic interactions with atom nuclei in a depth of d and of course excited X-rays.

the electron beam over a rectangular area of the sample surface. In contrast to conventional optics, the electromagnetic coils do not magnify the image. The magnification is simply achieved by focussing the beam and changing the scanned area. The electron beam typically has diameters of around 0.2–20 nm. The scan speed can be varied. Typically, the slower the scan speed, the better is the image, but this might lead in bad conducting materials to charging up effects. Therefore, one can also scan the area of interest at a high speed and integrate the resulting images. Then every point of the surface is charged for a short time only.

Different types of signals originate from elastic and inelastic scattering of the incident electrons. The most important signals are secondary electrons (SE), backscattered electrons (BSE), and characteristic X-rays used for energy-dispersive X-ray spectroscopy (EDX). For aerogels, most often only the SE detector is used, but sometimes also the BSE signal. Secondary electrons result from inelastic scattering of incident (primary) electrons at the sample atoms. Primary electrons may ionise sample atoms, excite electrons to higher energy states or stimulate an electron plasma. The ionisation energy is proportional to the nuclear charge. This interaction causes the beam electrons to slow down. They lose a large amount of energy. The SEs typically have energies < 50 eV. They can leave the sample only when their kinetic energy is sufficiently high, i.e., when they are generated in a thin layer beneath the surface, denoted in Figure 6.1 with t. The SEs are collected with a positively charged grid and accelerated towards a detector mounted at an off-axis position. The signal is amplified by a photomultiplier before the image is displayed. The topography of the sample is reproduced because the SE intensity depends on the orientation of the sample surface. More SEs are emitted from prominent edges which appear brighter than, e.g., grooves. A necessary condition for SEM pictures is simply that at least the sample surface is electrical conducting, otherwise the surface will charge up and deflect the electron beam.

Backscattered electrons result from the elastic interaction of electrons with the sample atoms or more specifically their nuclei. The incident electrons interact with the positively charged nuclei such that their trajectory is deflected and can leave the surface of the sample with a high energy. BSEs result from an electron diffusion cloud inside the material, having a raindrop shape and an extension d of around 0.1–1 μm.

The heavier an element is, the higher its atomic number, the more strongly an incident beam is deflected. Thus BSE electrons mirror the atomic number distribution in a material. BSE electrons typically have a high energy close to that of the incident beam. Therefore, they are easy to discriminate from secondary electrons. Today, typically a BSE detector has four circular segments placed above the sample surface, with a hole for the primary beam. For more details of the electron sample interaction, see [383, 384].

6.1.2 Transmission Electron Microscopy

The TEM is used to study microstructural features on a submicron to nanometre scale. Typical examples are phase distributions with a high spatial resolution, detection of grain boundary phases and segregations, evidence of orientation relationships between adjacent phases and analyses of dislocations. In contrast to SEM in TEM, a beam of electrons is transmitted through a specimen, to form an image. In order that an electron beam can be transmitted through a specimen, this has to be very thin, typically below 100 nm. In the simplest case, an image forms, since a sample might have different local varying electron densities (e.g., inclusions, precipitates) such that the electron beam is absorbed differently during passage through the specimen. This gives a picture with a contrast variation and is called a bright field image. The image can be magnified and focussed onto an imaging device, often a fluorescent screen or today a charge-coupled device. In other situations, one has to take into account that the electrons being accelerated up to a few hundred kV are no longer simple particles, but are waves whose wavelength can be calculated according to the de Broglie relation. Then the image is due to interference of electron waves and the intensity on a screen is a superposition of electron waves.

6.2 TEM Images of Aerogels

In contrast to metals and ceramics, which need very special techniques to prepare thin enough samples that can be penetrated by an electron beam, it is very easy to prepare samples of silica aerogels for TEM investigations, since one can make use of the nanostructured network and its brittleness. It is sufficient to scratch off from a sample a few particles or grains and collect them on a TEM sample grid and place it in the TEM. In this way, the pictures shown in Figures 6.2 were taken.

The two pictures show an example of a typical absorption contrast, meaning the brightness of a given area is determined by the thickness of the silica particle network. Very dark, almost black areas show that a few layers of particles are one over the other, whereas the very light grey regions in the upper-left corner look as if silica particles are connected like a string of pearls. This is easier to see in the right figure, which shows many single lines of aggregated particles. Looking at the scale bar in the left corner allows us to estimate the thickness of these lines of particles, namely a value of around 10 nm.

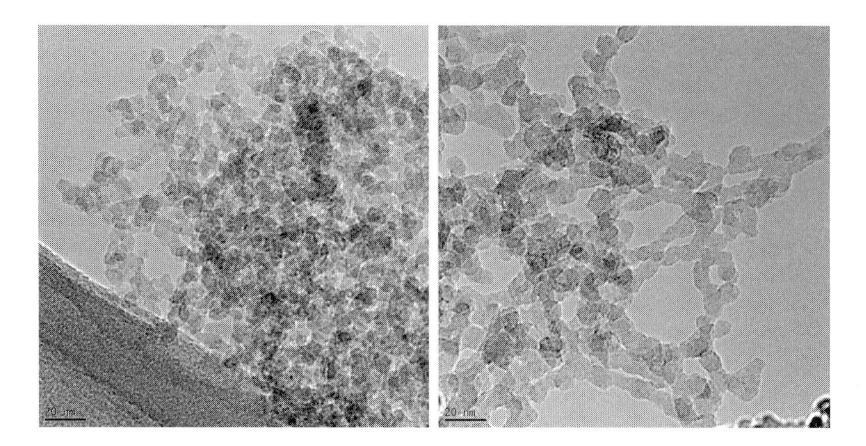

Figure 6.2 TEM pictures of a silica aerogel exhibiting the nanonetwork of colloidal silica particles having an amorphous structure itself.

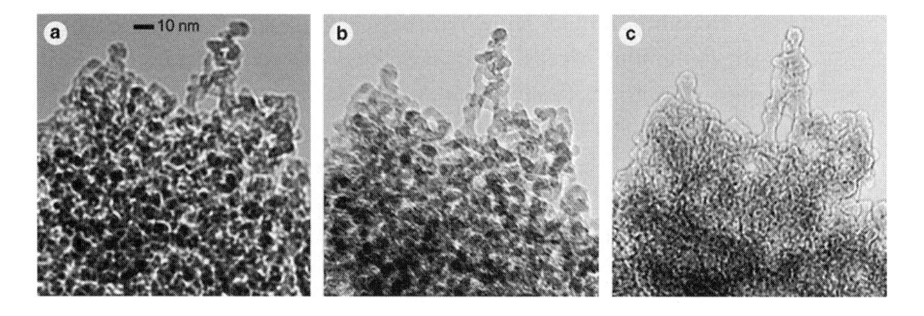

Figure 6.3 Bright field images of a silica aerogel after Stroud et al. [385]. Panel (a) shows the focussed image, (b) a Gaussian focus and (c) an overfocussed image. The pictures reveal that care must be taken to extract from a figure microstructural data, such as particle or neck size. The parameters of making the photos need to be taken into account. Reprinted from [385] with permission from Elsevier

Stroud et al. [385] recently wrote a practical guide for the application of TEM techniques to aerogels. From their work, we reproduce Figure 6.3, which shows an image of a silica aerogel and the effect of focussing.

Similar TEM images of silica aerogels were shown by Guangwu Liu and co-workers [386], showing that essentially all TEM figures look similar to Figure 6.2. TEM pictures of RF aerogels were already prepared by Pekala and co-workers [8] (see Figure 6.4). Although the quality of the picture is not that high, it demonstrates that the RF aerogels prepared 30 years ago by Pekala indeed have a nice particular structure with nanometre-sized particles connected in three dimensions.

Despetis and co-workers [387] made in 2012 the first approach to make a 3D reconstruction of the microstructure of silica aerogels. They used TEM pictures as shown in Figure 6.5. Their reconstruction unfortunately was only on a coarse level, and thus the real 3D structure could not be revealed as one would expect looking at the TEM figures.

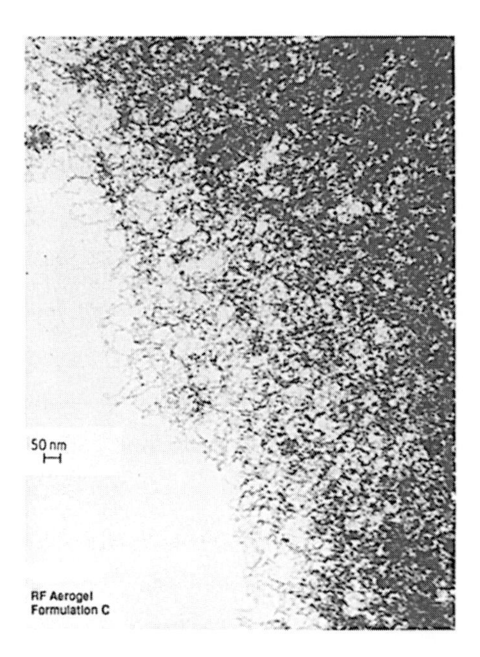

Figure 6.4 TEM of an RF aerogel presented by Pekala et al. [8], exhibiting a particulate microstructure of nanoparticles. Reprinted from [8] with permission from Springer-Nature

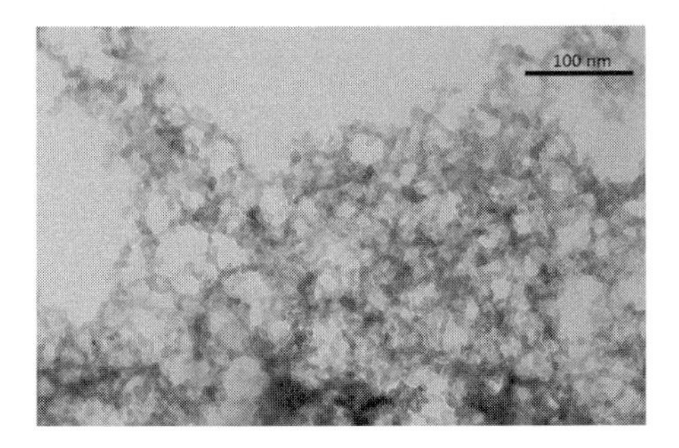

Figure 6.5 Bright field TEM image of base catalysed silica aerogel made by Despetis and co-workers. The structure looks like the TEM figures shown previously: particles connected like a string of pearls with relatively large necks. One also gets an impression of the pores and their size. Reprinted from [387] with permission from Elsevier

TEM pictures from carbon aerogels prepared by pyrolysis of Pekala RF aerogels look as if they would be silica aerogels: the nanostructure looks very similar, namely string of pearls being connected in 3D. Figure 6.6 shows two examples.

As discussed in the section on silica aerogels, one can prepare intrinsically flexible silica aerogels by replacing TEOS by a trifunctional silane, MTMS. Hedge and Rao [388] presented their results on MTMS aerogels and also showed that the variation of the ratio MTMS to methanol changes the microstructure considerably. Their TEM images are shown in Figure 6.7.

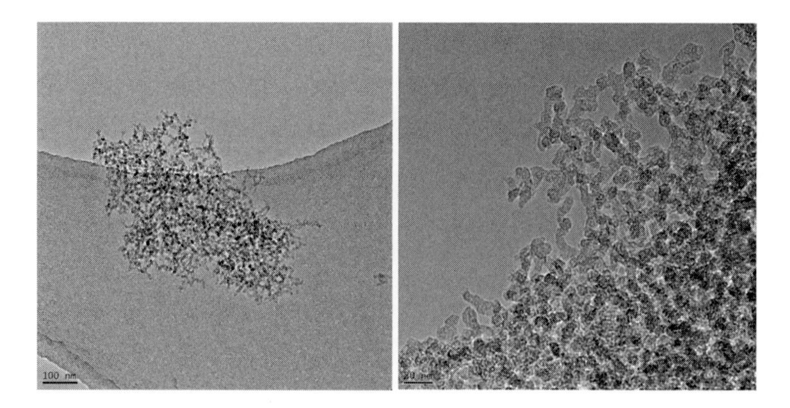

Figure 6.6 TEM pictures of a carbon aerogel exhibiting the nanonetwork of colloidal carbon particles having an amorphous structure itself. Starting material was a Pekala-type RF aerogel converted to carbon by pyrolysis.

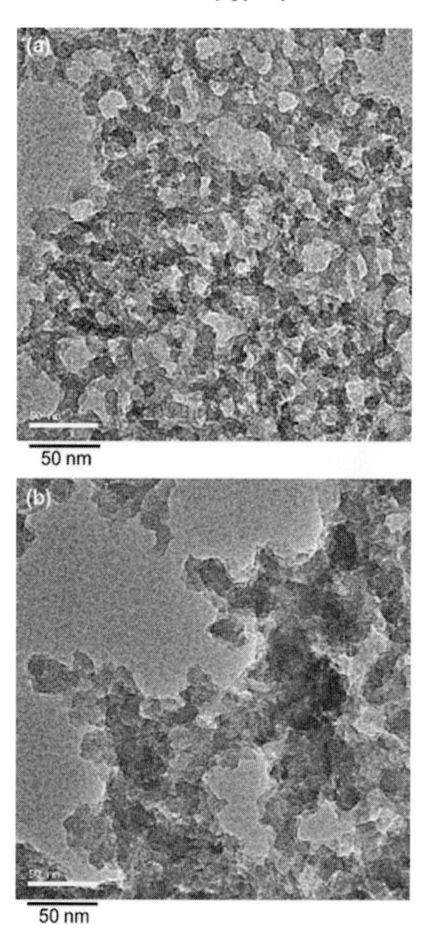

Figure 6.7 Bright field TEM images two silica aerogels made with different molar ratios of methanol to MTMS [388]. In (a), the molar ratio MeOH/MTMTS is 14, and in (b) 35. The more diluted the system is, the more irregular and compact is the microstructure. Reprinted from [388] with perimission from Springer-Nature

6.3 SEM Images of Particular Aerogels

Scanning electron microscopic images of silica aerogels are more difficult to realise due to the isolating character of the silica particles. This can lead to charging of the particles and therefore deviation of the electron beam, blurring and other image artefacts. Therefore, all aerogels are organic or inorganic in nature, and they are isolators that are sputtered with a thin film of either gold or platinum to yield a certain electrical conductivity. This can alter the image, because it leads, for instance, to spots on the surface of a particle, which are not the nanostructure of the aerogels but an artefact. In addition, one has to mention that sputtering means that a plasma is produced in an argon atmosphere between two electrodes. On one electrode, a thin disc of Au or Pt is fixed on the other sample to be coated. The plasma etches Au or Pt away from the source and deposits them onto the rough surface of the aerogel. There is absolutely no guarantee that the deposition is everywhere homogeneous on the nano- to micrometre-sized particles in the porous network, and the metal film is deposited everywhere with the same thickness. It even cannot be assured that a film has been established, and sputtering may lead to more than islands or patches of metal, as shown in Figure 6.8. Therefore, interpretation of SEM pictures always needs care, and one should vary the acceleration voltage and the beam current to see how these affect the image. An example of a sputter artefact is shown in Figure 6.8, illustrating possible misleading deposit structures.

The nanostructure of silica aerogels can also be looked at with SEM. But the structure interpretation is not always easy and straightforward, since one has to take into account that one always looks at a fracture surface and it has to be proven that this surface and the underlying sample reflects the real material structure inside an aerogel (see Figure 6.9). During breaking, the sample might undergo changes in morphology which are difficult to describe. This is well known for cellulose aerogels, since they are soft and a fracture at room temperature deforms the material considerably.

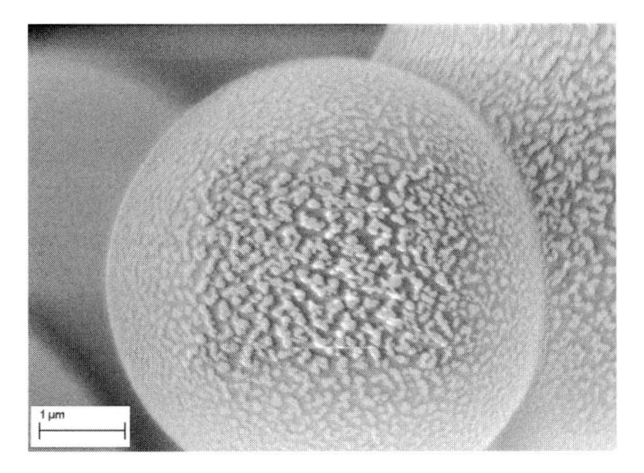

Figure 6.8 SEM picture of an RF aerogel synthesised with citric acid. The deposit is an Au sputtering. Spots on the spherical surface are Au particles stemming from sputtering.

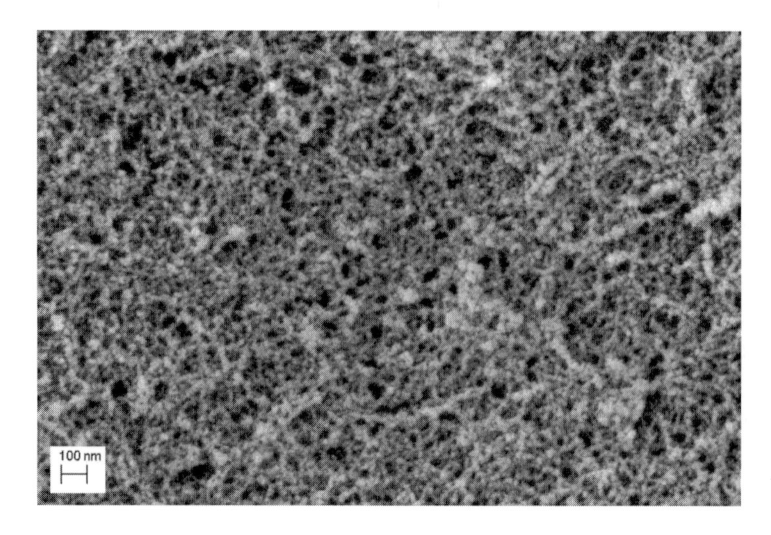

Figure 6.9 SEM picture of a silica aerogel fracture surface exhibiting in a different way the nanonetwork. This should be compared to the TEM pictures shown previously.

Performing breakage under liquid nitrogen helps and avoids such deformations. In addition, one has to carefully check that the sample does not change during irradiation with an electron beam. Especially in biopolymeric aerogels, one can observe that the electron beam can completely vaporise fibrils. The only way out might be a reduction of accelerating voltage and beam current density. As a first example, we use the SEM image of a silica aerogel prepared by a two-step routine from TEOS. Gurav and co-workers [389] analysed the physical properties of silica aerogels prepared from sodium silicate. The sodium silicate was treated with an acid and allowed to gel, followed by the step of passing water vapour through the hydrogel, and finally they performed a solvent exchange with pure methanol and a step of hydrophobisation with TMCS. The hydrophobisation was allowed to dry the materials under ambient conditions. Figure 6.10 shows a result this procedure on the microstructure.

Preparing RF aerogels according to the recipes given by Pekala leads to a particulate structure which mimics a mixture of a fibrillar network, but in reality one can observe with high magnification particles of polycondensed resorcinol lying on strings and looking like fibrils as shown in Figure 2.15.

As mentioned in Section 2.2, on RF aerogels, they can be prepared either along a base catalysed route, most often with sodium carbonate used as a catalyst, or an acid catalysed route. Two examples of aerogels prepared with two acids are shown in Figure 6.11. In both cases, spherical particles are obtained connected in 3D to a wonderful network of particles having almost the same diameter. Notice also that both aerogels do not really exhibit a nanostructure, but the scale of their particles is in the micrometre range. This, as mentioned previously, is due to the fast condensation reaction, the point that under acidic conditions fewer nucleation centres are formed and the fact that the fast growth by the condensation reaction then leads to precipitating of all monomers onto the small number of starting particles (nuclei).

Figure 6.10 SEM images of silica aerogels prepared from sodium silicate by Gurav et al. [389] after 30 min (a) and 2 h treatment (b) with water vapour in the state of a hydrogel. The longer treatment increased the density by a factor of 2 leading to a denser network, which is reflected in the SEM pictures. Reprinted from [389] with permission from Elsevier

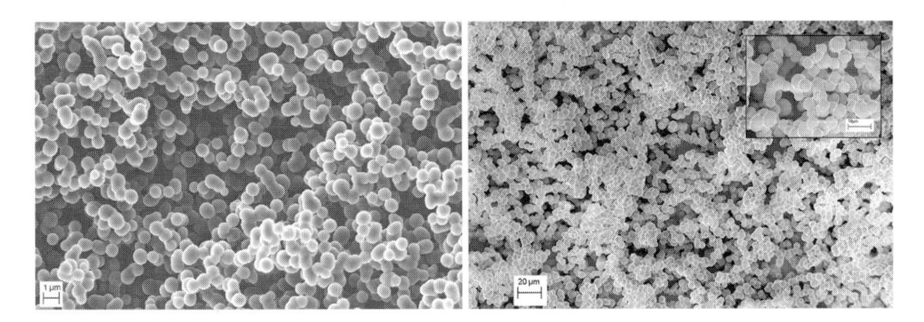

Figure 6.11 Microstructure of RF aerogels prepared with HCl (left panel) and citric acid (right panel).

Figure 6.12 SEM images of RF aerogels after Mulik et al. [103]. Panel (a) shows the RF aerogel (envelope density of 0.175 and 1:3 R:F molar ratio). Panel (b) shows for comparison a native silica aerogel of similar density $\rho_e = 0.169$. Courtesy American Chemical Society

A nanostructure can be prepared with an acid route if one follows the recipe of Mulik et al. [103]. Figure 6.12 shows two SEM images presented by Mulik et al. comparing an RF aerogel prepared according to their route with a silica aerogel of similar envelope density. In both cases, nanosized particles can be seen which look as if they contain smaller subparticles.

6.4 SEM Images of Fibrillar Structures

Biopolymeric aerogels often exhibit a different nanostructure. In many but not all cases, a so-called fibrillar structure is obtained. The effect of an increase in cellulose content makes the network more dense, but essentially keeps the fibrillar network. An example is shown in Figure 6.13. One should realise that in these aerogels the cross points of the fibrils are chemically bonded. Therefore, the network has no similarities with a felt, which is a mechanically loose mixture of fibres with no bonding at touching

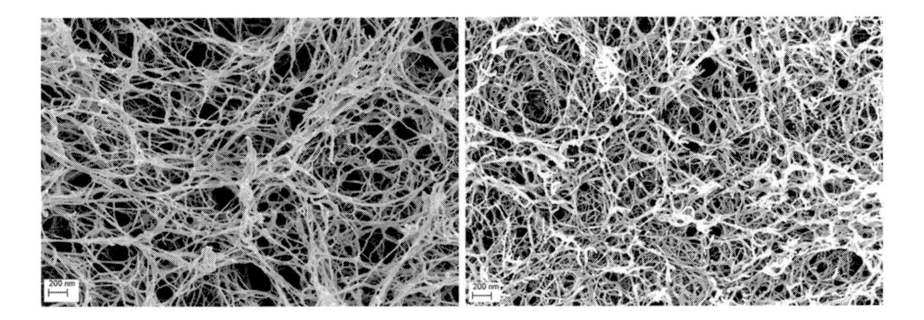

Figure 6.13 Left panel: nanostructure of cellulose aerogels prepared with the calciumthiocyanate route (2 wt% cellulose); right panel: cellulose aerogels with 4 wt% cellulose. Courtesy K. Ganesan, DLR

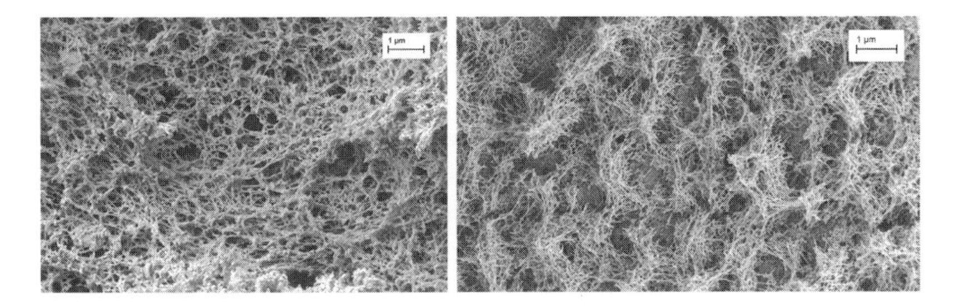

Figure 6.14 Nanostructure of cellulose aerogels prepared with a $ZnCl_2$–hydrate melt (3 wt% cellulose). washed with different fluids. Left panel after washing with water, right panel after washing with iso-propylalcohol. Both aerogels were dried supercritically with carbon dioxide. Reprinted from [150] with permission from Elsevier

points. The similarity on the macro-scale is more to that of a force transmitting fleece in which fibres are connected at the cross points by for instance a melt puddle in the case of polymers or a glue. If cellulose aerogels are prepared with a different solvent, such as a $ZnCl_2$-hydrate melt, and the gel fluid is washed out with different liquids (water, ethanol, acetone), different nanostructures result. These are a result of a complex interaction between the washing fluid and the cellulose complexed with salt hydrates. One example is shown in Figure 6.14. Details can be found in the paper by Schestakow et al. [150].

7 Density

Models and Measures

The seemingly simplest property of any material and aerogels especially is the density ρ, which is defined as the ratio of mass m and volume V, $\rho = m/V$. For any regularly shaped body such as a cube, sphere or cylinder, the volume is readily determined and the mass obtained by simply weighing the body. For a porous material, especially if the shape is not regular, the density is not that easy to determine. For aerogels, two different values are usually determined: the so-called envelope density ρ_e and the skeletal density ρ_s. Both density value definitions are schematically depicted in Figure 7.1. The envelope density is defined as the mass m divided by the total volume V_e enclosing the porous structure. The skeletal density instead is the density of the solid backbone of the aerogel, i.e., the sum of the volume of all nanoparticles making up the aerogel V_s, $\rho_s = m/V_s$.

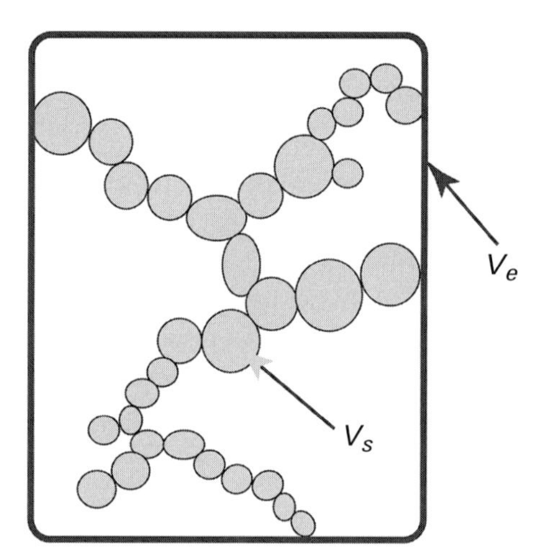

Figure 7.1 The envelope volume V_e of an aerogel with the embedded nanostructure of a total volume V_s defining the skeleton or backbone of the aerogel.

7.1 Measurement of Envelope Density

Let us take an arbitrary piece of aerogel or porous body in general. Its mass is easily measured on a balance. The volume of an irregularly shaped body needs special experimental techniques. For instance, one could think to use Archimedes' principle by submerging the body in a suitable fluid and measuring the buoyant uprise. For an open porous body, this technique cannot be used, since the fluid generally would be sucked into the pores. One can, however, use a solid equivalent realised in a facility called GeoPyc^{TM} [390]. In this measurement equipment, a glass tube of well-defined diameter is filled with a fixed amount of a calibrated sand and compressed by a vibrating piston with the preset force (50 N or less). After compression, the length of the sand pillar is automatically measured with a transducer. Then the irregular shaped piece of material is mixed with the same sand and again compressed and the new length determined. With the known diameter one also calculates the new volume. Subtracting the volume of the pure sand pillar gives the volume of the enclosed aerogel piece. The relation between sand volume and aerogel volume should be 70/30 to give accurate results. Using the weight of the aerogel finally yields the envelope density. The force for compression must be chosen to avoid a deformation of the aerogel. A very compressible or flexible aerogel cannot be measured with this technique. Best are stiff ones, which retain their shape on compression with a small load. In any case, one could perform measurements with various compression forces and extrapolate the apparent volume of the aerogel to zero force.

7.2 Measurement of Skeletal Density

The skeletal density of aerogels is typically measured with a gas pycnometer. The principle is simple: in a given volume of a container, a gas is filled and the pressure measured (see Figure 7.2). Putting into the container an open porous material and filling in the same amount of gas as in the empty container, a pressure difference is measured from which the volume of the material can be determined using the ideal gas law. Technically the gas displacement measurement is done a bit differently and more conveniently.

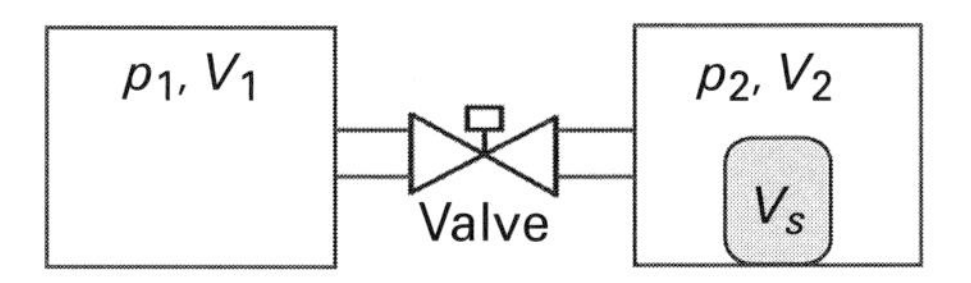

Figure 7.2 Principle measurement setup for determining the skeletal density of a porous body using two compartments of the same volume connected by a valve.

7.2.1 Measurement Principle

A schematic drawing of a gas pycnometer is shown in Figure 7.2. The system essentially consists of two compartments connected by a motorised valve and, of course, being at the same temperature for the whole time of the measurement! Initially, compartment 1 having a volume V_1 is evacuated and a certain time later filled with helium to a pressure p_1. The pressure is recorded and the valve is opened. Helium flows into container 2 with a volume V_2. In this container the aerogel to be measured also is inserted. After some time, the pressure in both connected compartments reaches a value p_2. Recording this pressure, one can calculate the unknown skeletal volume V_s and from this the skeletal density

$$\rho_s = \frac{m}{V_s}. \tag{7.1}$$

The volume V_s is calculated using the ideal gas law. The balance equation of both compartments reads

$$p_1 V_1 = p_2(V_2 + V_1 - V_s). \tag{7.2}$$

Let us assume for simplicity that both compartments have the same volume, $V_2 = V_1 = V_0$. Solving Eq. (7.2) yields the skeletal volume

$$V_s = V_0\left(2 - \frac{p_1}{p_2}\right). \tag{7.3}$$

This relation sets boundaries on the pressure p_2. If $p_2 = p_1/2$, we have $V_s = 0$ (empty compartment). If $p_2 = p_1$, we have $V_s = V_0$, meaning the second compartment is completely filled. Any measurement of p_2 should therefore fall in the interval $p_1/2 < p_2 < p_1$. With this sekeletal volume, the skeletal density is calculated as

$$\rho_s = \frac{m}{V_s} = \rho_e \frac{V_e}{V_0} \frac{1}{2 - \frac{p_1}{p_2}}. \tag{7.4}$$

Eq. (7.4) also shows, that always $\rho_s > \rho_e$. For a good measurement, one should have a large V_e in the chamber, or better a large V_s, and pressure values $p_2 > p_1/2$ to get reliable values (a filling of the chamber of about 10% solid is recommended, but is not easy to achieve with low-density aerogels).

One can also solve Eq. (7.2) for the final equilibrium pressure in the connected compartments, naming it now p_2^s:

$$p_2^s = p_1 \frac{1}{2 - \frac{V_s}{V_0}}. \tag{7.5}$$

This result tells us that any sample put into compartment 2 increases the final pressure above the empty value $p_1/2$. If inside the porous material the equilibrium pressure p_2^s is rapidly reached, the measured skeletal density is determined using Eq. (7.4).

Although this all looks fine, one has to be aware of a few issues:

1. Typically the skeletal volume is small. With porosities of aerogels typically in the range of 90–97% and a ratio of envelope to compartment volume of around $V_e/V_0 \approx 1/10$, the ratio V_s/V_0 is easily only 0.005 and then the pressure difference is small, and that might lead to errors.
2. Helium is used since it is the smallest noble gas atom and goes into all *open* pores. If the material contains also a certain amount of closed porosity, this is not detected.
3. The value for the skeletal density extracted from a measurement must always be smaller than the value of the fully crystalline or massive or bulky material having no pores, and the relation $\rho_s < \rho_{bulk}$ strictly holds.
4. One should bear in mind, that there might already be some adsorbates on the surface of aerogels, e.g., water, especially in hydrophilic silica or biopolymeric aerogels. This might change the skeletal density. The same is true if at the surface other molecules are bonded, such as ethoxy or trimethylsilyl groups.
5. What happens if the diffusion of helium (and maybe also nitrogen and oxygen left over from the evacuation of compartment 2 before filling with helium) out or into the aerogel is impeded and takes not seconds but minutes or hours? Then we expect that the pressure p_2 will be a function of time $p_2(t)$ and thus the apparent skeletal volume and the density. We will calculate the characteristic times for this equilibration process later.
6. Even helium as detecting molecules has a finite size, and therefore it might be possible that it cannot detect the necks between nanosized particles.

We first start with a simple consideration concerning a possible effect of adsorbates on the skeletal density. We then discuss the time dependence of pressure evolution in the measurement chamber (diffusion in or out the aerogel) and finally we come back to a possible size effect of the test gas.

7.2.2 Effect of Adsorbates

Before we do a calculation, let us think about a possible effect of adsorbates. Adsorbates are molecules which are attached to the pore surface of the aerogel. Therefore, they increase the mass and the measured skeletal volume. Thus any adsorbate on the surface could change the apparent skeletal density. How can this be calculated and related to known properties of the adsorbate and the aerogel?

Let S_m be the measured specific surface area in units m^2/g of the aerogel, ρ_e the envelope density measured in kg/m^3 and ϕ_s the fraction solid of the skeletal material in the aerogel body. We then have for the specific surface area per volume the relation

$$S_V = \rho_e S_m. \tag{7.6}$$

From this, we get the total inner surface of the aerogel as

$$A_s = S_V V_s = S_V \phi_s V_e. \tag{7.7}$$

Let Ω_m be the atomic volume of the adsorbate (measured in m^3). The number of adsorbed molecules at the inner surface is

$$N_m = \frac{A_s}{\Omega_m^{2/3}}. \tag{7.8}$$

These molecules occupy a volume at the inner surface of

$$V_{ads} = N_m \Omega_m = A_s \Omega^{1/3} = \Omega_m^{1/3} \phi_s V_e \rho_e S_m. \tag{7.9}$$

The additional mass of the adsorbate is calculated with the atomic mass M_a, given in grams, for instance, as

$$m_{ads} = N_m M_a. \tag{7.10}$$

Using the preceding relations, one arrives at

$$m_{ads} = \frac{M_a}{\Omega_m^{2/3}} \phi_s \rho_e V_e S_m. \tag{7.11}$$

The adsorbate modified skeletal density ρ_s^{ads} is defined as

$$\rho_s^{ads} = \frac{m_{true} + m_{ads}}{V_s^{true} + V_{ads}}, \tag{7.12}$$

where we have denoted with V_s^{true} the skeletal volume if there are no adsorbates and split the mass into two contributions, the true mass and the adsorbate one. This relation can be simplified a bit:

$$\rho_s^{ads} \cong \left(\frac{m_{true}}{V_s^{true}} + \frac{m_{ads}}{V_s^{true}} \right) \left(1 - \frac{V_{ads}}{V_s^{true}} \right). \tag{7.13}$$

Using Eq. (7.9) leads to

$$\rho_s^{ads} = \left(\rho_s^{true} + \frac{m_{ads}}{V_s^{true}} \right) \left(1 - \rho_e S_m \Omega_m^{1/3} \right) \cong \rho_s^{true} \left(1 - \rho_e S_m \Omega_m^{1/3} \right) + \frac{m_{ads}}{V_s^{true}} \tag{7.14}$$

and thus finally we arrive at

$$\rho_s^{ads} = \rho_s^{true} (1 - \rho_e S_m \Omega_m^{1/3}) + \frac{M_a}{\Omega_m^{2/3}} \rho_e S_m. \tag{7.15}$$

In accordance with our expectation, adsorbates at the inner surface change the measured skeletal density. The larger the specific surface area, the larger the effect. Molecules with a larger atomic volume, such as water or alcohol compared to nitrogen, have a larger effect. Therefore, it is recommended before performing a skeletal measurement to heat treat a sample in a vacuum chamber. Let us make an estimate of the effect. For cellulose aerogel, the specific surface area is 200 m^2/g. The envelope density is 100 kg/m^3 and water is the adsorbate. The molar volume of water is $18 \cdot 10^{-6}$ m^3/mol. With the Avogadro number, this gives an atomic volume of $3 \cdot 10^{-29}$ m^3/molecule. Inserting these values into Eq. (7.14) gives a correction of the order of 0.006 or approximately 0.01. Cellulose has in its crystalline state a density of 1.5 g/cm^3. The correction is then of the order of 1% and less. One has to keep in mind that a multilayer adsorption of water would increase the value by the number of layers. The change due to adsorbed mass can be calculated to be $\approx 0.08\ \rho_e$, and this means a change from 1.5 to 1.508 g/cm^3 and thus an effect of around 5%. Interestingly, Eq. (7.15) shows that there are two opposing effects. The additional mass of the adsorbent increases the skeletal density, and the additional volume of the adsorbent yields a larger displaced volume and thus a lower skeletal density. The effect of chemically bonded methoxy, ethoxy or trimethylsilyl groups at the surface is discussed later; see also Section 7.5.1.

7.2.3 Pressure Evolution

There is limited literature with respect to the time-dependent aspect of skeletal density. A measurement using aerogels was reported in the paper of Woignier et al. [391, 392]. Woignier and co-workers developed their own gas pycnometer. They report that equilibration can take a long time, due to the slow diffusion of gas into the nanopores. There are of course other papers on gas permeation through aerogels, an aspect being important to discuss the time necessary to either evacuate the aerogel pore space or fill it with helium [393–396]. There are two different problems that might occur in the measurement leading to two different model calculations:

Model 1: The impeding step in reaching the true equilibrium pressure p_2^s is the diffusion of gas molecules such as nitrogen and oxygen out of the aerogel into both compartments during the evacuation step of compartment 2. There is a leftover gas content in the aerogel, and thus the real equilibrium pressure cannot be reached.

Model 2: The impeding step is the filling of the aerogel pores with helium by diffusion and thus reducing the pressure in the connected compartments. This means the evacuation time and value are sufficiently fast compared to the diffusion of helium into the pores.

In any case, there would be a pressure difference, and the final equilibrium value p_s^s derived earlier will be reached later. Let us check how big the possible effect is.

Non-Equilibrium Pressure Difference

Let us assume that the pressure equilibration in the machine runs in two steps, since the helium diffusion into the aerogel pores takes much more time compared to the equilibration time of the pressure in both chambers. The final value in compartment 2 after a few seconds will be determined by the compartment volumes and the envelope volume of the aerogel V_e. The diffusion of helium into the evacuated aerogel pores would in a second step decrease the pressure in the machine. Let us calculate this. If the volume V_e displaces gas in compartment 2, the equilibrium pressure would be

$$p_2^e = p_1 \frac{1}{2 - \frac{V_e}{V_0}}. \tag{7.16}$$

This pressure differs from the final equilibrium pressure after the diffusion of the gas into the pore network of the aerogel by a pressure difference Δp, which can be calculated. Note that the skeletal and the envelope volume are related by the fraction of solid material in the envelope volume, ϕ_s (fraction of solid aerogel stuff in the volume):

$$V_S = \phi_S V_e. \tag{7.17}$$

We then have

$$\Delta p = p_2^s - p_2^e = p_1 \left(\frac{1}{2 - \frac{V_S}{V_0}} - \frac{1}{2 - \frac{V_e}{V_0}} \right) \cong -p_1 \frac{V_e}{2V_0} \frac{1 - \phi_s}{2 - \frac{V_e}{V_0}}. \tag{7.18}$$

The pressure difference always is negative, since $V_e < V_0$. The pressure difference can be estimated. Assume $p_1 = 1$ bar, compartment volumes $V_0 = 100$ mL and envelope volume of 1 mL and a porosity of 90%. Then the pressure difference would be -2.2 mbar on top of the empty volume pressure reading of 500 mbar or roughly 0.5 percent pressure difference. One can also argue that the larger the porosity, $\phi_p = 1 - \phi_s$, the stronger the second pressure reduction after the first rapid equilibration. This pressure difference and its time evolution is the basis of the dynamic gas expansion method developed by Reichenauer and Fricke [393, 394].

Characteristic Diffusion Time

Looking at the diffusion equations in Chapter 10, the characteristic time is quite generally defined as

$$\tau_c \approx \frac{\ell_c^2}{4D}, \tag{7.19}$$

with ℓ_c a characteristic length, like the sample size, and D the diffusion coefficient. As will be explained in more detail in Chapter 10, if the pores are large such that the random motion of gas molecules is not impeded by the pore walls, the diffusion coefficient differs from one on which the molecules collide mainly with the pore walls and not with themselves. For the case of narrow pores, the so-called Knudsen effect has to be taken into account. In Chapter 10, the diffusion coefficient for such a case $D = D_{Kn}$ is derived. The characteristic length is the shortest geometrical length scale of the sample under investigation. For a sphere, it is just the radius, for a cube it is half the edge length, and for a plate it is half the plate thickness. We then have two different characteristic times. Let us denote the time under Knudsen conditions as τ_c^{Kn} and under normal diffusive conditions without any free path constraint as τ_c^{μ}. Using Eqs. (10.22) and (10.27),

$$\tau_c^{Kn} = \frac{\ell_c^2}{4D_{Kn}} = 3\frac{\ell_c^2}{2R_p\phi_p}\sqrt{\frac{\pi M_a}{8k_B T}} \tag{7.20}$$

and

$$\tau_c^{\mu} = \frac{\ell_c^2}{4D_{\mu}} = 3\frac{\ell_c^2 p a_m^2}{2\phi_p T^{3/2}}\sqrt{\frac{\pi^3 M_a}{k_B^3}}. \tag{7.21}$$

Let us compare both times and estimate a few values to get an impression of how important the time-dependent evacuation of an aerogel sample or its filling with helium really is. The dimensionless ratio of both characteristic times is

$$\frac{\tau_c^\eta}{\tau_c^{Kn}} = \frac{\sqrt{8}p\pi a_m^2 R_p}{k_B T}. \tag{7.22}$$

Using that in the volume of the connected containers V_0 at equilibrium pressure the number of molecules N_m is constant, we can use the ideal gas law to simplify this expression to

$$\frac{\tau_c^\mu}{\tau_c^{Kn}} = \sqrt{8}\pi a_m^2 R_p N_A \frac{N_m}{V_0}. \tag{7.23}$$

First it is interesting to note that the ratio only depends on the pore size R_p, neither the temperature nor the pressure or the porosity. Second, we can conclude that the Knudsen time is essential for the evacuation of the sample, since the pores are typically small. It would thus be sufficient to calculate the characteristic Knudsen time. Let us make an estimate. If the characteristic length ℓ_c is the size of an aerogel sample, say 10 mm, takes the temperature as 295 K, the pore size as 10 nm and uses the well-known mass of helium and a porosity of 95%, one can calculate a characteristic Knudsen diffusion time of 50 seconds. Thus in a measurement it would be good to allow for evacuation of the compartment with the sample something like 200 seconds and for equilibration the same. One can then be sure that the slow diffusive motion of gas molecules out and into the aerogel does not affect the final pressure measurement and thus the skeletal density value.

Before we discuss a possible effect of the test gas size on the skeletal density measurement, let us first present some results. Job et al. [397] measured the skeletal density on RF aerogels. They used different R/C ratios and different dilution ratios. They also prepared not only supercritically dried RF aerogels but also freeze dried and ambient dried ones. Looking at their data, there is no clear trend that relates, for instance, skeletal and envelope density, or specific surface area and skeletal density nor any other trend. All data seem to scatter around a value of 1.5 ± 0.05 g/cm^3 as can be realised from Figure 7.3.

RF aerogels were also prepared by Schaefer and co-workers [252]. They used a standard recipe described earlier, but varied the amount of catalyst. Besides their interesting other results, which were treated in Chapter 4, they also measured the skeletal density. Their result is shown in Figure 7.4 Comparing the measurement by Schaefer et al. and Job and co-workers, it first is striking that the older measurements from Schaefer et al. yield much lower values, below 1.2 g/cm^3. In most papers on RF aerogels, the skeletal density is reported as having a value of around 1.5 g/cm^3, which is well in the range of Job's measurements. This could be a hint to slightly different gelation, ageing and washing procedures. Job et al. [397] prepared their RF aerogels according to the same procedure as Schaefer et al. [252] with respect to the chemical components and their ratios, but they left them in sealed flasks for about 72 h at 85°C, washed the RF gels with ethanol and performed supercritical drying

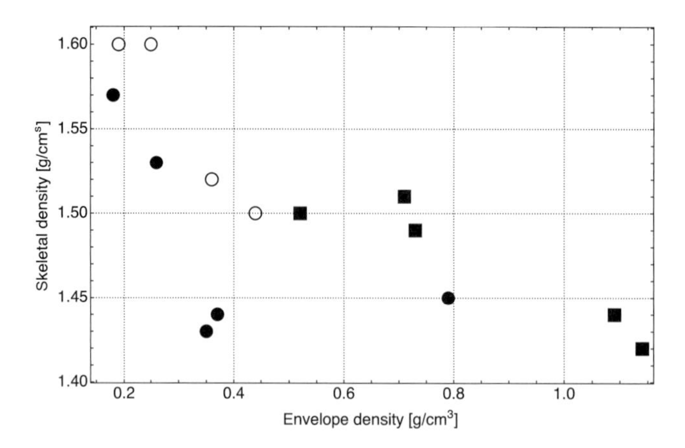

Figure 7.3 Skeletal density of RF aerogels after Job et al. [397]. Open circles are data from supercritically dried aerogels, full circles from freeze dried aerogels and full squares from xerogels.

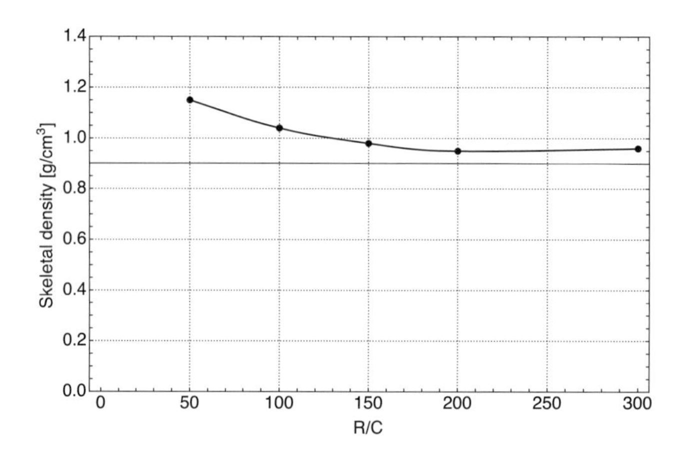

Figure 7.4 Skeletal density of RF aerogels. The supercritically dried RF aerogels were prepared with different ratios of resorcinol (R) to catalyst (C), here sodium carbonate. Reprinted and adapted from [252] with permission from Elsevier

with carbon dioxide. Schaefer instead left the gelling solution 7 days at 85–90°C in a sealed container, washed them in a dilute acid to increase the cross-linking by dimethylenether groups and exchanged the liquid in the pore with acetone before performing the same supercritical drying. This slight difference in procedure might lead in the case of Schaefer et al. [252] to closed pores, which give then an apparent smaller skeletal density.

For silica aerogels prepared from TMOS, Woignier and co-workers carefully measured the skeletal density and discussed aspects of supercritical drying and test gas size on the results [391, 392]. The aerogels were prepared from TMOS diluted in methanol and hydrolysed with 0.0001 molar nitric acid. The volume fraction of TMOS was varied between 6–33%. Directly after gelation, the alcogels were

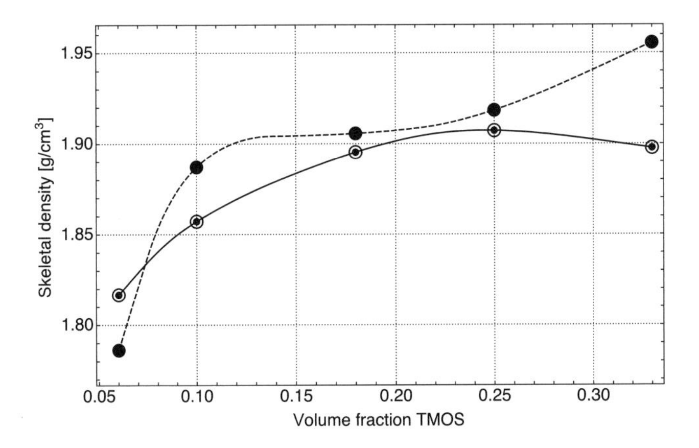

Figure 7.5 Skeletal density of silica aerogels prepared from TMOS after [392]. The full circles belong to aerogels supercritically dried with carbon dioxide, the open circles are silica aerogels supercritically dried with methanol.

put into an autoclave and the methanol liquid in the pores brought directly to a supercritical state. As mentioned by them and in the literature on silica aerogels, the direct utilisation of the methanol rich pore liquid for supercritical drying leads to an esterification of the particle surface (see Section 5.3). Instead of silanol groups, methoxy groups are at the surface. In order to avoid this esterification, they also exchanged the methanol with carbon dioxide and then performed the supercritical drying. These different procedures indeed lead to different skeletal densities, as shown in Figure 7.5 Looking at the figure, it is first interesting to note that the skeletal densities of both types of silica aerogels are below the value known for vitreous silica or fused quartz, namely 2.203 g/cm^3, as it should be. Generally, the skeletal density of carbon dioxide dried aerogels is higher, which the authors attribute to the presence of only silanol groups at the surface. The existence of methoxy groups at the surface in the methanol dried samples reduces the skeletal density, as is to be expected from Eq. (7.15).

7.2.4 Effect of Test Gas on Skeletal Density

Wognier and co-workers [392] discuss in depth a possible effect of the test gas size on the measurement of skeletal density. Their treatment uses a model presented in Iler's book on silica chemistry [45]. Iler describes on pages 483 ff. how the finite size of a test gas changes the measurement of specific surface area. We will discuss this later in Chapter 8. Here we use the illustration made already by Iler to show why the finite test size affects skeletal density. Consider two spherical particles of radius R which just touch in a point. The test gas shall be considered as spherical with a radius r_t. Then the test gas cannot reach all points of the two particle arrangement. A sketch of this situation is shown in Figure 7.6. It illustrates that the shaded volume cannot be mirrored by the test gas and thus the volume of the solid backbone of an aerogel is

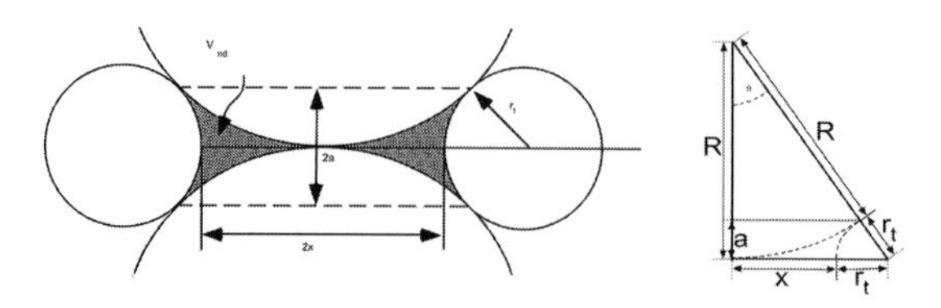

Figure 7.6 Schematic view of the undetectable volume by gas molecules or atoms of size r_t and particles touching having a size R (left panel). The right panel shows relations between particle radius R, neck radius x and radius of the test gas atoms or molecules r_t. Note: $R^2 + (x + r_t)^2 = (R + r_t)^2$.

overestimated, which will reduce the skeletal density. Together with the geometrical relations as they are depicted in Figure 7.6, one derives the following relationships:

$$a = \frac{x + r_t}{R + r_t} \tag{7.24}$$

and

$$R^2 + (x + r_t)^2 = (R + r_t)^2, \tag{7.25}$$

which can be solved for x and then inserted into Eq. (7.24) to express also a in terms of R, r_t. The shaded volume not reached by the gas molecules is calculated under simplifying assumptions. First we calculate the cylinder volume with the base $\pi(x + r_t)^2$, which reads

$$V_{cylinder} = 2\pi r_t (x + r_t)^2. \tag{7.26}$$

From this we must subtract half the volume of the torus:

$$V_{torus} = \frac{1}{2}\pi^2 r_t^2 (x + r_t). \tag{7.27}$$

We must also subtract the volume of the two sphere caps which have a height a:

$$V_{spherecap} = \frac{2\pi}{3} a^2 (3R - a). \tag{7.28}$$

The volume V_{nd} being not detectable is then

$$V_{nd} = V_{cylinder} - V_{torus} - V_{spherecap}. \tag{7.29}$$

The apparent volume detected by the gas is then $V = V_0 + V_{nd}$, with V_0 the volume if the gas molecules would have no radius at all and could reach any part of the surface. Since the mass of the aerogel is constant, we arrive at an apparent skeletal density of

$$\rho_s = \frac{m}{V_0 + V_{nd}} = \rho_s^0 \frac{1}{1 + \frac{V_{nd}}{V_0}} \cong \rho_s^0 \left(1 - \frac{V_{nd}}{V_0}\right). \tag{7.30}$$

The effect is shown in Figure 7.7. The apparent skeletal density decreases by a few percent depending essentially on the ratio of the gas used in the pycnometer to the particle size. For helium, the covalent radius is 0.028 nm and the van der Waals radius

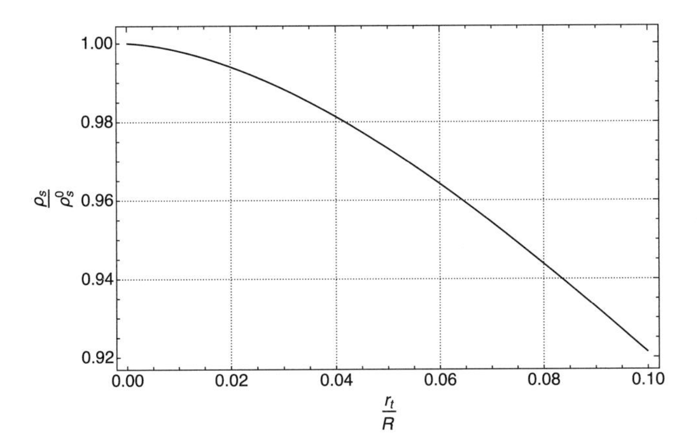

Figure 7.7 Relative change of skeletal density due to the size of the gas used in a pycnometer.

0.14 nm. If the particle radius would be 5 nm, the ratio is in the range 0.006–0.028. One could test this in a pycnometer by using different gases (helium, argon, xenon).

7.3 Porosity

Having both the envelope and skeletal density, one can calculate the porosity of an aerogel. The porosity is defined as the ratio of the pore volume divided by the envelope volume:

$$\phi_p = \frac{V_p}{V_e} = \frac{V_e - V_s}{V_e} = 1 - \frac{V_s}{V_e}. \tag{7.31}$$

Mass has the relation

$$m = \rho_s V_s = \rho_e V_e \tag{7.32}$$

and thus we have

$$\phi_p = \frac{V_p}{V_e} = 1 - \frac{\rho_e}{\rho_s}. \tag{7.33}$$

Often in the literature the specific pore volume is calculated. This can be done as follows. The specific pore volume v_p having the unit m^3/g is defined as

$$v_p = \frac{V_p}{m} = \frac{V_e - V_s}{m} = \frac{1}{\rho_e} - \frac{1}{\rho_s}. \tag{7.34}$$

Therefore, measurement of the envelope and skeletal density yields additional information on aerogels.

7.4 Rule of Mixtures for the Envelope Density

In the following, we will discuss general aspects of aerogel density and perform a few simple calculations on the density of aerogels. Quite generally, an aerogel is a multiphase material, and one phase at least is air. For a multiphase material, independent of the spatial arrangement of the phases, the average or envelope density ρ_e is simply a weighted superposition of the densities of the constituting phases yielding the so-called rule of mixtures (ROM) [398]:

$$\rho_e = \sum_{k=1}^{n} \rho_k \phi_k. \tag{7.35}$$

Here the ρ_k are the densities of phases k and ϕ_k are the volume fractions of these phases. For the volume fractions, we have the constraint

$$\sum_{k=1}^{n} \phi_k = 1. \tag{7.36}$$

One can express Eq. (7.35) also in terms of molar concentrations x_k or weight fraction w_k using the molar volume of the phases $V_{m,k}$ or the molar weight $M_{m,k}$, since the volume fraction can be expressed in both terms:

$$\phi_k = \frac{\frac{x_k}{V_{m,k}}}{\sum_{k=1}^{n} \frac{x_k}{V_{m,k}}} \tag{7.37}$$

$$\phi_k = \frac{\frac{w_k}{M_{m,k}}}{\sum_{k=1}^{n} \frac{w_k}{M_{m,k}}}. \tag{7.38}$$

For aerogels consisting of one solid skeleton with a density ρ_s, and only air as a second phase, having the density ρ_A, these relations simplify to

$$\rho_e = \phi_s \rho_s + \phi_A \rho_A = \phi_s \rho_s + \phi_p \rho_A = (1 - \phi_p)\rho_s + \phi_p \rho_A = \rho_s - \phi_p(\rho_s - \rho_A). \tag{7.39}$$

Since generally $\rho_A \ll \rho_s$, we obtain

$$\rho_e \cong \phi_s \rho_s = (1 - \phi_p)\rho_s. \tag{7.40}$$

The pore volume fraction ϕ_p (equal to the air fraction ϕ_A) can thus be determined by the measurement of the skeletal and the envelope density as discussed before regarding Eq. (7.33):

$$\phi_p = \frac{\rho_s - \rho_e}{\rho_s - \rho_A} \cong 1 - \frac{\rho_e}{\rho_s}. \tag{7.41}$$

Here one might ask whether an aerogel can be lighter than air. The general answer is *no*, since the pores are filled with air and thus the density of air is the lower limit. If an aerogel is, however, evacuated, the density of the remaining solid skeleton $\phi_s \rho_s$

can be smaller than the density of air. The critical amount of solid ϕ_s^c or pore volume fraction ϕ_p^c is

$$\phi_s^c \leqslant \frac{\rho_A}{\rho_s} \quad \text{or} \quad \phi_p^c \geqslant 1 - \frac{\rho_A}{\rho_s}. \tag{7.42}$$

For example, let the aerogel made from silica have the skeletal density of 2200 kg/m^3. Air has under standard conditions (STP, 25°C, 100 hPa pressure) a density of 1.184 kg/m^3. Then the required pore volume fraction would be 99.95%. Such aerogels can and have been produced.

7.5 Relation between Monomer Content and Final Density

One might ask the question: is it possible to calculate the density of an aerogel using known values of the components it is made from? In principle, yes, and preparing a lightweight aerogel, it generally is known that one can always make an estimate of the envelope density to be expected by simply calculating the mass of the constituents being used and the volume of the solution. Let the aerogel being made of p components (such as TEOS, resorcinol, water, ethanol, catalyst, etc.), then the total mass is given by

$$m_{total} = \sum_{k=1}^{p} m_k = \sum_{k=1}^{p} N_k^m M_k^m. \tag{7.43}$$

Typically recipes found in the literature give the composition used in terms of molar ratios. Therefore, we have written the mass as the sum of the number of moles N_k^m and the molar mass M_k^m. The volume of the solution can in principle be calculated as the sum of the molar volumes V_k^m weighted by the number of moles of each constituent. Here we have, however, to take into account that on mixing the components, the volume can increase or decrease. There is a so-called excess volume ΔV_{ex} which stems from the interaction between the solute and the solvent molecules (typically water or ethanol or in biopolymers it can be the sodium hydroxide solution or the salt hydrate melt). Thus the volume of the solution has to be written as

$$V_{total} = \sum_{k=1}^{p} m_k N_k^m V_k^m + \Delta V_{ex}, \tag{7.44}$$

which gives an envelope density expressed as

$$\rho_e = \frac{m_{\text{total}}}{V_{\text{total}}} = \frac{\sum_{k=1}^{m} N_k^m M_k^m}{\sum_{k=1}^{p} m_k N_k^m V_k^m + \Delta V_{ex}}. \tag{7.45}$$

The excess volume can be either negative or positive and in general it will even depend on the composition. For the estimates to be made here, we ignore such a complication and will either set it to zero or a constant value. The aerogel density is then calculated as the ratio of m_{total} and V_{total}, provided that, first, all components in the solution fully react with each other and are incorporated into the gel body;

and second, that the pore liquid can be extracted without any shrinkage and the solid network of the gel is perfectly intact.

7.5.1 Estimate of the Envelope Density of Silica Aerogels

Quite often silica aerogels are made from TEOS (tetraethyl orthosilicate), adding water to hydrolyse the molecule and adding ethanol to allow for a solubility of silica in the ethanol–water solution. Most often, then either an acid is added (such as HCl, HNO_3, etc.) or a base such as NH_4OH, sodium hydroxide or others. Most often, first a colloidal solution is formed under acidic conditions and the final step of condensation or polycondensation is performed under base conditions (for more details, see [45] and Chapter 2). In principle, each alkoxy group (C_2H_5–O) at the Si atom in the TEOS molecule is during hydrolysis replaced by an OH group while the alkoxy group becomes an ethanol molecule. Starting with TEOS, we thus have that 1 mole of TEOS is transformed into 1 mole of $Si(OH)_4$ and 4 moles ethanol (EtOH). The fully hydrolysed $Si(OH)_4$ molecules react with each other under formation of 2 moles water and form SiO_2. Thus 1 mole of TEOS is transformed into 1 mole SiO_2, 2 moles water and 4 moles EtOH. In order to make this reaction work, usually EtOH and water are used in excess. (In reality, typically the TEOS is not really fully hydrolysed and then reacts, but either it is hydrolysed partially and the partial hydrolysed molecules react with others or a hydrolysed site can directly react with an alkoxy group to form EtOH, etc. The real reaction scheme is much more complicated than used for the simple calculation here.) For our estimation of the principally possible aerogel density, we simplify the reactions by the following scheme:

$$\mathrm{TEOS} + 2\mathrm{H_2O} \rightarrow \mathrm{SiO_2} + 4\mathrm{EtOH}. \tag{7.46}$$

For a calculation, we have to add on the right side excess molar quantities of water $\Delta N_{\mathrm{H_2O}}$ and ethanol ΔN_{EtOH}, since the dilution determines the density considerably; both excess quantities are only spectators affecting the final density, but not the reaction itself. The density is then calculated in the following way:

$$\rho_e^{sil} = \frac{N_{\mathrm{TEOS}} M_{\mathrm{SiO_2}}}{N_{\mathrm{TEOS}} V_{m,\mathrm{SiO_2}} + (4N_{\mathrm{EtOH}} + \Delta N_{\mathrm{EtOH}}) V_{m,\mathrm{EtOH}} + \Delta N_{\mathrm{H_2O}} V_{m,\mathrm{H_2O}}}. \tag{7.47}$$

Here N_{TEOS} denotes the molar amount of TEOS used; $M_{\mathrm{SiO_2}}$ is the molar weight of SiO_2; and the $V_{m,\mathrm{SiO_2}}$, $V_{m,\mathrm{EtOH}}$ and $V_{m,\mathrm{H_2O}}$ are their molar volumes. Taking the molar weight of $M_{\mathrm{SiO_2}} = 60.1$ g/mol, the molar volume of ethanol $V_{m,\mathrm{EtOH}} = 58.3682$ $\mathrm{cm^3/mol}$, that of SiO_2 as $V_{m,\mathrm{SiO_2}} = 22.6792$ $\mathrm{cm^3/mol}$ and that of water 18 $\mathrm{cm^3/mol}$, we can calculate the possible envelope density as a function of dilution by varying the excess concentrations of water or ethanol. An example of the effect of ethanol and water content on the envelope density of silica aerogels is shown in Figure 7.8. The agreement between measured data and the estimate is quite good with a fit parameter for the molar volume of silica as 26 $\mathrm{cm^3/g}$. This value leads to a skeletal density of 2.3 and is thus in the range of various modifications of silica; see footnote 1. The envelope density can be varied also in a completely different way as shown by Wong and

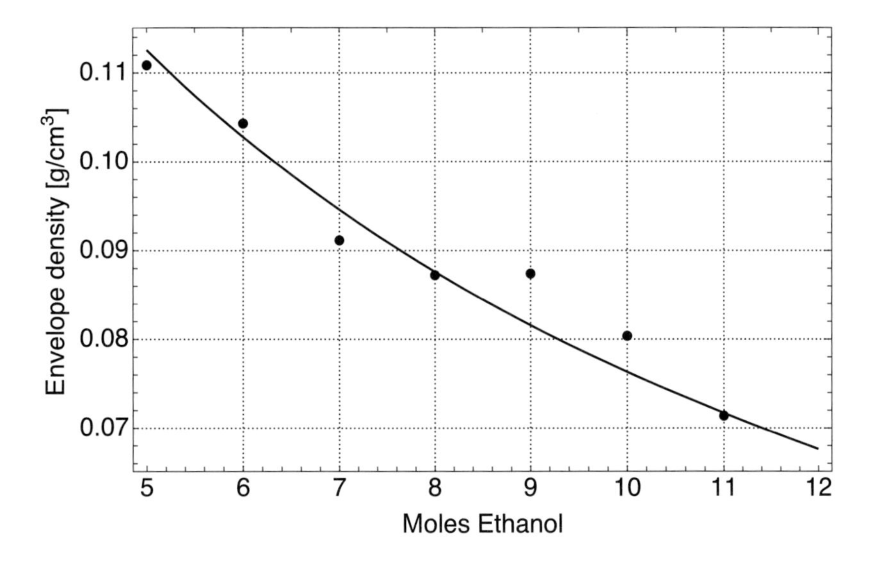

Figure 7.8 Measured envelope density of silica aerogels prepared from TEOS at a constant amount of water of 5.5 moles and the excess molar concentration of ethanol as stated on the abscissa. The points are the experimental data and the solid line is a fit with Eq. (7.47).

coworkers [399]. They did not use TEOS and prepared the gel from a mixture with water, ethanol and a catalyst, but used pre-polymerised TEOS, so-called PEDS, which is commercially available. It is a mixture of water and TEOS in a ratio of 1.5 and has a silica content of 20 wt% stored in ethanol. The PEDS used by Wong et al. had the designation PEDS-P_{750E20}. One can easily dilute such a stock solution with ethanol and induce gelation by adding, for instance, ammonium hydroxide. They prepared a huge quantity of differently diluted samples to vary the envelope density in order to investigate the dependence of several properties on envelope density (Chapter 13 presents a more detailed discussion of their results). Here we are concerned with the density alone. Figure 7.9 shows their result. The volume concentration silica of the stock solution can be calculated from the weight percentage of silica in the solution assuming a density of silica as 2.65 g/cm^3 and that of ethanol as 0.7893 g/cm^3 to be 6.93 vol% [1]. This would give an envelope density, replacing the ethanol by air without any shrinkage to 0.184 g/cm^3, which is the end point of the dashed line in Figure 7.9 at 100% PEDS. Wong and Koebel report a volumetric shrinkage of 30–40%. Correcting the dashed line with such a value, which is slightly dependent on the density itself, yields the dotted line (for further discussion of how the degree of shrinkage depends on density see Section 5.4.2.) The experimental values are still higher, which is attributed to the attachment of trimethylsilyl groups at the surface used to hydrophobise the silica aerogel (they used hexamethyldisilazane for hydrophobisation). The effect of

[1] The density value of silica used by Wong et al. is quite high. Fused silica has instead a density of 2.2 g/cm^3, cristobalite has a density of 2.33 and tridymite a density of 2.28. Amorphous silica has a broad density range between 2.19–2.66 g/cm^3.

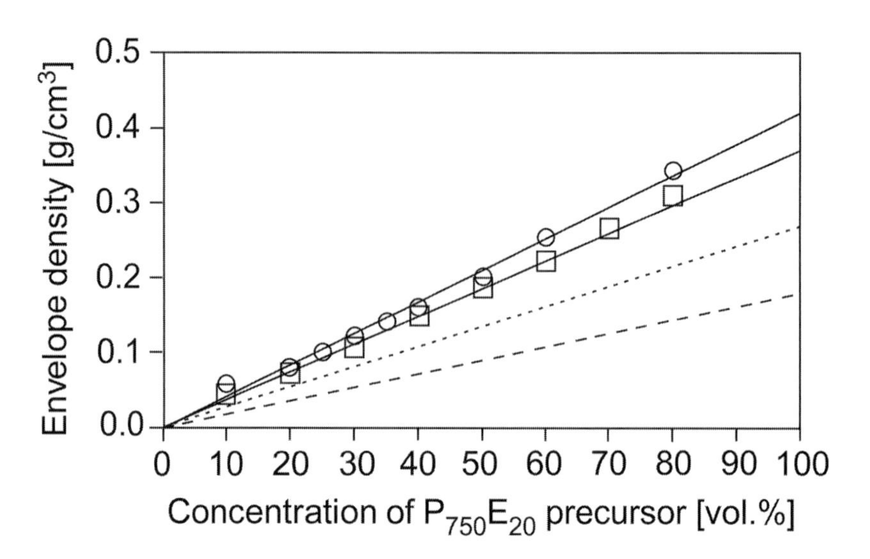

Figure 7.9 Measured envelope density of silica aerogels prepared from PEDS by dilution of two stock solutions (open circles and squares) as a function of volume concentration of the PEDS precursor. The dashed and dotted lines are explained in the text. Reprinted and redrawn from [399] with permission from Elsevier

the trimethylsilyl (TMS) groups at the surface can be calculated with the model given in Section 7.2.2 for adsorption of molecules at the inner surface of an aerogel. The mass bonded m_b to the inner surface is

$$m_b = M_a \frac{A_s}{\Omega_m^{2/3}} = M_a \frac{S_V \phi_s V_e}{\Omega_m^{2/3}}, \tag{7.48}$$

with Ω_m the molecule volume measured in m^3 and M_a measured in grams. Using the relation between surface area per unit volume and surface area per unit mass, we arrive at

$$\rho_e^{corr} = \rho_e^0 \left(1 + \frac{M_a S_m}{\Omega_m^{2/3}} \phi_s \right). \tag{7.49}$$

Using for the molar volume of the trimethylsilyl group a value of 111 cm^3/g and a mass of 73 g/mol, and using the specific surface area they measured to be around 740 m^2/g, one can calculate a corrected envelope density as

$$\rho_e^{corr} = \rho_e^0(1 + 0.171\phi_s). \tag{7.50}$$

Identifying ϕ_s with the concentration of the precursor allows us to correct the results of Wong et al. by at maximum 17%, which is still below the measured value but closer to it. As discussed by Wong and co-workers, their aerogels always contain more TMS groups than just a monolayer on the inner surface. Typically around 25% is built in the aerogels. This leads to an increase in density of around 33%. With the addition of the adsorbed TMS group at the surface, the measured values can be described as an effect of composition (SiO_2 + TMS + ethoxy groups), shrinkage and replacement of ethoxy and OH groups at the inner surface. These examples outline that the envelope density of silica aerogels can be estimated from known data of its constituents with an accuracy of around 10–20%.

7.5.2 Estimate of the Envelope Density of RF Aerogels

So-called RF aerogels are made from dissolving resorcinol (R) and formaldehyde (F) in water and adding a catalyst, such as sodium carbonate or citric acid. The reaction between R and F in solution is quite complex under base or acid conditions. Eventually a molecule made of two R molecules bonded by a methylene or dimethylene ether bridge is formed, and these can react further to produce polymers in either a so-called resol or novolake structure [83] (see also Section 2.2). On reaction of two molecules R with 1 formaldehyde, 1 mole of water is produced. We estimate the envelope density by calculating the mass of the reactant (R–CH_2–R) or (R–CH_2–O–CH_2–R) from the molar masses of R, methylene and dimethylene ether and calculating the volume of the solution from the molar volume of each constituent (R, F, water) before any reaction has happened. We neglect the catalyst, since it usually is used in very low amounts. This way of estimating the final density should give the ratio of the solid polymers $(R–CH_2–R)_n$ (n is the degree of polymerisation) confined in particles and connected to a 3D network with respect to the initial volume of the solution. Let $M_{m,R}, M_{m,F}, M_{m,CH_2}, M_{m,H_2O}, M_{m,CH_2-O-CH_2}$ be the molar masses of the constituents and ρ_R, ρ_F be the density of R and F. We then can calculate the mass of the dimer R–CH_2–R, from which the polymer is formed, as

$$m_{\text{dimer}}^{\text{methylene}} = N_{\text{resorcinol}} M_{m,R} + \frac{N_{\text{resorcinol}}}{2} M_{m,CH_2} \tag{7.51}$$

if only methylene bridges are formed and $N_{\text{resorcinol}}$ are the moles resorcin. If only dimethylene ether bridges are formed, we have

$$m_{\text{dimer}}^{\text{dimethyleneether}} = N_{\text{resorcinol}} M_{m,R} + N_{\text{resorcinol}} M_{m,CH_2-O-CH_2}. \tag{7.52}$$

In reality, both will appear in different amounts depending on the R:F ratio, the catalyst concentration, etc. Here we have used that typically RF aerogels are produced with exactly the stoichiometric necessary relation of R:F being 1:2 (in some cases an excess of F is used). The volume of the solution before any reaction has happened is calculated as

$$V_{sol} = N_{\text{resorcinol}} \frac{M_{m,R}}{\rho_R} + N_{\text{formaldehyde}} \frac{M_{m,CH_2}}{\rho_F} + N_{H_2O} \frac{M_{m,H_2O}}{\rho_{H_2O}}. \tag{7.53}$$

Typically in RF aerogels, molar ratios of R/water of 20–200 are used. A careful study of the effect of dilution was made by Ludwig in her bachelor thesis [400]. She studied RF aerogels catalysed with citric acid and dried under ambient conditions. In one series of experiments, the molar ratio of R:F:C was fixed at 1:1.3:0.061 (C = citric acid) and only the water amount was varied between 20–60 moles of water. The effect of dilution is shown in Figure 7.10 together with a fit using the

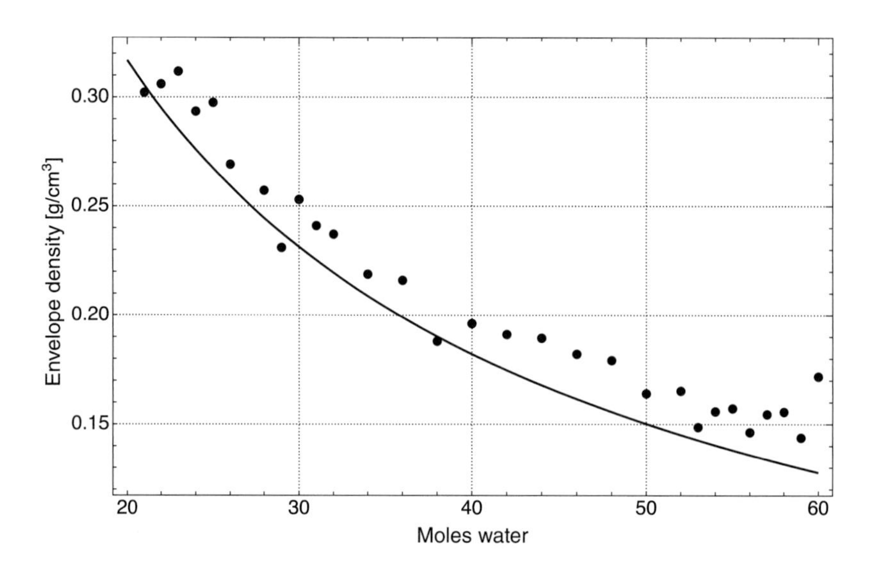

Figure 7.10 Envelope density of RF aerogels catalysed with citric acid having a constant ratio of R/F of 1:1.3 as a function of dilution with water in the range of 20–60 moles. The solid curve assumes that only dimethylene bridges are formed between two resorcin molecules.

relation derived previously. This comparison shows that the envelope density can be calculated close to the experimental values. These are higher than the predicted ones, but shrinkage could not be taken into account, since it was not measured. Another careful study was made by Schwan in her PhD thesis [89]. She prepared base catalysed RF aerogel using different R;F ratios (0.4, 0.5, 0.6) and diluted them with water. She also measured the shrinkage. Since the measured envelope density of all RF aerogels does only depend on the R:W ratio and not measurable on the R/F, we plot here all data together and fit them with the expressions derived earlier. The result is shown in Figure 7.11 together with a calculation assuming only dimethylene ether bridges are formed and the R/F ratio is 0.5.

7.5.3 Density of Cellulose Aerogels

Cellulose aerogels can be prepared, for instance, by dissolution of cellulose in salt hydrate melts as described in Chapter 3. Two salt hydrates are typically used: one based on $ZnCl_2 \cdot (3\text{–}4)H_2O$, and the other one rhodanide $Ca(SCN)_2 \cdot 4H_2O$. The salts are dissolved water (which is an exothermic reaction) at temperatures between 70–120°C, then cellulose is stirred in and dissolved until a clear solution is obtained. A wet gel forms on cooling to lower temperatures. In the case of zinc chloride hydrate, this happens around 30°C, and in the case of rhodanide at around 80°C. After washing out the dissolved salts with an appropriate medium, such as acetone, ethanol, isopropanol and supercritical drying, cellulose aerogels are obtained whose envelope

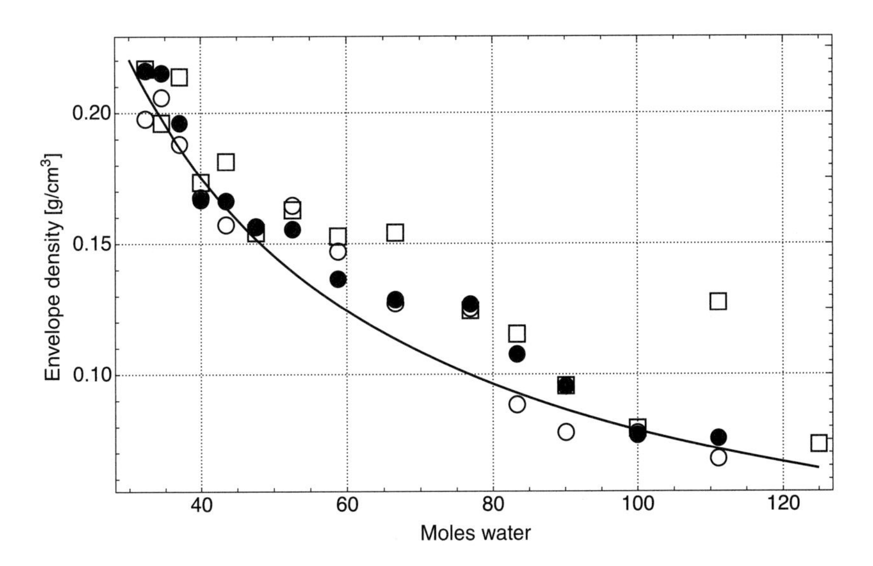

Figure 7.11 Envelope density of RF aerogels catalysed with sodium carbonate having ratios of R/F of 0.4, 0.5, 0.6 as a function of dilution with water in the range of 30–125 moles (the open circles are for R/F = 0.4, the full circles for R/F = 0.5, the open squares for R/F = 0.6). The solid curve assumes that only dimethylene bridges are formed between two resorcin molecules.

density varies proportional to the weight percentage of cellulose stirred into the salt hydrate melt. This behaviour can be calculated. We start with the definition of the volume fraction of cellulose introduced into the salt hydrate melt ϕ_c. As mentioned previously, this is related to the weight fraction w_c by

$$\phi_c = \frac{w_c/\rho_c}{w_c/\rho_c + w_S/\rho_S} = \frac{1}{1 + \frac{w_S \rho_c}{w_c \rho_S}}, \tag{7.54}$$

in which ρ_c, ρ_S denote the densities of the cellulose (typically around 1500 kg/m^3) and the salt hydrate melt (index S). w_S denotes the weight fraction of the salt hydrate melt. For small amounts of cellulose (typically weight fractions below 10% can be dissolved), we can simplify the relation to

$$\phi_c \cong \frac{w_c \rho_S}{w_S \rho_c} = \frac{w_c}{1 - w_c} \frac{\rho_S}{\rho_c} \cong w_c \frac{\rho_S}{\rho_c}. \tag{7.55}$$

The aerogel envelope density ρ_e^{cell} reads

$$\rho_e^{cell} = \phi_c \rho_c + (1 - \phi_c)\rho_A = \phi_c \rho_c (1 - \frac{\rho_A}{\rho_c}) + \rho_A \cong \phi_c \rho_c + \rho_A. \tag{7.56}$$

Using Eq. (7.55), we finally get

$$\rho_e^{cell} = w_c \rho_S + \rho_A. \tag{7.57}$$

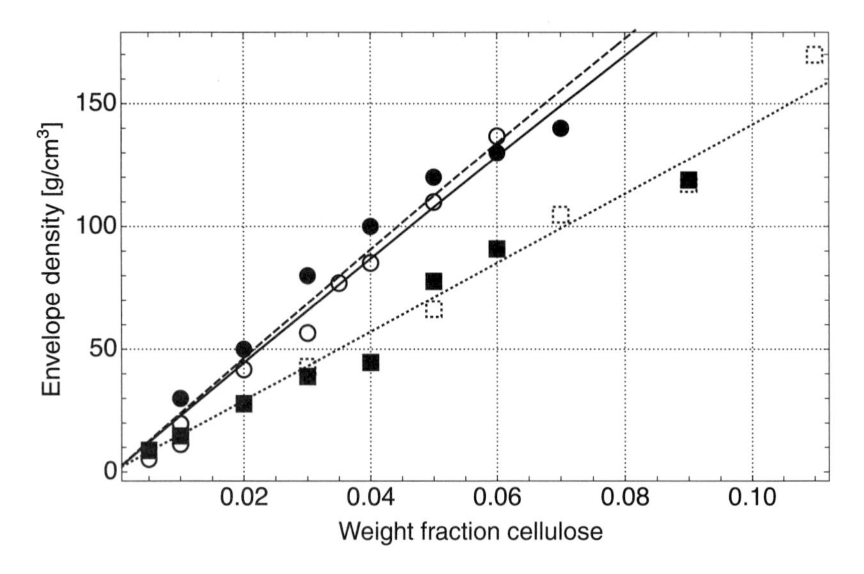

Figure 7.12 Envelope density of cellulose aerogels as measured by several authors [142, 148, 149, 218, 401]. All data were corrected for shrinkage if the total shrinkage from the wet gel to the dry aerogel was reported. The open circles are the data from [148, 149], who both used exactly the same procedures to prepare cellulose aerogels by dissolution with a rhodanide hydrate melt and supercritical drying with carbondioxide after solvent exchange with pure ethanol. The full circles are from the paper of Cai et al. [401], who used an aqueous LiOH mixed with urea to dissolve cellulose. The open squares are data from Buchtová and Budtova [218], who dissolved cellulose in a mixture of DMSO and an ionic liquid called EMImAC. Supercritical drying with carbondioxide was performed after solvent exchange with ethanol. The full squares are data from Innerlohinger et al. [142]. They used the well-known ionic liquid NMMO and supercritical drying with carbon dioxide.

Thus the density of cellulose aerogels depends linearly on the weight fraction of cellulose introduced into the salt hydrate melt. The slope of such a curve is the density of the salt hydrate melt (without cellulose dissolved in it). The density of the salt hydrate melt has to be measured. For instance, for a rhodanide melt the density was measured to be $\rho_S = 1.43$ g/cm^3 at 80°C [148]. Using Eq. (7.56) and data available in the literature on cellulose aerogels prepared by different routes yields a result shown in Figure 7.12. This diagram shows an interesting trend: both cellulose aerogels prepared with an ionic liquid fall essentially on one curve or follow one trend, whereas the aerogels prepared either from a salt hydrate route or an alkaline solution with urea follow another trend. In both cases, the fitted density of the solvent differ considerably. For the ionic liquids, a density of around 1400 kg/m^3 is obtained and for the other cases a density of around 2150 kg/m^3. Thus the salt hydrate melt as well as the LiOH–urea melt would have a considerably higher density than both ionic liquids. Unfortunately, there are no measurements available concerning the temperature-dependent density of such liquids.

7.6 Variables and Symbols

Table 7.1 lists some of the variable and symbols of importance in this chapter.

Table 7.1 List of variables and symbols in this chapter.

Symbol	Meaning	Definition	Units
a_m	Molecule diameter	–	m
A_s	Total inner surface of an aerogel	–	m^2
D	Diffusion coefficient	–	m^2/s
D_{Kn}	Knudsen diffusion coefficient	–	m^2/s
D_μ	Viscous gas diffusion coefficient	–	m^2/s
k_B	Boltzmann coefficient	1.38×10^{-23}	J/K
ℓ	Characteristic sample size	–	m
m	Sample mass	–	kg
m_{ads}	Adsorbed mass	–	kg
M_a	Atomic mass	–	kg
N_A	Avogadro's number	$6.022 \cdot 10^{23}$	1/mol
p	Pressure	–	Pa
R_p	Pore radius	–	m
r_t	Test gas molecule radius	–	m
Sh	Shrinkage	–	–
S_m	Specific surface area per sample mass	–	m^2/kg
S_V	Specific surface area per sample volume	–	m^2/m^3
T	Temperature	–	K
v_p	Specific pore volume	–	m^3/kg
V_e	Volume of the sample's envelope	–	m^3
V_s	Volume of the solid backbone	–	m^3
Ω_m	Atomic volume	–	m^3
ϕ	Volume fraction	–	–
ϕ_s	Volume fraction solid	–	–
ϕ_p	Pore volume fraction	–	–
ρ_A	Air density	1.204	kg/m^3 at 20 °
ρ_e	Envelope density	$\rho_e = m/V_e$	kg/m^3
ρ_s	Skeletal density	$\rho_s = m/V_s$	kg/m^3
τ_c	Characteristic time	–	s

8 Specific Surface Area

One important characteristic of all aerogels is their large specific surface area. Almost in every paper on aerogels, not only the envelope density is reported but also the specific surface area in terms of inner surface per unit mass, m^2/g. We first present some fundamental relations for the specific surface area of particulate aerogels, such as silica or RF aerogels and of fibrillar aerogels such as cellulose and define terms and present selected experimental results to compare the models with reality. Techniques to measure the surface area are dealt in Section 8.6.

8.1 Definitions and Relations

The specific surface area of any body with a surface A and volume V is defined as the ratio

$$S_V = \frac{A}{V} \quad \text{unit} \quad \frac{\mathrm{m}^2}{\mathrm{m}^3} = \frac{1}{\mathrm{m}}. \tag{8.1}$$

We use here the subscript V to denote the surface area per unit volume. The body might be included in a test or sample volume V_T (see Figure 8.1). The specific surface area per unit test or sample volume, superscript T, is then

$$S_V^T = \frac{A}{V_T} = \frac{A}{V}\frac{V}{V_T} = S_V \phi_s, \tag{8.2}$$

with ϕ_s the volume fraction solid in the test volume:

$$\phi_s = \frac{V}{V_T}. \tag{8.3}$$

Typically the specific surface area of porous bodies is measured by an adsorption technique, to be dealt with in more detail later in this chapter. Most often, nitrogen physisorption at 77 K is applied due to easy availability of liquid nitrogen and its low price. But also argon at 87 K is an attractive adsorptive since it has no quadrupole moment which might interfere with charged surface features (actually recommended over N_2 by the International Union of Pure and Applied Chemistry (IUPAC) since 2015 [402], and for materials with a low specific surface area krypton is recommended

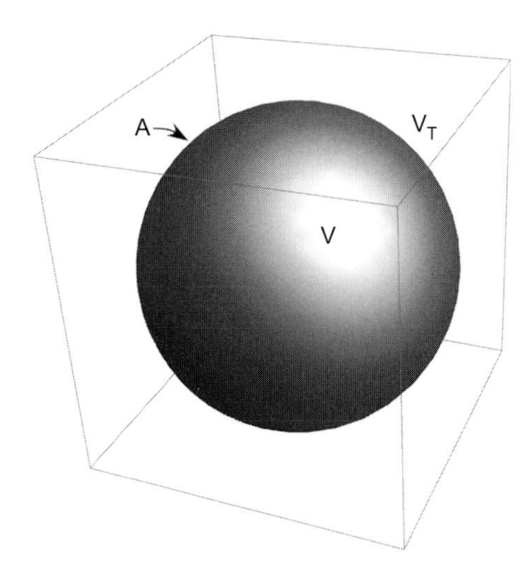

Figure 8.1 Scheme of a body in a test volume defining the two different surface areas in Eqs. (8.1), (8.2).

at 77 K. The evaluation of the adsorption isotherms leads to a surface area given in terms of surface per mass of a sample S_m, with the unit m^2/g. Both surface areas are related by

$$S_V^T = S_m \rho_e \quad \text{or} \quad S_m = \frac{S_V^T}{\rho_e}. \tag{8.4}$$

Using the relation derived in Eqs. (7.40) and (8.2), yields

$$S_m = \frac{S_V^T}{\rho_e} = \frac{S_V \phi_s}{\rho_s \phi_s} = \frac{S_V}{\rho_s}. \tag{8.5}$$

This relation shows clearly that the specific surface area of an aerogel sample is not defined by the porosity but *only* by the particle size forming the 3D network!

8.2 Surface Area of Simple Shapes

Let us calculate the specific surface area for simple bodies. A sphere of radius R has the surface area $A = 4\pi R^2$ and a volume of $V = 4\pi R^3/3$, yielding

$$S_V^{sphere} = \frac{3}{R} = \frac{6}{D} \tag{8.6}$$

if D denotes the diameter. For a cube with an edge length d, we have $A = 6d^2$ and $V = d^3$ and thus

$$S_V^{box} = \frac{6}{d}. \tag{8.7}$$

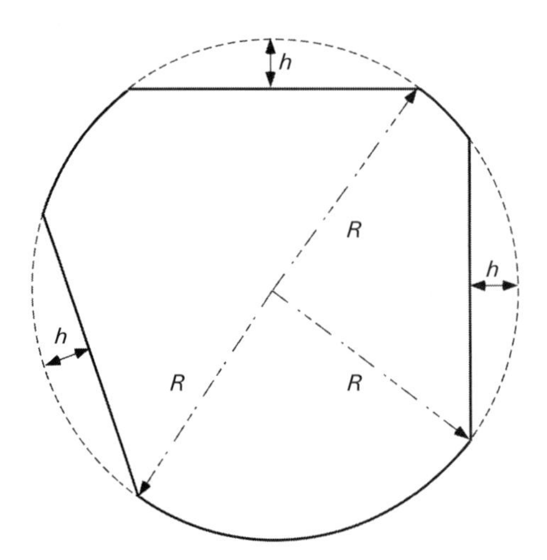

Figure 8.2 Scheme of a sphere which is truncated at three regions. This illustrates how an overlap of particles in a pearl-necklace structure of an aerogel changes the specific surface area.

For a cylinder with radius R and length ℓ, we have $A = 2\pi R^2 + 2\pi R\ell$ and $V = \pi R^2 \ell$ and thus

$$S_V^{cyl} = \frac{2}{R}\left(1 + \frac{R}{\ell}\right), \tag{8.8}$$

or in the case of a long cylinder $R \ll \ell$, we get

$$S_V^{cyl} = \frac{2}{R}. \tag{8.9}$$

For long cylinders, the radius determines the specific surface area and its length is not important. Let us treat a very special case, namely that of a truncated sphere as shown in Figure 8.2, which is relevant in the context of particulate aerogels. Quite often in such structures, the particles building the network overlap as if they would penetrate into each other. They establish so-called necks. An example of such a neck formation observed by Schwan and co-workers [89] is shown in Figure 8.3. Such features can be modelled as truncated spheres. With reference to Figure 8.2, the volume of each spherical cap can be calculated as

$$V_{cap} = \frac{\pi}{3} h^2 (3R - h) \tag{8.10}$$

and the mantle surface is

$$A_{cap} = 2\pi R h. \tag{8.11}$$

The rest volume of the original sphere $V_0 = 4\pi/3R^3$ is then

$$V_{rest} = V_0 - n_{NN} V_{cap}, \tag{8.12}$$

Figure 8.3 Formation of a string of pearls in RF aerogels exhibiting the feature of particles connected via necks of different size [89]. Courtesy M. Schwan, DLR

with n_{NN} the number of spherical caps cutting out volume of the original sphere, which will be equal to the number of nearest neighbours. The surface area of the remaining rest volume

$$A_{rest} = A_0 - n_{NN} A_{cap} \tag{8.13}$$

and thus the specific surface area is

$$S_V^{ts} = \frac{A_{rest}}{V_{rest}} = S_V^0 \frac{1 - n_{NN}\frac{A_{cap}}{A_0}}{1 - n_{NN}\frac{V_{cap}}{V_0}}, \tag{8.14}$$

with S_V^0 the surface area of the full sphere. Inserting the expressions derived earlier and assuming that $h/R < 1$, we obtain after some calculations

$$S_V^{ts} \cong S_V^0 \left(1 - n_{NN}\frac{h}{2R}\right)\left(1 + \frac{9}{4} n_{NN}\left(\frac{h}{R}\right)^2\right). \tag{8.15}$$

This equation tells us that the first term leads to a reduction in specific surface area, whereas the second term leads to an increase. The first term is determined by the surface area, whereas the second, and thus the increase, is dictated by the reduction in volume. In order to proceed, we must have an idea about the number of cutouts or, in other words, the number of neighbouring particles touching and merging such that a truncated sphere as shown in Figure 8.2 appears. If the particles building the aerogel would be arranged like a hexagonal closed packed (hdp) or face-centred cubic lattice (fcc), the number of nearest neighbours would be 12. On a body-centred cubic lattice it would be 8, and on a simple cubic lattice 6, diamond lattice 4. In a random loose packing of spheres, values below 6 are typical. Aerogels are definitively not densely packed, and even a random loose packing is too high. Values for n_{NN} should be in the range of 2–4. Iler [45] cites the following expression for next nearest neighbours n_{NN} in a random packing of spheres:

$$n_{NN} = 2\exp(2.4\phi_s). \tag{8.16}$$

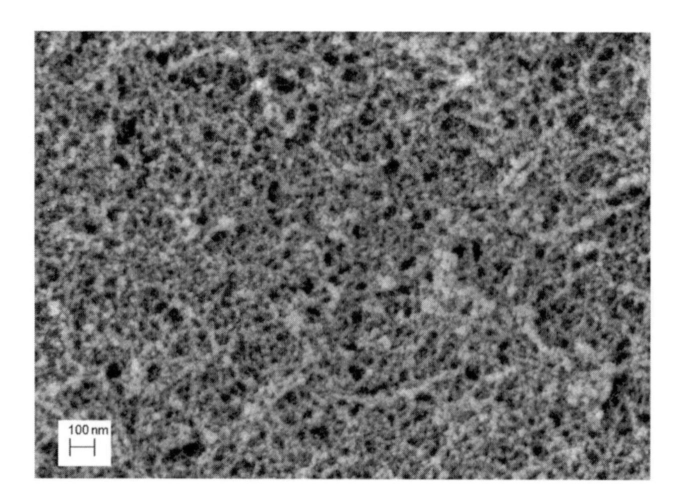

Figure 8.4 Scanning electron micrograph of a silica aerogel, showing a pearl-necklace structure. One could get the impression that the particles of the network are connected via two or three bonds to neighbouring particles. Courtesy M. Heyer, DLR

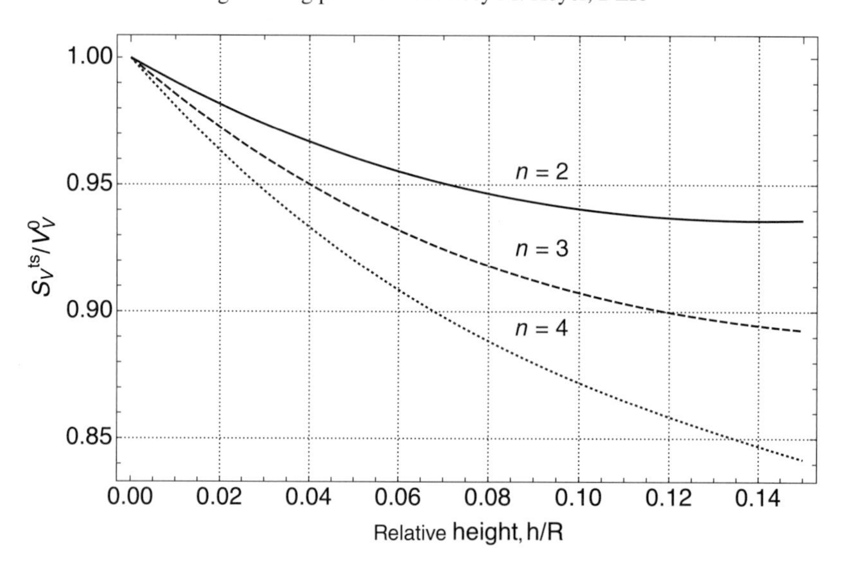

Figure 8.5 The specific surface area of particles connected via necks by two, three or four neighbours reduces between 5–15%, depending on the number of nearest neighbours $n = n_{NN}$.

In aerogels having typically solid fractions below 10%, the coordination or next nearest neighbour number would be below 2.5. In Figure 8.3, the coordination number clearly is 2. Figure 8.4 shows an SEM picture of a silica aerogel prepared from TEOS. Here the coordination number might be guessed as something between 2 and 3. Using for n_{NN} values between 2 and 4, one can calculate the ratio of the specific surface area of a truncated sphere arrangement compared to the value of a full sphere. The result is shown in Figure 8.5. Looking at the effect of neck formation between particles, which is enhanced during ageing of the wet gel, one has to conclude that inasmuch as particles overlap or build necks, the structure changes from spherical to cylindrical or rod-like, and as a consequence particulate aerogels with small necks will always have a larger specific surface area than fibrillar aerogels, even if the fibril diameter is the same as the particle diameter.

8.3 Surface Area of Irregular-Shaped Bodies

It remains a problem with all the preceding considerations: generally, aerogels are not made from perfect spheres or even intersecting spheres, but they exhibit irregular particle shapes. Most often, the particles or fibrils forming the 3D network have not one radius or length, but there is a distribution of radii or length or slenderness. To compare the shape with vegetables, one could say they can have a shape of a potato, zucchini, cucumber, melon or pumpkin, and fibrillar structures might look like an irregular spider web. We have to think about a possible description and especially how to measure the surface area using something like a diameter or length.

There is a way out of that dilemma, well described in the context of stereology [403]. Instead of a particle diameter or radius, we use the so-called mean intercept or chord length $\overline{L}$. How do we measure this and how is it related to a particle diameter or any other characteristic length of a body or phase inside a multiphase material?

Take an arbitrary porous body and make a cut through an arbitrary plane of it as shown in Figure 8.6. Draw on the plane a line and measure along that line either the average length in the pore space or in the solid phase (since both are related, it does not matter). One might do this for many lines on a cross section or even take another cut through the body and then simply average all measured intercept lengths. This gives

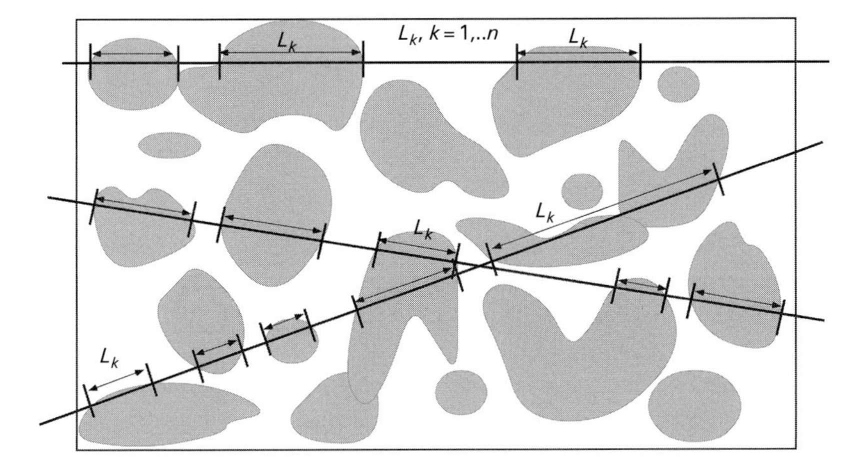

Figure 8.6 Schematic drawing of a two-phase material, such as a porous one, with drawn lines and the linear intercepts with these lines. The average of all intercepts $L_k, k = 1, \cdots, n$ defines the linear intercept used to determine the specific surface area.

Table 8.1 A few examples of the relation between a characteristic length of a body and the mean intercept length after Underwood [403].

Particle shape	Volume	Surface	$\overline{L}$
Cube	a^3	$6\,a^2$	$\frac{2}{3}a$
Sphere	$\frac{4\pi}{3}R^3$	$4\pi R^2$	$\frac{4}{3}R$
Cylinder	$\pi R^2 \ell$	$2\pi R(R+\ell)$	$\frac{2R\ell}{R+\ell}$
Rod	$\pi R^2 \ell$	$2\pi R\ell$	$2R$

just the average or mean intercept length $\overline{L} = 1/N \sum_{k=1}^{N} L_k$. The specific surface area per unit volume can be related to the mean intercept length. Stereology proves that for an arbitrary morphology of pores in a body, the following relation exactly holds (see [403]):

$$S_V^T = \frac{4\phi_s}{\overline{L}}, \tag{8.17}$$

with ϕ_s the volume fraction solid, which is for aerogels almost always ρ_e/ρ_s. But how do we relate the mean intercept length to, for instance, the particle size? Underwood provides a table for simple geometrical bodies (see [403], pages 90–92), which is reproduced here in Table 8.1 for the shapes previously discussed. Eq. (8.17) is quite often used in the literature on aerogels to calculate a particle or fibril size. The average intercept length can be converted to an average fibril or particle diameter using the relations given in Table 8.1. Let us reproduce here a few results:

$$\overline{D}_{sphere} = \frac{3}{2}\overline{L} \quad \text{sphere diameter} \tag{8.18}$$

$$\overline{D}_{fibril} = \overline{L} \quad \text{fibril diameter with} \quad \ell \gg R. \tag{8.19}$$

Using Eq. (8.5) one can easily see that the following equation yields a correct estimate of an average size:

$$S_m = \frac{S_V^T}{\rho_e} = \frac{4\phi_s}{\overline{L}\rho_e} \cong \frac{4}{\overline{L}\rho_s}, \tag{8.20}$$

and thus is independent of the volume fraction of solid phase. One must, however, take into account that the intercept length will depend on the volume fraction. With increasing volume fraction of solid phase, the particles forming the network will touch and inasmuch as they do the intercept increases and the surface area decreases. Eq. (8.17) is correct, but the transformation to a particle or rod radius according to Table 8.1 is only valid if the volume fraction is low and the particles do not touch. We have already seen that for a truncated sphere, the specific surface area reduces inasmuch as the barrel becomes shorter and shorter or, in other words, the radius estimated using a sphere approximation yields wrong results. Such effects were already discussed by John W. Cahn 50 years ago [404]. He suggested in his paper that the specific surface area per unit volume of the sample depends on the volume fraction of the dispersed or network forming phase like

$$S_V^T = K\phi_s^{2/3}(1-\phi_s)^{2/3}, \tag{8.21}$$

in which K is a constant. Since Eq. (8.17) rigorously holds, such a relation means that the mean intercept depends on the volume fraction of solid in the gel like

$$\overline{L} = \frac{4\phi_s^{1/3}}{K(1-\phi_s)^{2/3}}. \tag{8.22}$$

The exponent 2/3 in this relation can easily be understood. Assume a sphere is located in the centre of a surrounding box as shown in Figure 8.1. This sphere will have a radius R and the enclosing box will have a side length of $2h$. The sphere touches the box when its radius has become h. Let us call this radius a critical one, R_c, because if it were larger, it would touch, for instance, another sphere close by, and thus we have the situation of a truncated sphere. The fraction solid inside this box can then be expressed as $\phi_s = \pi/6(R/R_c)^3$. The maximum value possible without truncating the sphere is then given by $\phi_s^{\max} = \pi/6$. The specific surface area is defined as before, namely $S_V^T = 3/R\phi_s$. Expressing now R by ϕ_s yields $R = R_c(\phi_s/\phi_s^{\max})^{1/3}$. And thus we have $S_V^T = S_V^m(\phi_s/\phi_s^{\max})^{2/3}$ with $S_V^m = 3/R_c\phi_s^{\max}$. This explains why Cahn chose the exponent 2/3. The second term in his relation is an approximation to treat the overlapping or truncation induced by other surrounding spheres. One can calculate this form exactly if one assumes a sort of packing. We are not going to do this here, since the relation by Cahn was recently questioned in the context of solidification research and formation of specific surface area in dendritic growth. Other exponents than 2/3 were proposed or fitted to experimental data [405–407]. Independent of this debate, since in aerogels the fraction solid is rather small, the linear intercept changes slowly with fraction solid phase and thus the particle size. We could simplify Eq. (8.22) then to

$$\overline{L} \cong \frac{4\phi_s^{1/3}}{K}. \tag{8.23}$$

This means extraction of a particle radius from surface area measurement needs some caution. We come back to this point in the Section 8.6.5 at the end of this chapter.

8.4 Surface Area of Fibrillar Aerogels

Fibrillar microstructures are typically observed in aerogels prepared from polymeric precursors as shown in Chapter 6. These structures exhibit similarities with macrostructures well known from textile products and paper. Felts or fleeces from polymeric or natural fibres can be envisaged or modelled as long cylinders placed randomly in space. In non-woven textiles or paper, the fibres stick to each other at their points of contact, either mechanically due to some surface roughness, or by Van der Waals–type interactions or by a binder added to the fibre slurry before forming

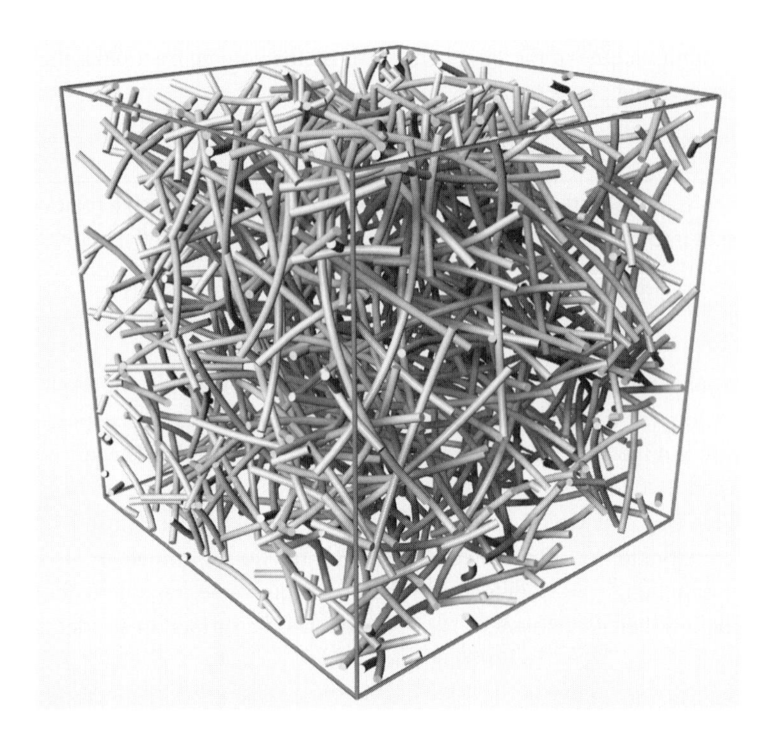

Figure 8.7 A 3D model of a random distribution of curly cylinders used by Heyden [408] to model papers made from a fluffy cellulose pulp. Reproduced with permission of S. Heyden, U Lund, Sweden and the Pulp and Paper Fundamental Res. Soc. (PPFRS), UK

into a sheet or any other shape. The microstructure description of aerogels with a fibrillar network is somehow similar to that of textile or paper, but first the structure is much finer in the nanometer range and secondly the fibres make at their points of contact real chemical bonds such that these nodes are able to transmit forces. From the literature on textile or paper structures we can, nevertheless, take some ideas, how to describe aerogel structures. An example of a modelled 3D structure used by Heyden [408] to model the mechanical response of a random network of curved cylindrical fibres is shown in Figure 8.7. Although fibrillar aerogels look somehow different, the general idea to just model them as elongated cylinders should work as a first approach. Let such cylinders have an average volume $\overline{v}_F$. The number of fibrils per unit volume is $n_F = N_F/V$. Then the volume fraction is

$$\phi_F = \overline{v}_F n_F \tag{8.24}$$

The average fibril volume is given by the fibril cross section and the average fibril length $\overline{l_F}$

$$\overline{v}_F = \pi \bar{R}_F^2 \bar{l}_F \tag{8.25}$$

with $\bar{R}_F$ the average fibril radius. The specific surface area per volume is the mantle surface multiplied with the number of fibrils per unit volume. Neglecting the front faces of the fibrils, since $l \gg \bar{R}$, we have

$$S_V = \frac{\bar{A}_F}{\bar{v}_F} = \frac{2}{\bar{R}_F}. \tag{8.26}$$

The specific surface area per unit sample volume V_T is

$$S_V^T = \frac{2\phi_F}{\bar{R}_F}, \tag{8.27}$$

with ϕ_F the volume fraction of fibrils in the sample volume. The specific surface area per mass is then

$$S_m = \frac{S_V^T}{\rho_e} = \frac{2}{\bar{R}_F \rho_s^F} \tag{8.28}$$

with ρ_s^F the skeletal density of the fibrils. As expected, the thinner the fibrils, the larger the surface area. One might ask, is there any way to experimentally control the fibril radius?

As already mentioned, one characteristic feature of fibrillar aerogels is that the nodal points are force transmitting; they are real bonded points, or in a sense there is a bond volume in a structure. There might also be some dangling fibrils, but in a first approach we neglect these. The points or areas of contacts reduce the specific surface area derived as described earlier. Let us make a simple model to get the essential parameters affecting the surface area. To model this, we denote with n_F^{con} the number of contacts per fibril and with $\bar{a}_{con}$ the area of one contact. The total number of fibrils in the sample volume V_T is $N_F = n_F V_T$. The total contact area in the sample volume is

$$A_{tot}^{con} = N_F \bar{A}_F = n_F^{con} \bar{a}_{con} n_F V_T \tag{8.29}$$

The specific surface area is then

$$S_V^{con} = \frac{A_{tot}^{con}}{V_T} = \bar{a}_{con} n_F^{con} n_F. \tag{8.30}$$

The specific surface area of the fibrils with contacts is therefore given as

$$S_V^{tot} = S_V^T - S_V^{con} = \frac{2\phi_F}{\bar{R}_F}\left(1 - \frac{n_F^{con}\bar{a}_{con}}{2\pi \bar{R}_F \bar{l}_F}\right). \tag{8.31}$$

We can play a bit around with this expression, if we assume for instance that $\bar{a}_{con} = \alpha \bar{R}_F^2$ with α an arbitrary but constant factor, probably $\alpha > 1$. We then would have

$$S_V^{tot} = \frac{2\phi_F}{\bar{R}_F}\left(1 - \frac{\alpha}{2\pi} n_F^{con} \frac{\bar{R}_F}{\bar{l}_F}\right). \tag{8.32}$$

To proceed we need the number of contacts per fibril. In the literature on paper felts, there is an expression for this [408], which we adopt for our nomenclature:

$$n_F^{con} = \frac{\phi_F \bar{l}_F}{2\bar{R}_F}. \tag{8.33}$$

Combining this with Eq. (8.32) yields

$$S_V^{tot} = \frac{2\phi_F}{\bar{R}_F}\left(1 - \frac{\alpha}{4\pi}\phi_F\right). \tag{8.34}$$

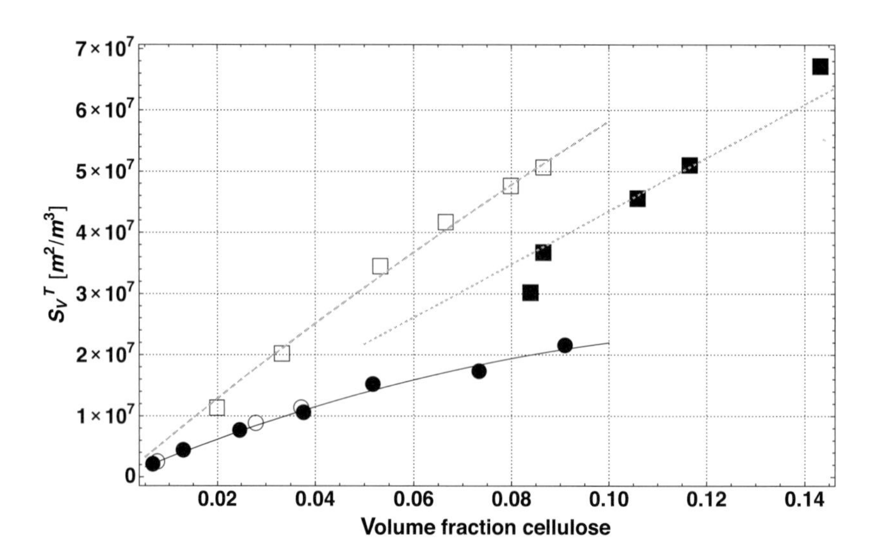

Figure 8.8 Specific surface area of cellulose aerogels prepared with different cellulose solvents all corrected for shrinkage. The open and full circles refer to the results of Hoepfner et al. [148] and Karadagli et al. [409]. The full squares are data from Buchtova and Budtova [218], who dissolved cellulose in a mixture of an ionic liquid and DMSO. The open squares are data from Cai et al. [401] using a LiOH–urea mixture to dissolve cellulose. Buchtova and Cai regenerated the wet gels in ethanol and performed a standard supercritical drying. The solid lines are a fit with Eq. (8.34).

This equation means that the larger the volume fraction of solid in an aerogel, the larger the reduction of surface area by fibril contacts. Yielding a fibrillar microstructure with a high surface area simply requires reducing the fibril radius by means of suitable processing during the gelation stage. The shrinkage during regeneration and washing steps (solvent exchange) can lead to an increase in the number of contacts per fibril via an increase in the effective volume fraction of fibrils and thus would lead to a reduction of the surface area.

There are fortunately some measurements in the literature on cellulose aerogels with a suitable range of cellulose compositions and surface area measurements. Hoepfner et al. [148] and Karadagli et al. [409] prepared cellulose aerogels using the same salt hydrate melt route and the same techniques to characterise the density and specific surface area (nitrogen adsorption; see Section 8.6). Combining their measurements with data from other authors who used different solvents for cellulose yields a result as shown in Figure 8.8. In all cases, a good fit is achieved, showing that the surface area is solely determined by the fibril diameter. The fit yields for the aerogels prepared by Hoepfner et al. and Karadagli et al. an average diameter of the cellulose fibril of 12 nm and a correction factor $\alpha = 42.78$. This means the contact area at points of intersection of fibrils is around 13 the cross section of a fibril. This sounds not unreasonable. For the materials prepared by Cai et al. [401], a diameter of 6 nm is obtained with a factor $\alpha = 14.6$, or in other words the area of contact is around five times the fibril diameter. In the case of Buchtova and Budtova's [218] aerogels, one obtains a fibril diameter of 9 nm with a correction factor $\alpha = 0$.

8.5 Surface Area of Aerogel Composites

Different types of composites can be made from and with aerogels. Aerogels of one kind can be dispersed in an aerogel matrix of another kind. Especially cellulose aerogels or other polysaccharide aerogels allow for another type, namely to fill the porous volume with another aerogel, for instance silica. Then we have two interconnected aerogel networks. A third variant would be possible, namely to disperse bulky granular materials (grains) in an aerogel matrix, either as nano-particles or as micron-sized ones. For the surface area of such composites, simple rules apply.

For a granular aerogel material embedded in another aerogel acting as a matrix, the total surface area in a composite of volume V can be calculated using the specific surface areas per volume. The matrix will occupy a volume of V_m and the particle phase all together a volume of V_P. We then have for the surface area of the composite S_V^c

$$S_V^c V = S_V^m V_m + S_V^p V_p \qquad \text{or} \qquad S_V^c = S_V^m \phi_m + S_V^p \phi_p \tag{8.35}$$

with ϕ_m, ϕ_p the volume fraction matrix or granular phase. Therefore, the specific surface area per volume of a composite is the weighted average of both components' specific surface area per volume and the weights are the volume fractions. If one measures instead the surface area per mass, another simple relation holds. Using the expressions for the volume fraction and, of course, that $m_p = \rho_e^p V_p, m_m = \rho_e^m V_m$ and using the definitions for weight percentages,

$$w_p = \frac{m_p}{m_p + m_m}. \tag{8.36}$$

We get after some calculations an expression:

$$S_m^c = w_p S_m^p + w_m S_m^m. \tag{8.37}$$

So the specific surface area per mass of a composite is the weighted average of both component's surface area per mass and the weights are the weight percentages of both components.

Let us treat another case, namely aerogels infiltrating a network of another aerogel (or a fibre batt or felt). Such aerogels were recently described by Budtova, Koebel and co-workers [31–33]. The porous material of the fibre phase will have a total envelope volume V_e, a density ρ_{fm} and a pore volume $V_p = \phi_{pf} V_e$ with ϕ_{pf} as the pore volume fraction of the fibrous material and $\phi_{fm} + \phi_{pf} = 1$. The specific surface area per volume of this material will be

$$S_V^{fm} = S_m^{fm} \rho_{fm}. \tag{8.38}$$

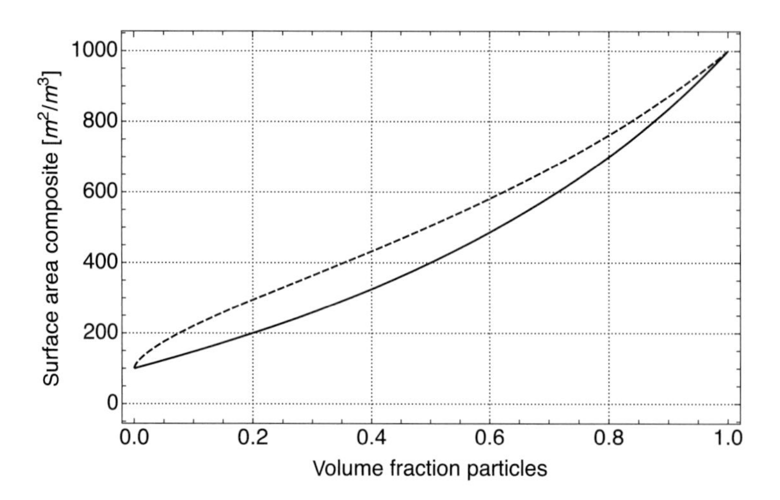

Figure 8.9 Possible effect of particles inside an aerogel matrix generating around themselves a region in which the surface area is different from the rest of the matrix (here assumed twice as large as the unaffected matrix). The total composite surface area increases especially at low-volume fractions of particles.

The aerogel filling will have a density of ρ^{fill} and a specific surface area per volume of $S_V^{\text{fill}} = S_m^{\text{fill}} \rho^{\text{fill}}$. The composite density can be calculated as

$$\rho_c = \rho_{fm}(1 - \phi_{pf}) + \rho_{\text{fill}} \phi_{pf}. \tag{8.39}$$

The total specific surface area per volume in the composite material is then

$$S_V^c = S_V^{\text{fill}} \phi_{pf} + S_V^{fm}(1 - \phi_{pf}). \tag{8.40}$$

Thus we again have that for the specific surface area a linear rule of mixtures exists in which, however, the pore volume of the fibrous material is the weighting factor. Measuring, as done usually the specific surface area with a BET technique, we get the specific surface area per mass. If we know it for both phases, we get again a linear relation

$$S_m^c = S_m^{fm} w^{fm} + S_m^{\text{fill}} w^{fill}. \tag{8.41}$$

In another case, it might, however, be possible that the matrix aerogel around the particles is affected by their presence leading to either smaller nanostructure around them or larger ones. The specific surface area per mass of an aerogel with small particles embedded and influencing the aerogel structure in their neighbourhood was derived by Laskowski and co-workers [410]. The effect is shown in Figure 8.9. At low-volume fractions, particle phase, the effect is large, becoming smaller at higher-volume fractions. Such effects were observed by Ratke for silica aerogels in which ceramic particles were dispersed under microgravity [411], and by Laskowski et al., who embedded aerogel particles in a cellulose aerogel matrix [410] and explained deviations from the simple rules of mixtures given previously with their model.

8.6 Measurement of Specific Surface Area

The most often used way to measure the surface area of an aerogel is nitrogen physisorption. Nitrogen sorption analysis is a method to determine nitrogen isotherms at the boiling point of liquid nitrogen (77 K). The shape and the height of the isotherm is characteristic for each porous sample and allows deriving quantitative information on the specific surface area and the pore size distribution of the porous material under investigation. It might also yield information on mechanical stability of nanosized structural features.

Figure 8.10 shows the typically adsorption isotherm of aerogels. The lower line is the adsorption path, indicated with an up arrow, and the upper line merging at lower pressure with the adsorption line is the desorption path. Typically a hysteresis between adsorption and desorption is observed. Almost all aerogels exhibit this so-called type IV isotherm according to the IUPAC classification scheme [402, 412] and may exhibit different forms of hysteresis curves. The type IV isotherms are only observed in aerogels with a pronounced mesoporosity. There are aerogels especially made, e.g., from carbon that show extreme microporosity and complete absence of mesoporosity and the corresponding hysteresis. Such aerogels might exhibit other isotherms, such as those of type Ia or Ib. For further details on the classification of isotherms, see the IUPAC reports [402, 412].

Prior to the measurement, adsorbed gases and vapours such as water vapour are removed from the sample by heating it under vacuum. Subsequently, starting from

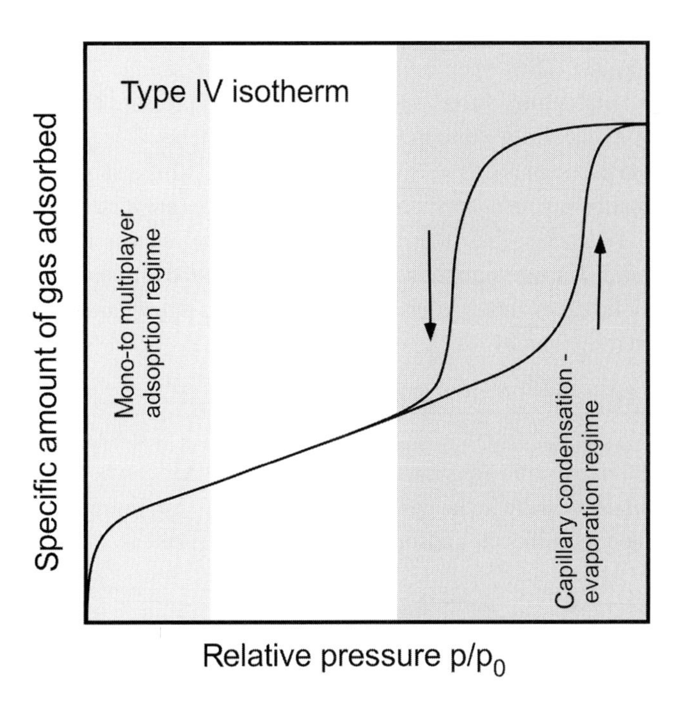

Figure 8.10 Schematic nitrogen sorption isotherm most often observed in aerogels.

the evacuated state, nitrogen is stepwise dosed, in a volumetric sorption setup, into the sample holder containing the porous material under investigation. The sample is hereby kept at a constant temperature of 77.3 K (−196 °C) by using immersion into a liquid nitrogen bath. In each dosing step, the gas pressure in the sample holder is monitored until equilibrium is reached. Equilibrium is defined in that context when the recorded pressure increase during a so-called equilibrium time interval t_{eq} divided by the mean pressure during the total time of measurement, or that specific measurement point is below a certain threshold, e.g., $\Delta p(t_{eq})/p(mean) \leqslant 0.01\%$. Due to adsorption of nitrogen within the sample, the pressure drop in the sample holder in each dosing step is larger than in the same sample holder with no sample being present. The mass specific amount of gas (per mass of the sample) adsorbed is calculated from this difference in gas pressure. To do so, the volume of the free space in the sample holder has to be known. This free volume is usually measured using the inert gas helium before or after analysis. For carbon-based samples or samples of significant benzene ring content, helium might be trapped in between those rings. Therefore, it is advisable to determine the free space after measurement. In general, the specific amount adsorbed is given in units of mg per gram of sample or in cm^3/g at STP.[1] The key parameter upon gas sorption is the relative gas pressure p/p_0, which is defined as the nitrogen partial pressure p over the saturation pressure p_0 at a given temperature.

In order to understand how such a measurement is made and the quantity of interest is derived, assume that in the sample holder with a volume V_0 being at a constant temperature T_0 there exists a defined pressure p_1. Then there are N_1 molecules in the gas phase above the sample which can be calculated by the ideal gas law as $N_1 = (p_1 V_0)/(k_B T_0)$. Then open a valve attached to the sample volume holder for a short but precisely defined time being at a well-defined pressure and temperature to let a number of molecules N_2 into the sample holder volume. This leads immediately to an increase in pressure $p_2(0) = (N_1 + N_2)/(V_0 k_B T_0)$. If the sample adsorbs (and this does not qualify any mechanism for how the molecules are attached to the surface of the sample) molecules, their number in the gas phase reduces from $N_1 + N_2$ to $N_1 + N_2 - N_{ad}(t)$. The pressure reduces with time, since the diffusion and adsorption into a porous solid takes time. After a sufficient period, a new equilibrium pressure establishes which is given by

$$p_2^e = \frac{N_1 + N_2 - N_{ad}(t \to \infty)}{V_0 k_B T_0} \quad \text{or} \quad p_2^e = p_2(0) - \frac{N_{ad}^{eq}}{V_0 k_B T_0}. \tag{8.42}$$

The pressure difference after equilibrium has established, Δp, can be measured as well as the new absolute pressure in the gas phase $p_a = p_2(0) - \Delta p$. From the measurement of Δp, the number of molecules adsorbed is simply calculated as

$$N_{ad} = \frac{\Delta p}{V_0 k_B T}. \tag{8.43}$$

[1] STP stands for standard temperature and pressure: 101.325 kPa, 0°C. The definition of STP is different in chemistry. IUPAC defines STP as 100 kPa for the pressure and the same temperature of 0°C

Since the amount adsorbed changes with the absolute pressure, one arrives after several increases of pressure to a so-called isotherm

$$N_{ad} = f(p_a, T, \text{gas, solid}). \tag{8.44}$$

The parameters gas and solid indicate that for a given solid surface, the isotherm will vary with the type of gas, and also for a given gas type, the isotherm changes with the kind of solid surface investigated. Most often in the literature, the amount of gas N_{ad} is calculated per gram of solid and the number of molecules expressed in terms of moles. As mentioned already, the pressure is given as a relative pressure p_a/p_0. The whole problem of adsorption, physisorption or chemisorption to any type of surface (smooth, rough, porous, etc.) is then reduced to derive theoretical models for adsorption isotherms. A few of these models, which are used especially in aerogel research to derive the specific surface area of a material, are outlined in the following subsections. Many other models are discussed in the literature [109, 402, 412–414] and are especially useful, for instance, if the gas molecules also dissociate at the surface or make a chemical reaction there or there is a mixed adsorption from a mixture of gases which requires special considerations [415].

8.6.1 Langmuir Isotherm

How can adsorption be understood? Principally, adsorption of a gas on a solid surface means that the gas molecules interact with the surface in a way that a part of them can be bounded to the surface. In adsorption or physisorption, in contrast to chemisorption, no covalent or metallic bonds are established between the gas molecules and the solid surface, but only Van der Waals–type interactions. These have interaction energies in the range of 4–40 kJ/mol in contrast to chemical bonds (ionic or covalent bonds), which are in the range of 100–400 kJ/mol (for a discussion of all kinds of bond mechanisms to a surface, see [117]). The Van der Waals interactions are due to molecules with, for instance, a permanent dipole inducing in a solid surface another dipole or the solid surface is polarised itself. The interaction is not only weak, but also its range of influence is rather small. Typically the dipole–dipole interaction potential $U(r)$ depends on distance r between the solid surface and the gas molecule like $U(r) \propto 1/r^6$. Inasmuch as the molecules come closer to the solid interface, the electron clouds of the interface and the gas molecule exert a repulsive force, such that a too close approach is almost impossible. Quite often this repulsive interaction is modelled as $U_{rep} \propto 1/r^{12}$. The total interaction force is then called as Lennard–Jones potential (exactly a (12,6)-LJ potential)

$$U_{LJ} = 4\epsilon \left[\left(\frac{\sigma}{r} \right)^{12} - \left(\frac{\sigma}{r} \right)^{6} \right], \tag{8.45}$$

in which ϵ is the interaction energy, which as stated previously is in the range of 4–40 kJ/mol and σ defines the minimum position of the particles, here the solid surface to gas molecule distance. Since the interaction energy in adsorption is rather small, it

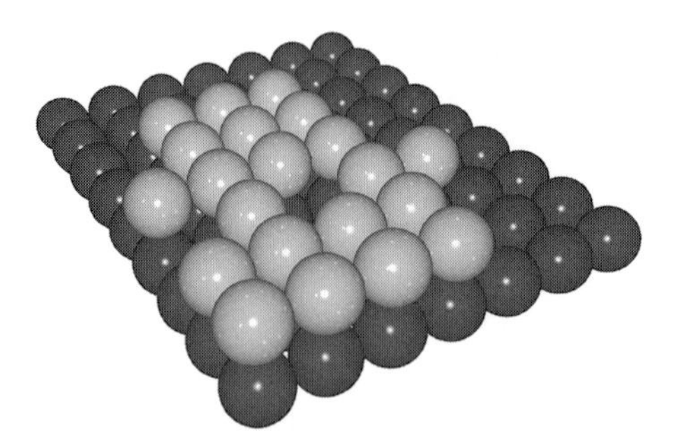

Figure 8.11 Schematic adsorption of molecules on a surface. The molecules are envisaged as simple spheres which attach to regular lattice places on the surface.

typically needs low temperatures to establish a layer or several layers of gas molecules at the surface. One then can imagine this layer of gas molecules as a surface phase on the solid structure.

The most simple way of adsorption is one single molecular layer of adsorbed gas (adsorbate) on an interface, which is assumed to be a flat one, and all positions on the surface have the same interaction potential to a molecule. This is illustrated in Figure 8.11. The adsorption isotherm for a single-layer adsorption was first calculated by Langmuir [416]. Assume that for a given pressure p of a single-component gas above the surface there are molecules adsorbed to the surface at a rate dn_{ad}/dt and desorbed at a rate dn_d/dt with n_{ad}, n_d the number of molecules adsorbed or desorbed. The surface exposed in the vacuum chamber to the gas will have an area A_S and the gas above the adsorbing surface will have a pressure p. The adsorption rate depends on the number of empty sites on the surface, S_0, since we assume that the gas molecules can only be bounded directly to the sample surface and not on areas where adsorption already has taken place. Desorption instead depends on the number of molecules already adsorbed on the surface, S_1, and not on the pressure above the surface. Kinetic gas theory shows that the number of molecules striking a surface per unit area and time is

$$Z(T) = \frac{p}{\sqrt{2\pi M_a k_B T}}, \tag{8.46}$$

with M_a being the mass of a molecule. The probability that a molecule hitting the surface is bonded is denoted by a factor $k_{ad}(T)$, which might depend on temperature. In the desorption case, the number of molecules being already adsorbed is relevant, since these only can leave the surface due to thermal activation, which is described by a rate constant $k_d^* e^{-\frac{E_{ad}}{RT}}$ with E_{ad} the energy of adsorption. We then can write for both rates the following equations

$$\frac{dn_{ad}}{dt} = k_{ad}(T)\, p\, S_0 \tag{8.47}$$

$$\frac{dn_d}{dt} = k_d^* e^{-\frac{E_{ad}}{RT}} S_1 = k_d(T) S_1, \tag{8.48}$$

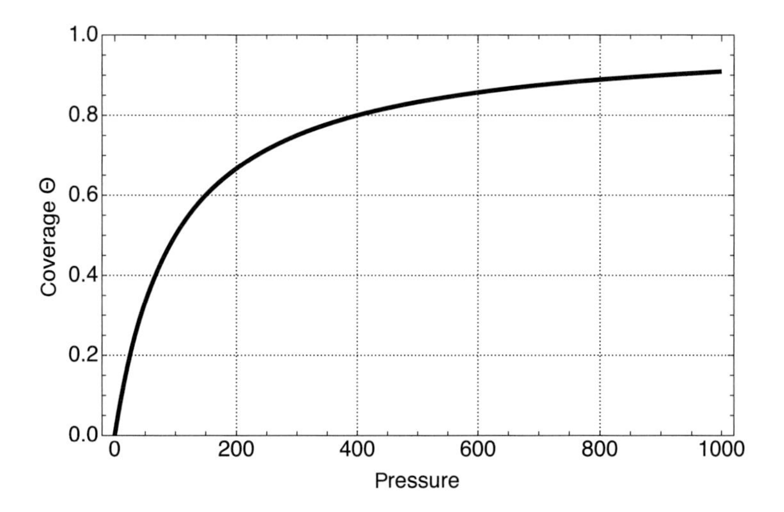

Figure 8.12 Langmuir isotherm for monolayer adsorption at constant temperature. The surface coverage depends in a non-linear way on the pressure.

where we have put the weak square root dependence on temperature of Eq. (8.46) into the rate constant $k_{ad}(T)$. Note the exponential dependence on temperature means that with increasing temperature, the desorption rate increases. The number of empty sites is related to the number of adsorbed sites by

$$S_0 = A_S - S_1. \tag{8.49}$$

We work at constant temperature and wait until the pressure has reached a constant value, assumed then to be a state where equilibrium between the gas above the surface and the surface with its given coverage of molecules has established. In that equilibrium situation, the rate of adsorption and desorption are equal. We define

$$\frac{k_{ad}}{k_d} = K_{eq}. \tag{8.50}$$

Here we call the ratio of both reaction rates K_{eq}. Using Eq. (8.49) and defining the relative coverage of the surface as

$$\theta = \frac{S_1}{A_S}, \tag{8.51}$$

we finally can write after some simple manipulations

$$\theta = \frac{K_{eq}\,p}{1 + K_{eq}\,p} \tag{8.52}$$

at the famous Langmuir isotherm for a monolayer adsorption on a surface. The behaviour is illustrated in Figure 8.12. Initially one has a linear increase of relative coverage with pressure which levels off at higher pressures. Note that K_{eq} has the dimension of an inverse pressure. Typically for gases the amount of adsorbed gas is measured and plotted as a function of pressure (as outlined earlier). The coverage of a surface can be expressed as the volume adsorbed v_a divided by the volumetric monolayer capacity v_m, i.e., $\theta = v_a/v_m$, as long as only a monolayer is formed. One should make a few comments regarding this isotherm and its derivation. The interaction energy of gas molecules with the surface is assumed to be independent of the coverage or the position where a gas molecule is bounded (all places on the surface are equal). This is a drastic oversimplification, especially in aerogels.

One can easily imagine that a gas molecule being alone on a surface just establishes one Van der Waals interaction with the solid surface, but if a second comes close to the first one, it not only establishes an interaction with the solid, but also a side interaction with the molecule being there. Inasmuch as the first layer develops, molecules might attach to patches on the surface or vacancies inside the layer, and the interaction energy with the solid is different from the state of being alone on the surface. Another point is that a molecule loosely bound to the interface will move along it, make surface diffusion and once it approaches a patch of other molecules it will stay there due to interactions with this patch of molecules. Such a surface diffusion in building up of a layer is not taken into account. We also have to mention that in aerogels as well as in other porous materials, the surface is not atomically flat and smooth, but rough, having steps, kinks and jogs and, probably most important, functional groups on the surface that bear dipoles, such as an OH group or in biopolymers amine groups etc. Molecules are preferably adsorbed to such rougher areas, since the interaction with the solid is stronger there. Finally we have to note that a monolayer adsorption is rather rare in aerogels; typically a second, third or fourth layer already builds up before the first layer is closed and within the nanosized pores space the gas phase can condense into the liquid state. Such a behaviour is often described with the so-called BET equation. In addition, one has to take into account that adsorption and desorption change the local interface temperature (heat of adsorption and desorption) and in a gas mixture such as air the desorption rate also depends on the pressure above the surface.

8.6.2 A Bit of Thermodynamics of Adsorption

Before we proceed to derive the BET adsorption equation, let us think a bit about thermodynamics of adsorption and adsorption layers, especially clarify the meaning of the energy of adsorption used previously. Generally the Gibbs free energy G of a system is a function of pressure p, temperature T, surface area A and number of molecules of species k, n_k, namely $G = G(p, T, A, n_k)$. The total differential of Gibbs free energy reads

$$dG = \left(\frac{\partial G}{\partial p}\right)_{T,A} dp + \left(\frac{\partial G}{\partial T}\right)_{p,A} dT + \left(\frac{\partial G}{\partial A}\right)_{T,p} dA + \left(\frac{\partial G}{\partial n_k}\right)_{T,p,A} dn_k. \tag{8.53}$$

Using that

$$\left(\frac{\partial G}{\partial p}\right)_{T,A} = V \quad \text{volume} \tag{8.54}$$

$$\left(\frac{\partial G}{\partial T}\right)_{p,A} = -S \quad \text{entropy} \tag{8.55}$$

$$\left(\frac{\partial G}{\partial A}\right)_{T,p} = \sigma \quad \text{surface energy} \tag{8.56}$$

$$\left(\frac{\partial G}{\partial n_k}\right)_{T,p,A} = \mu_k \quad \text{chemical potential,} \tag{8.57}$$

we can write this differential as

$$dG = Vdp - SdT + \sigma dA + \sum_k \mu_k dn_k. \tag{8.58}$$

Let us define a new quantity namely the so-called surface excess $\Gamma = n_{ad}/A$ defined as the ratio of molecules adsorbed at the surface per unit surface area. This is constant if the number of moles arriving at the surface and being attached there is equal to the number of moles desorbed from the surface. The surface layer is in equilibrium with the gas phase above, if the chemical potential of the adsorbed phase is equal to the chemical potential of the gas phase, $\mu_{ad} = \mu_{gas}$ or $d\mu_{ad} = d\mu_{gas}$. The change of chemical potential of the adsorbed surface layer can be calculated as

$$d\mu_{ad} = -S_{ad}dT + V_{ad}dp + \left(\frac{\partial \mu_{ad}}{\partial \Gamma}\right) d\Gamma. \tag{8.59}$$

The chemical potential of the gas phase is simply given as

$$d\mu_{gas} = -S_{gas}dT + V_{gas}dp. \tag{8.60}$$

If the surface excess Γ is constant, we have $d\Gamma = 0$ and thus arrive at

$$-S_{ad}dT + V_{ad}dp = -S_{gas}dT + V_{gas}dp. \tag{8.61}$$

This equation can be rearranged to

$$\frac{dp}{dT} = \frac{S_{gas} - S_{ad}}{V_{gas} - V_{ad}} \tag{8.62}$$

Since one generally has $V_{gas} \gg V_{ad}$ and for the gas phase one can assume the ideal gas law to hold, $V_{gas} = R_g T/p$, we arrive after some manipulations at

$$\frac{1}{p}\frac{dp}{dT} = \frac{S_{gas} - S_{ad}}{R_g T}. \tag{8.63}$$

Using that in equilibrium there is a simple relation between entropy and enthalpy, namely $\Delta S = \Delta H/T$, this equation reads

$$\frac{1}{p}\frac{dp}{dT} = \frac{H_{gas} - H_{ad}}{R_g T^2} = \frac{q_{st}}{R_g T^2}. \tag{8.64}$$

The term q_{st} is called the differential isosteric heat of adsorption and represents the energy difference between the state of the system before and after adsorption of a differential amount of molecules on the surface. It can be calculated by plotting $\ln p$

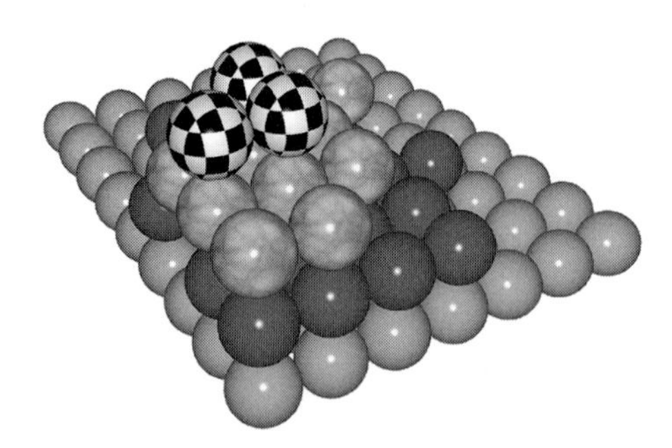

Figure 8.13 Drawing of a multilayer adsorption.

as a function of $1/T$ as long as the surface coverage Γ is constant, which has to be checked experimentally. This short derivation clarifies that the heat of adsorption is an enthalpy. In principle, one can do the same calculation for the second layer as it is adsorbed on the first one and would of course get another isosteric heat of adsorption. Let us now calculate the famous BET isotherm.

8.6.3 BET Isotherm

BET stands for the names of the researchers who developed the method, i.e., S. Brunauer, P. H. Emmett and E. Teller [417]. Starting from the evacuated state, the nitrogen dosed into the sample holder is initially adsorbed at the inner surface of the porous material (adsorbent) under investigation. The BET equation used for the evaluation of the surface area is based in the assumption that in parallel to the monolayer adsorption, statistical adsorption processes also take place in higher layers (multilayer adsorption; see Figure 8.13). This leads to the following relationship between the amount of gas adsorbed v and the relative gas pressure p/p_0, in which p_0 is the so-called saturation pressure at which all pores are completely filled with adsorbent:

$$\frac{p/p_0}{v\left(1-p/p_0\right)} = \frac{1}{c \cdot v_m} + \frac{c-1}{c \cdot v_m}\frac{p}{p_0}. \tag{8.65}$$

Here v_m is the specific monolayer capacity, i.e., the specific amount of gas necessary to cover the adsorbent with a monolayer measured in cm^3/g, or one states the moles/g, mostly in mmol/g, and then denotes this capacity as Q_m. The parameter c describes the strength of the interaction between the adsorbents and the adsorbate interface. The larger c the higher the interaction.

Although this equation is routinely used in almost all papers on aerogels and the evaluation software for an isotherm analysis in modern sorption equipments, there is rarely a derivation of it, showing the implications and assumptions behind it. We therefore follow the original paper of BET and derive this equation.

Let $S_0, S_1, S_2, \ldots, S_i$ represent the surface areas on the sample surface with i layers on it. So S_0 denotes the surface area with no molecule (empty area), S_1 the area covered with a monolayer, S_2 the area covered with two layers and so on. We assume that the situation with i layers is an equilibrium situation. This means, first, that the adsorption and desorption rates for the first layer are equal:

$$a_1 p S_0 = b_1 S_1 e^{-E_{ad}/RT}. \tag{8.66}$$

In this equation a_1, b_1 are constants. This is exactly the Langmuir isotherm equation. We solve this equation for the surface coverage by one molecular layer and write

$$S_1 = \frac{a_1}{b_1} p e^{E_{ad}/RT} S_0 = y S_0. \tag{8.67}$$

We must, however, take into account that other areas of the surface are covered with layers of multiple thickness. Therefore, let us look to the areas with a second layer. Again, at equilibrium the area covered with one layer S_1 must be constant. there are four ways S_1 can change: (a) adsorption processes on the empty surface, (b) evaporation from the first layer, (c) adsorption on the first layer (becoming then part of the surface with two layers) and (d) desorption from the area with two layers. In equilibrium, all four processes balance each other and thus we have

$$a_2 p S_1 + b_1 S_1 e^{-E_{ad}/RT} = b_2 S_2 e^{-E_2/RT} + a_1 p S_0, \tag{8.68}$$

where we have noticed that the interaction of molecules with a monolayer at the surface is different from the adsorption energy and have given it a new symbol E_2. Using Eq. (8.66) leads to

$$a_2 p S_1 = b_2 S_2 e^{-E_2/RT}. \tag{8.69}$$

This relation means that the rate of adsorption on the first layer is equal to the rate of desorption from the second layer. We also rewrite this equation as

$$S_2 = \frac{a_2}{b_2} e^{E_2/RT} S_1 = x_2 S_1. \tag{8.70}$$

Extending this way of argumentation to the next layer, one yields a set of equations having the following general form:

$$a_i p S_{i-1} = b_i S_i e^{-E_i/RT}, \tag{8.71}$$

which we again solve for the S_i:

$$S_i = \frac{a_i}{b_i} p e^{E_i/RT} S_{i-1} = x_i S_{i-1}. \tag{8.72}$$

The next two steps in argumentation are important in the derivation of the BET equation. First the sample area relates all S_i to each other:

$$A_s = \sum_{i=0}^{\infty} S_i. \tag{8.73}$$

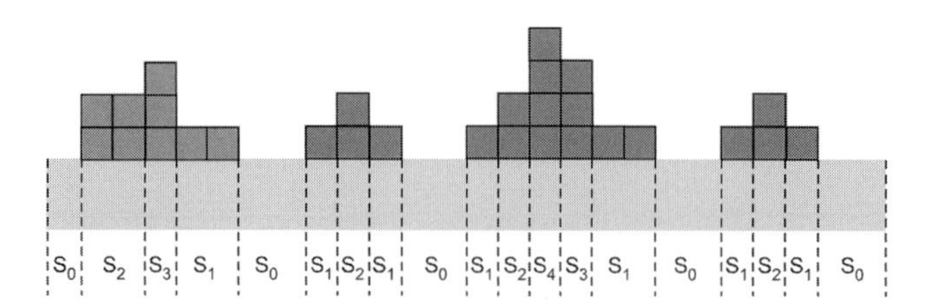

Figure 8.14 Scheme of a multilayer adsorption and definition of the areas covered by different number of layers. All area indicated with S_0 belong to areas with no adsorption, all with S_1 belong to the layer with a unimolecular adsorption and so on.

A sketch of a multilayer adsorption with the definitions of the different areas is given in Figure 8.14. This illustrates Eq. (8.73) and also the next equation, which calculates the amount of gas adsorbed:

$$v = v_0 \sum_{i=0}^{\infty} i\, S_i, \tag{8.74}$$

in which v_0 is the amount of gas adsorbed per area if it is covered just with a monolayer. If we calculate now the ratio of the adsorbed volume and the volume if the sample surface would be covered only with a monolayer, we obtain

$$\frac{v}{A v_0} = \frac{\sum_{i=0}^{\infty} i\, S_i}{\sum_{i=0}^{\infty} S_i} \tag{8.75}$$

with $A = A_s/a_m$ and a_m the area a single molecule occupies on the surface. In principle we solved the problem of multilayer adsorption, but to go along with this equation, Brunauer, Emmett and Teller made a few interesting assumptions which allow us to simplify this expression and eventually lead to the famous BET equation. First, they assume that the interaction energies with the second layer of gas molecules on the monolayer of gas and that of all other layers with each other (gas gas interactions) are equal, meaning

$$E_2 = E_3 = \cdots E_i = E_L, \tag{8.76}$$

and are equal to the liquefaction or condensation energy (we implicitly assume herewith that all layers above the first are in the liquid state). Brunauer, Emmett and Teller have then made a further simplification, namely

$$\frac{b_2}{a_2} = \frac{b_3}{a_3} = \cdots \frac{b_i}{a_i} = g, \tag{8.77}$$

meaning that all constants are related and their ratio is a constant, g. With these simplifications, the x_i's in Eq. (8.72) are constant and we denote the value simply as x. Then the area of all layers except the first one are related to each other:

$$S_i = x S_{i-1} = x^{i-1} S_1 = y x^{i-1} S_0 = c x^i S_0, \tag{8.78}$$

in which we introduce the abbreviation

$$c = \frac{y}{x} = \frac{a_1 g}{b_1} e^{(E_{ad} - E_L)/RT}, \tag{8.79}$$

assuming later that this is a constant. If we substitute these relations into Eq. (8.75), it can be rewritten as

$$\frac{v}{v_m} = \frac{c \sum_{i=1}^{\infty} i x^i}{1 + c \sum_{i=1}^{\infty} x^i} \tag{8.80}$$

with $v_m = A v_0$. The sums are easily evaluated, keeping in mind that they are closely related to the geometric series. We have that $\sum_{i=1}^{\infty} x^i = x/(1-x)$ and for the sum in the denominator use $x dx^i/dx = i x^i$. Insertion into Eq. (8.80) leads to

$$\frac{v}{v_m} = \frac{cx}{(1-x)(1-x+cx)}. \tag{8.81}$$

We now use that at saturation, all pores are filled with liquid gas, $p = p_0$, and thus $x = 1$. This yields

$$\frac{p_0}{g} e^{E_L/RT} = 1 \quad \text{or} \quad x = \frac{p}{p_0}. \tag{8.82}$$

Substituting the expression for x into Eq. (8.81) leads to the famous BET equation

$$v = \frac{v_m c p}{(p_0 - p)(1 + (c-1)\frac{p}{p_0})}. \tag{8.83}$$

The volume adsorbed depends on the pressure and the surface area A_s included in the term v_m and of course on the interaction of the gas with the solid surface and the energy of condensation (see Eq. (8.79)). We plot the pressure dependence in Figure 8.15 to show the effect of c on the shape of the curve. The plot shows that essentially the constant c dictates the behaviour at low relative pressures. Close to saturation, the curves merge. In the range at lower relative pressures, the difference between c-value has a clear effect. This was used by Brunauer, Emmett and Teller to rearrange the equation to allow a linear plot, as already shown in Eq. (8.65). Plotting $(p/p_0)/(v(1 - p/p_0))$ vs. p/p_0 results in a straight line, making the intersect $y_0 = 1/(v_m c)$ with the ordinate and $m = (c-1)/(c \cdot v_m)$ as its slope. Reading of the intersection and the slope allows us then to calculate $c = m/y_0 + 1$. The specific monolayer capacity v_m is given by $v_m = 1/(m + y_0)$. The linear relationship is schematically shown in Figure 8.16. This linear plot is the basis of all evaluation programs used in commercial adsorption equipment. Typically a plot is made in the range of relative pressures from 0.05 to 0.3. Let us make another point on the meaning of the c-value. From Eq. (8.79), it becomes clear (looking also at the derivation of the Langmuir equation) that the c-value becomes large when the activation energy for desorption from a monolayer is large compared with that from the other layers. One can interpret this also in the following sense: if the heat of vaporisation (mirrored in the liquefaction energy E_L) is smaller than the heat of adsorption, than the c-value is large.

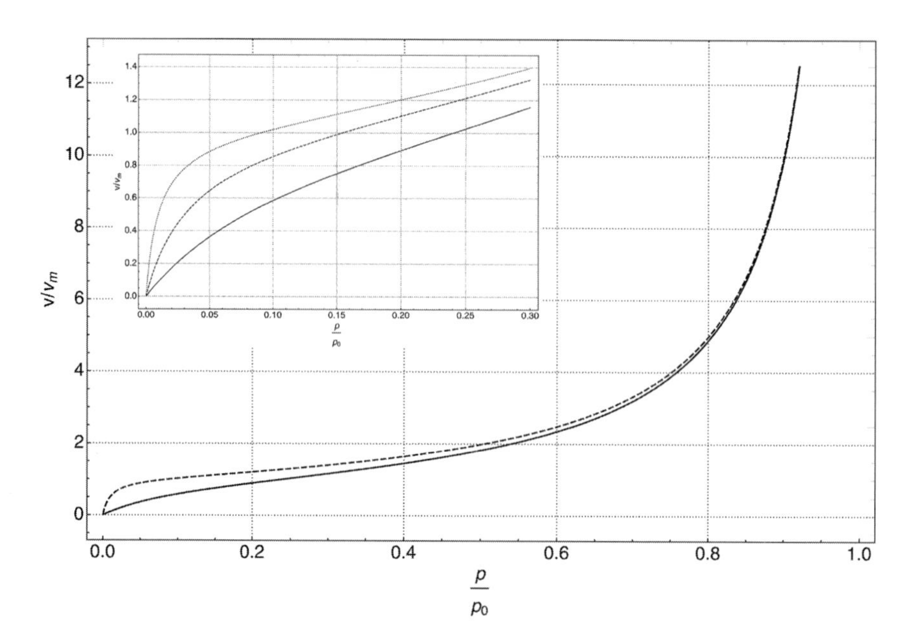

Figure 8.15 Plot of the BET Eq. (8.81). Shown is the adsorbed volume v/v_m as a function of relative pressure $x = p/p_0$ for two values of the constant c. The solid line is for $c = 10$ and the dashed one for $c = 100$. The inset magnifies the behaviour at low pressures. The solid line is for $c = 10$, the dashed one for $c = 30$ and the dotted one for $c = 100$.

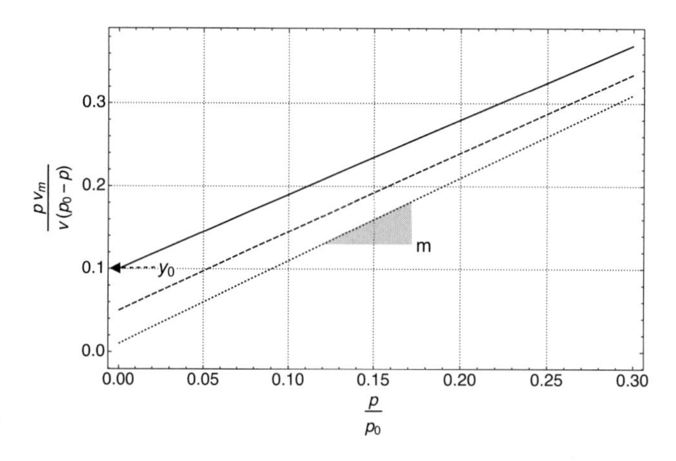

Figure 8.16 Scheme of the linearised BET equation with the slope and intercept indicated for three different values of c. Solid line $c = 10$, dashed line $c = 20$ and dotted line $c = 100$.

The specific surface area can be calculated from the specific monolayer capacity by using the following relationship:

$$v_m = A v_0 = \frac{A_s}{a_m} \frac{\Omega_m}{N_A}. \tag{8.84}$$

In this equation N_A denotes Avogadro's number and Ω_m the molar volume of the gas used at standard conditions. a_m is the surface area taken, for instance, by a single

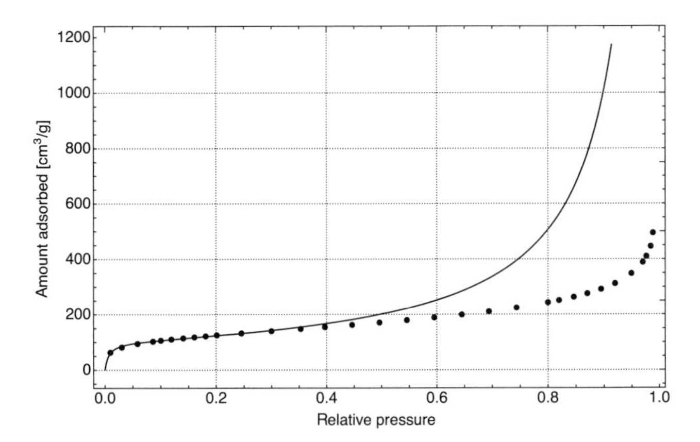

Figure 8.17 Nitrogen adsorption isotherm on a Pekala-type aerogel with the fit of the BET equation. The fit yields a c value of 148.7 and a monolayer capacity of 101. The BET surface area is 435 m^2/g. Courtesy M. Schwan, DLR

nitrogen molecule ($a_m = 0.16\ nm^2$). Solving this equation for A_s and calling it now S_{BET}, we arrive at the standard equation used in many textbooks:

$$S_{BET} = a_m \frac{v_m}{\Omega_m} N_A, \tag{8.85}$$

indicating with the subscript BET that it was determined using the BET model. One example will illustrate the application of the BET adsorption isotherm to aerogels. Figure 8.17 shows the adsorption branch of a nitrogen adsorption on a Pekala-type RF aerogel together with the fit of the BET equation over the whole pressure range. The comparison shows that the BET equation fits well up to around 0.4 relative pressure and then would drastically overestimate the adsorption.

In the literature on adsorption, physisorption and chemisorption, a large variety of isotherms are used to often describe special sorption processes, such as the Temkin or Freundlich isotherm. Only very few adsorbents/adsorptive systems fulfil Langmuir's assumptions. The vast majority of surfaces are strongly heterogeneous, meaning also they have a variety of interaction energies with the adsorbing gas. Therefore, a lot of research was invested to develop isotherm equations targeting certain heterogeneities of certain material classes, such as the t-plot discussed in the next subsection. Thus many of the more than a dozen isotherm equations that can be found in the literature are more or less empirical and in a certain sense only used to fit experimental data. We are not going to discuss these. For a first reading, see the textbook of Atkins et al. [109] and for a deeper discussion the book of Rouquerol et al. [414].

8.6.4 T-plot

Quite often an adsorption isotherm taken on aerogels does not fit to the BET model, as shown in the preceding subsection. At best there is a good agreement in the pressure range of $0.05 < p/p_0 < 0.3$. The deviation from the BET-like behaviour can have various origins. As an example, there could be a certain amount of micropores, pores in the range below 2 nm, surface roughness in the nanometer range, the aforementioned patches of adsorbed layer with a variety of layers, ranging from mono- to multilayer etc. This led in the decades after the seminal paper of BET to a variety of different approaches to adsorption. One concept is that of a so-called statistical or average thickness $\bar{t}$ of adsorbed gas. This thickness then is a fractional multiple of a complete monolayer. This statistical thickness is defined by the following relation. Let the mass adsorbed be m_a and m_m be the mass of a complete monolayer on the sample with a surface area A_s. Then we have the following relations:

$$\frac{m_a}{m_m} = \frac{V_a \rho_a}{V_m \rho_a} = \frac{\bar{t} \cdot A_s}{d_a A_s}, \tag{8.86}$$

and thus the simple equation results

$$\bar{t} = d_a \frac{V_a}{V_m}. \tag{8.87}$$

In these equations ρ_a is the density of the adsorbents and d_a is the thickness of a monolayer. If V_a/V_m would be identified with the BET isotherm, one directly gets the average thickness of the adsorbed molecules, provided the adsorption follows the BET model. This equation also implies that on a smooth flat surface the adsorption isotherm would yield a straight line passing the origin. During decades of studying the adsorption of gases on non-porous surfaces, it was observed, especially by de Boer from the TU Delft, that for many materials, particularly oxides, the statistical thickness follows a universal behaviour as a function of relative pressure. This led to the development of the so-called t-plot model. In principle, this is only another approach to describe adsorption, as the BET equation or other classical equations have been. Intensive investigations on non-porous materials have shown that there is not really a universality, but several material classes seem to follow the same adsorption rules (and thus a t-plot) [413]. Nevertheless, many models for the statistical thickness have been developed (and are still being developed). The standard procedure is to measure adsorption on a reference material being non-porous and extract the statistical thickness from it, $\bar{t} = f(p/p_0)$, using Eq. (8.87). Having the isotherms measured on the porous material $V_{ads} = g(p/p_0)$ then allows us to plot $V_{ads} = h(\bar{t})$. This is illustrated in Figure 8.18.

The scheme relies on having a suitable non-porous reference material. This is not easy for aerogels in general. For silica aerogels, one could take, for instance, a vitreous silica glass substrate. For carbon, several references are available in the literature. For the biopolymers and RF or resorcinol-melamine-formaldehyde (RMF) aerogels such a reference model substance is not really available and therefore a t-plot could, for instance, use a theoretical model for the statistical thickness. For different materials, thickness equations are suggested in the IUPAC report [402]. For convenience, we write some of the most often used equations:

$$\text{Halsey} \qquad \bar{t} = 0.354 \left(\frac{-5}{\ln \frac{p}{p_0}} \right)^{1/3} \tag{8.88}$$

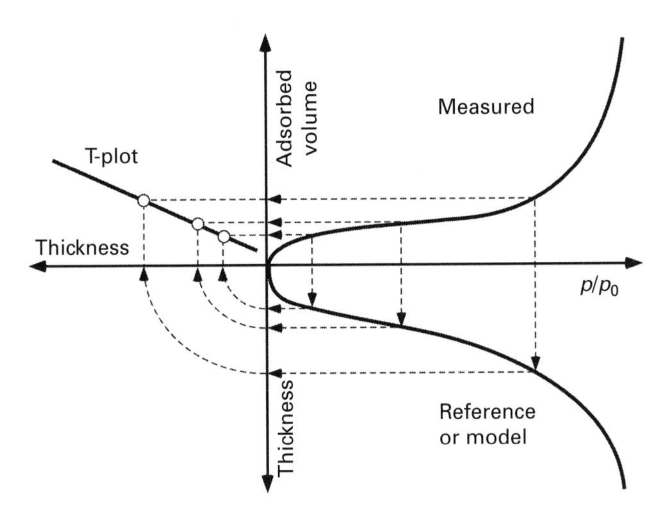

Figure 8.18 Scheme of the transformation of an adsorption isotherm, $\bar{t} = f(p/p_0)$, here a BET isotherm with c = 100, a plot of the average thickness as a function of relative pressure and the adsorption isotherm as a function of this average thickness.

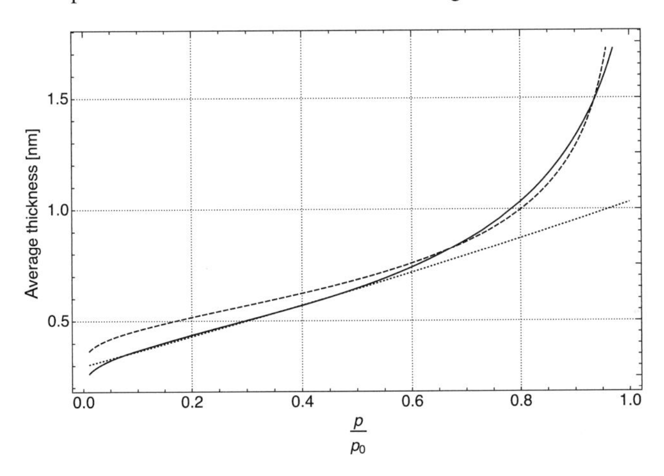

Figure 8.19 Comparison of the average adsorbed layer thickness as given by the Harkins and Jura Eq. (8.89), solid line, and the Halsey Eq. (8.88), dashed line and the Magee Eq. (8.90) (dotted line)

$$\text{Harkins and Jura} \quad \bar{t} = 0.1\left(\frac{13.99}{0.34 - \lg\frac{p}{p_0}}\right)^{1/2} \tag{8.89}$$

$$\text{Magee} \quad \bar{t} = 0.088\left(\frac{p}{p_0}\right)^2 + 0.645\left(\frac{p}{p_0}\right) + 0.298. \tag{8.90}$$

The statistical thickness in these equations is always given in nm. A plot of these curves is shown in Figure 8.19. The Halsey and Harkins–Jura equations differ in the low pressure range, whereas at higher pressures both are more or less equal. Since these equations are routinely built in modern software, one can check the difference

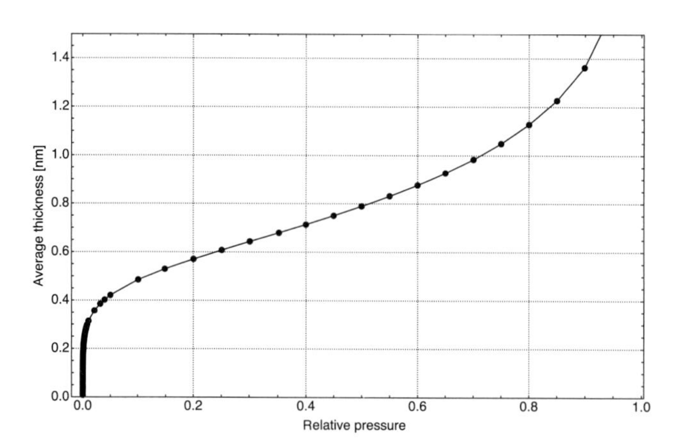

Figure 8.20 Statistical or average thickness of the nitrogen adsorption on a silica-alumina non-porous reference material.

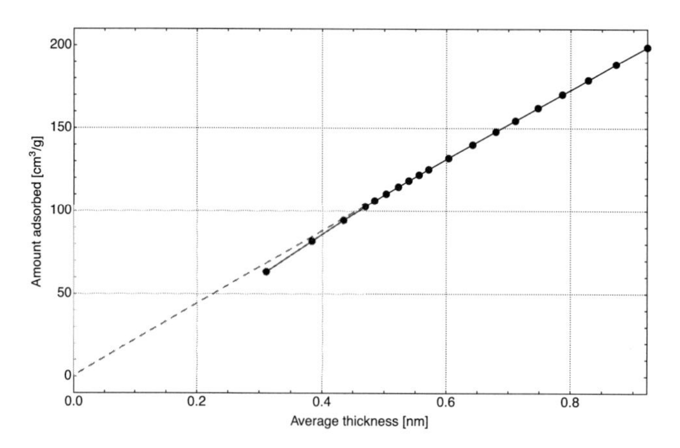

Figure 8.21 T-plot of the Pekala type aerogel. The dashed line extrapolates linearly to the origin and shows that this aerogel does not contain micropores.

easily. The Magee equation is a special one used as a reference material for carbon (non-porous carbon black).

The t-plot method will be illustrated with a material already shown before, the Pekala-type RF aerogel. The average thickness as a function of pressure of a reference material built in the software of the adsorption device was used. That reference material, an amorphous silica-alumina non-porous material, yields the statistical or average thickness as a function of relative pressure shown in Figure 8.20. A fit of this curve, for instance by a spline function, allows us then to extract the thickness at the discrete relative pressure values of the measured adsorption isotherm (see Figure 8.17). These thickness values are then related to the amount adsorbed, and thus a plot of the amount adsorbed as a function of thickness is possible. The result is shown in Figure 8.21. This t-plot can be used and is quite often used to determine microporosity, since in that case the plot shown in Figure 8.21 would not extrapolate to zero but a finite ordinate value would be reached, which allows then to determine the amount of microporosity. As mentioned previously (see Eq. (8.87)), the amount adsorbed increases linearly on a smooth substrate; however, if there are micropores, the situation changes. Even at very low relative pressures, these micropores are rapidly filled without any increase in average thickness. Once these are filled, the average thickness increases according to, for instance, one of the previously described theoretical models. This means a plot of the amount adsorbed against the thickness starts with a finite amount already adsorbed in the micropores, and then an isotherm follows which might be described by a BET isotherm, for instance.

8.6.5 Small Angle X-Ray Scattering (SAXS)

A completely different method to determine the specific surface area of a porous material uses expressions derived by Debye, Guinier and Porod for the evaluation of small angle X-ray scattering data (SAXS) of porous materials [418–421]. The method is not that often applied in aerogel research compared to BET, which mainly is due to the availability of the relatively expensive SAXS instruments compared to nitrogen adsorption equipment. In Appendix C, we briefly review all relevant equations to determine the specific surface area. Most often used is a two-phase model, one phase being the solid, one the void or pore space and a random uncorrelated structure is described by an autocorrelation function (or radial distribution function), allowing calculation of the scattered X-ray intensity as a function of wave vector $\vec{q}$ or equivalently the scattering angle 2θ. In typical SAXS devices, the detector is far away from the sample (around a metre) and angles up to 10° are looked at [422]. One can perform SAXS measurements with lab devices or at synchrotron radiation facilities. As shown in Appendix C, the specific surface area is determined using the following equation:

$$S_V = \pi\phi_s(1-\phi_s)\frac{K}{Q}, \tag{8.91}$$

in which K is the limit of $I_s(q)q^4$ as $q \to \infty$ and Q is defined as $\int q^2 I_s(q)dq$ and $I_s(q)$ denotes the scattered intensity (for more details, see Section C.3). Examples of such results can be looked at in [94–97, 99, 101, 102, 151, 153, 253, 423–426].

8.7 Variables and Symbols

Table 8.2 lists some of the variable and symbols of importance in this chapter.

Table 8.2 List of variables and symbols in this chapter.

Symbol	Meaning	Definition	Units
A	Surface area of a body	–	m^2
A_{cap}	Mantle surface of a spherical cap	–	m^2
A_s	Surface area of a body	–	m^2
a_k, b_k	Constants in the BET equation	–	–

Table 8.2 Continued

Symbol	Meaning	Definition	Units
$\bar{a}_{con}$	Average area of one contact	–	m^2
$\bar{D}_{sphere}$	Diameter of a sphere	–	m
$\bar{D}_{fibril}$	Diameter of a fibril	–	m
E_{ad}	Adsoprtion energy	–	J
E_L	Liquefaction energy	–	J
$\overline{L}$	Mean intercept or chord length	–	m
$\bar{\ell}_F$	Average fibril length	–	m
k_{ad}	Adsorption rate	–	–
k_{des}	Desorption rate	–	–
K_{eq}	Ratio of reaction rates	–	1/Pa
n_{ad}	Number of adsorption events	–	–
n_F	Number of fibrils per volume	–	–
n_F^{con}	Number of contacts per fibril	–	–
n_{NN}	Number of next nearest neighbours	–	–
N_A	Avogadro's number	$6.022 \cdot 10^{23}$	1/mol
p	Pressure	–	Pa
p_0	Saturation pressure	–	Pa
$\bar{R}_F$	Average radius of a fibril	–	m
S_0	Area of empty sites on a surface	–	m^2
S_1	Area of occupied sites on a surface	–	m^2
S_k	Area of sites on a surface occupied with k layers	–	m^2
S_m	Specific surface area per sample mass	–	m^2/kg
S_V	Specific surface area per body volume	$S_V = A/V$	m^2/m^3
S_V^T	Specific surface area per sample volume	$S_V^T = A/V_T$	m^2/m^3
S_V^{sphere}	Specific surface area of a sphere	–	m^2/m^3
S_V^{cyl}	Specific surface area of a cylinder	–	m^2/m^3
S_V^c	Specific surface area of a composite aerogel	–	m^2/m^3
S_V^m	Specific surface area of the matrix in a composite aerogel	–	m^2/m^3
S_V^p	Specific surface area of the second phase "p" in a composite aerogel	–	m^2/m^3
$\bar{t}$	Average thickness of the adsorbed layers	–	m
T	Temperature	–	K
U_{LJ}	Lennard–Jones potential	–	–
$\bar{v}_F$	Average volume of cylindrical fibrils	–	m^3
v	Adsorbed volume	–	m^3
v_m	Specific monolayer capacity	–	–
V	Volume of a body	–	m^3

Table 8.2 Continued

Symbol	Meaning	Definition	Units
w_p	Weight percentage of second phase in a composite	–	–
$Z(T)$	Number of molecules striking a surface	–	moles/(m^2 s)
α	Proportionality constant	$\frac{\bar{a}_{con}}{\bar{R}_F^2}$	m^3
Ω_m	Atomic or molar volume	–	m^3
ϕ	Volume fraction	–	–
ϕ_F	Volume fraction of fibrils	–	–
ϕ_s	Volume fraction solid	–	–
ϕ_p	Pore volume fraction	–	–
ρ_A	Air density	1.204	kg/m^3 at 20°
ρ_e	Envelope density	$\rho_e = m/V_e$	kg/m^3
ρ_s	Skeletal density	$\rho_s = m/V_s$	kg/m^3
ρ_s^F	Skeletal density of a fibril	–	kg/m^3
θ	Relative coverage of a surface with adsorbates	$\frac{S_1}{A_s}$	–

9 Pores and Pore Sizes

Since in aerogels the pores are as essential as the solid network, one should think about ways to describe or classify them. A description of the pores is difficult and the situation is not comparable with, for instance, closed cell foams. Especially aerogels have till now not been analysed by in situ techniques in 3D to give a picture of the nanostructure. Scanning electron microscopy gives an image, but the images are typically taken at fracture surfaces and it is hard to evaluate how the fracture process changed the arrangement of the particles or fibrils and thus the pores. Most often, experimenters simply use the nitrogen adsorption measurement technique described in Chapter 8 and use the so-called BJH formalism (see Section 9.2.1) from the desorption curve to derive the pores size distribution using a simple model of pores.

Nevertheless, there are measures of pore sizes possible, which are well defined in stereology, namely the mean free distance between particles or fibrils in a network or the next nearest neighbour distance. In addition, scattering methods allow us to extract chord length in the pore and the solid phase, assuming a suitable model of the two-phase structure. Before we present both, let us think about simple geometrical pictures of aerogels.

9.1 Simple Geometrical Models

First we will look into particulate aerogels. We take for simplicity as a model of an aerogel a box whose edges are made by quadratic bars of thickness t_w. These bars represent the string of pearls of particles building an aerogel network in reality. The quadratic box has a cube edge length of t_c. Thus the pore size is given by $t_p = t_c - t_w$. This arrangement is shown in Figure 9.1. Let us first calculate the volume fraction of solid material inside the cell. The volume of all bars is calculated as

$$V_s = 4t_w^2 t_c + 8t_w^2(t_c - t_w). \tag{9.1}$$

The cell or box volume is

$$V_c = t_c^3 \tag{9.2}$$

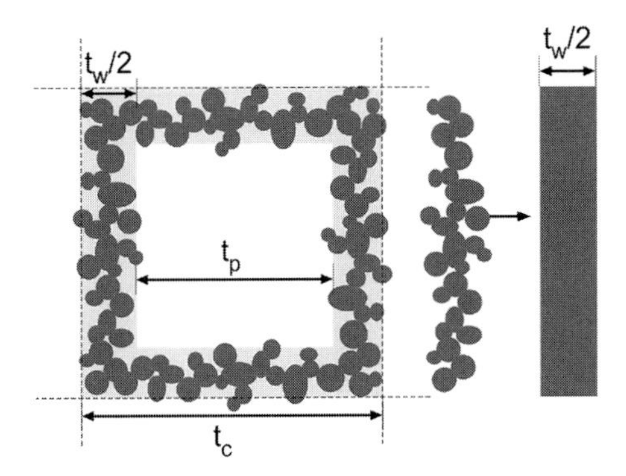

Figure 9.1 Scheme of an aerogel cell of size t_c and wall thickness t_w. The cell walls are made of aerogel particles and will be thought of as if they were bars. The bars will be quadratic prisms of thickness t_w. The rest of the box is empty: the porous space. The dashed lines illustrate that the boxes are extended space filling in 3D.

and thus the volume fraction solid material in the aerogel is

$$\phi_s = \frac{V_s}{V_c} = \frac{4t_w^2 t_c + 8t_w^2(t_c - 2t_w)}{t_c^3}. \tag{9.3}$$

Assuming $t_w/t_c \ll 1$ we can write

$$\phi_s = 12\frac{t_w^2}{t_c^2} \tag{9.4}$$

or

$$t_w = \frac{\sqrt{\phi_s}}{2\sqrt{3}} t_c. \tag{9.5}$$

Since we are here interested in pore measures, we rearrange the equation

$$t_p = t_c - t_w \tag{9.6}$$

and arrive at

$$t_p = t_w \left(1 - \frac{\sqrt{\phi_s}}{2\sqrt{3}}\right). \tag{9.7}$$

In this simple picture the pore size is almost equal to the cell size and decreases only a few percentage point with increasing volume fraction solid. The wall thickness increases also only slightly as the fraction solid increases. This is shown in Figure 9.2. Fibrillar networks can also be treated by this simple cube model with the only difference being that then the wall thickness is not an average of the particles building them as for particulate aerogels, but is identical to the fibril thickness.

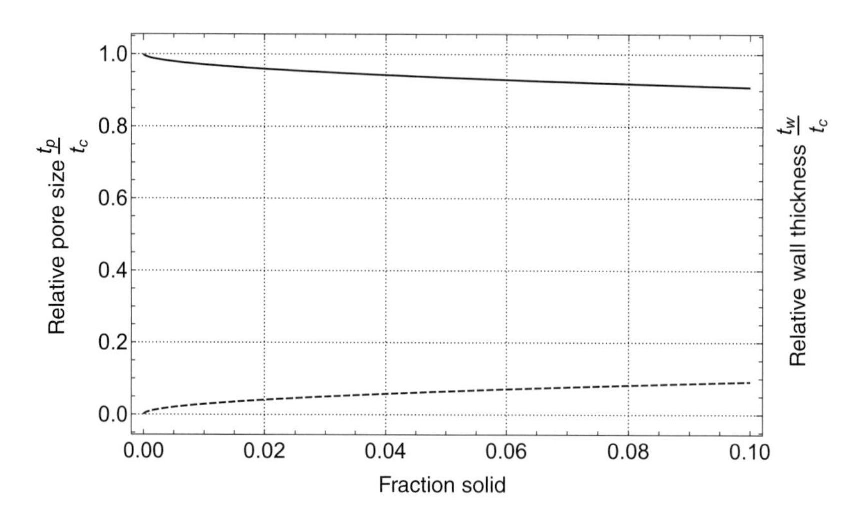

Figure 9.2 Reduction of average pore size according to Eq. (9.7) with volume fraction solid (solid line) and increase of wall thickness with fraction solid (dashed line).

9.2 Stereological Pore Size Description

Since pores in aerogels usually do not have such a regular appearance, meaning neither their size nor their thickness are uniform, they are irregular, and we need another description of pores and porosity. As done before, stereology provides methods to describe such structures. One of these descriptors is the *mean free path*. The mean free path or distance between particles is the a mean edge-to-edge distance. It represents the uninterrupted interparticle distance through the pore space averaged between all possible pairs of chains of particles or fibrils. This mean free distance is defined as

$$\lambda = 4\frac{1-\phi_s}{S_V^T} = \overline{L}\frac{1-\phi_s}{\phi_s}. \tag{9.8}$$

This expression shows as the volume fraction solid phase goes to zero, the mean free distance goes to infinity and at volume fraction solid phase equal to 1 the mean free distance is zero. Since the mean intercept length (see Figure 8.6 in Section 8.3) enters proportionally, bigger particles reduce at the same volume fraction the mean free distance. Using Eq. (7.40), one can also write

$$\lambda \cong \overline{L}\frac{\rho_s}{\rho_e}, \tag{9.9}$$

meaning inasmuch as the envelope density of an aerogel decreases, the mean free distance between particles or string of particles or fibrils becomes larger and the mean free path is inversely proportional to the envelope density. If the particles constituting an aerogel structure could be modelled as a system of impenetrable hard spheres and thrown randomly into space, one can calculate the mean free path in terms of the particle radius. Let the particle size distribution $n_V(R)dR$ denote the number of particles per unit volume in the size range $R, R+dR$. Then all moments are defined as

$$\overline{R^m} = \int_0^{\infty} R^m n_V(R) dR. \tag{9.10}$$

Binglin Lu and Torquato have shown [427] that the mean free path in such a hard sphere system λ_{hs} is then

$$\lambda_{hs} = \frac{1-\phi_s}{\phi_s} \frac{\overline{R^3}}{\pi \overline{R^2}}. \tag{9.11}$$

If one would write $\overline{R^3} \approx \overline{R}^3$ and the same for the second moment, we would arrive at

$$\lambda_{hs} = \overline{R} \frac{1-\phi_s}{\pi \phi_s}, \tag{9.12}$$

which is the same as Eq. (9.8), besides the factor π, which would allow us to say $\overline{L} = \overline{R}/\pi$. One can make the expression a bit more explicit using a special size distribution. We take here for the sake of generality a so-called Schulz distribution defined as

$$n_v(R) = \frac{1}{\Gamma(p+1)} \left(\frac{p+1}{\overline{R}} \right)^{p+1} R^p \exp\left(-\frac{(p+1)R}{\overline{R}} \right) \tag{9.13}$$

for the particle radii with $\Gamma(x)$ the gamma function; see [428]. The Schulz distribution has only one control parameter p. By increasing p, the variance decreases, and the distribution becomes sharper. At $p \to \infty$, a monodisperse size distribution is obtained. Evaluating the moments leads to

$$\overline{R^m} = \overline{R}^m \frac{(p+1)^{-m}}{p} \prod_{i=0}^{m} (p+i). \tag{9.14}$$

Insertion into Eq. (9.11) gives

$$\lambda_{hs} = \overline{R} \frac{1-\phi_s}{\phi_s} \frac{(3+p)}{\pi(1+p)}. \tag{9.15}$$

In the limit $p \to \infty$, Eq. (9.12) is recovered. The ratio $(3+p)/(1+p)$ falls from $p = 1$ rapidly from 2 rapidly to values below 1.5, or in other words, the sharper the size distribution, the more the approximate expression of Eq. (9.12) is correct. We can use the result to relate the mean free chord or intercept length to the average particle size. Comparing Eq. (9.8) with Eq. (9.15) yields

$$\overline{L} = \overline{R} \frac{(3+p)}{\pi(1+p)}. \tag{9.16}$$

This stereological analysis can be compared with computer simulations of an aerogel. Halperin and co-workers [236] simulated the microstructure of an aerogel using a so-called diffusion-limited cluster aggregation (DLCA) model of particles. Details were already discussed in Section 4.2.2. In their three-dimensional DLCA simulation, they measured the mean free path. Their results suggest that the inverse of the mean free path is directly proportional to the simulated aerogels density as shown in Figure 4.16 in full agreement with Eq. (9.9).

Another expression useful for aerogels might be the centre-to-centre distance Δ_3 of particles in the network or the surface-to-surface distance δ_3. The centre-to-centre distance was already calculated by Chandrasekhar [429] in the context of spatial distribution of stars. Taking particles as points (their centres) and assuming that these particles are distributed randomly (or more precisely by a Poisson process) in space, he derived that the probability to find a particle in a shell between radii $R, R + dR$ is given by

$$P(R) = 4\pi R^2 n_V \exp(-4\pi/3R^3 n_V), \tag{9.17}$$

with n_V the density of particles per unit volume. The average centre-to-centre distance is then

$$\Delta_3 = 2\int_0^\infty RP(R)dR = 2\frac{\Gamma\left(\frac{1}{3}\right)}{6^{2/3}\pi^{1/3}n_V^{1/3}} \cong 1.108 n_V^{-1/3} \tag{9.18}$$

Here $\Gamma(z)$ is the Γ-function. Thus the more particles exist per unit volume, the shorter the average centre-to-centre distance. The particle density is related to the volume fraction and the average particle size as

$$\phi_s = n_V \overline{v} \cong n_V \overline{L}^3 \tag{9.19}$$

and thus we arrive at

$$\Delta_3 = 2.078 \frac{\overline{L}}{\phi_s^{1/3}}. \tag{9.20}$$

The distance is directly proportional to the particle size (measured by the mean linear intercept). The relation derived by Chandrasekhar is, however, only valid for points randomly distributed in space and sufficiently far away. This definitively is not a good description of aerogels. A bit better description was given by Steele [430] for particles with a particle size distribution $f(R)dR$. He calculated the distance between faces of a Voronoi polyhedron constructed around each particle and arrived at

$$\Delta_{14} = \overline{R}\Delta_{1-14}, \tag{9.21}$$

with $\overline{R}$ the average particle size and

$$\Delta_{1-14} = \frac{1}{14}\sum_{k=1}^{14}\left(\frac{m_3}{m_1\phi_s}\right)^{1/3}\frac{\Gamma(k+1/3,x)}{\Gamma(k,x)} \quad \text{with} \quad x = \frac{8m_1^3}{m_3} \tag{9.22}$$

the distance being an average of the first nearest neighbour up to the 14th next neighbour. m_1, m_3 are the two moments of the size distribution

$$m_1 = \int_0^\infty Rf(R)dR \qquad m_3 = \int_0^\infty R^3 f(R)dR, \tag{9.23}$$

and here $\Gamma(x, z)$ is the incomplete Γ-function. Assuming $m_3^{1/3} \cong m_1$ one can calculate Δ_{14} as

$$\Delta_{14} = \beta \frac{\overline{R}}{\phi_s^{1/3}}, \tag{9.24}$$

with $\beta = 2.2$. Since this equation was derived for spherical particles, we can use the results in Table 8.1 to obtain

$$\Delta_{14} \cong 1.65 \frac{\overline{L}}{\phi_s^{1/3}}. \tag{9.25}$$

This distance is shorter than that of Chandrasekhar's estimate. According to Steele, this relation is valid for volume fractions solid below 30% and thus would be applicable to particulate aerogels. Let us check Steele's for the nearest neighbour and compare it to Chandrasekhar's result. This can only be done if $m_3 = m_1^3$, for particles with an equal size. Performing the calculations yields

$$\Delta_3 = 2.078 \frac{\overline{L}}{\phi_s^{1/3}} \text{Chandrasekhar – points} \tag{9.26}$$

$$\Delta_3 = 1.558 \frac{\overline{L}}{\phi_s^{1/3}} \text{Steele–Voronoi} \tag{9.27}$$

The estimate of Steele is smaller by around 25%. This is due to the method used: spherical shells around points in space and asking for the average nearest neighbour distance to the next neighbour is different from constructing a Voronoi polyhedron around a point and asking for the size of this polyhedron. In any case, these estimates of the centre-to-centre distance is the maximum pore size possible. As in the simple geometrical pictures described earlier, one has to reduce this spacing by the average particle diameter. We thus arrive at the following expression:

$$\delta_3 = \Delta_3 - 2\overline{R} = \overline{L} \left(\frac{2\alpha - 3\phi_s^{1/3}}{2\phi_s^{1/3}} \right) \tag{9.28}$$

with $\alpha = 1.558$ or 2.078 according to Eqs. (9.26) and (9.27). A fit using the numerical constant value of Steele leads to Figure 9.3. Note that expression of the centre-to-centre distance in Eq. (9.28) is similar to the mean free path expression of Eq. (9.8) with the essential difference that the volume fraction enters only as its cube root and thus the effect of increasing the volume fraction on the centre-to-centre distance is smaller compared to its effect on the mean free path between particles or fibrils randomly distributed in space.

9.2.1 Pore Size Distribution: The BJH Method

The pore size distribution of mesoporous materials can be determined using the BJH method. BJH stands for Barrett, Joyner and Halenda, who developed this evaluation method [431]. Their approach is based on the Kelvin equation, which relates relative

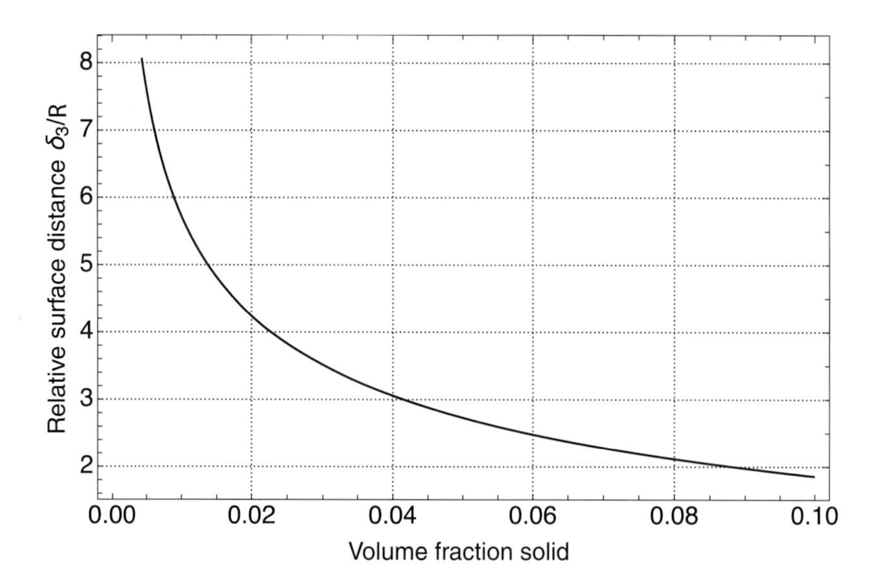

Figure 9.3 Surface-to-surface distance in a random arrangement of particles taken as a measure for the pore size according to Eq. (9.28).

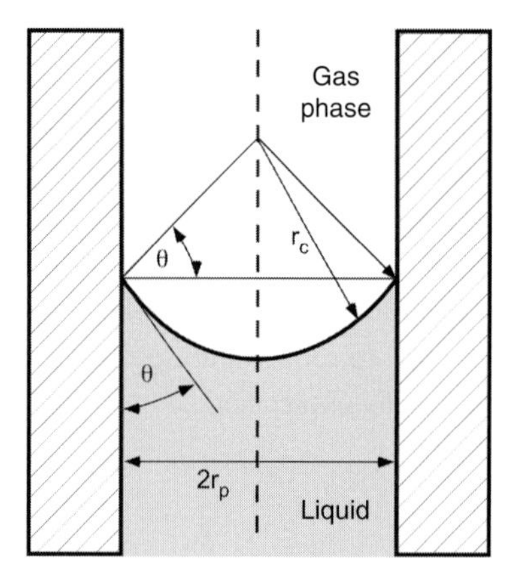

Figure 9.4 Schematic representation of a meniscus in a partially filled pore with a radius of curvature r_c and a pore diameter $2\ r_p$.

pressure to the curvature of a liquid droplet. Inside a pore, the relative pressure over a liquid meniscus with a radius of curvature r_c is related to the pore radius r_p by the wetting angle θ the liquid has with the pore wall. Assuming that the liquid–gas surface can be described by a sphere, the relation reads $r_c = r_p / \cos\theta$. This is illustrated in Figure 9.4. The essential equation used in the BJH analysis is the so-called Kelvin equation. Here we give a short derivation.

Imagine the liquid surface shown in Figure 9.4 to have a surface area A exposed to its vapour above. Then the Gibbs free energy is changed from a reference state G_0 by the surface energy σ_{LV} to a value

$$G = G_0 + \sigma_{LV} A. \tag{9.29}$$

As usual, equilibrium between the liquid and its vapour is defined as the equality of the chemical potential of both phases, liquid and vapour. The chemical potential μ of a pure component is defined as the partial derivative of the Gibbs free energy with respect to the number of molecules n:

$$\mu = \frac{\partial G}{\partial n}. \tag{9.30}$$

Rewriting this in terms of the particle volume, we obtain the following:

$$\mu = \frac{\partial G}{\partial n} = \frac{\partial G}{\partial V}\frac{\partial V}{\partial n} = \frac{\partial G}{\partial V}\Omega_m, \tag{9.31}$$

with Ω_m as the molar volume. The derivative of the Gibbs free energy with respect to volume is, however, the following:

$$\frac{\partial G}{\partial V} = \sigma_{LV}\frac{\partial A}{\partial V} = \sigma_{LV}\left(\frac{1}{r_1} + \frac{1}{r_2}\right), \tag{9.32}$$

where we have made use of the general relationship between the principal radii of curvature r_1, r_2 of an arbitrary shaped surface and its curvature

$$\mathcal{H} = \frac{\partial A}{\partial V} = \left(\frac{1}{r_1} + \frac{1}{r_2}\right). \tag{9.33}$$

Therefore, the chemical potential of the liquid μ_l can be written as

$$\mu_l = \mu_l^0 + \sigma_{LV}\Omega_m\mathcal{H}, \tag{9.34}$$

with μ_l^0 the chemical potential of the liquid with no curvature effects, meaning $\mathcal{H} = 0$. If the vapour can be described as an ideal gas, its chemical potential can be written as

$$\mu_g = \mu_g^0 + k_B T \ln p \tag{9.35}$$

and μ_g^0 a reference state of the gas phase. In equilibrium, the chemical potential of the liquid and the vapour phase are equal:

$$\mu_g = \mu_l. \tag{9.36}$$

Insertion leads to

$$\mu_g^0 + k_B T \cdot \ln p = \mu_l^0 + \sigma_{LV}\Omega_m\mathcal{H}. \tag{9.37}$$

If the curvature is zero, i.e., we have a flat surface, then a reference pressure is defined by $\mu_g^0 + k_B T \cdot \ln p_0 = \mu_l^0$. With this relation, we finally arrive at the so-called Kelvin equation:

$$p(K,T) = p_0 \exp\left(\frac{\sigma_{LV}\Omega_m\mathcal{H}}{k_B T}\right), \tag{9.38}$$

which describes the equilibrium vapour pressure above a curved surface compared to a flat surface (curvature $\mathcal{H} = 0$). The physical meaning of the equation can be visualised by imagining that a convex surface has a positive curvature and thus the vapour pressure is higher compared to a flat surface and a concave surface has a lower one. It is also worthwhile to notice that the derivative of G with respect to volume represents a pressure ([G]=[J]=[Nm]; [dG/dV]=[N/m^2]), i.e.:

$$p = \frac{\partial G}{\partial V} = \sigma_{LV} \cdot K = \sigma_{LV} \cdot \left(\frac{1}{r_1} + \frac{1}{r_2} \right). \tag{9.39}$$

For spherical surfaces, this expression reduces to the well-known Laplace pressure, since then the two radii of curvature are identical, $r_1 = r_2 = r_c$:

$$p_{Laplace} = \frac{2\sigma_{LV}}{r_c}. \tag{9.40}$$

If the curvature is negative, the pressure $\partial G/\partial V$ is called a tension; otherwise, it is called a pressure. One can visualise this definition of tension and pressure by considering water rising up in a capillary. Then its surface is concave with a negative curvature. The liquid is drawn into the capillary against gravity. Thus the surface tension with the wetting of the capillary walls and the curvature act like a tension.

Writing the Kelvin equation a bit different is useful to analyse the pore size distribution:

$$\ln \frac{p}{p_0} = \frac{2\Omega_m \sigma_{LV} \cos\theta}{r_p RT}. \tag{9.41}$$

The idea behind the BJH analysis is that during desorption, liquid in the pores continuously evaporates with a pressure given by Eq. (9.41). Evaluating the desorption branch allows us to convert the pressures to pore radii, if one assumes a pore geometry. For a description of adsorption and desorption isotherms, see Chapter 8. Solving Eq. (9.41) for the radius then yields the following relationship between the pore radius and the relative pressure, assuming a contact angle of zero degrees (perfect wetting):

$$r_p = \frac{2\sigma_{LV}\Omega_m}{\ln\left(p/p_0\right) RT}. \tag{9.42}$$

Since a film of adsorbed nitrogen is always present at the pore walls when capillary condensation starts, the thickness of this film has to be added to the radius of the pore. The film thickness $\bar{t}$ can be calculated by the empirical relationship derived by Halsey in Eq. (8.88) or Harkins and Jura in Eq. (8.89). With the film thickness, the overall pore size $2r_{total}$ is given by the following:

$$2r_{total} = 2\left(r_p + \bar{t}\right). \tag{9.43}$$

It has to be noted that the BJH evaluation should be applied for mesopore sizes <100 nm only. The basic assumption in the BJH formalism is that the pores are cylindrical and that the Kelvin equation for the capillary pressure as it depends on pore size and wetting angle can be applied. The applicability of the Kelvin equation describing the desorption branch of an adsorption curve is questioned in the literature since decades. In macroscopic capillaries, it is accepted that the Kelvin equation is valid; in nano- to micrometre-sized capillaries it was questioned if the meniscus can

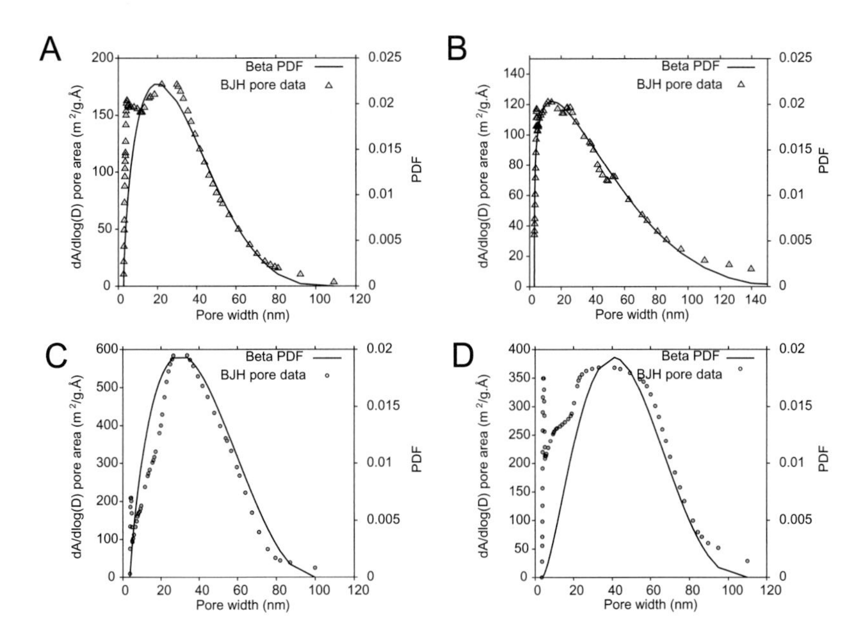

Figure 9.5 Pore size distribution (obtained through BJH data) of cellulose aerogels prepared from $Ca(SCN)_2 4H_2O$ (A: 3 wt% and B: 5 wt%) and $ZnCl_2 4H_2O$ (C: 3 wt% and D: 5 wt%) vs. the best fit of the beta distribution function. Reprinted from [433] with permission of the Royal Society of Chemistry

be described as a spherical cap as drawn in Figure 9.4. The effect of a thin, almost liquid, film on the surface was investigated [432]. In almost all aerogel literature, the BJH analysis is applied although it is known to have its shortcomings. The essential drawback is that independent of the nature of aerogels, the pores are never cylindrical, not even as a coarse approximation. In biopolymeric aerogels, the network consists of fibrils connected in 3D, in silica or RF or carbon aerogels particles, to build branched chains, and these are interconnected to form a spanning cluster.

For cellulose aerogels prepared from microcrystalline cellulose with a degree of polymerisation of DP = 211 using either rhodanide or $ZnCl_2$ hydrate as dissolving agents, ethanol as regeneration liquid and supercritical drying with carbon dioxide, Rege et al. performed a BJH analysis [433]. They have fit the experimental pore size distribution with a generalised beta distribution function

$$p(l) = \frac{(l - l_{\min})^{\alpha-1}(l_{\max} - l)^{\beta-1}}{B(\alpha,\beta)(l_{\max} - l_{\min})^{\alpha+\beta-1}}, \tag{9.44}$$

where α and β denote the shape parameters of the distribution function. They are estimated from the shape of the BJH pore data. The fit and the experimental results shown in Figure 9.5 indicate a reasonable agreement.

9.2.2 Pore Size Distribution: Thermoporosimetry

The BJH analysis relies on the validity of the Kelvin equation and uses the fact that the equilibrium gas pressure above a curved liquid–gas interface depends on its curvature. The melting or solidification point of a single component or pure material depends on curvature, but in this case also that of the solid–liquid interface. A standard procedure to measure the melting or solidification temperature is to heat or cool a sample in a differential scanning calorimeter (DSC) [434]. During cooling, the onset of solidification (crystallisation) is detected, since the latent heat of solidification evolves at the solid–liquid interface. The process of solidification is exothermic. During heating, the situation is just the opposite. Typical experiments, however, use the cooling process to detect the liquid-to-solid phase transition temperature. A way to measure the pore size distribution is therefore to infiltrate a liquid, being inert and non-reactive with the porous material backbone, into the pore space and performing, for instance, a DSC run with it. Instead of one exothermic event at one temperature, a broad curve of solidification temperatures will be observed. Using a model for the depression or undercooling below the equilibrium melting temperature as it depends on the pore size allows us then to extract the pore size distribution. The model used is based on the so-called Gibbs–Thomson effect.

Gibbs–Thomson Effect

We start with a simple arrangement shown already in Figure 9.4 in which, however, the gas phase is replaced by a liquid and the liquid by a solid phase. We thus have a solid–liquid two-phase equilibrium in a cylindrical pore of radius r_p, and the curved interface between solid and liquid will be again approximated as a spherical cap with radius r_c. We first derive generally the equilibrium melting temperature of an interface and then that of a sphere.

Let G denote the free energy per mole as composed of a enthalpic term H and an entropic one, $T \cdot S$:

$$G = H - T \cdot S. \tag{9.45}$$

We can write for the liquid and the solid phase

$$G_s = H_s - T \cdot S_s \quad \text{and} \quad G_L = H_L - T \cdot S_L. \tag{9.46}$$

Thermodynamics shows that at a given temperature the phase with the lower free energy is stable: the free enthalpy is minimised (whereas the entropy is maximised). The free energy of the liquid and the solid phase differ as a function of temperature, since their specific heats are different. But there is a temperature at which both free energies are equal, the equilibrium melting temperature T_m. This means

$$H_s - T \cdot S_s = H_L - T \cdot S_L \quad \text{or} \quad T_m = \frac{\Delta H_m}{\Delta S_m} \tag{9.47}$$

with the obvious abbreviations $\Delta H_m = H_L - H_s$ and $\Delta S_m = S_L - S_s$, which have the names enthalpy of melting or latent heat of melting ΔH_m and entropy of melting ΔS_m.

The situation changes if the solid phase has a curved interface to the surrounding liquid. Then its free enthalpy is increased due to the solid–liquid surface tension σ_{LS} spent in building up the interface. Let A_m be the interface area per mole. We then can write for the solid and the liquid phases:

$$G_{solid} = H_{solid} - TS_{solid} + \sigma_{LS} A_m = H_{solid} - TS_{solid} + \sigma_{LS} \frac{\partial A_m}{\partial n} \tag{9.48}$$

$$G_{liquid} = H_{liquid} - TS_{liquid}. \tag{9.49}$$

The temperature at which both free energies are equal, T^*, defines the new equilibrium melting temperature of a sphere with radius r_c. Let us first look what we can do with the derivative of the surface area with respect to the number of moles, n. Here we can use

$$\frac{\partial A}{\partial n} = \frac{\partial A}{\partial V}\frac{\partial V}{\partial n} = \Omega_m \frac{\partial A}{\partial V} \tag{9.50}$$

with Ω_m the molar volume.

$$\begin{aligned} G_{solid}(T = T^*) &= G_{liquid}(T = T^*) \\ H_{liquid} - H_{solid} - T^* S_{liquid} + T^* S_{solid} &= \sigma_{LS}\Omega_m \frac{\partial A}{\partial V} \\ \Delta H_m - T^* \Delta S_m &= \sigma_{LS}\Omega_m \frac{2}{R}. \end{aligned} \tag{9.51}$$

Using the relation for a planar interface in Eq. (9.47), we can rewrite Eq. (9.51) as

$$\Delta T_m(r_c) = T^* - T_m = T_m \frac{\sigma_{LS}}{\Delta H_m}\frac{2}{r_c}. \tag{9.52}$$

Using the relation $r_c = r_p/\cos\theta$, we arrive at

$$\Delta T_m(r_p) = T_m \frac{\sigma_{LS}\cos\theta}{\Delta H_m}\frac{2}{r_p}. \tag{9.53}$$

This equation is the standard form of the Gibbs–Thomson equation for the melting point depression of a single component liquid as it solidifies in a cylindrical pore. It shows that the smaller the pore size, the larger the deviation of the measured solidification temperature from its value obtained on massive bulky materials. Since the pores in aerogels are in the mesoporous range, one can additionally take into account that the solid–liquid interface tension depends on curvature itself. This was done by Grosse et al. [309] and leads to a correction of the Gibbs–Thomson relation:

$$\Delta T = \frac{2\sigma\cos\Theta}{\Delta S_m[r_p - 2\Omega_m \Gamma_a \cos\Theta]}, \tag{9.54}$$

with $\Gamma_a = n/A$ being the number of moles at the interface per unit area. According to Grosse, the correction is of the order of a nanometer.

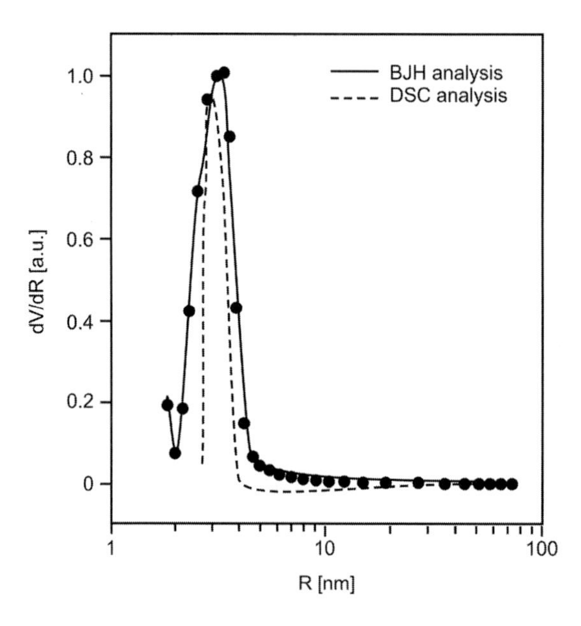

Figure 9.6 Pores size distribution derived from the DSC measurement (dashed line) compared with one derived from nitrogen adsorption and BJH (solid line with full circles) analysis. Reprinted with permission of Springer-Nature from [435]

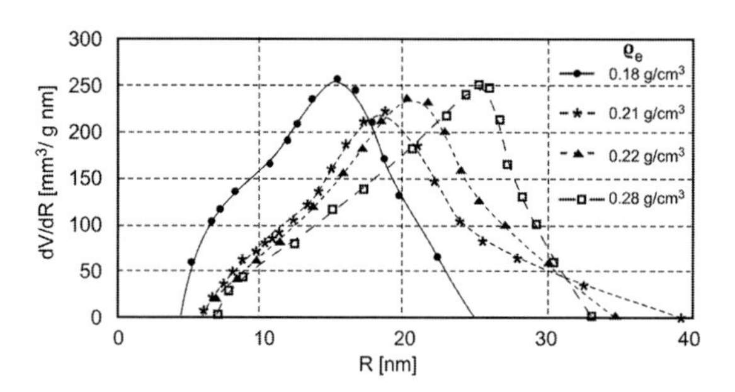

Figure 9.7 Pore size distribution of silica aerogels with different envelope densities as determined from thermoporosimetry after [436].

Pore Size Distribution from Thermoporosimetry

Using measured DSC curves, Nedelec and co-workers [435] extracted from them with help of the Gibbs–Thomson equation a pore size distribution and compared this with a pore size distribution obtained from nitrogen adsorption and a BJH analysis. A result is shown in Figure 9.6.

The technique of thermoporosimetry was used already in the early 1990s by Woignier and co-workers [436]. They prepared a silica aerogel by a standard two-step routine including supercritical drying. The aerogel had initially a density of 0.18 g/cm^3 and was sintered at 950°C to increase the density up to 0.84 g/cm^3. The meso- and macropore size distribution determined is shown in Figure 9.7. The curves always show a broad size distribution, which shifts its maximum to larger sizes with

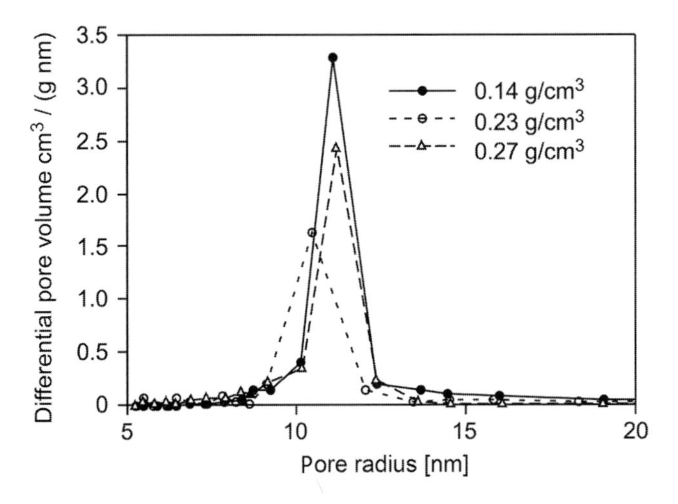

Figure 9.8 Pore size distribution as determined from thermoporosimetry using light scattering after [309] for supercritically dried silica aerogels made from TMOS.

increasing envelope density. Sintering of aerogels leads to a reduction of small pores and a redistribution of the nanoparticles to a coarser network. In the same decade, Grosse and co-workers determined also with a special variant of thermoporosimetry the pore size in silica aerogels made by Airglass^{TM} in Sweden [309]. They did, however, not use a DSC facility to extract the pore size distribution but used light scattering and analysed the scattering curves. For a liquid, they used high-purity succinonitrile, which has an equilibrium freezing point of 58°C. They recorded the scattering intensity as a function of undercooling below the equilibrium melting temperature, and from a comparison of calculated scattering intensity curves they determined the pore size distribution shown in Figure 9.8. Pircher and co-workers made a careful study of the pores' size distribution in cellulose aerogels prepared using different solvents [151]. For a cellulose base material, they used cotton linters and dissolved them in NMMO hydrate, $Ca(SCN)_2$ hydrate with an excess of water and LiCl as an addition. A third solvent was tetrabutylammonium fluoride trihydrate (TBAF) and finally 1-Ethyl-3-methyl-1H-imidazolium acetate (EMIm). The hot solutions were always after gelation washed with pure ethanol and supercritically dried with carbon dioxide. The thermoporosimetry was performed in samples infiltrated with o-xylane and the onset of crystallisation as it depends on pore size analysed. They derived for the pore size distribution an equation based on the previous derived Gibbs–Thomson relation

$$\frac{dV_p}{dr_p} = \frac{Y(T)\Delta T^2 c}{\Delta H_m(T) r_p} \tag{9.55}$$

with c an experimentally determined constant for o-xylene and $\Delta H_m(T)$ the temperature-dependent crystallisation enthalpy of o-xylene. $Y(T)$ is the heat flow of the sample as preset in the DSC facility. The pore size distributions obtained for cellulose aerogels prepared with different solvents are shown in Figure 9.9.

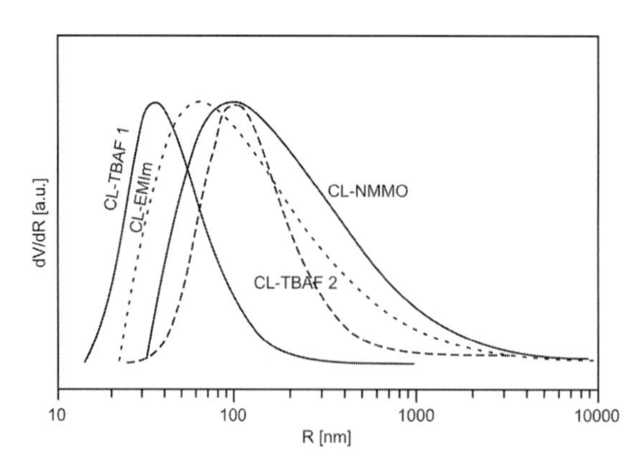

Figure 9.9 Pore size distributions of cellulose aerogels obtained from thermoporosimetry. The pore size spectrum differs considerably for different cellulose solvents. After [151] reprinted under Creative Commons license.

All distributions are typically skew symmetric with a shift to smaller sizes. Interestingly, in contrast to a BJH analysis, thermoporosimetry allows us to determine pore size up to 1 μm, and it is interesting that cellulose aerogels show a broad size spectrum.

As mentioned in Chapter 8 on surface area, aerogels usually are characterised by a type IV isotherm with a pronounced hysteresis. However, for aerogels with densities well below 200 kg m^{-3}, the type IV isotherms can be largely deformed, since capillary forces are compressing the sample. In case of significant deformation on desorption, an evaluation via BJH yields erroneous pore size distributions. The effect of solid network deformation of aerogels during desorption was intensively studied by Reichenauer and Scherer [437–439].

A final remark is necessary: all methods we have described for pore sizes and their distribution make the assumption that the pores are cylindrical. Looking at the morphology described in Chapter 6, this looks like an oversimplification. Currently, there are, however, no other methods available.

9.3 Variables and Symbols

Table 9.1 lists some of the variables and symbols used in this chapter.

Table 9.1 Variables and symbols.

Symbol	Meaning	Definition	Units
G	Gibbs free enthalpy	–	J
G_S	Gibbs free enthalpy of a solid	–	J
G_L	Gibbs free enthalpy of a liquid	–	J
H	Enthalpy	–	J

Table 9.1 Continued

Symbol	Meaning	Definition	Units
H_s	Enthalpy of a solid	–	J
H_L	Enthalpy of a liquid	–	J
$\mathcal{H}$	Mean curvature	–	1/m
$n_v(R)dR$	Particle size distribution, radius R	–	–
n_v	Particle density	–	$1/m^3$
N_A	Avogadro's number	$6.022 \cdot 10^{23}$	1/mol
p	Pressure	–	Pa
$p_{Laplace}$	Laplace pressure	–	Pa
$\overline{R}$	Average particle size	–	m
$\overline{R^m}$	mth moment of the particle size distribution	–	–
S_m	Specific surface area per sample mass	–	m^2/kg
S_V	Specific surface area per body volume	$S_V = A/V$	m^2/m^3
S_V^T	Specific surface area per sample volume	$S_V^T = A/V_T$	m^2/m^3
r_1, r_2	Principal radii of curvature	–	1/m
r_c	Capillary size	–	m
r_p	Pore size	–	m
r_{total}	Total pore size	$r_{total} = 2r_p + \bar{t}$	m
S	Entropy	–	J/K
S_s	Entropy of a solid	–	J
S_L	Entropy of a liquid	–	J
t_p	Pore size in a simple cubic lattice	–	m
t_w	Wall size in a simple cubic lattice	–	m
t_c	Size of a simple cubic cell	–	m
$\bar{t}$	Thickness of adsorbed layer	–	m
T	Temperature	–	K
$\Delta T_m(r_p)$	Melting point supression	–	K
$\bar{v}_F$	Average particle volume	–	m^3
v	Adsorbed volume	–	m^3
v_m	Specific monolayer capacity	–	–
V	Volume of a body	–	m^3
w_p	Weight percentage of second phase in a composite	–	–
δ_3	Surface-to-surface distance in a dispersion	–	m
Δ_3	Centre-to-centre distance in a dispersion	–	m
ΔH_m	Enthalpy of melting	–	J
ΔS_m	Entropy of melting	–	J/K
λ	Mean free path	–	m
λ_{hs}	Mean free path in a dispersion of hard spheres	–	m
μ	Chemical potential	–	–
Ω_m	Atomic or molar volume	–	m^3/mol

Table 9.1 Continued

Symbol	Meaning	Definition	Units
ϕ	Volume fraction	–	–
ϕ_s	Volume fraction solid	–	–
ϕ_p	Pore volume fraction	–	–
ρ_e	Envelope density	$\rho_e = m/V_e$	kg/m^3
ρ_s	Skeletal density	$\rho_s = m/V_s$	kg/m^3
σ_{LV}	Interfacial energy between liquid and vapour	–	J/m^2
σ_{LS}	Interfacial energy between solid and liquid	–	J/m^2
θ	Wetting angle liquid to pore wall	–	–

10 Diffusion in Aerogels

In synthesis, processing and applications of aerogels, transport of liquids and gases in and through a gel or aerogel quite often determines the time scale and the properties. Having a liquid solution of precursors and solvent, the molecular precursors move in the solvent, form dimers, trimers, tetramers and so on and eventually build particles, which can further grow. The particles themselves will move in the solvent and can aggregate and form a network. Once the network is formed, monomers and oligomers are still in the solvent and move until they are captured by the network and stiffen it. This process has been called ageing. The transport of molecules is a so-called diffusive process, meaning that the molecules move randomly in the solvent biased by concentration or chemical potential gradients. They typically diffuse from point of higher potential (mostly also higher concentration) to regions of lower potential or concentration. Similarly, a diffusion process occurs, when a wet gel is washed, for instance, in an ethanol bath to exchange the pore fluid after gelling and ageing to one which is miscible with, for instance, carbon dioxide. In such a situation, the wet gel is overlaid by ethanol, which then diffuses into the pore space. Inasmuch as it diffuses inwards, the gel fluid moves outwards into the ethanol layer. During adsorption and desorption studies on aerogels, treated in Chapter 8, nitrogen diffuses into the pore system and will then be adsorbed or desorbed there. This transport takes time, and we will discuss in this chapter characteristic times for such a process. Gas transport through an aerogel not only occurs due to concentration differences, but typically due to a pressure difference. Such a pressure difference applied over a piece of aerogel will drive the gas through the pore system. The speed of transport or the amount of transported gas depends on a property called permeability, which essentially reflects the pore size and the volume fraction of pores. For any type of filter application, this property is decisive. We discuss this type of gas transport in Chapter 11. The transport of heat through an aerogel has a part determined by diffusive gas transport through the pore space. This will be treated later in Chapter 12. We therefore discuss in this chapter concepts of diffusion of species in general and in aerogels especially.

10.1 A Phenomenological Approach to Diffusion

Consider the following experimental setup, which also can be done in every kitchen. Take a glass, fill in water and add some amount of balsamic vinegar, since it has a nice dark brown colour, with a syringe and try to place the vinegar at the bottom. If you do not stir, it takes days to dissolve vinegar such that the solution has the same colour everywhere. From the bottom, the vinegar seems to move upwards, slowly, and ahead of the dark brown pool of vinegar is a diffuse interface with varying colour. This slow dissolution occurs by diffusion. Why does it occur? The answer has two parts. First, vinegar and water are miscible in each other (if one would do the same experiment with olive oil, nothing would happen, even if one would vigorously stir). Second, distributing the vinegar molecules homogeneously in water reduces the free energy of the system and increases the mixing entropy. The state of complete dissolution is energetically favourable compared to a system with two layers, water above or below vinegar (in the case of oil, this is the energetically favourable situation). How can one describe the diffusion of one fluid into another?

If we place at a distance ahead of the interface a plane and measure the number of vinegar molecules passing this plane per second, we can calculate a current density $\vec{j}$, which is measured in mass/(m^2 s). This current density is related in the concept of diffusion to the concentration difference (vinegar to water or more generally B into A) or more precisely not the difference but the local gradient, dc/dx or in 3D grad$c = \nabla c$, in a linear way. The relation

$$\vec{j} = -D\nabla c \tag{10.1}$$

is called Fick's first law.[1] It states that the current moving through a plane is directly proportional to the driving force, and the constant of proportionality is called the diffusion coefficient. In principle, this already is a simplification, since the real driving force cannot be the concentration difference. Imagine a glass filled with water and oil. They will build a two-layer system and neither water nor oil molecules will diffuse into the layer of the other. The generalisation of Fick's first law therefore states that the driving force is the chemical potential gradient and the constant of proportionality is then called the mobility M:

$$\vec{j} = -M\nabla\mu. \tag{10.2}$$

Both constants are related as already shown in Chapter 4. If $G(c, T)$ denotes the Gibbs free energy of a binary system, we have shown that M and D are related by

$$M = \frac{D}{\partial^2 G(c, T)/\partial c^2}. \tag{10.3}$$

Let us perform a calculation for an ideal solution. In that case, Gibbs free energy $G_{id}(c, T)$ is fully determined by the mixing entropy and can be written as (see also Appendix A)

[1] The concentration c can be measured in several ways: molar fractions, mass per volume etc. Although the general equation for diffusion will look similar, the definition of the concentration measure defines a coordinate system and during diffusion of species into each other the centre of mass or the centre of volume etc. can change considerably, giving rise to an additional so-called advection term in the diffusion equation. This is in detail discussed in the book of Rosenberger on crystal growth and is especially important in gas diffusion [302].

$$G_{id} = R_g T[c \ln c + (1-c)\ln(1-c)]. \tag{10.4}$$

Insertion into Eq. (10.3) leads to

$$M = \frac{Dc(1-c)}{R_g T}. \tag{10.5}$$

If the solution is diluted $c \ll 1$, we would have

$$M = \frac{Dc}{R_g T}. \tag{10.6}$$

Since the diffusion of species into or out of a volume element V changes there its local concentration, we have to consider that any time change in local concentration in a volume element is balanced by an appropriate flux of species through the surface of the volume element. This is mathematically written in Fick's second law:

$$\frac{\partial c}{\partial t} + \vec{j} = 0 \quad \text{or} \quad \frac{\partial c}{\partial t} = \nabla \cdot D\nabla c = D\nabla^2 c. \tag{10.7}$$

The last equality holds only if the diffusion coefficient is independent of concentration. Eq. (10.7) is a parabolic partial differential equation of second order. As such it requires additional conditions to solve it, like an initial condition, stating the concentration field being there at time zero and boundary conditions defining values of the concentration or the gradient at the boundary. Many analytical solutions can be found in the textbook of Crank [290] and, since the equation is similar to the heat conduction equation, also in the famous textbook of Carslaw and Jaeger [440] on conduction of heat in solids.

We are not going to discuss this equation in depth, but would like to extract from it the so-called characteristic time of a diffusion process. Let us first take a look at the equation in a one-dimensional form:

$$\frac{\partial c}{\partial t} = D\frac{\partial^2 c}{\partial x^2}. \tag{10.8}$$

There are many fundamental solutions possible. The first one is a scaling solution. We write as a new coordinate $\xi = x/(2\sqrt{Dt})$ and can show by insertion and performing the necessary differentiations that

$$c(x,t) \propto \frac{1}{t^{1/2}} \exp\left(-\frac{x^2}{4Dt}\right) \tag{10.9}$$

is a solution of Eq. (10.8). Another possible solution is obtained by integrating this solution, leading to

$$c(x,t) \propto \operatorname{erf}\frac{x}{2\sqrt{Dt}}, \tag{10.10}$$

with $\operatorname{erf}(x)$ the error function. The essential issue in both solutions is, that

$$\frac{x}{2\sqrt{Dt}} \tag{10.11}$$

is a single dimensionless parameter dictating all solutions at least for diffusion in an infinite or semi-infinite medium. As Crank pointed out, it follows that

- the distance of penetration of any given concentration is proportional to the square root of time;
- the time required for any point to reach a given concentration is proportional to the square of its distance from the surface and varies inversely as the diffusion coefficient;
- the amount of diffusing substance entering the medium through unit area of its surface varies as the square root of time.

Therefore one can always make a first guess of the time it takes that a certain concentration level is reached. Let us say we have a slab of thickness h and on one face a constant concentration. Then the time it requires to reach a certain concentration level at a distance h away from the source would be

$$\tau_c = \frac{h^2}{2D}. \tag{10.12}$$

Looking at real solutions in finite media, one can see that this is only a gross estimate, but the order of magnitude always is right and this might be helpful.

The preceding diffusion equations are valid for non-porous, bulky materials. What does change, if the material is porous as aerogels are? The most simple answer would be, almost nothing. The cross section through which a flow takes place is reduced by the porosity ϕ_p and thus the flux density changes to

$$\vec{j}_p = -D\phi_p \nabla c. \tag{10.13}$$

This changes Fick's second law to

$$\frac{\partial c}{\partial t} = -D\phi_p \nabla^2 c \tag{10.14}$$

as long as the diffusion coefficient does not depend on the porosity itself.

10.2 Diffusion Coefficients

The speed of diffusion differs orders of magnitude in liquids and gases. One can easily imagine that in liquids, due to the dense packing of molecules, the motion is much smaller than in gases. It also is intuitively clear that movement of molecules in a highly viscous liquid is slower than in a low viscosity liquid. If one treats a molecule as a ball of radius R_m, the motion in a liquid is balanced by the Stokes drag on such a sphere. Einstein showed that the diffusion coefficient is inversely proportional to the friction coefficient on a sphere and proportional to the thermal energy $k_B T$. Combining this result with the hydrodynamic result for friction on a sphere, which is $6\pi\mu_L R_m$, the diffusion coefficient of liquids with viscosity μ_L is

$$D = \frac{k_B T}{6\pi \mu_L R_m}. \tag{10.15}$$

This is the famous Stokes–Einstein relation. The diffusion is faster in low viscous liquids, and the smaller a molecule, the faster its random motion. One might question the applicability of continuum hydrodynamics to the motion of molecules, but this relation at least always gives the correct order of magnitude. The dynamic viscosity of a liquid μ_L is strongly temperature dependent given by an Arrhenius expression

$$\mu_L = \mu_0 e^{\frac{Q}{k_B T}}, \tag{10.16}$$

with Q an activation energy of the order of 100 kJ/mol. The temperature dependence of the diffusion coefficient in Eq. (10.15) essentially reflects the temperature dependence of viscosity. Typical values for a diffusion coefficient in liquids cluster around 10^{-9} m^2/s.

In gel forming solutions, the liquid continuously changes from a solution of monomers dissolved in a solvent to one of dimers, trimers, tetramers, pentamers etc. In many cases, polymers are formed and inasmuch as oligomers or polymers develop in a solvent, two things happen: first, the viscosity slightly increases; and secondly, oligomers have another diffusion coefficient. If one takes the Stokes–Einstein equation to be valid even in the case of polymers, one would have to calculate the friction coefficient of polymer chains. There are several models developed in the literature on polymer solutions [54]. One is the so-called Rouse model describing the centre-of-mass diffusion:

$$D \cong \frac{k_B T}{N_P \pi \mu_L a_m}, \tag{10.17}$$

with a_m as the molecule diameter and N_P the degree of polymerisation. For a rod-like molecule, there is an additional weak, logarithmic dependence on the rod length [441, 442]. In this model, the polymers move independently of all others and they are treated as stiff rods. In the Zimm model, hydrodynamic interaction between the molecules along the polymer chain are allowed. Then the diffusion coefficient is

$$D \cong \frac{k_B T}{\mu_L N_P^{1/2} a_m}. \tag{10.18}$$

The difference between both models is obvious. Diffusion coefficients in polymer solutions have values of around 10^{-12} m^2/s.

The diffusion coefficient for gases is derived in the kinetic theory of gases as a product of the mean free path ℓ and the average molecule velocity $\overline{v}$:

$$D_\eta = \frac{1}{3} \ell \overline{v}. \tag{10.19}$$

The mean molecular velocity is a function of temperature which can be calculated in the framework of kinetic theory of gases to be ([443, p. 214], [212])

$$\overline{v} = \sqrt{\frac{8k_BT}{\pi M_a}} \tag{10.20}$$

and the mean free path between two molecule collisions can be calculated as

$$\ell = \frac{k_BT}{\sqrt{2}\pi a_m^2 p}, \tag{10.21}$$

with p the pressure and a_m the molecule diameter.[2] This gives for the diffusion coefficient the expression

$$D_\mu = \frac{2}{3}\sqrt{\frac{k_B^3}{\pi^3 M_a}\frac{T^{3/2}}{pa_m^2}}. \tag{10.22}$$

Typical values for diffusion coefficients of gases are in the range of 10^{-5} m^2/s to 10^{-4} m^2/s. Thus diffusion in gases is orders of magnitude faster than in liquids or polymer solutions.

10.2.1 Knudsen Diffusion

Diffusion or flow of a gas into or out of the pores of aerogels needs special consideration. We have to take into account that the pore size spectrum of aerogels typically is in the range of a few ten of nanometers and thus it is under standard conditions in the range of the mean free path of gas molecules. For air, the mean free path is calculated to be 69 nm. This means that gas molecules do not flow through the pore space in a viscous-like manner, but they may have a lot of collisions with the pore walls. This case is denoted in the literature as Knudsen diffusion through small orifices and characterised by the so-called Knudsen number:

$$Kn = \frac{\ell}{2R}. \tag{10.23}$$

Here ℓ is the mean free path, being dependent on pressure and temperature, and $2R$ denotes the diameter of a pore through which the gas has to move. If $Kn < 1$, the gas flow through the pore is described by conventional fluid dynamics and is of viscous type. If $Kn \gg 1$, the gas transport through the pores is determined by collisions between the pore walls and follows new laws, described later in this chapter and also in Chapter 11. A thorough discussion of Knudsen effects can be found in the textbook of Landau and Lifshitz on theoretical physics [210].

For a discussion of Knudsen diffusion, let us start with a simple arrangement. There is a hole in a wall having a radius R and a length L; see Figure 10.1. Let us consider a

[2] It might not be obvious that Eqs. (10.20) and (10.21) have the correct units. But let us check this. Insert the unit for Boltzmann's constant, which is J/K. k_BT thus has a unit J or Nm. A N is just 1 kg m/ s^2. Since the molecular weight is measured in kg, we have that k_BT/M_a has the unit m^2/s^2 and thus the square root indeed gives the unit of a velocity. The mean free path is simple: since k_BT has the unit Nm, $a_m^2 p$ has the unit N, and thus the result is indeed a length.

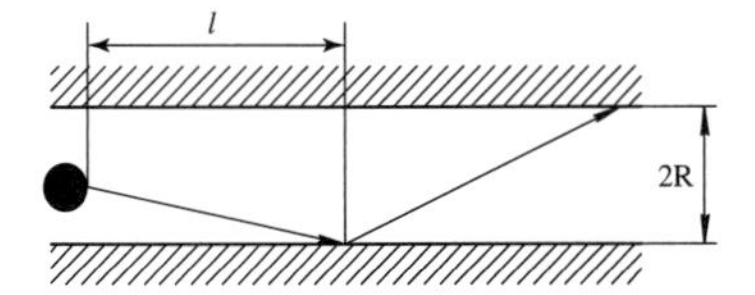

Figure 10.1 Schematic drawing of a molecule in a cylindrical pore of diameter 2R. The collision with the wall defines the mean free path (ℓ).

gas with a density of n_1 molecules per cubic metre at one side of a hole and $n_2 < n_1$ on the other side. Then the free molecular gas flow density j_K through the hole is

$$j_K = w(n_2 - n_1)\overline{v}. \tag{10.24}$$

Here $\overline{v}$ is the mean gas velocity, to be calculated with kinetic gas theory, and w characterises a probability factor taking into account that only a limited number of molecules entering the hole in the wall are able to run through it (calculate all molecules that enter the hole and are able to run through by specular or diffusive reflections inside the cylindrical or otherwise shaped hole, and compare this number with the number of molecules colliding with the cross section of the hole as if this would be an impenetrable surface). This has to be calculated using assumptions on scattering of gas molecules at the pore walls' (diffusive, specular) ratio of the diameter to the length of the pores. Exercise 8 in the textbook of Landau and Lifshitz solves this problem for a cylindrical hole (see [210], page 68)). Calculation of the probability factor is quite cumbersome, and therefore we take from literature that for a thin orifice $w = 1/4$ and for long straight tubes ($L \gg R$) $w = 2/3\, R/L$.

Inserting the last expression into Eq. (10.24) and using Eq. (10.6), we obtain

$$j_K = \frac{2R}{3L}(n_1 - n_2)\sqrt{\frac{8R_gT}{\pi M_m}}. \tag{10.25}$$

Replacing now the difference in density by a differential $(n_1 - n_2)/L \approx -\partial n/\partial z$ leads to

$$j_K = -\frac{2}{3}R\sqrt{\frac{8R_gT}{\pi M_m}}\frac{\partial n}{\partial z}. \tag{10.26}$$

Comparison with the general expression of diffusion in one dimension reveals that

$$D_{Kn} = \frac{4}{3}R\sqrt{\frac{2R_gT}{\pi M_m}} \quad \text{tubular pores} \tag{10.27}$$

is the *diffusion coefficient* under the condition of $Kn \gg 1$ (see also [444]) having the unit m^2/s. Thus this value depends on the pore radius and not on the pressure, and it increases with the square root of the temperature. One could also derive a similar expression directly from the definition of the diffusion coefficient of gases given in

Eq. (10.19). Since the mean free path in that equation has to be replaced by the pore diameter $2R$, we directly get

$$D_{Kn} = \frac{2}{3}\overline{v}R = \frac{4}{3}R\sqrt{\frac{2R_gT}{\pi M_m}} \tag{10.28}$$

as before. Note that this is only valid if the pores can be described as long cylinders, which might be acceptable for dense particulate aerogels. It will not be a good approximation for all types of fibrillar aerogels. In these aerogels, a molecule does not move into a more or less cylindrical pore with a smooth or rough surface, and once hitting the surface, the molecule is scattered and then moves forwards from scatter event to scatter event and eventually leaves the pore. One can imagine such a pore as a sieve or filter itself made of the fibrils and between them there is enough space for molecules to escape. Thus a flight through a fibrillar aerogels has much fewer scatter events with the fibrils than in a particulate aerogel. Since the molecules will pass such a microstructure almost as if fibrils would not be there, a good guess is that Eq. (10.8) applies. One could make as a rough estimate that in fibrillar aerogels the diffusion coefficient is a mixture of unhindered gas diffusion and Knudsen diffusion:

$$D_{fibrillar} = D_{\mu}\phi_p + D_{Kn}(1 - \phi_p). \tag{10.29}$$

As mentioned several times, aerogels do not possess straight, smooth pores. The pores are typically rough, especially if they consist of particles. The roughness could have an effect on the diffusion of gas molecules. It can easily be imagined that the gas molecules will make many more collisions with a rough wall than with a smooth one and that they will move back and forth in such a pore, and they even can be trapped in holes and reside there for some time. There is ample literature on the effect of surface roughness on diffusion in general and Knudsen diffusion especially, see for example [445–448]. The analysis shows that although the diffusion coefficient is reduced by surface roughness compared with a smooth pore, this effect is cancelled exactly by the additional pore cross section and therefore transport current density. The effective transport current through a rough and a smooth pore is only determined by the cross section.

10.3 Measurement of Gas Diffusion in Aerogels

The diffusion equations sketched earlier in this chapter and solutions of Fick's second law give in principle hints how to measure the diffusion coefficient. As stated by Cussler [139], there are numerous method available to measure diffusion in solids, liquids and gases, and the interested reader should take a look at the cited book. We will not give a description of the techniques here, but describe only those used to study diffusion of gases in aerogels. As far as we know, there were two techniques used.

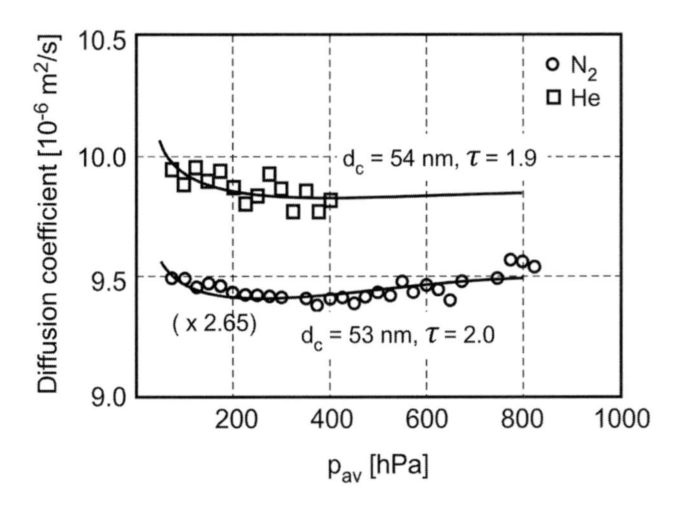

Figure 10.2 Helium and nitrogen diffusion coefficients for a sintered SiO_2 aerogel with a density of 270 kg/m^3 as a function of average gas pressure. The lines are fits of Eq. (10.30) with the pore diameter and the tortuosity as fit factors. Reprinted from [394] with permission from Elsevier

Reichenauer et al. [394] developed a new technique almost 25 years ago, which was sketched in Chapter 7. They place in a test chamber an aerogel sample, evacuate this and after a sufficient time of evacuation they apply small pressure jumps and measure the relaxation of the pressure due to diffusion in (or out) of a test piece. Recording the pressure relaxation as a function of time allows us to extract the volumetric flow in or out the sample. By this dynamic gas expansion method, they measured the diffusion coefficient of helium and nitrogen in sintered silica, RF and carbon aerogels. The basic equation they used to analyse their experimental data relies on the following equation in which d is the average pore diameter and p_{av} is the average pressure:

$$D = \frac{\phi_p}{\tau}\left(\frac{d^2}{32\mu}p_{av} + \frac{1+\sqrt{\frac{R_gT}{M_m}}d\frac{p_{av}}{\mu}}{1+1.24\sqrt{\frac{R_gT}{M_m}}d\frac{p_{av}}{\mu}}\frac{2}{3}\sqrt{\frac{2R_gT}{\pi M_m}}d\right). \tag{10.30}$$

The first term stems from a simple analysis of a Hagen–Poiseuille like flow through cylindrical pores (a derivation is given in Chapter 11). The second term describes Knudsen diffusion with a slightly different pre-factor, which is close to unity. The porosity of the sample enters as shown previously and also the tortuosity τ, which also is defined in Chapter 11. Figures 10.2 and 10.3 show some results. One should notice that the diffusion coefficient is for both gases in the range of 10^{-5} m^2/s, with helium having the higher diffusion coefficient by around a factor of 3, which is essentially due to the smaller molecular mass. If the viscous flow analog to a Hagen–Poiseuille flow would be important, the diffusion coefficient should increase with pressure.

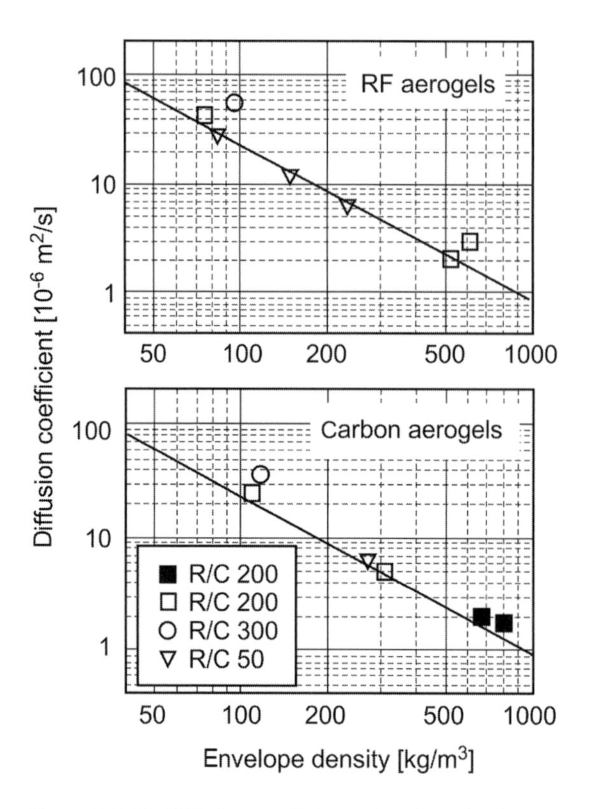

Figure 10.3 He diffusion coefficients for RF and carbon aerogels as a function of envelope density. The lines in the figures are a fit to a scaling function $D \propto \rho_e^{-1.44}$. Reprinted from [394] with permission from Elsevier

These measurements show that in silica aerogels the diffusion coefficient for both gases is almost independent of pressure and thus mainly determined by a Knudsen diffusion in pores. For RF and carbon aerogels, Reichenauer and co-workers studied the influence of envelope density on the diffusion of helium as shown in Figure 10.3.

Hosticka and co-workers [395] also studied a few years later the gas flow through a silica aerogel of a density of 0.06 g/cm^3, with a specific surface area of 800 m^2/g, an average pore size of 24 nm and a pore size spectrum of 2–200 nm measured by the BJH method (see Chapter 9). Although in principle they have measured the permeability (see Chapter 11), one can derive from such a measurement the diffusion coefficient (see Eqs. (11.40 and 11.41)). They calculate a diffusion coefficient for He and nitrogen in the range of 10^{-5} m^2/s and thus in the range of free molecular gas diffusion. In contrast to Reichenauers et al.'s results shown earlier, they observe a strong pressure dependence, indicating that the gas flow setup they have used as well as the porosity of their sample allows us essentially to study gas permeation. Only at very low pressures, a Knudsen diffusion regime was achieved.

10.4 Variables and Symbols

Table 10.1 lists many of the variables and symbols used in this chapter.

Table 10.1 Variables and symbols.

Symbol	Meaning	Definition	Units
a_m	Molecule diameter	–	m
c	Concentration	–	kg/m^3
D	Diffusion coefficient	–	m^2/s
G	Gibbs free energy	–	J
G_{id}	Gibbs free energy of an ideal solution	–	J
$\vec{j}$	Diffusion current density	–	kg/m^{2}s
k_B	Boltzmann coefficient	1.38×10^{-23}	J/K
M	Mobility	–	–
M_a	Atomic mass	–	kg
M_m	Molar mass	–	kg/mol
N_A	Avogadro's number	6.022×10^{23}	1/mol
N_p	Degree of polymerisation	–	–
p	Pressure	–	Pa
R	Pore radius	–	m
R_m	Radius of a ball-like molecule	–	m
R_g	Universal gas constant	8.314	J/(mol K)
T	Temperature	–	K
$\overline{v}$	Average molecule velocity	–	m/s
ℓ	Mean free path	–	m
μ	Chemical potential	–	–
μ_L	Viscosity of a liquid	–	Ns/m^2
ϕ_p	Pore volume fraction	–	–

11 Permeability for Gases

The flow through porous media is a well-investigated subject of fluid and gas dynamics, since it is of great practical importance. Many applications need to know the physics behind flow through porous materials, such as extraction of oil or gas from soils, filtration in the chemical lab or wastewater treatment, filtration of dirt in a vacuum cleaner using filters, filters in air conditioning systems and many more. Porous media can be made from many materials, such as paper fibres, cloth, woven and non-woven fabrics, loosely or dense packed sand, gravel, concrete, metals powders, open porous polymeric foams (sponges) and many more. Here we are interested in flow through aerogels. Since aerogels and xerogels possess a nanostructure (pore sizes between 10 nm and 1 μm), and porosities ranging from 50% for xerogels to 99.9% for aerogels, the flow of a fluid through its pores needs special considerations. Fluids are either liquids, such as the multicomponent solutions in which the wet gels are prepared, or gases, which pass through an aerogel during operation (heat exchange, Knudsen pump, filtration, gas chromatography, catalysis and others).

Here we are only concerned with gas flow through aerogels. This makes it in a certain sense more difficult, since gases are compressible and thus the density is not constant, but depends on pressure and might vary from point to point in a sample. There are several modes a gas can flow through a porous body. Three essential ones are depicted in Figure 11.1. First there is of course the conventional viscous flow determined mainly by the pressure gradient and the viscosity, like in Hagen–Poisseuille flow [274]. In such a flow situation, the molecules interact or collide with each other much more frequently than with the pore walls. Knudsen flow instead is determined by the interaction of molecules with the pore walls, meaning they collide more often with pore walls, and collision events between themselves are negligible. The third possibility is a sliding of molecules along the surface of the pore walls determined by the friction coefficient between molecules and the pore surface. All three modes are discussed in some detail in this chapter. Independent of the way that molecules travel through the pores, the essential characteristic property or parameter determining the flow through a porous body is the so-called permeability. The basics of the porous media are described in the classical textbooks of Scheidegger [449] or Churchill [274].

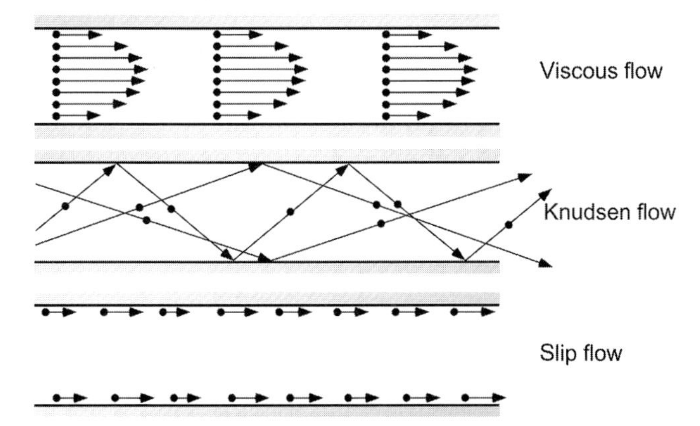

Figure 11.1 Scheme of possible gas flows through a pore in an aerogel. Top: viscous flow determined mainly by the pressure gradient and the viscosity, like in a Hagen–Poisseuille flow. Middle: Knudsen flow, determined by the interaction of molecules with the pore walls. Bottom: slip of molecules along the surface of the pore walls determined by the friction coefficient between molecules and the pore surface.

11.1 An Experimental Approach

We define permeability as a parameter or characteristic property of a porous body not theoretically, but by a measurement rule. Imagine an open porous body, like a stiff sponge, is subjected to a pressure difference. Then the gas will flow from the high- to the low-pressure side. If the pressure difference Δp is kept constant and the flux Q is measured, having the units mass or volume per time, one observes that both the pressure difference and the flux are linearly related:

$$Q \propto \Delta p. \tag{11.1}$$

Making experiments with a suitable facility ensuring that the gas flows only through the porous body, one finds out that the flux is proportional to the area of the body exposed to the gas, inversely proportional to its thickness and also to the viscosity of the gas (or generally a fluid). Making experiments with different porous solids, one observes that the proportionalities stated are always observed only the constant of proportionality changes. This constant is then called the permeability and taken as a property of the porous material. The mathematical relation between flux and pressure difference is described by the famous Darcy law for porous bodies. There are different notations of this law in the literature used. Here we first state the fundamental one, which says the flow velocity $\vec{u}_D$ of the gas through a porous body is proportional to the pressure gradient:

$$\vec{u}_D = -K\frac{1}{\mu}\nabla p = -K\frac{1}{\mu}\frac{dp}{dx}. \tag{11.2}$$

The last equality holds for one-dimensional flow. The proportionality constant K is the so-called *permeability* having the unit *length*^{2} and μ is the dynamic viscosity of

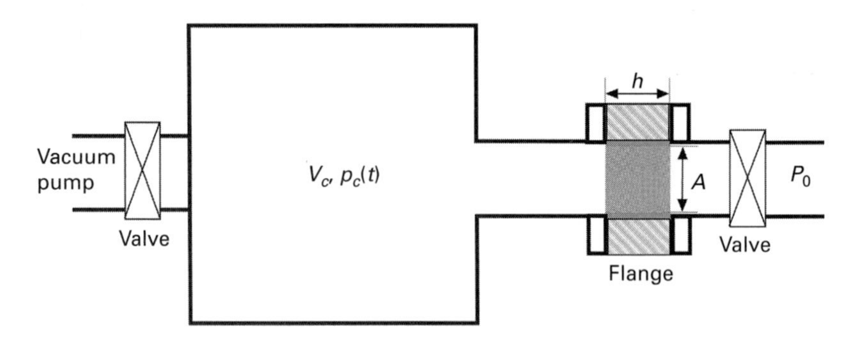

Figure 11.2 Sketch of the experimental setup for the measurement of the permeability of a porous body sketched by the dotted area. The body has a thickness of h and a cross section A.

the gas (unit Ns/m^2). The mass flow through the body of cross section A is simply the product of the velocity, the gas density ρ and of course the cross section:

$$Q = \frac{dm}{dt} = \rho u_D A = -K A \frac{\rho}{\mu} \frac{dp}{dx}. \tag{11.3}$$

Q has the unit of kg/s. Instead of the mass flow, one could also measure the volumetric flow, leading to

$$Q_V = \frac{dV}{dt} = u_D A = -K A \frac{1}{\mu} \frac{dp}{dx}, \tag{11.4}$$

with Q_V having the unit of m^3/s. Both equations give directly a measurement principle: measure the mass or volume flow rate at a constant pressure gradient (or pressure difference, since the thickness of the body h is a constant and $dp/dx = \Delta p/h$) and measure the pressure difference. This measurement idea is the basis of many experimental setups to determine the permeability of porous materials, especially if a liquid is used. It also has been applied to gas flow through aerogels. Before we discuss this, we would like to present another method, which is a dynamic one and does not need to measure a flow rate, just pressures (which today is a much easier task and can be done very precisely).

11.1.1 A Dynamic Measurement Setup

The permeability of a porous body such as an aerogel is measured dynamically using a simple setup shown in Figure 11.2. The test piece or sample to be measured is mounted in a stainless steel ring and glued therein with a special silicon gel such that no gas can pass the porous body aside but has to pass through its pores. The stainless steel ring with the sample is connected to a flange that itself is connected to a valve. A cylindrical chamber that can be evacuated serves as a test volume.

For the measurement the test volume, V_c is evacuated to a preset pressure p_c^0 and the valve in front of the aerogel piece is opened. Thus a constant ambient pressure p_0, which is measured, is applied on one side of the test piece, whereas on the other side initially a defined lower pressure $p_c^0 < p_0$ exists, which increases inasmuch as gas flows into the chamber. A vacuum gauge measures the pressure increase with time. The measurement principle is similar to the technique described by Torrent [450] for permeability measurements of concrete and by Tanikawa and Shimamoto [451] for experiments on the Klinkenberg effect on rocks.

The time-dependent pressure increase can be modelled in a simple way calculating the mass flow Q through the body. We assume that the gas can be treated as an ideal one, meaning $pV = NR_gT$ with the symbols having their usual meaning. Defining the gas density as

$$\rho = \frac{NM_m}{V}, \tag{11.5}$$

with M_m the molar weight, measured in kg/mol and N the number of moles in volume V, the ideal gas law is written as

$$\rho = \frac{pM_m}{R_gT}. \tag{11.6}$$

Insertion of Eq. (11.5) into Eq. (11.3) and using Eq. (11.2) yields

$$Q = -\frac{M_m AK}{R_gT\mu} p\frac{dp}{dx}. \tag{11.7}$$

Integration leads to a Darcy equation valid for a compressible gas

$$Q = \frac{dm}{dt} = \frac{M_m AK}{2\mu R_g Th}\left(p_0^2 - p_c^2\right), \tag{11.8}$$

with $p_0 > p_c(t)$. The pressure change in the chamber volume V_c with time can then be expressed as

$$\frac{dp_c}{dt} = \frac{Q}{M_m}\frac{R_gT}{V_c} = \frac{AK}{2\mu hV_c}\left(p_0^2 - p_c(t)^2\right). \tag{11.9}$$

This ordinary differential equation can be solved directly by separation of variables and integration. If we assume that initially the chamber has a pressure $p_c(t = 0) = p_c^0$ we obtain as a solution

$$p_c(t) = p_0 \tanh\left(\frac{AKp_0}{2\mu hV_c}t + \operatorname{artanh}\left(\frac{p_c^0}{p_0}\right)\right) = p_0 \tanh\left(\frac{t}{\tau_c} + \operatorname{artanh}\left(\frac{p_c^0}{p_0}\right)\right). \tag{11.10}$$

Thus there will be essentially an exponential increase of pressure in the chamber starting with an initial pressure p_c^0. Some real measurements on porous glass bodies are shown in Figure 11.3. The characteristic time τ_c can be expressed as

$$\tau_c = \frac{2\mu hV_c}{AKp_0}. \tag{11.11}$$

The relation shows that the larger the ambient pressure, the shorter the time to raise the pressure in the chamber to its ambient value. The larger the chamber volume, the larger the time. Thus by increasing the volume V_c or the sample thickness and

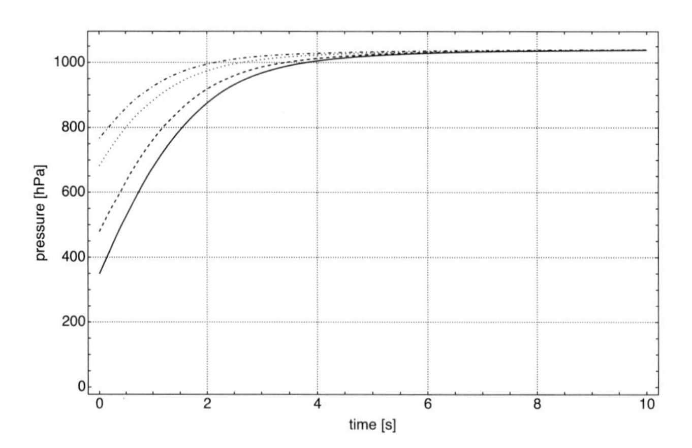

Figure 11.3 Some experimental pressure curves measured with the device sketched in Figure 11.2 on a porous glass body using different initial pressure values in the test chamber. The time dependence follows exactly the predicted behaviour of Eq. (11.10).

reducing the ambient pressure p_0 (but keeping it constant), the measurement time can be increased. Measurement of the pressure increase as a function of time and fitting the resulting data points with the expression of Eq. (11.10) to determine the characteristic time τ_c then allows us to determine the permeability as

$$K = \frac{2\mu h V_c}{A p_0 \tau_c}. \tag{11.12}$$

Let us state here: this is a one-parameter fit; since only τ_c is unknown, all other values are either preset or measured. Some evaluations with respect to the permeability are shown in Section 11.5.2. From an experimenter's point of view, the chamber volume has to be measured accurately, and this includes all dead volumes in the valves. In addition, the surface area A and the thickness h should be measured as accurately as possible. In addition, one has to ensure that the aerogel body does not deform under a given pressure difference. Looking at Eq. (11.8) one could also rewrite it in terms of pressure difference and average pressure:

$$Q = \frac{dm}{dt} = \frac{M_m A K}{\mu R_g T h}(p_0 - p_c)\frac{(p_0 + p_c)}{2} = \frac{M_m A K p_{average}}{\mu R_g T h}(p_0 - p_c), \tag{11.13}$$

with $p_{average} = (p_0 + p_c)/2$. This formulation is quite often used in the literature.

11.1.2 Stationary Measurement of Permeability

The standard procedure to measure permeability is the direct application of Eq. (11.4). A scheme of such an equipment is shown in Figure 11.4. The test chamber is divided into two parts, which are connected by a piece of an aerogel. The dividing plate has a hole of a well-defined area A. In both compartments, the pressure is measured. The gas flows into the upper chamber. The gas flow rate is measured and regulated

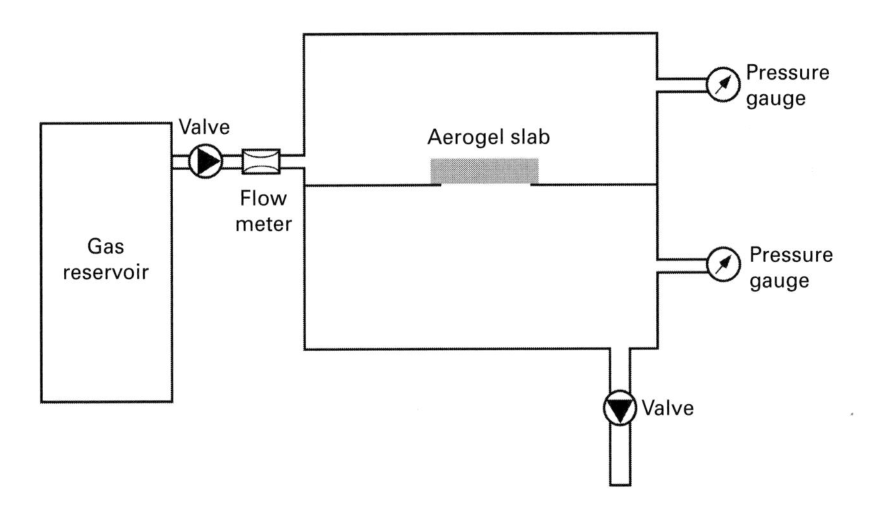

Figure 11.4 Sketch of an experimental setup for the measurement of the permeability of a porous body by a stationary method.

by an adjustable valve. The source of gas will typically be a gas bottle such that a well-defined gas, such as nitrogen, argon or helium, can be used. At the bottom chamber is also a valve is attached which regulates the outflow, if necessary. Both chambers can be evacuated initially. A measurement runs in a simple way: after evacuation, gas flows from the reservoir into the upper chamber, establishing a certain pressure p_2, which is recorded. The gas flows through the sample and the pressure in the lower compartment p_1 is recorded. Once a stable and constant pressure is established in both compartments, the flow rate dV/dt and the two pressures together with the material parameters A, h allow one to determine the permeability. If a volumetric flow rate is measured, Eq. (11.4) can be solved and the parameters inserted:

$$K = \frac{\mu h \frac{dV}{dt}}{A(p_2 - p_1)}. \tag{11.14}$$

If the mass flow rate is measured solving Eq. (11.8) for K yields

$$K = \frac{2\mu R_g T h \frac{dm}{dt}}{M_m A \left(p_2^2 - p_1^2\right)}. \tag{11.15}$$

The precision of both measurements depends essentially on the accuracy with which the flow rate can be measured and that a stationary state is achieved.

11.2 Full Mathematical Model of Porous Media

The models presented in Section 11.1 rely on several assumptions, that will be discussed with the help of a full mathematical model of the gas flow through a porous medium. A complete description of the process does allow for a time-dependent pressure variation in the porous medium itself. Initially there is no gas in the test piece on opening the valve. After establishing in a very short time the pressure p_0 on one side, the pressure difference used in Darcy's law has to build up and generally will be time dependent. It might be that a stationary state cannot be reached, meaning inflow and outflow will not balance. A more general model allows for a spatial and temporal variation of gas density ρ inside the test piece. We therefore start with the continuity equation

$$\frac{\partial \rho}{\partial t} + \nabla \cdot \vec{j} = 0, \tag{11.16}$$

in which $\vec{j}$ is the gas current density flowing through the porous sample as follows:

$$\vec{j} = \rho \vec{\bar{u}} = \rho \vec{u}_s / \phi_p. \tag{11.17}$$

Here $\vec{\bar{u}}$ is the local average velocity related to the flow inside the pores as $\vec{\bar{u}} = \vec{u}_s / \phi_p$ with ϕ_p the porosity of the test piece.[1] Using Darcy's law, Eq. (11.2), the continuity equation reads

$$\frac{\partial \rho}{\partial t} = \nabla \cdot \left(\rho \frac{K}{\mu} \nabla p \right), \tag{11.18}$$

and using the relation for the ideal gas, Eq. (11.6), we rewrite the equation as

$$\frac{\partial p}{\partial t} = \frac{K}{\mu} \nabla \cdot (p \nabla p). \tag{11.19}$$

Using $\nabla p^2 = 2p\nabla p$, we can transform this equation to

$$\frac{\partial p^2}{\partial t} = \frac{Kp}{2\mu\phi_p} \nabla^2 p^2 = K_p \nabla^2 p^2. \tag{11.20}$$

This equation looks like a diffusion equation for the square of the pressure with a pressure-dependent diffusion coefficient. One should notice that K_p has the dimension m^2/s. We define dimensionless variables

$$\phi = p/p_0 \tag{11.21}$$

$$\tau = \frac{K p_0}{2\mu\phi_p h^2} t. \tag{11.22}$$

$$\vec{r'} = \vec{r}/h \tag{11.23}$$

[1] To have a better idea about these velocities, let us make the following consideration: let q be the volumetric flow rate, $q = \bar{u}A$, with $\bar{u}$ the average flow velocity inside the pores and A the total area of the pores inside a cross section through the body. The area A is related to the total area of the sample A_{total} by the pore volume fraction ϕ_p, and thus we have $A = A_{total}\phi_p$. We then can write for the flow rate $q = \bar{u}\phi_p A_{total} = u_s A_{total}$ or $\bar{u} = u_s/\phi_p$. The so-called superficial velocity u_s can be measured using $u_s = q/A_{total}$, whereas the local average velocity is hard to measure directly.

The dimensionless time in Eq. (11.22) also defines a characteristic time for the relaxation of the gas flow inside the porous medium, namely

$$\tau_r = \frac{\mu\phi_p h^2}{2Kp_0}, \tag{11.24}$$

to obtain the continuity equation in a simple form

$$\frac{\partial\phi}{\partial\tau} = \nabla\cdot(\phi\nabla\phi), \tag{11.25}$$

which reads in one dimension

$$\frac{\partial\phi}{\partial\tau} = \frac{\partial}{\partial\xi}\left(\phi\frac{\partial\phi}{\partial\xi}\right), \tag{11.26}$$

with $\xi = x/h$. Eq. (11.25) is known as the 'porous media equation', and solutions of it can be found in the literature [452]. Unfortunately, Eqs. (11.25) and (11.26) are nonlinear partial differential equations of parabolic type and therefore not easy to solve by the well-established methods applicable to linear parabolic equations with a constant coefficient, such as the heat or mass diffusion equation [290, 440]. Depending on the initial conditions, solutions can be as simple as

$$\phi = -\frac{\lambda}{6(1+\lambda\tau)}\xi^2 \quad \text{with} \quad \lambda \quad \text{a constant} \tag{11.27}$$

$$\phi = \frac{(b+2a\xi)^2}{4a(1-6a\tau)} \quad \text{with} \quad a, b \text{ constants} \tag{11.28}$$

$$\phi = \frac{\xi^3}{2(1-3\tau)^2} \tag{11.29}$$

$$\phi = f(\xi/\sqrt{\tau}), \quad f(\xi) = Ce^{-\xi} - \frac{\xi}{2} + \frac{\xi^2}{4} + B, \quad C, B \text{ constants}, \tag{11.30}$$

and more complicated and scaling solutions can be found (see [452]). We will, however, not discuss any solution, but try to draw some general conclusions for the dynamical measurement setup. This can be done without solving the equation.

11.2.1 Characteristic Time of Permeation

First we can estimate the characteristic time τ_r for a bed of particles. The viscosity of air at ambient conditions is around $\mu = 2\cdot 10^{-6}$ Ns/m^2. Let the sample thickness be $h = 10$ mm and the initial pressure be $p_0 = 1$ hPa. Typical porosities are something like 40% for a loose packed bed of a granular material, for dense materials it can be as low as 10%. In aerogels, which are treated later, the porosity is typically between 90–99.9%. Then the permeability is of the order of 10^{-11} m^2. The characteristic time τ_r is of the order of a few milliseconds. Concrete in contrast to beds of particles has a permeability in the range of 10^{-16} m^2 [450], and thus the relaxation time is of the order of 100–1000 seconds. Therefore, the simple model derived earlier would not be applicable to concrete. In the dynamic experimental setup, the relaxation time of the gas flow inside the porous sample is much smaller than the measurement time for the pressure increase in the test chamber, which typically is of the order of a few tens to 100 seconds. In any case, one has to adjust sample thickness and pressure difference and test volume in a suitable way to be able to measure the permeability of aerogels.

11.2.2 Stationary State

The preceding considerations also show that we can simplify Eq. (11.20) in setting the time derivative zero, leading to

$$\nabla^2 p^2 = 0. \tag{11.31}$$

In one dimension, this simplifies to

$$\frac{d^2 p^2}{dx^2} = 0. \tag{11.32}$$

This equation can be solved straightforwardly by direct integration to yield

$$p(x) = p_1 \sqrt{1 + \left(\frac{p_0^2}{p_1^2} - 1\right)\frac{x}{h}} \cong p_1 \left(1 + \frac{1}{2}\left(\frac{p_0^2}{p_1^2} - 1\right)\frac{x}{h}\right), \tag{11.33}$$

where p_0 is the pressure on the test chamber side of the sample and p_1 the outside pressure. This shows that in a stationary state the pressure variation inside the sample varies with position non-linearly. If the pressure difference is small or the ratio p_0/p_1 is close to one, a linear variation is observed (expansion of the square root to first order).

11.2.3 Reynolds Number

In order to apply Darcy's law, the flow through the porous medium must be a creeping flow. To ensure this, we estimate the Reynolds number Re

$$\mathrm{Re} = \frac{2\overline{u}R}{\nu}, \tag{11.34}$$

in which $2R$ is the pore diameter and $\overline{u}$ the average flow velocity of the gas. Assume that the pore diameter is around 0.1 mm and less, and the kinematic viscosity ν of air is around $1.4 \cdot 10^{-5}$ m^2/s. The local average velocity is less than 1 mm/s, and thus we have

$$\mathrm{Re} \approx 0.01. \tag{11.35}$$

The flow inside the pores is a creeping flow, and Darcy's law can be applied. Inside an aerogel, the pore diameter is in the range of 10–100 nm, and thus the Reynolds number is even smaller in the range of

$$\mathrm{Re}_{\mathrm{aerogel}} \approx 10^{-5} \tag{11.36}$$

and less. Although this would define that the average flow in aerogels always is a creeping flow, one should bear in mind that often pores are not straight channels but complex tortuous entities, and thus locally it might not be a simple flow.

11.3 Flow through an Aerogel

Since the pores in aerogels are typically very thin, the flow might not simply follow Darcy's equation. The molecules are scattered at the pore walls quite often compared to scattering events between themselves such that this changes the motion through the pores with different pressures on both sides of the test piece. We therefore have to look for new relations between pressure and flow rate. As discussed within the context of diffusion, the new situation in aerogels is described by the so-called Knudsen number (see Eq. (10.23)), whose definition is repeated here:

$$Kn = \frac{\ell}{2R}. \tag{11.37}$$

Here ℓ is the mean free path of a gas molecule being dependent on pressure and temperature and $2R$ denotes the diameter of a pore through which the gas has to move. If $Kn < 1$, the gas flow through the pore is described by conventional fluid dynamics and is of viscous type. If $Kn \gg 1$, the gas transport through the pores is determined by collisions with the pore walls and follows new laws. The situation is schematically shown in Figure 10.1, but is extended here, since the pores are typically not straight cylindrical ones as normally used to calculate, for instance, the diffusion of molecules through narrow pores. We therefore take up a point made in Section 11.2.3. The pores in aerogels are interconnected, worm-like channels whose pore walls are not smooth but rough, since they are built up by more or less spherical particles or wiggled fibrils. Since molecules pass a narrow hole or capillary in a plate or a pore in an aerogel only by diffusion and not viscous flow, the permeability under this condition has to be redefined. To do so, we consider that the pore spectrum is broad, such that there might a viscous, diffusive and slip flow of a gas, and this leads to a situation in which all modes of transport combine.

Pure Knudsen flow through a capillary of diameter $2R$ can be calculated with a technique described in detail in the famous textbook of Landau and Lifshitz [210]. We do not repeat the derivation made there, but state the important assumptions, that there is a full accommodation between the gas in the pore and the wall, such that the gas has a Maxwellian velocity distribution and the gas molecules are reflected at the pore wall following a cosine law, meaning they are reflected diffusively. Then the flow through a pore, which is assumed to be much longer than its diameter (such that it can be treated as infinitely long, which makes the calculations very easy). The mass flux measured in kg/s is then under isothermal conditions given by

$$Q_{Kn} = \frac{4\pi}{3}\frac{R^3}{L}\sqrt{\frac{2\pi M_m}{R_g T}}\Delta p, \tag{11.38}$$

with Δp the pressure difference along the pore to be assumed linear and L the length of the pore. The gas mass current density (measured in kg/(m^{2}s)) is this value divided by the cross section of the cylindrical pore:

$$j_{Kn} = \frac{4}{3}\frac{R}{L}\sqrt{\frac{2\pi M_m}{R_g T}}\Delta p = D_{Kn}\frac{\pi M_m}{2R_g T}\frac{\Delta p}{L}. \tag{11.39}$$

The combined effect of Knudsen gas diffusion and permeation was already treated by Knudsen, who gave a semi-empirical equation which reads in the nomenclature used here:

$$j_{comb} = \left(K\frac{p_2 + p_1}{2\mu} + 0.89 D_{Kn}\right)\frac{M_m}{R_g T}\frac{p_2 - p_1}{L}, \tag{11.40}$$

with p_2, p_1 the pressure at the inlet and outlet sides of the pore of length L. The first term in this relation has a pore radius dependence through the permeability (which depends essentially on the square of the radius, as explained later) [444]. The second term depends linearly on the pore radius via the Knudsen diffusion coefficient. The factor 0.89 is empirical. If we compare the second term with that derived by Landau and Lifshitz, there is a small difference. The term given by Knudsen is smaller (compare 0.89 to $\pi/2$).

For the combined effect of all three kinds of motion, a semi-empirical equation was derived 60 years ago by Cunningham and Williams [453] describing the gas flow through a pore system. They suggest the following equation for the mass flux:

$$Q_{all} = -\frac{K}{\mu}\frac{pM_m}{R_g T}\frac{dp}{dx} - \frac{4}{3}\left(1 - \frac{Kn}{1 + Kn}\right)D_{slip}\frac{M_m}{R_g T}\frac{dp}{dx} - \frac{Kn}{1 + Kn}D_{Kn}\frac{M_m}{R_g T}\frac{dp}{dx}. \tag{11.41}$$

The first term describes viscous flow of a gas and is identical with Eq. (11.7), the second term flow is due to slip of molecules along the pore walls and the third term Knudsen flux is as described earlier (although formally it differs, especially looking to Eq. (11.38)). Let us check the meaning of the second term. Usually in fluid dynamics there is a famous boundary condition applying to all kinds of flow at a solid wall. At such a wall, the tangential flow velocity is always zero. This is called the no-slip boundary condition. It turns out that especially in nanopores it is possible that there is motion of atoms or molecules at the surface of the pores parallel to it along the surface. Such slip flow can be measured and depends on many factors. It is handled by a slip boundary condition. The flow along the surface of a pore is treated as a viscous flow but corrected by a friction factor; for more details, see [454]. The slip self-diffusivity is then

$$D_{slip} = \frac{\pi R \bar{u}_m}{8}, \tag{11.42}$$

with R the pore radius and $\bar{u}_m$ the average flow velocity in the pore. Let us make an estimate of the permeability of aerogels. The pore radius is 5 nm, the gas nitrogen and the temperature 300 K. We then have for D_{Kn} a value of $2.2 \cdot 10^{-6}$ m^2/s, and from this a permeability K_{Kn} of 0.4 10^{-12} m^2 . Therefore, the Knudsen permeability is rather small at standard conditions.

11.4 Knudsen Effect

Here we should add another relation often used in the aerogel literature, namely that for the so-called *Knudsen effect*. Assume we have a plate with a hole of cross section A_h in it. The number of moles colliding with the plate per unit of time and area is given by the kinetic theory of gases:

$$Z(p, T) = \frac{p(T)}{\sqrt{2\pi M_m R_g T}}. \tag{11.43}$$

The number of moles per second leaving the hole is therefore

$$Q(p, T) = A_h \frac{p(T)}{\sqrt{2\pi M_m R_g T}}. \tag{11.44}$$

Assume now that two gas reservoirs being at different pressures and temperatures p_1, T_1 and p_2, T_2, are connected by a hole having a diameter much smaller than the mean free path, $\sqrt{A_h} \ll \ell$. In contrast to the situation in which the hole is large in diameter compared to the mean free path, then a mechanical equilibrium of both gas reservoirs is only possible if the number of molecules going from the one to the other side (and vice versa) is equal, meaning $Q(p_1, T_1) = Q(p_2, T_2)$. This leads to a situation that

$$\frac{p_1}{\sqrt{T_1}} = \frac{p_2}{\sqrt{T_2}}. \tag{11.45}$$

The pressures are on both sides different, and their ratio is determined by the square root of their temperatures. This can be used to build so-called Knudsen pumps, and aerogels are excellent candidates to build very efficient Knudsen pumps [455], meaning essentially one has two containers connected by an aerogel and applies a temperature differential and a gas flow sets in according to Eq. (11.45). Thus gas is pumped without any moving part.

11.5 The Meaning of Permeability

By applying Darcy's law, a numerical value for permeability can be calculated, but what does it actually mean? This is similar to knowing the electrical resistance given in ohms but not knowing why different materials with identical dimensions exhibit different values. Thus, the questions is: Is it possible to correlate the permeability with characteristic properties of porous bodies? From numerous investigations carried out by various scientific disciplines (geology, hydrology, physics of building materials, hydrodynamics), several models were developed. Let us start with a very simple one.

11.5.1 Parallel Cylindrical Pore Arrangement

Assume all pores are cylinders with different diameters $d = 2R$. These cylinders are all aligned parallel. The flow through a single cylindrical pore of diameter d is described by Hagen–Poiseuille's law [274], namely the average flow velocity in the pore u_p is

$$u_p = -\frac{d^2}{32\mu}\frac{dp}{dx}. \tag{11.46}$$

Taking into account the porosity ϕ_p, defined here as

$$\phi_p = n\pi d^2/4 \tag{11.47}$$

with n the number density of cylindrical pores per unit area, we can express the pore velocity through the arrangement of pores as

$$u_D = \phi_p u_p = -\frac{nd^4}{128\mu}\frac{dp}{dx}. \tag{11.48}$$

We compare this expression with the Darcy equation Eq. (11.2) and get an expression for the permeability

$$K = \frac{\phi_p d^2}{32}. \tag{11.49}$$

Important in this relation is that the permeability is proportional to the square of the pore size. This relation was used to determine the diffusion through aerogels, as mentioned in Section 10.3, Eq. (10.30). The first term in that equation is almost identical to this expression, with the difference that the permeability is multiplied with the average pressure divided by the viscosity. In this way, a diffusion constant with the dimension of m^2/s is achieved. This manipulation implies that the diffusion and the Hagen–Poisseuille flow are for gases identical physical processes. Although, as shown in Section 10.3, the experimental results support that in aerogels gas diffusion is mainly due to Knudsen diffusion or permeation, one should bear in mind that a Hagen–Poisseuille flow is a viscous flow through a tube and fully determined by hydrodynamics, whereas diffusion in a pressure gradient is a biased random process. The bias comes from the pressure difference that drives the gas from the high- to the low-pressure regions (in ideal gases, concentration and pressure are related by the state equation). Therefore, although the results mentioned fit to the expectations, caution must be taken not to mix a random process with a gradient-driven fluid flow. The simple cylindrical model is very often used in the aerogel literature because it is so simple. But it probably does not describe reality, and any way to describe gas permeation through aerogels by this equation leads to errors. The permeability is a more complex function of porosity. Many other models have been developed; see [274]. Most famous and often used is a model developed 80 years ago by Karman and Kozeny.

11.5.2 Karman–Kozeny Permeability

Starting point of this and other concepts to relate the measured permeability value to geometrical features of the porous body is the so-called *hydraulic diameter* D_h. Let us start with a value named in stereology [403] the mean intercept length $\overline{L}$ as was defined in Section 8.3. How can we measure the hydraulic length? As mentioned before in Chapter 8, the specific surface or interface area can be related to the mean intercept length. Stereology proves that for an arbitrary morphology of pores in a body, the following relation holds (see [403]) as already shown in Chapter 8:

$$S_V^T = \frac{4\phi_s}{\overline{L}}. \tag{11.50}$$

The mean intercept length is related to the particle size, at least for regular-shaped bodies in a simple manner. The hydraulic length D_h characterising the flow through the porous body is besides a factor similar to the mean free path λ_p in a porous body. The mean free path in a two-phase mixture of arbitrary morphology has the meaning of the average distance one can travel in such a material without leaving the phase one is in. It is related to the average particle diameter $\overline{D_p}$ and the solid fraction by

$$\lambda_p = \overline{D_p}\frac{\phi_s}{1-\phi_s}. \tag{11.51}$$

If we use the equations given before, we get for the hydraulic radius the standard definition

$$D_h = \overline{L}\frac{1-\phi_s}{\phi_s}. \tag{11.52}$$

This can also be written as

$$D_h = \frac{4(1-\phi_s)}{S_V^T\phi_s}. \tag{11.53}$$

For uniform spheres one can show, after some calculations,

$$D_h = \frac{2}{3}\frac{\phi_p D_p}{1-\phi_p}. \tag{11.54}$$

In the next step we have to calculate the so-called *tortuosity*. As pointed out by Churchill [274], tortuosity τ_h is usually defined as ratio of the average distance travelled by a particle or gas molecule through the pore space L_e before it leaves the porous body to the real thickness or thickness of the porous body L:

$$\tau_h = \frac{L_e}{L}. \tag{11.55}$$

Typically this factor is hard to determine theoretically, and thus one might use it as a fudge factor. Normally, one can imagine that in aerogels $\tau_h \gg 1$. The typically velocity through a porous body is then in reality

$$u_g = \frac{u_0}{\phi_p}\tau_h \tag{11.56}$$

if u_0 is the velocity in the absence of pores. Let us now calculate Darcy's law modified with tortuosity:

$$u = -\frac{\phi_p D_h^2}{16 k_0 \mu \tau_h} \nabla p = -\frac{K}{\mu} \nabla p. \tag{11.57}$$

Solving this equation for K, we get

$$K = \frac{\phi_p D_h^2}{16 k_0 \tau^2} = \frac{\phi_p^3}{k_k (1 - \phi_p)^2 S_{Vs}^2}. \tag{11.58}$$

here we used Eq. (11.53) and of course set $k_0 \tau^2 = k_k$ constant. Without going into further detail, one can show that for spherical particles making the porous body Eq. (11.58) reads

$$K = \frac{\phi_p^3}{180(1 - \phi_p)^2} D_p^2. \tag{11.59}$$

This is the famous permeability relation often named as Karman–Kozeny's law:

$$K = \frac{(1 - \phi_s)^3}{5 \phi_s^2 S_{VT}^2}. \tag{11.60}$$

Here we denoted $S_V^T = S_{VT}$ to avoid any misreading with squaring this value. Since this permeability reflects the morphology of the porous body, it should be independent of the pressure gradient or the absolute values of the pressures used. Such behavior is reflected by the sintered glass bodies described earlier. Figure 11.5 shows an evaluation for different glass bodies, whose pressure-time diagrams are shown in Figure 11.3.

These curves clearly show that the permeability is a geometrical feature of porous bodies and that also the permeability does not depend on the average pressure. Such a behaviour is under discussion, especially also for aerogels (see also [453]).

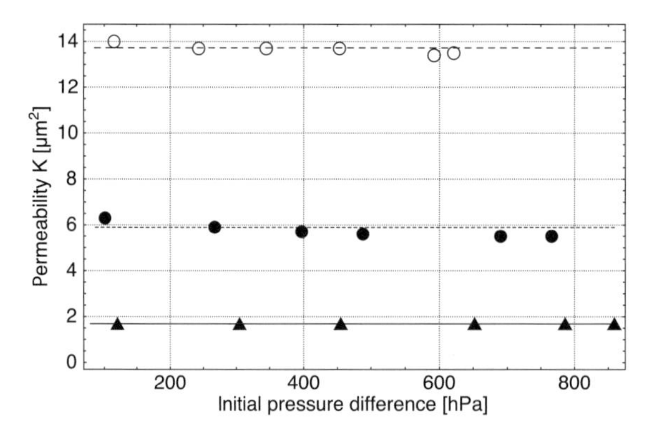

Figure 11.5 Permeability of sintered porous glass bodies; see Figure 11.3, showing that the permeability is a function of the glass particle size (d_p) but not a function of pressure difference and therefore also the average pressure.

Coming back to Eq. (11.60), one should notice, since S_V^T is related to the BET surface area, $S_V^T = \rho_e S_m$. We therefore can relate the BET surface area to the permeability. Let us do an estimate for an aerogel. Typically the porosity is 95%, the surface area 500 m^2/g and the envelope density 100 kg/m^3. Then the Karman–Kozeny permeability is calculated as $2.7 \cdot 10^{-14}$ m^2. If this would be a correct value, the characteristic time to achieve a stationary flow would be of the order of a few seconds and a measurement possible without taking into account an instationary pressure development in an aerogel sample. Nevertheless, one should in most cases better take a thin sample (if it is mechanically possible, plate bending of the sample and bending strength should be considered) and use a small test volume, so that one always measures the steep increase in the pressure time curves, shown in Figure 11.3.

It must be noted that permeability is not a scalar value but a tensor like electrical or thermal conductivity. In Darcy's law, the flow velocity as well as the pressure gradient are vectors and the mediating factor is a second-rank tensor, i.e., a matrix. Thus, the permeability is not necessarily isotropic. This is illustrated using the example of a body containing cylindrical pores. The permeability along the cylinder axis is very high but becomes zero in perpendicular direction.

11.6 Permeability of Some Aerogels

The permeability of aerogels was measured in two states, either in the wet liquid state or the dry one. Whereas the first case is important in processing, since replacement of a gel liquid by another one is often a critical step, the second one is important in any application of aerogels as filters or catalysts, loading of an aerogel with a substance that is able to condense or precipitate inside the aerogel. We are not going to discuss experimental results on liquid permeability, such as the work done by example the group of Wognier and co-workers [456], but concentrate here on gas permeability measurements. In a recent paper, Woignier et al. [457] studied again the permeability of silica aerogels with fumed silica added, comparing liquid and gaseous permeability. Their results show an enormous difference between liquid and gaseous permeability. Figure 11.6 shows a direct comparison. Whereas the liquid

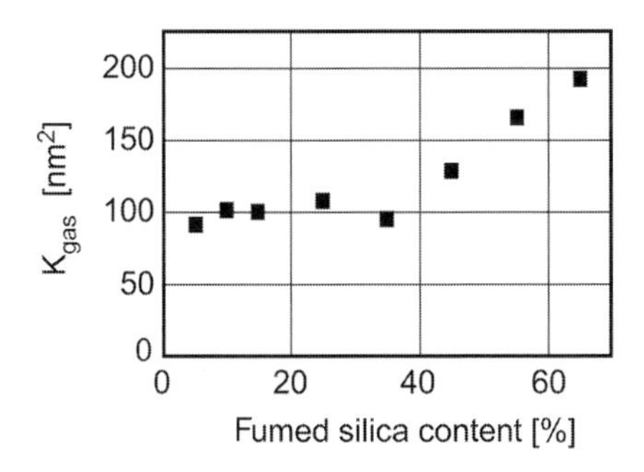

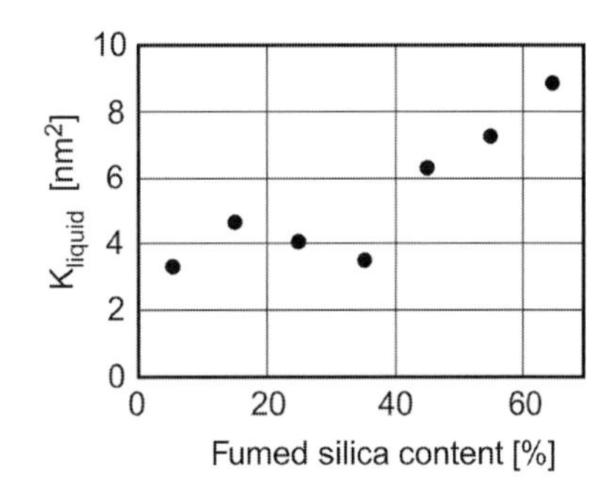

Figure 11.6 Gaseous and liquid permeability as a function of fumed silica added. Reprinted from [457] with permission from Elsevier

Table 11.1 Some properties of cellulose aerogels with three different amounts of cellulose after [458].

Cellulose content	Envelope density kg/m^3	BET surface area m^2/g	Permeability K μm^2
2 wt%	63	287	0.0465
4 wt%	109	314	0.005
6 wt%	138	282	0.0036

Table 11.2 Permeability of some selected aerogels measured by the dynamical test setup described at DLR, Cologne

Type of aerogel	Envelope density kg/m^3	BET surface area m^2/g	Permeability K μm^2
RF (R/C 1500)	610	1.9	1.9
Carbon	330	540	33
Silica	138	540	0.038
Cellulose 3 wt% ($ZnCl_2$)	57	170–240	0.465

permeability is in the range of a few nm^2, the gaseous permeability is in the range of 100–200 nm^2. Woignier and co-workers show that this difference is essentially due to the Knudsen effect valid for gases. They also show that slip along the pore surface is an additional effect.

Recently, Ganesan and co-workers [458] published some data on cellulose aerogels prepared with the rhodanide route explained in Chapter 3. Some properties of his cellulose aerogels are given in Table 11.1. These data reveal that with an increasing amount of cellulose, the density increases and the permeability decreases to values of around $0.36 \cdot 10^{-14}$ m^2 or 3600 nm^2. The values of cellulose aerogels are thus a factor of 10 higher than those reported by Woignier for silica. This points to the difference in microstructure between particulate and fibrillar aerogels. Table 11.2 gives some unpublished data measured at DLR in recent years. These data exhibit no clear trend of the permeability with specific surface area. Although the cellulose aerogels prepared with the rhodanide route have only a slightly higher specific surface area then that prepared with the zinc-chloride route, the permeability differs by an order of magnitude. These data also disagree with the values of Wognier, which might be due to different experimental method in determining the permeability. Independent of that, we expect a Knudsen permeability of the order of 10^{-12}–10^{-14} m^2. The measured data fit to this expectation. The data currently available do not, however, allow us to draw any conclusion on the applicability of, for instance, a Karman–Kozeny model or give a sufficient motive to develop another model. Unfortunately, no systematic investigations were performed measuring permeability as a function of envelope density or specific surface area and characterising carefully the pore structure of aerogels.

11.7 Variables and Symbols

Table 11.3 lists many of the variables and symbols used in this chapter.

Table 11.3 Variables and symbols.

Symbol	Meaning	Definition	Units
A	Cross section of a sample	–	m^2
A_h	Cross section of a hole in a plate	–	m^2
D_{slip}	Slip diffusion coefficient	–	m^2/s
D_{Kn}	Knudsen diffusion coefficient	–	m^2/s
$\overline{D}_p$	Average pore diameter	–	m
$\overline{D}_h$	Hydraulic diameter	–	m
h	Height or thickness of a sample	–	m
$\vec{j}_{Kn}$	Knudsen current density	–	–
k_B	Boltzmann coefficient	1.38×10^{-23}	J/K
K	Permeability	–	m^2
K_{Kn}	Knudsen permeability	–	m^2
Kn	Knudsen number	$\ell/2R$	–
K_p	Pressure-dependent permeability	$Kp/(\mu\phi_p)$	m^2/s
L	Length of a pore	–	m
L_e	Real distance travelled by a molecule in a porous body	–	m
M_m	Molar mass	–	kg/mol
N	Number of moles	–	
p	Pressure	–	Pa
[1.5pt] p_c^0	Initial pressure in a test volume	–	Pa
p_0	Ambient pressure	–	Pa
Δp	Pressure difference	–	Pa
∇p	Pressure gradient	–	Pa/m
Q	Mass gas flux	–	mass/time
Q_V	Volumetric gas flux	–	volume/time
$2R$	Pore radius	–	m
R_g	Universal gas constant	8.314	J/(mol K)
Re	Reynold's number	$2\overline{u}/\nu$	–
T	Temperature	–	K
$\overline{u}$	Average flow velocity	–	m/s
$\overline{u}_D$	Gas flow velocity	–	m/s
$\overline{u}_p$	Average flow velocity in a cylindrical pore	–	m/s
V_c	Test volume	–	m^3
ℓ	Mean free path in a gas	–	m
λ_p	Mean free path in a dispersion of particles	–	Ns/m^2
μ	Viscosity of a liquid	–	Ns/m^2

Table 11.2 Continued

Symbol	Meaning	Definition	Units
ν	Kinematic viscosity	μ/ρ	m^2/s
ϕ	Relative pressure	p/p_0	–
ϕ_p	Pore volume fraction	–	–
ϕ_s	Solid volume fraction	–	–
ρ	Gas density	–	kg/m^3
τ	Characteristic time	–	–
τ_c	Characteristic time	–	–
τ_r	Relaxation time	–	–
τ_n	Tortuosity	L_e/L	–

12 Thermal Properties

Aerogels are famous for their low thermal conductivity making them the super insulation materials of the future. The extreme small conductivity, which can be even close to the conductivity of vacuum isolation panels, poses, however, problems to many conventional measurement techniques, since even smallest heat leaks might give erroneous results by easily 20–30%. This chapter therefore is divided into several sections: first, general aspects of heat conduction are treated, and second, models are discussed explaining the thermal conductivity of porous materials, especially aerogels. Several techniques to measure thermal conductivity are described in Section 12.5.

12.1 Heat Conduction Equation

For simplicity, take a piece of material as shown in Figure 12.1 and apply to one side a temperature T_{hot} and the other side a smaller temperature T_{low}. Assume that the piece of material has on both sides the same cross section area A and that the hot and cold sides are separated by a distance h. Then there will be a heat flux from the higher to the lower temperature.

The heat flux is an energy flux per unit time measured therefore in J/s, and this is denoted with the symbol $\dot{q}$, where the dot indicates differentiation with respect to time. The heat flux can be calculated from the temperature difference, the cross section and the sample thickness h as

$$\dot{q} = \frac{dq}{dt} = -\lambda \frac{T_{hot} - T_{low}}{h} A. \tag{12.1}$$

Here we have introduced a constant of proportionality λ, which is called the thermal conductivity and is measured in W/mK. On the right side, we have the driving force for heat transport, the temperature gradient $(T_{hot} - T_{low})/h$, and on the left side we have the resulting flux. Both are related linearly to each other, establishing Fourier's first law. This relation defines also a method to measure the thermal conductivity: One establishes in a sample of cross section A a constant temperature difference, and measures at two positions in the sample, separated by some distance h, the resulting temperatures and the energy flux through the sample. This could be done, for instance,

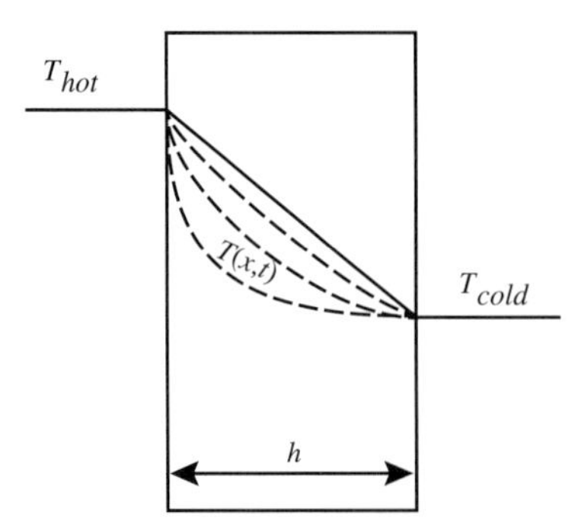

Figure 12.1 Scheme of the conduction through a slab of thickness h. The full lines characterise the stationary state. The dashed lines sketch time and space-dependent temperature profiles if initially the sample would have been at temperature T_{cold} and then brought into contact at $t = 0$ with a hot plate of temperature T_{hot}.

with electrical heaters and measuring voltage and current used to establish the constant temperature profile. Eq. (12.1) is the basis of a lot of measurement devices available on the market.

Let us extend the discussion of heat flux (in isotropic solids) a bit. Eq. (12.1) can be generalised. Assume there is an arbitrarily shaped body with a non-homogeneous temperature distribution. Let us consider a small volume element through which heat will flow. There is an input and output of heat in this element, and the difference between heat flowing into the volume element and out defines the temperature change within that element (provided there is no heat source inside the body, like a phase change, radioactively decaying elements, Joule heating due to an electrical current etc.). The heat flux per unit area and time will now be denoted by a vector $\vec{j}$, since the heat flux generally will be different in different directions. This local heat flux is then proportional to the acting temperature gradient, denoted ∇T, which is the vectorised and the limiting form of the temperature difference $\lim_{h\to 0}(T_{hot} - T_{low})/h$, or better, the derivative with respect to the spatial coordinates represented by a vector $\vec{r}$, namely $\partial T/\partial \vec{r}$:

$$\vec{j} = -\lambda \nabla T. \tag{12.2}$$

The temperature change in a volume element can be calculated from the change in energy content (enthalpy). The enthalpy H per unit volume is defined by

$$H = \rho c_p T, \tag{12.3}$$

in which ρ is the material density (in principle a function of temperature) and c_p is the specific heat at constant pressure (being also a function of temperature in general). If density is measured in kg/m^3 and the specific heat (or heat capacity) in J/(kg K), the unit of H is J/m^3. This also means $\rho c_p = c_V$ is the specific heat per volume. The net flux of heat in and out of the volume element is balanced by the rate of change of energy in that volume element, which is just the divergence of the flux noted as $div\vec{j} = \nabla \cdot \vec{j}$. We therefore arrive at the following equation:

$$\frac{\partial H}{\partial t} + \nabla \cdot \vec{j} = 0, \tag{12.4}$$

and assuming that the density and the specific heat are not explicitly dependent on time and that the thermal conductivity is almost constant, we arrive at the following

$$\rho c_p \frac{\partial T}{\partial t} = \lambda \nabla^2 T. \tag{12.5}$$

Introducing the thermal diffusivity as

$$\kappa = \frac{\lambda}{\rho c_p}, \tag{12.6}$$

the equation reads

$$\frac{\partial T}{\partial t} = \kappa \nabla^2 T. \tag{12.7}$$

This is Fourier's second law. The thermal diffusivity κ has the unit m^2/s.[1] This differential equation can be solved for many geometries and initial and boundary conditions. The interested reader may consult the standard textbook written almost 80 years ago by Carslaw and Jaeger [440]). Today this heat transport equation can be solved numerically in arbitrary complex geometries with any kind of material, composite materials in any shape and with any kind of interface. Many commercial computer codes are available for such a task. Typically modern measurement devices are not only tested rigorously by experiments but also optimised with numerical analyses of the heat fluxes inside using the real geometry and the real materials with their temperature dependent properties. We are not going into any detail.[2] Analytical methods and solutions in simplified geometries are still of great value to get insight into the physics of heat transport and essential parameters acting. This especially is true for a principal understanding of measurement techniques and a crosscheck of possible measurement errors and validity of values obtained as well as validating new numerical implementations.

12.2 Heat Conduction in Porous Materials: Aerogels

In this section, we give a brief overview of physical modes of heat transports, describe transport paths in porous materials, make simple models for the heat conduction in porous materials and apply that to aerogels (more details can be read in [14, 459]).

[1] And thus the same unit as mass diffusivity in Fick's second law and kinematic viscosity in the Stokes equation for laminar viscous flow.

[2] Without any claim of completeness, here a few possible hints for where to find such numerical codes for heat transfer and more: www.cfd-online.com/, Comsol Multiphysics, Flow3D, ANSYS, Fluent, Magmasoft™, ProCast™ and of course MATLAB™, Mathematica™, Maple™and many others.

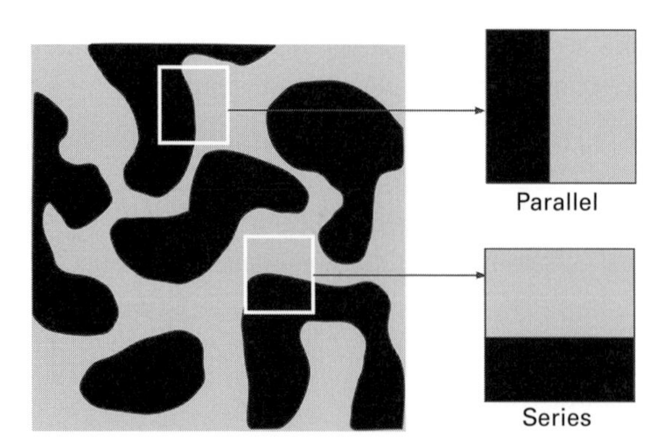

Figure 12.2 Sketch of a section through a two-phase microstructure. The small figures to the right show geometrically simplified arrangement of the two phases, most often called a series or parallel arrangement.

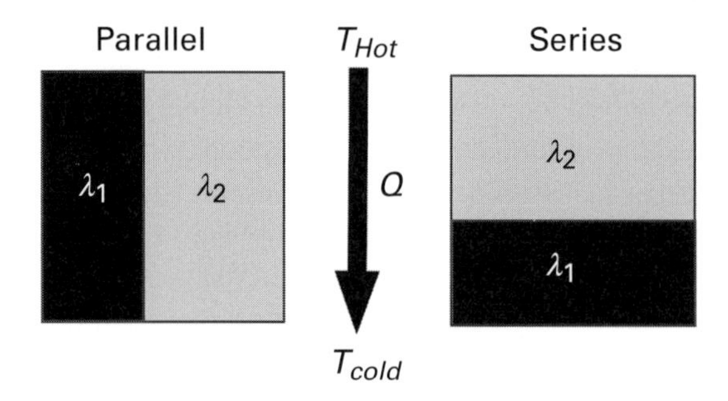

Figure 12.3 The two most simple arrangements of two phases: parallel to the heat flux or perpendicular.

12.2.1 Porous Media in General

A porous medium has at least two distinct phases: typically one solid phase and a second fluid phase, which might be just empty or filled with air or a liquid. Porous materials can have arbitrary complex microstructures such as open cell foams, sponges, beds of particles of arbitrary shape, networks of fibres, felts and fibre batts, filters and many more [1]. A sketch of a section through such a two-phase microstructure is shown in Figure 12.2. A description of the properties of a porous material requires knowledge of the 3D arrangement of the phases. The mathematical description of the microstructure is rather complex and needs in general stochastic methods; see Torquato [398] and Underwood [403]. Here we just try to discuss a few simple models, especially suitable to discuss the thermal conductivity of aerogels. For the two simple cases sketched in Figure 12.3, one can easily derive that the thermal

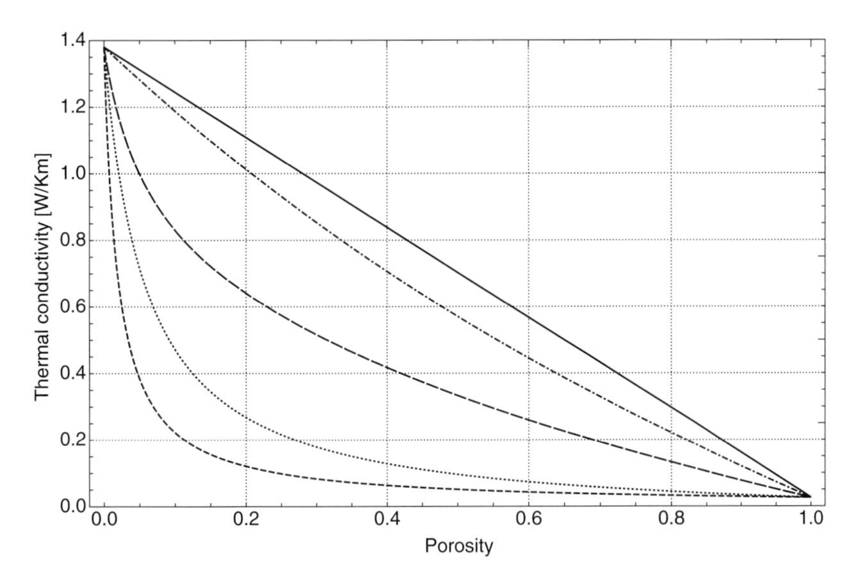

Figure 12.4 Wiener bounds, Eqs. (12.8) and (12.9), solid and dashed, and Hashin–Shtrikman (HS) bounds, Eqs. (12.12) and (12.13) dotted and dot-dashed, for a porous material. The pores are filled with still air (thermal conductivity 0.026 W/mK) and the solid is Silica with a conductivity of 1.38 W/mK. Also shown is the arithmetic average of the HS bounds as a line with long dashes.

conductivity of an arrangement of phases being parallel or perpendicular to the heat flux is simply given by two very extreme laws (called Wiener bounds):

$$\text{parallel} \quad \lambda_c^p = \phi_1 \lambda_1 + \lambda_2 \phi_2 \tag{12.8}$$

$$\text{series} \quad \frac{1}{\lambda_c^s} = \frac{\phi_1}{\lambda_1} + \frac{\phi_2}{\lambda_2}. \tag{12.9}$$

Since the two extreme relations for the thermal conductivity, Eqs. (12.8) and (12.9), are far apart from each other, if the thermal conductivities differ appreciably, these bounds are rarely of any use. In addition, as sketched in Figure 12.2, a real microstructure is possibly a complex mixture of local elements of a series and parallel arrangement, and therefore these two bounds are the absolute minimum and maximum. All real values for the thermal conductivity of a porous solid must be within the area they span. A calculation of both together with other bounds is shown in Figure 12.4. Therefore, many attempts have been undertaken to give more rigorous bounds [398, 460, 461]. Most famous in this context are the so-called Hashin–Shtrikman (HS) bounds. These bounds can be written more generally: assume to have a multi-phase material with M phases. Each phase has a volume fraction ϕ_k and a thermal conductivity λ_k. Let $\lambda_{\min}, \lambda_{\max}$ be the lowest and the highest thermal conductivities of all phases. Then the lower HS bound is

$$\lambda_{\text{low}}^{HS} = \frac{1}{\sum_{k=1}^{M} \frac{\phi_k}{\lambda_{\min}+\lambda_k}} - \lambda_{\min} \tag{12.10}$$

and the upper bound reads

$$\lambda_{\text{up}}^{HS} = \frac{1}{\sum_{k=1}^{M} \frac{\phi_k}{\lambda_{\max}+\lambda_k}} - \lambda_{\max}. \tag{12.11}$$

They read as follows for aerogels in which the pore space has the lower conductivity than the solid backbone material:

$$\lambda_{\text{low}}^{HS} = \lambda_p \frac{2\lambda_p\phi_p + \lambda_s(3-2\phi_p)}{\lambda_p(3-\phi_p)+\lambda_s\phi_p} \tag{12.12}$$

$$\lambda_{\text{up}}^{HS} = \lambda_s \frac{\lambda_p + 2\lambda_s(1-\phi_p) + 2\lambda_p\phi_p}{\lambda_s(2+\phi_p)+(1-\phi_p)\lambda_p}. \tag{12.13}$$

Typical aerogels have porosities larger than 90% and thus aerogels would be at the right side of Figure 12.4. These bounds predict that in all cases, the thermal conductivity of a porous material in which the pores are filled with air is higher than that of still air. Therefore, these models are not sufficient to describe the super-insulating properties of aerogels. Two things are directly striking:

- If the pores would not be filled with air but be evacuated, meaning we artificially set the thermal conductivity to zero (which is nonsense, since there always is radiative heat transfer), what would be the result?
- If we take into account, that the solid skeleton of an aerogel is not just a compact material but has typically a necklace-like structure in 3D and therefore only point or nanosized area contacts exist, and these transport the heat and the fraction solid $\phi_s = 1 - \phi_p$ and then overestimates drastically the cross section that transports heat. One could for instance guess that this nanostructure reduces the thermal conductivity of the skeleton by some factor.

Taking both effects into account would lead to a much lower effective thermal conductivity. In principle, one can therefore imagine that aerogels can have very low thermal conductivities, since their nanostructure induces strong deviations from conventional porous media theory. We are not further following such considerations, but discuss models of thermal conductivity developed for aerogels going back to fundamental work by Lu et al. [78], Hrubesh and Pekala [462] and others three decades ago. Meanwhile, improved and more detailed models were developed; see [463].

12.3 Thermal Conductivity of Aerogels

We first repeat the model developed by Lu et al. and Hrubesh and Pekkala [78, 462]. They treat an aerogel as an effective medium and derive an effective thermal conductivity composed of three contributions, heat conduction through the solid backbone of the aerogel λ_s, gaseous heat transport through the pore space λ_g and of course radiative heat transport λ_r, which always is there and needs very special consideration. They finally add up these contributions to yield an expression for the thermal conductivity of an aerogel:

$$\lambda_{\text{eff}} = \lambda_s + \lambda_g + \lambda_r. \tag{12.14}$$

Let us briefly review their considerations for each kind of heat transport mode in aerogels (for more detail, see [459]).

12.3.1 Heat Conduction of the Solid Backbone

Heat is transported through a solid by two modes: electrons or phonons. Phonons in any material if crystalline or amorphous are vibrations of the atom or molecule chains in a solid material (which generally are also quantised in a crystalline material). In isolating aerogels (and besides carbon aerogels, all aerogels treated in this book are isolators), any electronic contribution can be neglected. Then the thermal conductivity can be written as [464]

$$\lambda_s = \frac{1}{3} c_V(T) v(T) \ell(T), \tag{12.15}$$

with $v(T)$ the average phonon velocity, $\ell(T)$ the phonon mean free path and c_V the specific heat capacity per unit volume. This is converted into the specific heat capacity per unit mass by division with the density of the aerogel:

$$c_m = c_V(T)/\rho_e \cong \frac{c_V}{\rho_s \phi_s}. \tag{12.16}$$

The last approximation holds if $\rho_e \ll \rho_s$. The free path of phonons is determined by two things: collisions or scattering between lattice vibrations themselves (as in gases the collision between two free moving molecules) and of course scattering at lattice imperfections. The theory of phonons in solids shows that $\ell \propto 1/T$ and thus at low temperatures the mean free path might exceed the crystal size, or in aerogels, having a nanostructure, this can already occur at room temperature. Then the mean free path might be simply given by the size of the nanoparticles or fibrils making up the aerogel network. This is in agreement with experiments and theory, as already shown a 90 years ago by Peierls [464, 465]. If we would like to calculate the thermal conductivity, we need to have a model for the specific heat, the average phonon velocity and the mean free path as they depend, for instance, on the envelope density (which somehow characterises the nanostructure on a coarse level). Figure 12.5 shows a sketch of the heat transport through a particulate aerogel backbone. Many models have been presented in the literature for the three factors or better for the thermal conductivity of porous and fractal materials. The simplest one assumes the sound velocity to be constant, the mean free path given by the size of the nanoparticles and the specific heat as being a constant. In more elaborate models, the thermal conductivity of an aerogel is expressed by the ratio of the longitudinal sound velocities of the aerogel v' and that of the solid backbone v_s as it depends on the envelope and skeletal densities ρ_e, ρ_s. As a scaling factor, one takes the thermal conductivity of

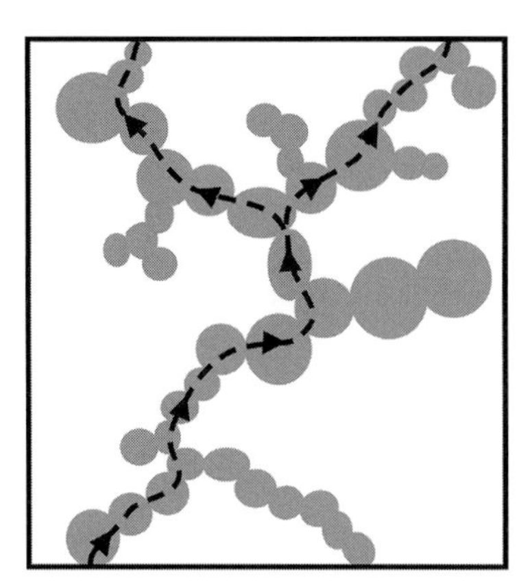

Figure 12.5 Scheme of the heat transport through the solid backbone of an aerogel.

the skeletal material λ_s. According to Fricke and co-workers, the conduction of heat through the tortuous solid structure of an aerogel can then be written as

$$\lambda_s^{aer} = \lambda_s \frac{\rho_e v'}{\rho_s v_s}. \tag{12.17}$$

The sound velocity of the aerogel depends on the gas pressure p_g, elastic modulus, gas density and porosity. To obtain the thermal conductivity of the solid backbone alone, one has to evacuate an aerogel [422]. A detailed experimental and theoretical analysis of this equation performed by Gross et al. [466] and extended by Weigold [422] showed that the sound velocity can be expressed in terms of simple power laws with respect to the aerogel density:

$$\rho_e \leqslant 100\text{kg/m}^3 \quad v'(p_g) = C_1 \sqrt{\frac{\lambda_s}{\rho_s v_s}} \tag{12.18}$$

$$\rho_e \geqslant 100\text{kg/m}^3 \quad v'(p_g) = C_2 \left(\frac{\rho_e}{\rho_s}\right)^{0.88} v_s. \tag{12.19}$$

They determined experimentally values for the constants C_1, C_2 for silica, RF, MF and carbon aerogels (see [462]). With these equations, the solid heat conduction can be calculated. Figure 12.6 shows a calculation of the solid conductivity of silica and RF aerogels based on Eq. (12.17) and the scalings in Eqs. (12.18) and (12.19). Other approaches take into account that aerogels are open porous nanostructured materials having in many cases a mass fractal structure. It was extensively discussed a few decades ago that lattice vibrations, phonons, in fractal objects have a different energy spectrum which mimic the fractal structure, and the term fractons was coined for lattice excitations in such materials. In a review on transport properties of fractal structures, including aerogels, Nakayama et al. discussed especially the temperature dependence of the thermal conductivity [467]. They showed that the thermal conductivity is described by four processes: first, fractons can be strongly localised to sizes

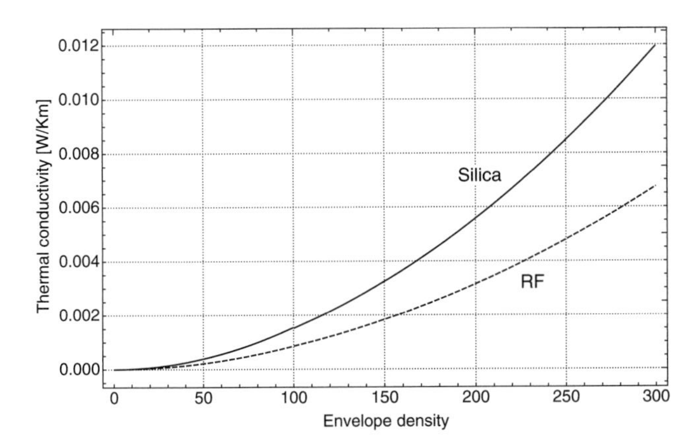

Figure 12.6 The variation of the solid conductivity of silica and RF aerogels with envelope density measured in kg/m^3.

of the order of two to three lattice constants at larger wave lengths; second, at shorter wavelengths the vibrations are extended in space; third, there are mesoscopic modes in the range of 10–100 lattice constants; and fourth, there are particle modes in which the particles forming the network vibrate. At very low temperature, the strongly localised modes lead to a linear temperature dependence of the thermal conductivity, the bit more extended modes lead to a constant thermal conductivity, followed by a hopping modes in which the localised modes hop from one region to another, while at higher temperatures the mesoscopic modes lead to a non-linear dependence on temperature. With this idea, the authors could describe the range of lower temperatures by a simple linear law fitting to silica aerogels of different densities in the temperature range below 10 K.

A completely different approach can be taken looking at Eq. (12.15) and assuming, as is often done in solid state physics, that the specific heat is almost constant and the phonon mean free path also (which is at least true at constant temperature). Then the only property left is the average phonon velocity, which can be of course interpreted as the sound velocity (see Eq. (12.17)). In Chapter 13, it is shown that the sound velocity is related to the elasticity of a material, Young's modulus and the envelope density of the material. Let us assume that the aerogel is plate-like and one has measured the propagation of longitudinal waves. Their sound velocity v_l is then given by Eq. (13.42), which we write here for convenience of the reader:

$$v_l^2 = \frac{E}{\rho_e(1-\nu^2)}, \tag{12.20}$$

with ν as Poisson's ratio. Inserting this equation into Eq. (12.15) leads to

$$\lambda_s = \frac{1}{3} c_V(T)\ell(T)\sqrt{\frac{E}{\rho_e(1-\nu^2)}}. \tag{12.21}$$

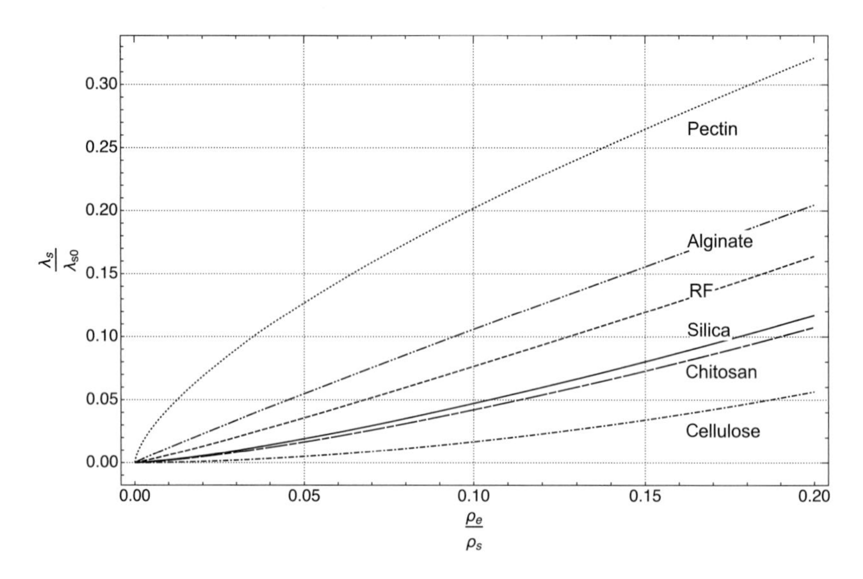

Figure 12.7 Variation of the solid backbone conductivity of various aerogels with relative density according to Eq. (12.24). In this plot, a Poisson ratio for silica and RF aerogels of 0.2 was taken and that for all biopolymers was set to zero. For all materials, the solid bulk Poisson ratio of 0.33 was assumed. The power-law exponents were taken from the data provided in Section 13.3.3.

We show in Chapter 13, that Young's modulus depends in many aerogels in a power-law fashion on density and express this as $E = E_0(\rho_e/\rho_s)^m$. This leads then to an expression for the thermal conductivity of the solid backbone:

$$\lambda_s = \frac{1}{3} c_V(T)\ell(T)\sqrt{\frac{E_0 \rho_e^m}{\rho_e \rho_s^m (1-\nu^2)}}. \tag{12.22}$$

Expressing the thermal conductivity of the skeleton material as

$$\lambda_s^0 = \frac{1}{3} c_V(T)\ell(T)\sqrt{\frac{E_0}{\rho_s (1-\nu_s^2)}} \tag{12.23}$$

with E_0, ρ_s, ν_s as Young's modulus, density and Poisson's ratio of the bulk material, we can rewrite this equation as

$$\lambda_s = \lambda_s^0 \sqrt{\frac{1-\nu_s^2}{1-\nu^2}} \left(\frac{\rho_e}{\rho_s}\right)^{\frac{m-1}{2}}. \tag{12.24}$$

Using the experimentally determined values for the power-law dependence of Young's modulus (see Section 13.3.3), one can plot the relative solid-state thermal conductivity of aerogels as shown in Figure 12.7. According to this model, the thermal conductivity of the solid backbone in aerogels is greatly reduced in cellulose, chitosan and RF and silica, whereas in alginate and pectin aerogels the solid backbone transports

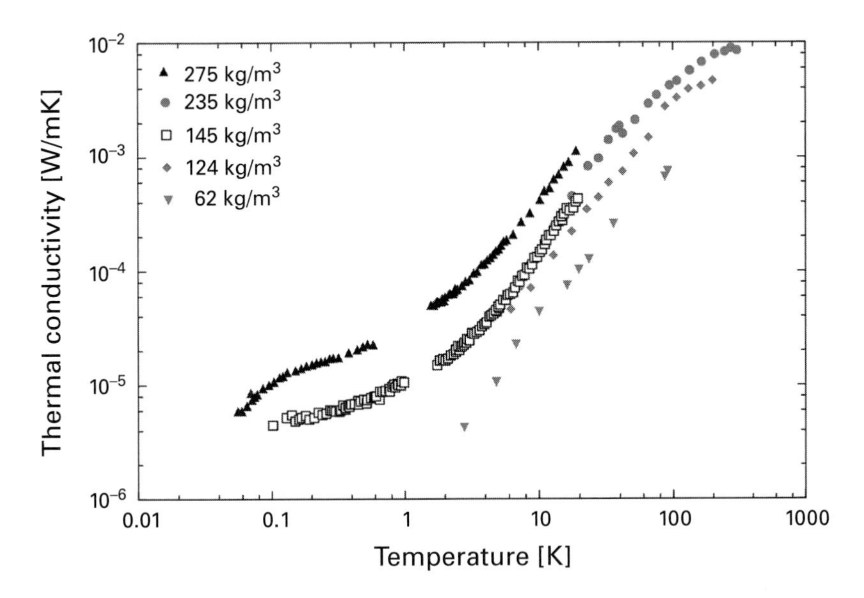

Figure 12.8 Temperature dependence of the solid backbone conductivity of silica aerogels with different envelope densities stated in the legend. Note that at very low temperature the conductivity goes to zero, due to the temperature dependence of the specific heat, followed by a plateau (localised fractons), a linear increase in the range of 2–20 K (hopping of localised fractons) and a levelling off to a non-linear temperature dependence. Reprinted from [459] with permission from Springer-Nature

relatively large amounts of heat. The relation between solid-state conductivity and elastic constants was recently carefully investigated by Weigold in her PhD thesis [422] in which she also took radiative heat transport in evacuated aerogels into account. The temperature dependence of the solid-state conductivity of aerogels of various density was carefully measured. Figure 12.8 shows a result taken from Ebert [459]. At very low temperature, the conductivity goes to zero due to the temperature dependence of the specific heat, followed by a plateau (localised fractons), a linear increase in the range of 2–20 K (hopping of localised fractons) and a levelling off to a non-linear temperature dependence. As outlined previously, this temperature behaviour is in full agreement with a fracton–phonon picture of transport of heat through a body with a fractal structure.

12.3.2 Gas Phase Transport of Heat

The transport of heat via the gas phase, shown schematically in Figure 12.9, through the pore space of an aerogel is tricky, since the pores have a broad size spectrum depending on their processing. For pores smaller than the mean free path of an air molecule, transport of heat via the gas phase is limited by Knudsen diffusion. This is characterised by the Knudsen number Kn (see Chapter 11). Since in many aerogels very small pores (around 10 nm) exist also with larger ones up to 100 nm (and even

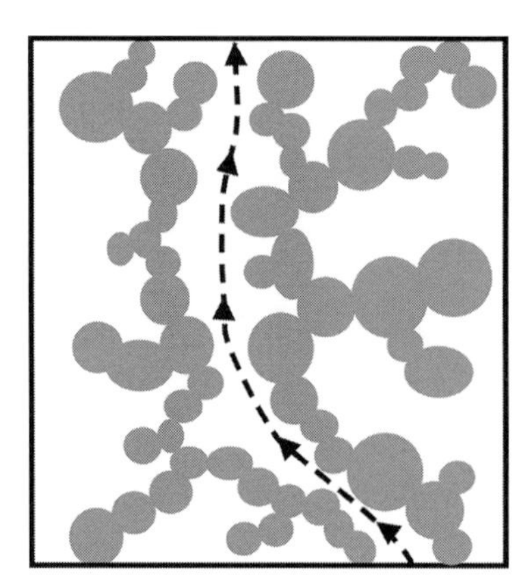

Figure 12.9 Scheme of the heat transport through the porous space by a gas inside an aerogel.

larger as is seen when compared the specific pore volume, Eq. (7.34) with specific pore volume from nitrogen desorption), an equation for the gaseous thermal conductivity is needed covering the range of large Knudsen numbers to small ones (conventional heat transport through a gas phase). Fricke and co-workers give an expression [78, 459, 466] as

$$\lambda_g = \frac{\lambda_{g0}\phi_p}{1 + \alpha Kn}. \tag{12.25}$$

Here λ_{g0} is the thermal conductivity of the gas in the pores not taking into account any Knudsen effect, α is a constant depending on the gas, $\alpha \approx 3$ for air. The last approximation stems from a detailed theoretical model [468, 469].

$$\alpha = \frac{5\pi}{16}\frac{9\gamma - 5}{\gamma + 1}, \tag{12.26}$$

with γ a factor depending on the type of gas

$$\gamma = \frac{2 + f}{f} \tag{12.27}$$

which is reflected in the degree of freedom f. For the diatomic molecules nitrogen and oxygen, $f = 5$ and thus $\gamma = 7/5$, taking only the translational and rotational degrees into account. This yields for α a value of approximately 3.1. The Knudsen number scales with the envelope density according to Hrubesh:

$$Kn = \frac{\ell_g}{d_p} \approx \ell_g \rho_e^{\omega}, \tag{12.28}$$

with ω a fit constant and ℓ_g the mean free path in the gas (the subscript g is used to differentiate it from the phonon mean free path). The effect of the Knudsen number on the gas phase thermal conductivity is illustrated in Figure 12.10. This figure clearly shows that a large Knudsen number reduces drastically the heat conduction via the gas phase, but also that even Knudsen numbers around two do already good. Using

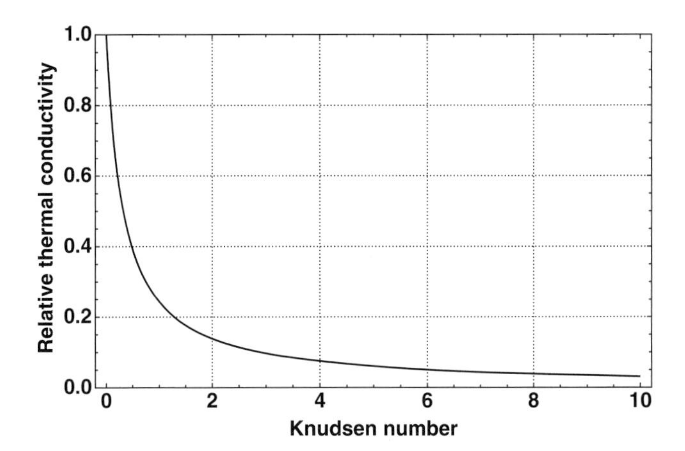

Figure 12.10 Effect of the Knudsen number on the heat conduction via the gas phase for diatomic molecules.

the expression for the density dependence of the pore size, one finally arrives at an equation for the gaseous conductivity in aerogels:

$$\lambda_g = \lambda_{g0} \frac{1 - \phi_s}{1 + 2C\ell_g \rho_e^{\omega}} \cong \lambda_{g0} \frac{1}{1 + 2C\ell_g \rho_e^{\omega}}. \tag{12.29}$$

The last relation holds since typically in aerogels $\phi_s = \rho_e/\rho_s \ll 1$. The constants C, ω vary from aerogel to aerogel; see [78, 466].

In this context, one can also make use of the numerical simulations of aerogel microstructures made by Halperin and co-workers [235] and the general expression for the mean free path in a porous material described in Chapter 9. From their simulations, they got that the inverse of the mean free path in the aerogel network or the pore size is directly proportional to the simulated aerogel's density in full agreement with the general relation of the mean free path in an arbitrarily structured porous body; see Eq. (9.9). In other words, we have that the Knudsen number varies like

$$Kn \propto \ell_g \rho_e. \tag{12.30}$$

This behaviour is similar to the assumption of Hrubesh fixing the exponent ω to a value of 1. Since the expression of Eq. (9.9) is generally valid, one can take the factor ω of Hrubesh's approach as a fudge factor. Experimental results on the thermal conductivity only by gas phase transport for RF aerogels prepared with different R/C ratios were given by Ebert [459]. Figure 12.11 shows the result. There is a clear decrease of thermal conductivity via the gas phase with envelope density. It also shows that different catalytic conditions, R/C ratio, yield a different porous microstructure. At small R/C ratios, a huge number of small particles exist and aggregate (or appear on stage in the solution via a polymerisation induced phase transformation), leading to smaller pores compared with larger R/C ratios.

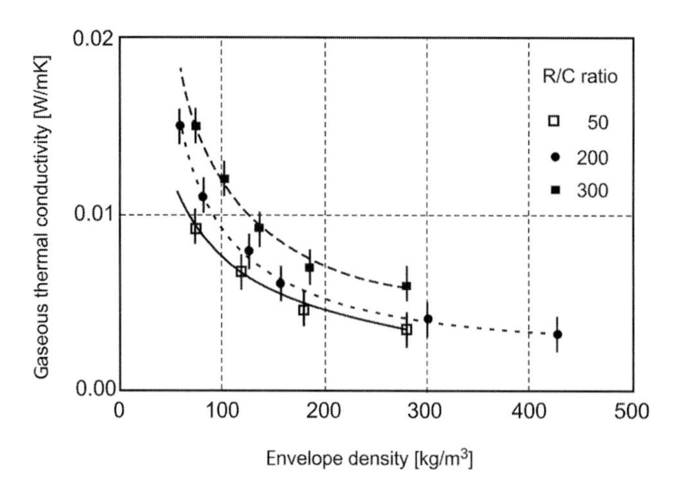

Figure 12.11 Variation of the thermal conductivity via the gas phase only, as determined experimentally in RF aerogels with envelope density measured in kg/m^3 after [459]. Reprinted with permission from Springer-Nature

The preceding description allows us also to take into account pressure and temperature. The thermal conductivity of a gas depends on temperature like [212, 443, 470]

$$\lambda_{g0} = \frac{75k_B}{64\pi a^2}\sqrt{\frac{\pi k_B T}{M_a}}, \tag{12.31}$$

where k_B is the Boltzmann constant in J/K and a is the diameter of the gas molecules in metres and M_a is the atomic or molecular mass; here we assumed a diatomic gas, such as nitrogen or oxygen. (Note: the thermal conductivity of a gas and its dynamic viscosity μ_g are related $\lambda_g = f\mu_g c_v$ with f a factor of the order of unity depending much on the theoretical model used to describe real gases and c_v is the specific heat per volume.) Interesting in this equation also is that the thermal conductivity (as the viscosity) does not depend on the gas density or the pressure [212, 443], a fact, that has already been derived by Maxwell 150 years ago and verified experimentally. Since the Knudsen number depends on the mean free path, we have to analyse its dependence on temperature. Kinetic theory of gases shows that the following relation holds:

$$\ell_g = \frac{k_B T}{\sqrt{2}\pi a^2 p}, \tag{12.32}$$

in which p is the pressure in Pascals and a is the diameter of the gas molecule assumed to be a hard sphere. We therefore get for the gas phase contribution in aerogels the expression

$$\lambda_g = \frac{\lambda_{g0}(T)}{1 + \alpha\frac{k_B T}{p\sqrt{2}\pi a^2 d_p}}. \tag{12.33}$$

This relation shows that at very low pressures the thermal conductivity goes linearly to zero and saturates at large pressure to the values of $\lambda_{g0}(T)$. At low temperatures, the thermal conductivity behaves like $\lambda_g \propto \sqrt{T}$ and does not depend on pressure. For high temperatures, the thermal conductivity goes hyperbolically to zero and depends inversely on pressure. This equation does not describe the whole regime of pressure dependence. Zheng and co-workers derived another expression [471], which was discussed and checked experimentally by Swimm and co-workers [472]:

$$\lambda_g = \frac{9\gamma - 5}{12}\sqrt{\frac{8M}{\pi R_g T}} c_V \frac{\phi_p}{\frac{S_{BET}\rho_e}{4\phi_p} + \frac{1}{\ell_g}} \frac{p}{\sqrt{T}}. \tag{12.34}$$

The pressure and temperature dependence are essentially similar to that given in Eq. (12.33). The thermal conductivity is generally not pressure dependent according to the kinetic theory of gases (and this was taken as one of the great breakthroughs for the kinetic theory of gases). The derivation leading to Eq. (12.31), however, makes several assumptions. First, at very low pressures the thermal conductivity must go to zero (vacuum). Second, when the mean free path is comparable to the container size, there is a modification also with pressure. Third, at high pressure the classical kinetic theory does not hold and the theory of dense fluids has to be applied, showing a dependence of thermal conductivity on pressure [470]. The effect of pressure on thermal conductivity has been discussed at length experimentally and theoretically for decades. Comings and Nathan [473] and Guildner [474] already investigated and analysed it with the use of the Chapman–Enskog model for non-ideal gases [212]. An early summary of the theoretical improvement of classical kinetic theory of gases was made by Kennard, Kleemann and Loeb et al. [475–477], showing that there can be a pressure dependence if the pressure is far off from normal pressures. Typically, therefore, one rewrites Eq. (12.33) as

$$\lambda_g = \frac{\lambda_{g0}(T, p)}{1 + \alpha \frac{k_B T}{p\sqrt{2}\pi a^2 d_p}} \tag{12.35}$$

if one would like to investigate the thermal conductivity at pressure above 1 MPa. Eq. (12.35) has been widely used for aerogels and beds of small particles, [459, 472, 478], and was used to fit experimental data successfully. Figure 12.12 shows the pressure dependence of Eq. (12.33).

There is another issue one should take into account. The pore size is not uniform, but there always is a spread in the size distribution. Let us calculate the effect of varying pore size or Knudsen number in a sample. For the sake of simplicity, we assume that all pores can be described as long straight rods with a diameter d_p and are either parallel or perpendicular to the heat flow. The pore size distribution will be described by a one-parameter Maxwellian size distribution:

$$f(d_p) = \frac{\sqrt{2}}{\sqrt{\pi \sigma^3}} d_p^2 \exp\left(-\frac{d_p^2}{2\sigma^2}\right). \tag{12.36}$$

This distribution has three advantages: first, the only parameter is the variance of the distribution, σ; second, it is right-sided, meaning there are no pores below zero size

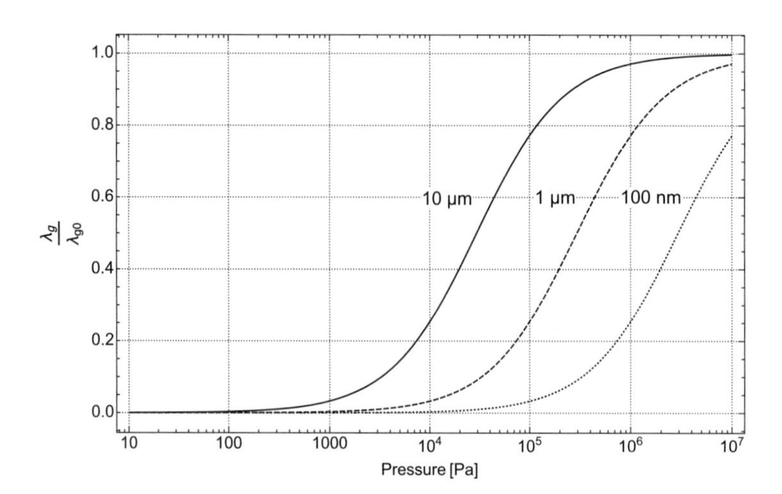

Figure 12.12 Pressure dependence of the relative thermal conductivity according to Eq. (12.33) at 300 K for three different pore sizes for a nitrogen gas. Interestingly, at lower pressures and thus larger mean free paths, the reduction in thermal conductivity is large even in pores of a few micron in size (as usual in polymeric foams).

(in contrast to, for instance, a Gaussian distribution); and third, the distribution is normalised, i.e., the integral over all sizes is one. The mean pore size $\overline{d_p}$ is defined by the variance alone:

$$\overline{d_p} = 2\sqrt{\frac{2}{\pi}}\sigma. \tag{12.37}$$

Calculating the Knudsen number under STP conditions leads to $Kn = 69/d_p$. The two situations, a parallel arrangement of pores or their arrangement perpendicular to the heat flow, are reflected in

$$\lambda_g^p = \int_0^\infty \lambda_g(d_p) f(d_p) dd_p \tag{12.38}$$

or

$$\lambda_g^s = \int_0^\infty \frac{1}{\lambda_g(d_p)} f(d_p) dd_p. \tag{12.39}$$

Although both integrations can be performed analytically, we omit the results here. We just state a few numerical examples. If the average pore diameter would be 50 nm, then most pores are in the range of 20–70 nm. For the parallel arrangement, one obtains an effective thermal conductivity of around 5 mW/mK. If the average pore size would be 100 nm, such that more than 50% of all pores are below 100 nm, the largest thermal transport is again given by the parallel arrangement and would have a value of 8 mW/mK. Both results show that not a few big pores determine the gas phase thermal transport, but the huge amount of small pores with large Knudsen numbers are important. Therefore, aerogels with a low thermal conductivity should possess a size spectrum with pores below 100 nm and the biggest number of pores should be below

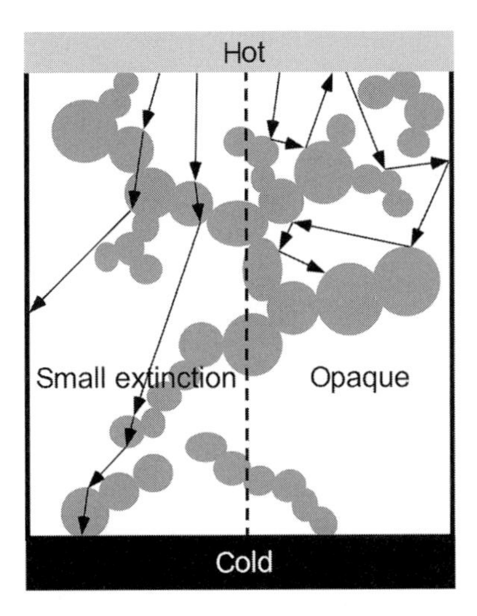

Figure 12.13 Scheme of the radiative heat transport through an aerogel.

the mean free path of air under STP conditions. If, for instance, one would take a silica aerogel and assume that the solid backbone has a thermal conductivity of 1.2 W/mK and at room temperature the radiation contribution would be around 1 mW/mK (see the next subsection), then the total conductivity of an aerogel would vary between 6–12 mW/mK in the porosity range of 90–99% for an average pore size of 50 nm. An increase to 100 nm makes a shift of 2.5 mW/mK. Therefore, it is important in aerogels to keep the largest pore size below 100 nm to have a super-insulating material.

12.3.3 Radiative Heat Transport

A more complicated and complex matter is the radiative heat transport in a porous body. Figure 12.13 shows a schematic of radiative heat transport in a porous body for two different situations: transmitting particles (small extinction) and reflecting, opaque particles. A detailed discussion can be found in [454]. Here we just adopt the simplified model derived by Caps and Fricke [479], which basically assumes that the photons are transmitted diffusively in the pore space of an aerogel. The amount of scattering, absorption or transmission is described by the Beer–Lambert law, which relates the absorption of light in a medium to the concentration of absorbers, the thickness of the material and a proportionality factor called the extinction coefficient. Electromagnetic wave theory, based on Maxwell's equation, shows that the extinction coefficient is inversely proportional to the refractive index of the absorbers, n. For silica aerogels, $n \cong 1$. They finally arrive at the following equation for the radiative contribution to the thermal conductivity:

$$\lambda_r = \frac{16 n^2 \sigma T_r^3}{3 \rho_e K_s / \rho_s}, \tag{12.40}$$

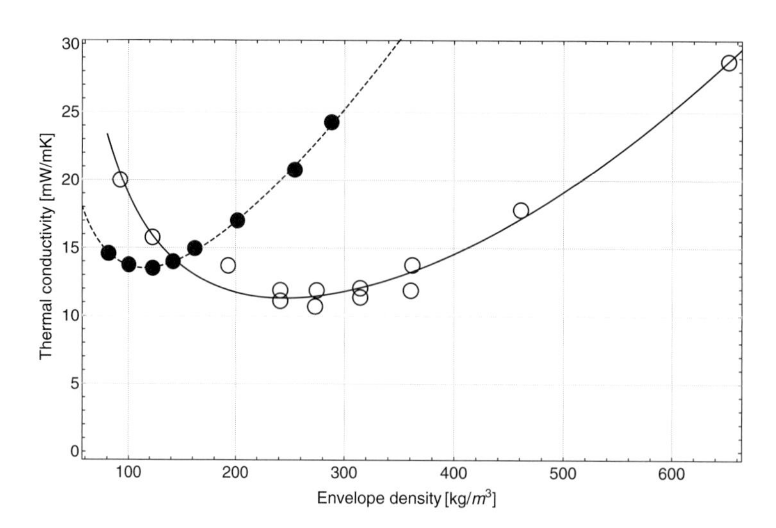

Figure 12.14 Total thermal conductivity of a silica (full circles) and RF aerogels (open circles) as it varies with envelope density measured in kg/m^3. Data extracted from Ebert [459] and Wong et al. [399]. Note the minimum occurs by silica in the density range of around 100–120 kg/m^3, whereas the minimum in RF aerogels is observed in the range of 250–270 kg/m^3.

with T_r the temperature of the aerogel. The specific extinction coefficient K_s/ρ_s was determined experimentally for many types of aerogels and is tabulated by Hrubesh and Pekkala [462]. One can use this equation to calculate the radiative contribution to heat transfer.

Adding all three contributions together leads to a curve with a minimum. In principle, this is to be expected: the solid-state thermal conductivity increases with increasing density, and the radiative should always decrease with increasing density and the gaseous too. Adding them up must give curve with a minimum. The exact location depends on many parameters, which are not very well known and difficult to measure precisely. Such a behaviour is always observed in aerogels. Figure 12.14 shows two measurements of an RF aerogel and a silica aerogel. The data points were extracted from publications by Ebert [459] and Wong et al. [399]. These experimental data suggest that aerogels exhibit a minimum of the thermal conductivity around 0.01 W/mK. Whether this minimum is an absolute one is open for investigation by better theoretical models and of course experiments [463].

12.4 Thermal Diffusivity

Aerogels not only are materials with a low thermal conductivity, but they also exhibit a low thermal diffusivity. Whereas the thermal conductivity is important for all insulation situations in which a stationary temperature profile is established, the thermal diffusivity determines the heating and cooling of a body, its time-dependent exchange of heat with the environment and thus is important in all applications where instationary temperature profiles prevail. Typical polymers have a thermal diffusivity in the range of 0.1–0.2 mm^2/s and oxides such as silica 0.9, glass wool around 0.58, cork 0.12 and concrete 0.54. Metals exhibit thermal diffusivities in the range from 3 to 160 mm^2/s, and even air has a thermal diffusivity of 20 mm^2/s. Thermal diffusivity is defined as

$$\kappa = \frac{\lambda_a}{\rho_e c_a}, \tag{12.41}$$

with λ_a the thermal conductivity of the aerogel, ρ_e its envelope density and c_a its specific heat capacity. The importance in any exchange of heat with an environment becomes clear looking at the characteristic time it takes to change the heat content of a body. Given a characteristic length of the body, like thickness or diameter as ℓ_c, the characteristic time τ is a multiple of

$$\tau \sim \frac{\ell_c^2}{\kappa}. \tag{12.42}$$

In this time, typically the temperature of a body has changed by a fraction 1/e, with e as Euler's number. The task to make at least a rough estimate of the thermal diffusivity of aerogels can be divided into two: first, modelling of the specific heat capacity; and second, modelling the thermal conductivity.

12.4.1 Specific Heat Capacity

The specific heat capacity in general is either the isobaric c_p or isochoric one c_V. Both are related by the following equation:

$$c_p = c_V(1 + \gamma_G \alpha T), \tag{12.43}$$

in which γ_G is the Grüneisen constant (typically with a value around 2) and α is the volumetric thermal expansion coefficient [107]. Both constants are related:

$$\gamma_G = \frac{V\alpha}{\chi_T c_V}, \tag{12.44}$$

with

$$\chi_T = -\frac{1}{V}\left(\frac{\partial V}{\partial p}\right)_T \tag{12.45}$$

as the isothermal compressibility. Eq. (12.43) shows that $c_p > c_V$. With this preliminary statement, let us try to calculate the specific heat of an aerogel c_a.

The specific heat capacity of an aerogel can first be written as a weighted sum of its components, the solid backbone and the air in the pores. Using that the sensible heat content in a material of mass m is given as $dH = mc_p dT$, we have at constant temperature the relation

$$m_a c_a = m_s c_s + m_L c_L \quad \text{with} \quad m_a = m_s + m_L, \tag{12.46}$$

in which c_s, c_L are the specific heat capacities of the solid backbone and air, m_a, m_s, m_L are the masses of the aerogel in total and its components. Denoting ρ_e, ρ_s, ρ_L

as the density of the components and dividing both sides by m_a using the general definition of density $m = \rho V$ and the rule of mixtures for any composite made of different phases $\rho_e = \rho_s\phi_s + \rho_L\phi_L$, and keeping in mind that $\phi_L = 1 - \phi_s$ holds for the volume fractions, we obtain

$$c_a = \frac{\rho_s}{\rho_e}\phi_s c_s + \frac{\rho_L}{\rho_e}(1-\phi_s)c_L \approx \frac{\rho_s}{\rho_e}\phi_s c_s = \frac{\rho_s\phi_s c_s}{\rho_s\phi_s + \rho_L\phi_L} \cong c_s, \tag{12.47}$$

since generally $\rho_L \ll \rho_s$ and $\rho_e \cong \rho_s\phi_s$[3]. Thus the specific heat of an aerogel is mainly determined by that of the solid backbone, which in itself is an astonishing result, since we would have expected a result like $\phi_s c_s$. We now have to specify in more detail the specific heat capacity. The isochoric specific heat capacity of solids $c_{s,V}$, indexed with an additional subscript V, is at high temperatures a constant given by the Dulong–Petit rule, expressed in units of kJ/(kg K) as

$$c_{s,V} = 3R_g/M_m, \tag{12.48}$$

with R_g the universal gas constant and M_m the molar weight. Insertion into Eq. (12.43) leads to an expression:

$$c_{s,p} = \frac{3R_g}{M_m}(1 + \gamma_G\alpha T). \tag{12.49}$$

Most often, the isobaric heat capacity $c_{s,p}$ is experimentally determined. This generally depends in a more complicated way on temperature as shown in Eq. (12.49). Most often it is described by a polynomial:[4]

$$c_{s,p}(T) = a + bT + cT^2 + \frac{d}{T^2}. \tag{12.50}$$

For many substances, the coefficients a, b, c, d are tabulated. Typically we have in polymers that the isobaric specific heat capacity increases almost linearly in the range from room temperature to 100°C. This linear behaviour also holds for RF aerogels on Pekala-type aerogels, shown in Figure 12.15. A fit to the data reveals that we can neglect the T^2 and the T^{-2} terms and write for the solid-state specific heat capacity the simplified expression

$$c_{s,p}(T) = c^0_{s,p} + bT. \tag{12.51}$$

According to Eq. (12.47), this also is the specific heat capacity of the aerogel.

Inserting Eq. (12.51) and Eqs. (12.18), (12.19), (12.29) and (12.40) into the definition of the thermal diffusivity in Eq. (12.41), we finally arrive at a simple expression

[3] Care most be taken. Cellulose aerogels can have an envelope density not much larger than air, a factor of 3–5, and then the density of air and its contribution to the specific heat capacity should be taken into account.

[4] Other representations are also possible and might be especially useful for certain temperature ranges, like the expression suggested by Horch [480], $c_{s,p} = 3R_g F(\theta_D/T) + bT + dT^3$. Here b, d are constants to be determined experimentally and $F(\theta_D/T)$ denotes the Debye function with θ_D the Debye temperature.

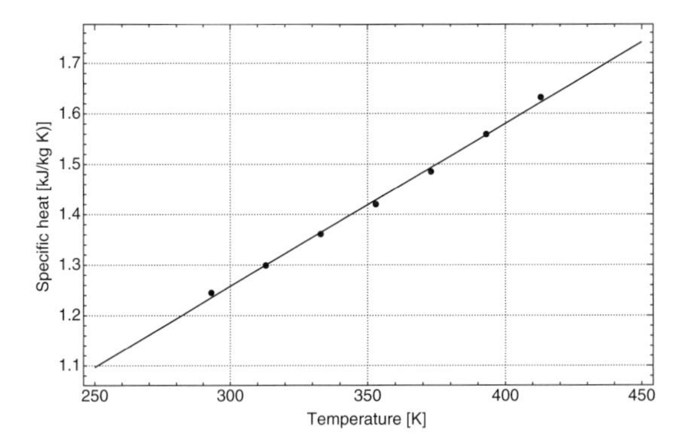

Figure 12.15 Specific heat capacity as a function of temperature for a Pekala aerogel.

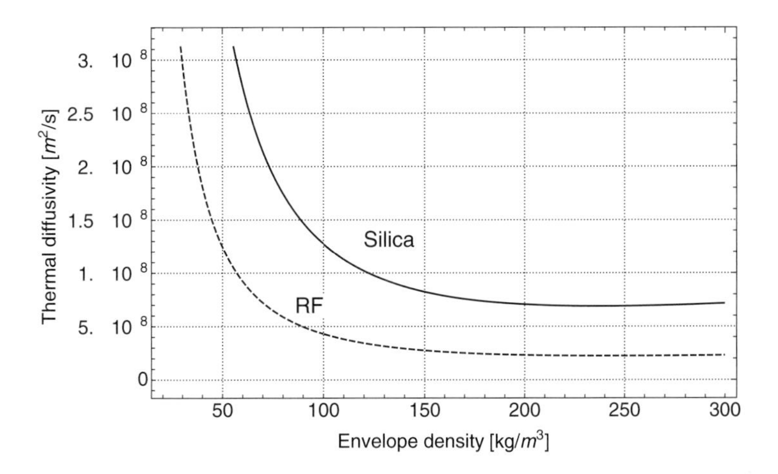

Figure 12.16 Thermal diffusivity as a function of envelope density for silica and RF aerogels.

for it. The dependence on density is much different. With increasing envelope density the thermal diffusivity goes to a constant value. The result is depicted in Figure 12.16. Note the values for typical aerogels above a density of 100 kg/m^3 are of the order of 10^{-7} m^2/s. For the characteristic time needed to relax a thermal field (see Eq. (12.42)), this means for a sample of, for instance, 1 cm diameter or thickness and a time of the order of 1000 s or 20 min. Apply this to, for instance, a tube, like an exhaust of a car or a long-distance heating pipe. Then this value means that isolating such a tube with an aerogel of 1 cm thickness shields the tube from temperature fluctuations outside. For exhaust tubes in cars, an aerogel isolation would be excellent to keep the temperature high in the exhaust. Especially in urban traffic, where combustion motors are never really in a steady state, the long characteristic time would help to transport the hot gas from the engine at a high temperature to the catalyst converter, which greatly improves its efficiency.

12.5 Measurement Techniques of Thermal Properties of Aerogels

From the short sketch of the fundamentals of heat conduction given in the previous section, two principally different methods to measure the thermal conductivity can be derived:

- Stationary methods, based on Eq. (12.1), or stationary solutions of Eq. (12.7)
- In-stationary methods, based on solutions of Eq. (12.7)

These then fall into many different types of experimental realisations. The most often used stationary ones are the following:

- Hot-plate methods, guarded and unguarded
- Hot-wire methods

The in-stationary ones are the following:

- Transient plane source (TPS) methods (circular and rectangular disks, one-sided and two-sided samples or sensors)
- Hot-wire methods (HWM) with alternating electrical heating
- Hot Bridge Method (HBM)

In the following paragraphs, we give a sketch of the most important methods used for aerogels too.

12.5.1 Stationary Measurement Methods

All stationary methods in principle take a sample, put on one side a heater and place on the other side a suitable heat sink. Both have to be regulated to establish a stationary and linear temperature field in a sample. Just this last point is the tricky one (and in reality any experimenter is faced with many other problems). There are two main geometries used: the plate method (one- or two-sided) and the cylindrical method with a heated wire in the centre and a cold wall at the outside. In any case, to make a good measurement, one has to determine accurately the heat flux $\dot{q}$ through the system by measuring the electrical power needed to keep the temperature at the hot side constant while it is of course constant at the cold side. The electrical measurement can be done with great accuracy. The second issue is to measure the temperature gradient $(T_{hot} - T_{cold})/h$. Here one can use the temperature of the hot and cold plates (wire or wall) and has of course to accurately measure the slab thickness h or hollow cylinder thickness. From these values, one could yield all data to get the thermal conductivity for a sample of area A from

$$\lambda = \frac{\dot{q}h}{A(T_{hot} - T_{cold})}. \tag{12.52}$$

Steady-state measurement methods can provide accurate and reliable results; however, they have the disadvantage of being time consuming. Achieving the stationary state can take considerable time; the characteristic time is $\tau_c = h^2/\kappa$ (see Eq. (12.7)). Let the thermal diffusivity be around 1 mm^2/s, $h = 100$ mm, then the characteristic time is 10,000 seconds. This means within 2.5 h the profile of being initially a unit step function inside the sample changes a bit (see Figure 12.1). The temperature in the centre, for instance, is around 1/3 of its final stationary value. It takes at least a factor of 10 more to establish stationary conditions. This means around one day. Aerogels can have a thermal diffusivity of around 0.01 mm^2/s and then all times become a factor of 100 larger or the sample thickness has to be reduced by, for instance, a factor of 100.

Guarded Hot Plate (GHP)

The guarded hot plate method is a commonly used method for determining the thermal conductivity of isolating materials. There are two different types of guarded hot plate instruments available: one-sided and two-sided. Figure 12.17 shows a schematic drawing of a two-sided GHP apparatus. Details for this standardised facility can be found on webpages of many manufacturers and in all necessary details are described in the ASTM C177 and ISO standard, how such devices have to be built, calibrated and operated [481, 482]). Figure 12.17 illustrates the main components of the system: two isothermal cold surface units and a guarded hot plate, being composed of a metred section thermally isolated from a concentric primary guard by a gap. Sometimes more than one guard is used. The two test specimen pieces are sandwiched between these three units. Thus a measurement produces a result that is the average of the two pieces, and therefore the two pieces should be closely identical. The guarded hot plate provides the power (heat flow per unit time) for the measurement (Q = voltage times current) and defines that portion of the specimen that is actually being measured. The function of the primary guard is to provide the proper thermal conditions within the test volume to reduce lateral heat flow within the apparatus. The three heating/cooling units create isothermal surfaces on the faces of the specimens at least below the central hot plate. The two cold surfaces are adjusted to the same temperature for the double-sided mode operation (typically a thermostat unit is used, water-cooled for instance) and the guard heater temperature does not deviate from that of the central heater by more than 0.2 K. There are additional heat fluxes in the lateral direction, since the outside temperature of the whole facility will typically be lower than the central part of the sample. To minimise these, a secondary guard is wrapped around the sample (rectangular box of heaters or a cylindrical tube with heaters attached).

Its temperature is usually taken as the average between the hot and cold temperature. The experimental method requires establishing steady-state conditions and measuring the heat flow Q in the metred section, the metred section area A and the temperature gradient across the specimen, in terms of the temperature T_{hot} of the hot surface and the temperature T_{cold} of the cold surface. The measurement requires electrical measurement of the flux Q, temperature measurement of the hot and cold plate and precise measurement of the sample thickness (not trivial for flexible materials such as even glass fibre batts or rock wool). In many cases, additional temperature sensors

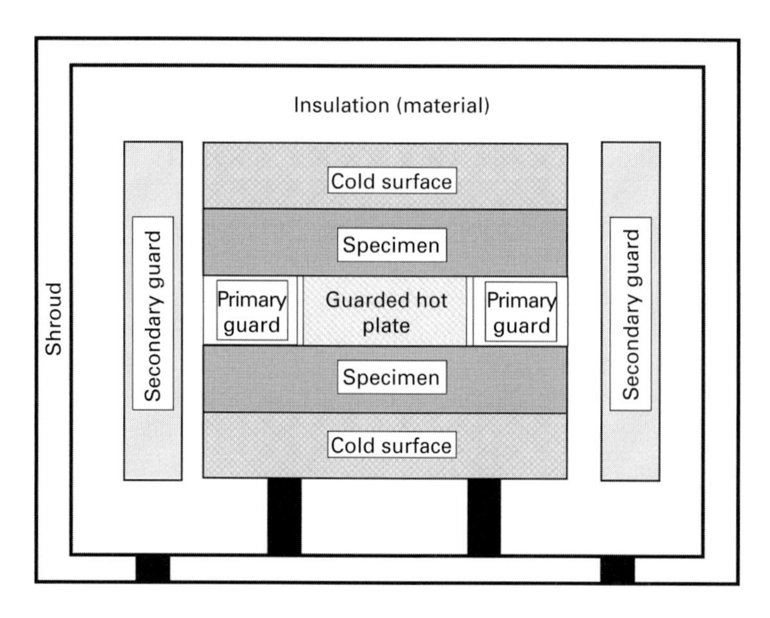

Figure 12.17 Sketch of the two-sided guarded hot plate method according to the ASTM norm [481].

are built in to control the guards, measure external temperature at the shroud etc. This allows one to calculate heat currents that may lead to errors. As temperature sensors, anything is possible that fits the temperature range of interest: Pt-resistors, thermocouples etc.

Finally, the thermal conductivity is determined using Eq. (12.52) with a modification of the area, now taken as

$$A = A_m + A_g/2, \tag{12.53}$$

with A_m, A_g, the area of the hot plate and the guard heater respectively. Today such machines, being electronically much improved and usually also numerically modelled to have a good impression of the quality of the temperature field in the sample, are readily available for sample size from 150×150 mm^2 to 500×500 mm^2, even with variable thickness and varying load on a specimen, to investigate the effect of compressibility and its effect on the thermophysical properties. Machines are available from various companies. There have been, however, many improvements made on the conventional setup described in the ASTM and the ISO norm. Dubois and Lebeau suggested a one-sided guarded hot plate using for the measurement of the heat flux a thermoelectric heat flux-metre [483]. A different approach is used by Jannot and co-workers [484]. They designed a new two-sided method, shown in Figure 12.18. They modelled the heat flow in their facility to ensure that super-isolating materials can really be measured. The essential issue of this facility are two aluminium blocks being held at constant temperature, two equal samples of small thickness (3–6 mm), a heater element being of exactly the same size as the sample and being thin (0.2 mm). The temperatures of the heater and the Al block are measured after heating with a

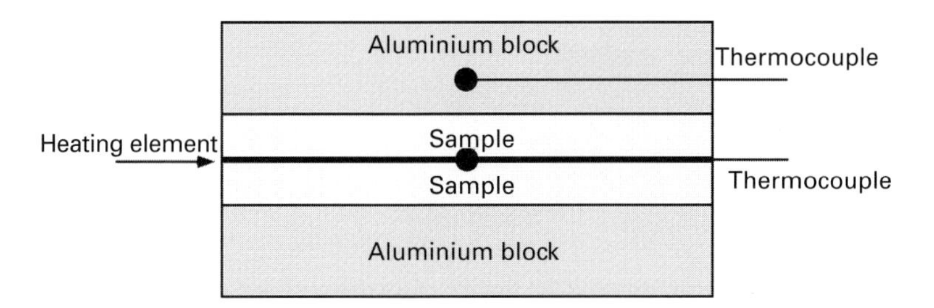

Figure 12.18 Sketch of the new two-sided method designed especially to measure materials with low-thermal conductivity. The two aluminium blocks serve to keep the temperature at the cold side constant, and the thin heating element between the two sample parts is used to establish a temperature gradient. The temperature difference between the centre of the sample and the aluminium blocks is measured as well as the power to maintain it. From the stationary-state temperatures and the geometry of the device, the thermal conductivity can be calculated.

suitable heating power and the temperature change as a function of time of both thermocouples measured and compared to both a 1D model and a 3D model. This allows one to determine the thermal conductivity once the stationary state is reached and the ρc_p value from the in-stationary solution. The accuracy of this method allows determination of thermal conductivities down to 0.01 W/mK.

There are other apparatuses available: plate arrangements with a cold and hot side, measuring the heat flux and therefrom determining the thermal conductivity. Axial or cylindrical arrangements, as briefly mentioned earlier, are especially suitable for measuring powders or compact beds of particles [485].

12.5.2 In-Stationary Measurement Methods

The advancement of electronics and the transformation from analogue to digital data acquisition and processing in the last two decades allowed researchers to realise in-stationary methods yielding the thermal conductivity, the diffusivity and the volumetric specific heat within a short measurement time, simple setups and small samples. We will briefly explain the different approaches available today. All in-stationary methods rely on solutions of Fourier's second equation, Eq. (12.7). Without going too much into detail (for greater detail, see, for instance, Carslaw and Jaeger [440]), one of the fundamental solutions of Eq. (12.7) in one-dimensional Cartesian coordinates reads

$$T(x,t) = \frac{1}{\sqrt{t}} \exp\left(-\frac{x^2}{4\kappa t}\right). \tag{12.54}$$

From this point source solution, it follows that another function also solves the heat conduction equation, namely

$$T(x,t) = T_c 2\sqrt{\kappa} \int_0^{x/2\sqrt{\kappa t}} \exp(-\xi^2) d\xi = T_c \sqrt{\pi\kappa} \operatorname{erf}\left(-\frac{x}{2\sqrt{\kappa t}}\right), \tag{12.55}$$

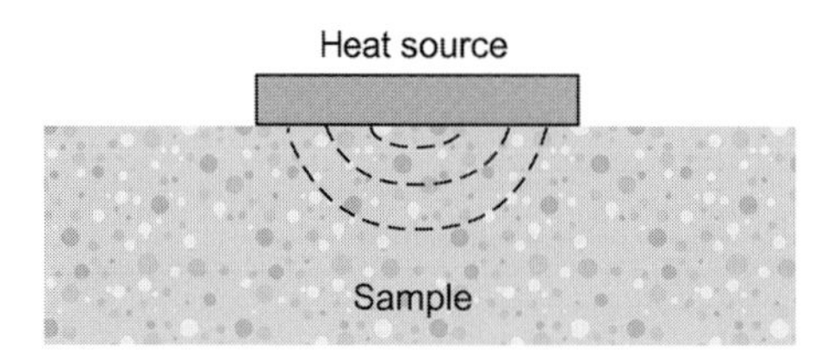

Figure 12.19 Plane heat source at the surface of a sample giving a heat pulse or wave into the sample.

which also defines the so-called error function erf, and where T_c is an arbitrary constant. The basis of most in-stationary measurement techniques is a situation described by Carslaw and Jaeger [440] by an additional heat flux at a surface (putting a heat source at the surface of a body), in which this heat flux f is determined by the heat conduction in the underlying solid:

$$f = -\lambda \frac{\partial T}{\partial x}. \tag{12.56}$$

A sketch of this idea is shown in Figure 12.19. This flux satisfies the heat conduction equation

$$\kappa \frac{\partial^2 f}{\partial x^2} = \frac{\partial f}{\partial t}. \tag{12.57}$$

Let us assume that $f = F_0$ =constant for $x = 0, t > 0$, then the solution is simply given by (check by insertion)

$$f = F_0 \operatorname{erfc}\left(\frac{x}{2\sqrt{\kappa t}}\right) \tag{12.58}$$

noting erfc $= 1 -$ erf. Inserting into Eq. (12.56) and integrating with respect to x gives the temperature in the underlying specimen as (after some rearrangements with the error function):

$$T(x,t) = \frac{2F_0}{\lambda}\left[\left(\frac{\kappa t}{\pi}\right)^{1/2} - \frac{x}{2}\operatorname{erfc}\left(\frac{x}{2\sqrt{\kappa t}}\right)\right]. \tag{12.59}$$

If the flux at the surface is time-dependent $f = f(t)$, the solution is given using Duhamel's theorem as

$$T(x,t) = \frac{\kappa^{1/2}}{\lambda \pi^{1/2}} \int_0^t \frac{1}{\sqrt{\tau}} f(t-\tau) \exp\left(-\frac{x^2}{4\kappa\tau}\right) d\tau. \tag{12.60}$$

This solution shows that within one measurement for a given heat pulse $f(t)$ at the surface, one can determine the thermal conductivity and diffusivity and thus also the volumetric specific heat and, if the density of the specimen is known, the mass specific heat capacity.

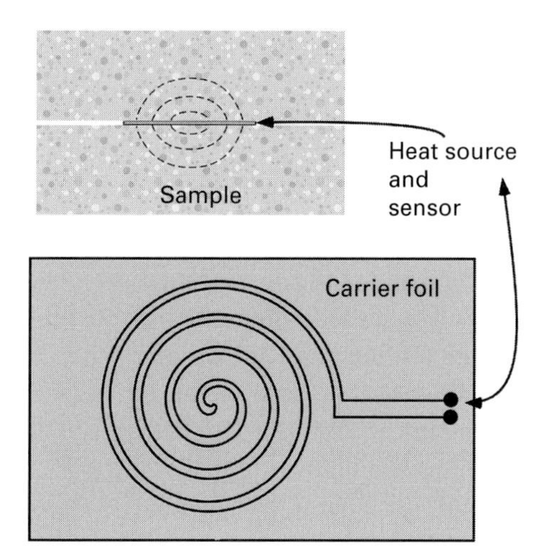

Figure 12.20 Schematic of the measurement geometry of a Hot-Disk^{TM}sensor at the bottom and the experimental setup above. A meander or spiral of a metallic wire is deposited on a carrier foil (Kapton, ceramic) and used for sending a short heat pulse and as a resistance sensor to the heat wave going into the sample on top and below the sensor.

There are three different methods currently available on the market as commercial devices: a transient plane source method, using a disk-like plane source, TPS, a hot bridge method, THB, and variants of hot wire methods. A completely different in-stationary method is the Laser Flash method, briefly discussed at the end.

The Transient Plane Source (TPS) Method

At the KTH in Stockholm a so-called TPS technique was developed in the last two decades which is designed for convenient thermal conductivity and thermal diffusivity measurements on various sample types [486–489]. The apparatus is available from HotDisk^{TM}. A sensor in the shape of a meandered spiral is applied to a sample. The double spiral of electrically conducting nickel (Ni) metal is sputtered on a thin foil and sandwiched between two layers of Kapton. Although only 0.013–0.025 mm thick, the Kapton provides both electrical insulation from the sample and mechanical stability for the probe. This sensor, shown at the bottom of Figure 12.20, is sandwiched between two identical pieces of the sample or applied only to one sample side. The sensor has a well-known temperature coefficient of resistivity, and is used simultaneously as a resistance heater and a resistance thermometer. Basically, a stepwise heating of the sensor is applied, which results in a transient temperature increase of the sample beneath the sensor and the sensor itself. By recording the temperature increase of the sensor element, typically of the order 1 K, the thermal transport properties of the sample can be estimated. The sensor is electrically part of a so-called Wheatstone bridge. The pulse heating of the meander (short characteristic time) and its much slower response due to the thermal response of the sample (longer characteristic time) brings this bridge out of balance. This is recorded with a sensible voltmeter (voltages in the range of microvolts and less). The temperature distribution originating from the plane source is modelled with a source solution (see Eq. (12.60)) and transferred into a resistance and voltage change in the Wheatstone bridge.

The method has its advantages and disadvantages: it is fast, taking just a few minutes to get a good result, if the thermal conductivity of the sample is not too low (say larger than 0.1 W/mK). There is not much sample preparation, besides the surface should be planar and smooth. The sensor is put on its surface, the other sample piece on it and a small pressure applied to increase the thermal contact. The technical disadvantages are that the sensor is only one piece of the bridge and thus prone to all errors of such a bridge measurement; different wire length and electromotive forces (emfs) due to soldering and contacts between different materials might affect the results negatively. (Emfs are voltages induced by any kind of electrical energy. Many sources are possible for that, for instance, contact potentials between materials of different kinds.) Often it is, however, possible to design the experimental setup so that one can measure and eliminate at least the direct influence of any thermal contact resistance between the sensor probe and the surface of the solid sample material. An effective way is to set up the measurement device so that the temperature response of the sensor, ΔT, is the sum of two clearly separable components, one component which is essentially constant, A, representing the total thermal contact resistance influence, and the second component a transient component, B, not influenced by any thermal contact resistance:

$$\Delta T = A + B(P_0/\lambda\ell)f(\tau_c). \tag{12.61}$$

In Eq. (12.61), constant B incorporates the bulk thermal conductivity λ of the sample, the heating power P_0 and a characteristic length dimension ℓ of the probe. The dimensionless time function f can be expressed in terms of the dimensionless time $\tau_c = \sqrt{(t-t_c)/\theta}$, where $\theta = \ell^2/\kappa$ represents the characteristic time of the measurement and t_c a time correction (heat pulse generation). The function f depends on the probe design and size, as well as the sample geometry. It is, however, not accessible to the experimenter, and the evaluation of the signal response of the Wheatstone bridge has to be judged by the experimenter. To measure very good insulators requires a special evaluation software, since any heat flow, even in the Kapton foil of the sensor, may lead to thermal short circuits below the foil. A slow response of the material and very small voltages in the bridge make the measurement difficult. Great care and experience are necessary to chosen proper parameters. A crosscheck with other measurement methods and calibration with known materials are necessary to become familiar with the measurement procedures or to introduce correcting factors. Nevertheless, it is possible with such a device to measure thermal conductivities down to 0.02 W/mK. A one-sided TPS facility is available for isolating materials and thermal conductivities $\geqslant$0.03 W/m and needs a small sample of a few 10 mm diameter. It is placed onto the sensor (similar to a HotDisk sensor) and pressed onto it with a small weight and the pulse heating started.

Figure 12.21 Top view of the transient hot bridge sensor layout. Each of the four tandem strips is 100 mm in length and 2 mm in width. Solder pads are at A, B, C, D: A–D, current source; B–C, digital voltage metre measuring the resistance change due to heat loss into the sample parts.

Transient Hot Bridge (THB) Method

The so-called transient hot bridge method was recently developed by the Physikalisch-Technische Bundesanstalt (PTB) in Braunschweig, Germany, to overcome certain problems of the TPS and the transient hot wire method described shortly [490]. The principle is the same as in the TPS system: a metallic wire is sputtered onto a polyimide substrate, which also works as a electrical insulating material. They do not use a meandering, but use thin strip elements as resistors, shown in Figure 12.21. The layout of the sensor consists of four tandem strips in parallel. Each tandem strip comes in two individual strips, a short and a long one. Two of the tandems are located very close to each other at the centre of the sensor and one additional tandem on either edge. All eight strips are symmetrically switched for an equal-resistance Wheatstone bridge. At uniform temperature, the bridge is initially balanced. An electric current makes the pairwise unequally spaced strips to establish a predefined inhomogeneous temperature profile that turns the bridge into an unbalanced condition. The sensor produces an almost offset-free output signal of high sensitivity. This voltage rise in time is a measure of the thermal conductivity, λ , thermal diffusivity, κ, and the volumetric specific heat, ρc_p, of the surrounding specimen. The signal is virtually free of thermal emfs because no external bridge resistors are needed. This is the essential difference from other transient methods. All resistors of the bridge are on the sensor and only the asymmetric heating induces an unbalancing of the bridge and a voltage. The method requires as the TPS method two pieces of material. The sensor is sandwiched between them. There are special sensors available that can cut directly into a soft material, improving the thermal contact.

Hammerschmdit and Meier [490] show that the voltage of the bridge is directly proportional to the time-dependent temperature evolution of the temperature field in the sample. They developed an analytical model based on a strip source solution (integration of the solution given earlier and using that the strip lays on the surface of a semi-infinite solid; see [440], Eq. §10.5(3)) and compared it with a numerical modelling, which showed that the analytical model works quite well. They especially discuss how even a small sample has to be isolated from the environment (isothermal or adiabatic isolation) to get reliable results. Their experimental investigation of Rockwool^{TM}samples and other low conducting materials showed that this could be an interesting method yielding quickly accurate and reproducible results also for aerogels. As with the TPS method, no big samples are required and since several sensor on polyimide, ceramic etc. are available, the device can be used also at higher temperatures. A commercial apparatus under license of the PTB is available.

Hot Wire and 3-Omega Method

As mentioned previously, the hot wire method is also a transient technique, and as it name already states, a wire is brought into close contact with a specimen and heated with a heat pulse (in-stationary or transient) and the response of the sample on the wire resistance recorded. The technique is in principle easy to apply, but requires the resolution of low changes in resistance or, since the wire is always a part of a bridge arrangement, the resolution of nanovolts. Although this is not a real issue, low voltages are prone to many errors or erratic measurements. They are not easy to handle, and the layout of such a facility is anything but simple. In good experimental conditions, an accuracy of less than 5% can be achieved for conductivity measurements. There are a lot of special thermal leak currents to be dealt with to achieve this accuracy. A new measurement method was recently developed by Dalton et al. [491]. Originally, the 3-omega technique was developed to measure the thermal diffusivity of metal filaments used in incandescent light bulbs [492]. The first reported use of the 3ω method to measure the thermal conductivity of solids was by Cahill and Pohl in 1987 [493]. A microfabricated metal line deposited on the specimen acts as a heater and thermometre. When an alternating current (AC) voltage signal is used to excite the heater at a frequency ω, the periodic heating generates oscillations in the electrical resistance of the metal line at the heater frequency of 2ω. Since voltage is given as the product of the current and the resistance of the wire, $V_0 = R_0 I_0$, multiplication of the sinusoidal current at the input frequency ω by the heater resistance fluctuations at 2ω results in a sum and a difference voltage signal at the first and third harmonics, which are used to infer the magnitude of the temperature oscillations [492]. Comparing the input root mean square (RMS) power per unit heater length to the frequency-dependent voltage oscillations, the amplitude and phase can be analysed to obtain the thermal conductivity and diffusivity of the specimen [493]. Cahill's and Dalton's model calculations show that the voltage amplitude depends inversely proportional on the thermal conductivity, and there is a logarithmic term that depends on the thermal diffusivity. Evaluating the third harmonic part of the voltage oscillations then allows us to determine both parameters. Dalton et al. showed recently [491] that indeed super-insulating materials can be measured with this setup.

Laser Flash Method

In the last decade, Laser Flash methods to measure thermophysical properties became popular, since they promise to measure thermal conductivities and other thermophysical properties up to 2800°C. The principle idea of Laser Flash is simple: a suitable laser beam is shot onto the surface of a sample and the heat pulse generated in the sample detected on the backside with an infrared sensor. The heat wave penetration of such a point heating and its time-dependent behaviour can be analysed, in principle with the source solution of Fourier's equation given earlier, and then of course the thermal diffusivity is determined. If from an independent experiment the specific heat is known the thermal conductivity can be calculated using the known envelope density. They have been used, especially to measure the thermal properties of carbon aerogels [494–497].

12.6 Application of Aerogels to Insulation Tasks

In this section, we calculate in detail the insulation of a plate by an insulation layer and that of a tube as two examples of general importance for aerogel applications. The insulation by a plate is important for all applications of aerogels in buildings, and the tube is important for insulation of long-distance heating pipes, car engine exhausts, oil pipelines and all tubes transporting hot gases or fluids. An excellent review of aerogel super-insulation was recently published by Koebel et al. [40].

12.6.1 Transfer of Heat from a Plate into the Surrounding Medium

The starting point of the analysis is Eq. (12.7), which we have to solve. In a building situation, we have a warm room being, let us say, at 25°C and let the outside be cool as it may be in winter, let us say −10°C. If we assume that the temperature in the room is constant (meaning our heaters have enough power to keep it constant), we can model that situation as if we have a hot plate at constant temperature and on top of it an insulation material, here an aerogel. In order to solve the heat conduction equation, we calculate the surface temperature of the aerogel plate, which will be in contact with a surrounding medium (air at rest or in convective motion, like a cold wind). The situation is depicted in Figure 12.22. The aerogel insulation will have a thickness of l and at the contact side of the wall it will have a constant surface temperature (imagine this as a brick wall with plaster on it). The heat conduction equation then has to be solved with the boundary condition:

$$T(0,t) = T_s. \tag{12.62}$$

The aerogel insulation layer is cooled with a so-called Newtonian cooling (for example, a sum of radiation and diffusive or convective cooling) leading to the boundary condition on the aerogel side:

$$-\lambda \left.\frac{\partial T}{\partial x}\right|_{x=l} = h(T - T_0). \tag{12.63}$$

Here T_0 is the temperature far away from the interface and of course will be constant, and h is the heat transfer coefficient having the unit W/m^{2}K. We will calculate the

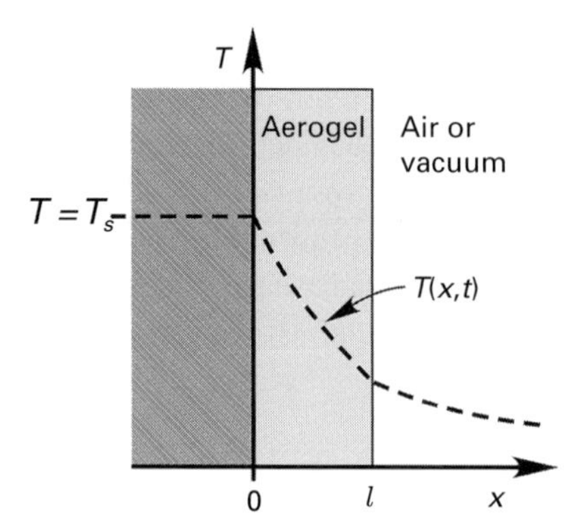

Figure 12.22 An aerogel plate of thickness l will have intimate contact to a warm source at a constant temperature T_S. At the cold side, heat is transported away from the aerogel by Newtonian cooling.

surface temperature on the cold side of the aerogel, at $T(l,t)$, for various thicknesses using a simple analytical model. In addition to the boundary conditions, we need an initial condition. Here we simply assume that initially the temperature inside the sample is constant:

$$T(x,0) = T_0. \tag{12.64}$$

12.6.2 The Solution

The solution of this problem was already obtained by Carslaw and Jaeger in their seminal book on heat conduction [440]. They, however, further simplify the boundary conditions. They assume that a region $0 < x < l$ has initially zero temperature and the surface $x = 0$ is kept at T_s and radiation will take place at $x = l$ into a medium (Newton cooling) at zero. These conditions differ from the real ones written earlier. We therefore correct the boundary conditions by introducing a new scale for the temperature:

$$T' = T - T_0. \tag{12.65}$$

We then have

$$T'(0,t) = T_s - T_0 \tag{12.66}$$

and

$$-\lambda \left. \frac{\partial T'}{\partial x} \right|_{x=l} = -hT' \tag{12.67}$$

and of course as the initial condition

$$T'(x,0) = 0. \tag{12.68}$$

Now the conditions of the solution given by Carslaw and Jaeger fit exactly and we can take their solution. This reads, retransformed to the real-world temperature scale:

$$T(x,t) = T_0 + (T_s - T_0)\frac{1 + \mathrm{Nu}(1 - x/l)}{1 + \mathrm{Nu}} - \sum_{n=1}^{\infty} \frac{2(\beta_n^2 + \mathrm{Nu}^2)\sin(\beta_n x/l)}{\beta_n(\mathrm{Nu} + \mathrm{Nu}^2 + \beta_n^2)} e^{-\frac{\beta_n^2 \kappa}{l^2} t}. \tag{12.69}$$

Here we introduced a new dimensionless number, the Nusselt number Nu[5].

$$\mathrm{Nu} = \frac{hl}{\lambda}, \tag{12.70}$$

which measures the ratio of heat transfer by conduction, given by λ to the actual heat transfer, given, for instance, by air at rest or convective heat transfer through a gas or a fluid. The coefficients β_n are solutions of the non-algebraic equation

$$\beta_n \cot \beta_n + \mathrm{Nu} = 0. \tag{12.71}$$

This equation has to be solved numerically. There are an infinite number of non-identical solutions of this equation.

12.6.3 Stationary Case

The meaning of the Nusselt number becomes clearer if we look at the boundary condition at the surface $x = l$ in more detail and solve the same problem for the stationary case, meaning after infinite time. In this problem, we have two interfaces, the aerogel–warm wall at a temperature $T_s - T_0$ located at $x = 0$, and the aerogel–air interface located at $x = l$. At the latter interface, we have the boundary condition

$$-\lambda \left.\frac{\partial T}{\partial x}\right|_{x=l} = h(T_l - T_0). \tag{12.72}$$

In this equation, T_l denotes the surface temperature $T(x = l) = T_l$ and T_0 is the temperature far away from the interface. The temperature T inside the aerogel plate is a function of position and time as stated in Eq. (12.69). In the stationary case at $x = l$ and $t \to \infty$, the heat energy current on the left side balances exactly the right side independent of time, and we can solve Eq. (12.72) by replacing the gradient by $\frac{\partial T}{\partial x} = -(T_s - T_l)/l$. The surface temperature then reads

$$T_l = T_s \frac{1 + \mathrm{Nu}\frac{T_0}{T_s}}{1 + \mathrm{Nu}}. \tag{12.73}$$

This relation can be interpreted as follows: if there is no heat transfer Nu $= 0$ at the aerogel–air interface, then in the stationary case the surface temperature at the plate

[5] In their book, Carslaw and Jaeger use another expression for this, since they typically express the heat transfer coefficient h as a value h/λ and then define a new dimensionless symbol $L = lh$. To avoid any misunderstanding we take the German way of naming it and then h is the heat transfer coefficient as given previously, and L is then the Nusselt number.

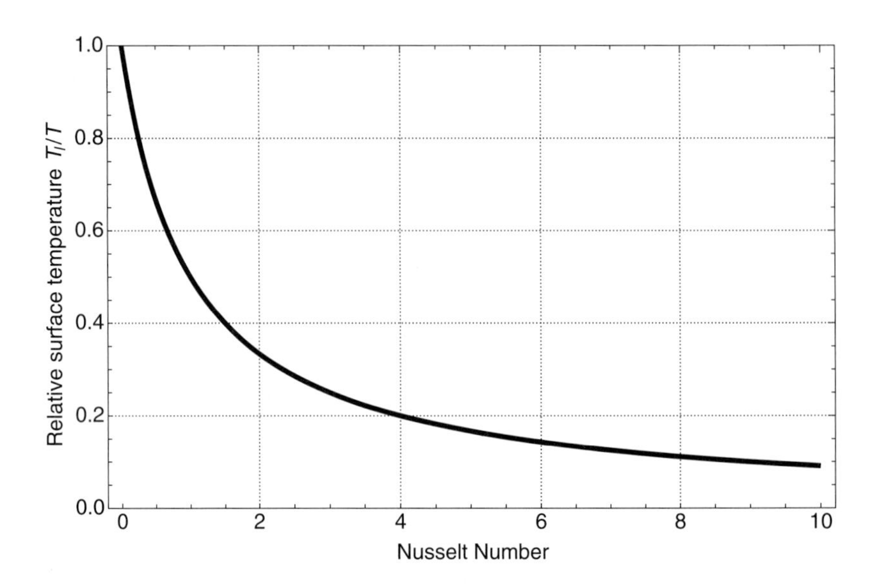

Figure 12.23 Relative surface temperature T_l/T_s as a function of Nusselt number Nu. Typical values for aerogel insulations are in the range of $1 < \mathrm{Nu} < 10$.

surface $x = l$ would be equal to that of the warm side $T_l = T_s$. If the heat transfer at the aerogel–air interface is infinite, $\mathrm{Nu} \to \infty$, the surface temperature would always be $T_l = T_0$, just a perfect insulation or a perfect heat transport away from the interface. As already mentioned, the Nusselt number characterises the heat transport through the plate in relation to the transport away from its surface. The case $T_0 = 0$ is that case discussed by Carslaw and Jaeger in their solution given earlier. This simplifies the expression Eq. (12.73) to

$$T_l = \frac{T_s}{1 + \mathrm{Nu}}. \tag{12.74}$$

Figure 12.23 shows the ratio T_l/T_s as a function of Nusselt number. Typical values for aerogels with a thermal conductivity of 0.01–0.03 and a heat transfer coefficient of around 10 would give Nusselt numbers ranging from 1 to 10. The lower the thermal conductivity of the aerogel plate, the higher the Nusselt number can be, and thus the lower the surface temperature on the aerogel side. From Eq. (12.73), it is clear that the surface temperature of the cold side always is smaller than that at the hot side. It might also be worth noteing that the thickness l_c needed to achieve a certain ratio between the temperatures at both sides depends only on the ratio λ/l:

$$l_c = \frac{\lambda}{h}\left(\frac{T_s - T_l}{T_l - T_0}\right). \tag{12.75}$$

One might conclude that under given conditions of heat transport on the aerogel side, characterised by l, the thickness needed for a given insulation ratio $\theta_{sl} = (T_s - T_l)/(T_l - T_0)$ depends linearly on the aerogel conductivity! This confirms the necessity to reduce in passive insulation situations the heat conductivity of the isolator material used. Let us calculate an example: the hot side is at 500°C, the cold side at 50°C, the heat transfer coefficient is that of air at rest, $h = 10$ W/m^{2}K and the outside air temperature is 20°C. Let the aerogel have a thermal conductivity of 0.02 W/mK. Then the thickness needed to achieve the insulation from 500 to 50°C is 30 mm. If the aerogel has a thermal conductivity of around 0.01, the thickness would be 15 mm. If the cold side of the aerogel should only be down to 100°C, a thickness of 10 or 5 mm is sufficient.

The Nusselt number is not a simple constant in typical insulation situations. Let us say the air is at rest, then the Nusselt number is low, whereas in a windy situation it is much higher and in a stormy situation even more. The Nusselt number generally depends on the Reynolds number and the Prandtl number.

The Reynolds number characterises the fluid flow (fluid = gas or liquid). It is defined as

$$Re = \frac{ul_c}{\nu}, \tag{12.76}$$

in which u is the flow velocity, l_c a characteristic length (for instance, the length or height of a building) and ν the kinematic viscosity. Let us calculate a value, using the kinematic viscosity of air being around $13 \cdot 10^{-6}$ m^2/s, and let the wall to be cooled by a flow from one side have a length of 10 m. Then for a wind of 10 knots ($\approx$5 m/s), we would have a Reynolds number of already $3.8 \cdot 10^6$. The Prandtl number is the ratio of kinematic viscosity to thermal diffusivity $Pr = \nu/\kappa$ and thus characterises the heat transport by flow in relation to heat transport by conduction. For air, the Prandtl number is approximately 0.72. The Nusselt number can be related to the Reynolds and Prandtl number in a power-law form

$$\mathrm{Nu} = 0.664 Re^{1/2} Pr^{1/3}. \tag{12.77}$$

This relation allows us to calculate heat transfer coefficient for any windy situations around a house.

12.6.4 Instationary Case

The time-dependent surface temperature on the cold side can be derived from Eq. (12.69) as

$$T(l,t) = T_0 + \frac{T_s - T_0}{1+\mathrm{Nu}} \left(1 + \sum_{n=1}^{\infty} \frac{2\mathrm{Nu}(\mathrm{Nu}+1)\sec\beta_n}{\mathrm{Nu}(\mathrm{Nu}+1)+\beta_n^2} \exp\left(-\frac{\beta_n^2 \kappa}{l^2} t \right) \right). \tag{12.78}$$

One can immediately see that the stationary solution of Eq. (12.74) is identical to this solution if we take $t \to \infty$. We use a Mathematica^{TM}notebook to look into solutions of this equation (especially solving the transcendental equation for the coefficients β_n). Before we do so, we have to evaluate some physical constants (see Table 12.1).

As mentioned previously, the heat transfer coefficient varies over orders of magnitude, depending especially on the heat transport by flow. For an illustration, we take a

Table 12.1 Properties of aerogels

Property	Symbol	Smallest value	Largest value	Unit
Thermal conductivity	λ	0.005	0.03	W/(mK)
Specific heat	c_p	415		J/(kg K)
Thermal diffusivity	κ	1	7	10^{-7} m^2/s
Density	ρ	30	300	kg/m^3
Heat transfer coefficient	h	1	100	W/m^{2}K

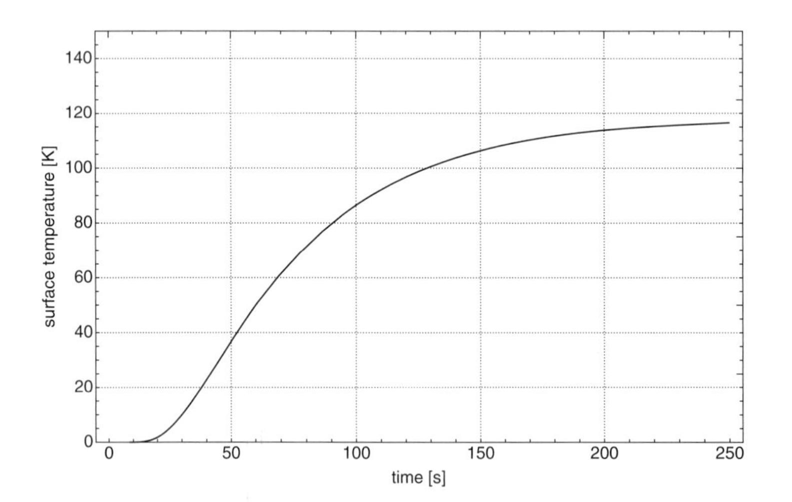

Figure 12.24 Surface temperature at the cold side of the aerogel as a function of time for $\lambda = 0.01$ W/mK and h $= 10$ W/m^{2}K. At the hot side the temperature is kept constant at $T_s = 1300$ K and we have set the initial temperature to zero, $T_0 = 0$ for convenience.

value of 10. The density of the aerogels varies and with it the thermal conductivity. We simply take a density of 100 kg/m^3. Figure 12.24 shows the time-dependent variation of the surface temperature at the interface aerogel–air for an assumed thermal conductivity of 0.01 W/mK and a heat transfer coefficient of 10 W/m^{2}K, which is a value for still air. The temperature reaches after some time the value given by Eq. (12.74). The characteristic or relaxation time can be estimated from the exponential part in Eq. (12.78), namely

$$\tau_n = \frac{l^2}{\beta_n^2 \kappa} = \frac{\rho c_p l^2}{\beta_n^2 \lambda}. \tag{12.79}$$

From all these times, we take just the first one of the series expansion with a β value of $\beta_1 = 2.65366$:

$$\tau = \frac{\rho c_p l^2}{\beta_1^2 \lambda} \cong \frac{\rho c_p l^2}{7.042\lambda}. \tag{12.80}$$

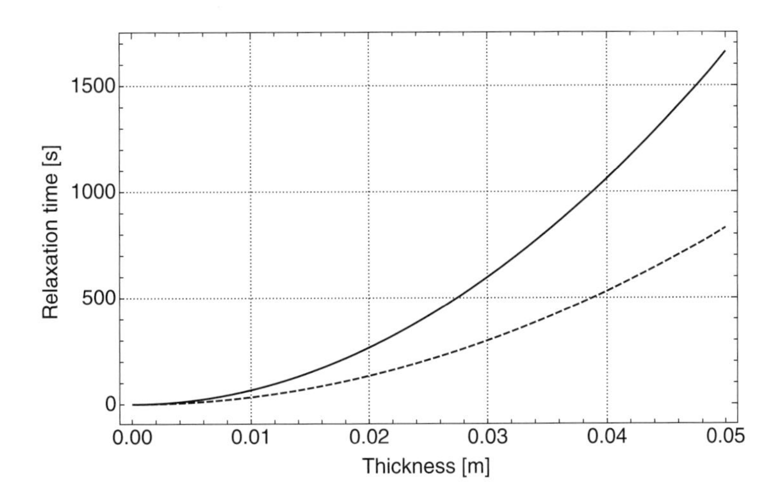

Figure 12.25 Relaxation time for the surface temperature in an aerogel plate as a function of its thickness for two different thermal conductivities. The solid line is for $\lambda = 0.01$ and the dashed line for $\lambda = 0.02$ W/mK.

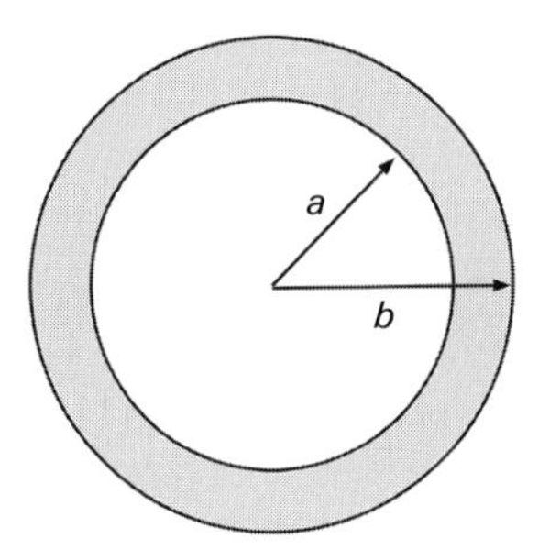

Figure 12.26 Sketch of the heat conduction through an insulation wrapped around a tube of inner radius $r = a$ having a thickness $l = b - a$.

Figure 12.25 shows the characteristic time for two different thermal conductivities as a function of aerogel thickness. As to be expected, the larger the thickness and the lower the thermal conductivity, the larger the time for the surface temperature to reach its stationary value.

12.6.5 Heat Transfer from a Tube through an Insulation

The heat conduction through a hollow cylinder manufactured from an aerogel material is in a certain sense similar to that of a plate, as long as the cylinder is taken as infinite, so that neither a heat loss nor heat transfer through the front faces occurs. Such a situation is sketched in Figure 12.26. The basic equation to be solved is Eq. (12.7) and is repeated here for convenience in cylindrical coordinates:

$$\frac{\partial T}{\partial t} = \kappa \left(\frac{\partial^2 T}{\partial r^2} + \frac{1}{r} \frac{\partial T}{\partial r} \right). \tag{12.81}$$

The boundary conditions are quite similar as in Eqs. (12.62–12.64). We first start with the stationary solution. The time-dependent solution is presented in Appendix E.

12.6.6 Stationary Solution for a Cylindrical Shell

For the stationary solution, we set the left side of Eq. (12.81) zero and obtain

$$\frac{1}{r}\frac{\partial}{\partial r}\left(r\frac{\partial T}{\partial r}\right) = 0. \tag{12.82}$$

This equation has the general solution

$$T(r) = C_1 + C_2 \ln r. \tag{12.83}$$

From the boundary conditions

$$T(a) = T_s \tag{12.84}$$

$$-\lambda \left.\frac{\partial T}{\partial r}\right|_{r=b} = h(T(b) - T_0), \tag{12.85}$$

we obtain the two coefficients C_1, C_2 as

$$C_1 = \frac{T_s + \mathrm{Nu}(T_s \ln b - T_0 \ln a)}{1 + \mathrm{Nu}\ln(b/a)} \tag{12.86}$$

$$C_2 = -\mathrm{Nu}\frac{T_s - T_0}{1 + \mathrm{Nu}\ln(b/a)} \tag{12.87}$$

with the Nusselt number

$$\mathrm{Nu} = \frac{hb}{\lambda}. \tag{12.88}$$

The stationary temperature field reads finally

$$T_{stat}(r) = \frac{T_s(1 - \mathrm{Nu}\ln(r/b) + \mathrm{Nu}\frac{T_0}{T_s}\ln(r/a)}{1 + \mathrm{Nu}\ln(b/a)}. \tag{12.89}$$

It is easy to prove that this solution fulfils the boundary conditions. The temperature at the aerogel side to the surrounding medium, $r = b$, is

$$T(b) = T_s\frac{1 + \mathrm{Nu}\frac{T_0}{T_s}\ln b/a}{1 + \mathrm{Nu}\ln(b/a)}. \tag{12.90}$$

Figure 12.27 shows how this temperature depends on the Nusselt number for some values of b/a and $T_s = 1300$ K and $T_0 = 300$ K. Naming $T(b) = T_b$ and introducing the thickness as $l = b - a$ allows us to obtain the relative thickness of the insulation needed to get a certain temperature ratio between inner and outer temperature as

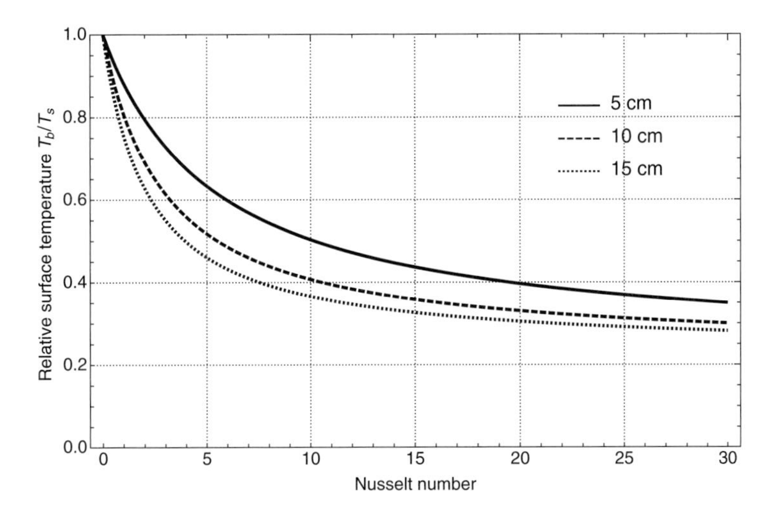

Figure 12.27 Stationary temperature at the cold side of the aerogel wrapped around a tube with diameter $a = 25$ mm. The aerogel has a thickness of 5, 10 and 15 cm. The variable here is the Nusselt number. The inner side of the tube is at $T_s = 1300$ K, and the cold side at $T_0 = 300$ K; see Eq. (12.90).

$$\frac{l}{a} = \exp\left(\frac{1}{\mathrm{Nu}}\frac{T_s - T_b}{T_b - T_0}\right) - 1 = \exp\left(\frac{\theta}{\mathrm{Nu}}\right) - 1. \tag{12.91}$$

This again shows that the larger the heat transfer at the aerogel side, the smaller the aerogel thickness has to be for a given ratio $\theta = \frac{T_s - T_b}{T_b - T_0}$. It also shows that the smaller the thermal conductivity of the aerogel or generally the insulation material, the smaller the insulation layer must be. In contrast to the plate solution, here we have an exponential dependency. Only in the case that the argument of the exponential function is small compared to unity can one expand the e-function and obtain again – as for plates – a linear relation. However, keep in mind that the Nusselt number also depends on the thickness, by its definition. A full solution for the thickness reads

$$l = \frac{\theta\lambda - ahW\left(\frac{\theta\lambda}{ah}\right)}{hW\left(\frac{\theta\lambda}{ah}\right)}, \tag{12.92}$$

in which $W(x)$ denotes the Lambert W-function, here taking the principle value [498]. Using Eq. (12.90) and rearranging it, we can use the measured stationary temperatures to obtain the Nusselt number as

$$\mathrm{Nu}_{exp} = \frac{1}{\ln(b/a)}\frac{T_s - T_b}{T_b - T_0}. \tag{12.93}$$

This relation is important for the time-dependent temperature at the aerogel surface treated in the Appendix E.

12.7 Variables and Symbols

Table 12.2 lists many of the variables and symbols used in this chapter.

Table 12.2 Variables and symbols.

Symbol	Meaning	Definition	Units
a	Diameter of a gas molecule	–	–
A	Cross section of a sample	–	m^2
c_p	Specific heat	–	J/(kg K)
c_V	Specific heat	–	$J/(m^3\ K)$
$\overline{d_p}$	Average pore diameter	–	m
E	Young's modulus	–	Pa
E_0	Young's modulus of a pore free material	–	Pa
h	Height or thickness of a sample	–	m
$\vec{j}$	Heat current density	–	W/m^2
k_B	Boltzmann coefficient	1.38×10^{-23}	J/K
H	Volumetric enthalpy	–	$J\ /m^3$
Kn	Knudsen number	$\ell_g/2R$	–
K_S	Specific extinction coefficient	–	–
l_c	Critical thickness of an insulation layer	–	–
M_a	Atomic mass	–	kg/mol
Nu	Nusselt number	hl/λ	–
Pr	Prandtl number	ν/κ	–
p_g	Gas pressure in an aerogel	–	Pa
R_g	Universal gas constant	8.314	J/(mol K)
$\overline{d_p}$	Average pore size	–	m
T	Temperature	–	K
$T(x,t)$	One-dimensional temperature field	–	K
T_{cold}	Temperature at the cold side of a sample	–	K
T_{hot}	Temperature at the hot side of a sample	–	K
T_s	Temperature at the warm side of a sample	–	K
T_0	Outside temperature	–	K
T_r	Temperature of an aerogel	–	K
$v(T)$	Average phonon velocity	–	m/s
v_s	Sound velocity	–	m/s
v_l	Longitudinal sound velocity	–	m/s
χ_T	Isothermal compressibility	–	–
ℓ	Phonon mean free path	–	m
ℓ_g	Mean free path in a gas	–	m
γ_G	Grüneisen constant	–	–
κ	Thermal diffusivity	$\lambda/(\rho c_p)$	m^2/s
λ	Thermal conductivity	–	W/mK

Table 12.2 Continued

Symbol	Meaning	Definition	Units
λ_s	Solid state thermal conductivity	–	W/mK
λ_s^{aer}	Solid state thermal conductivity for an aerogel	–	W/mK
λ_g	Gas phase thermal conductivity	–	W/mK
λ_{g0}	Thermal conductivity of a gas	–	W/mK
λ_r	Radiative thermal conductivity	–	W/mK
λ_{eff}	Effective thermal conductivity	–	W/mK
λ_{low}^{HS}	Lower Hashin-Shtrikman bound thermal conductivity	–	W/mK
λ_{up}^{HS}	Upper Hashin-Shtrikman bound thermal conductivity	–	W/mK
ν	Poisson ratio	–	–
ν	Kinematic viscosity	μ/ρ	–
ϕ_1, ϕ_2	Volume fraction of phase 1 and 2	–	–
ϕ_p	Pore volume fraction	–	–
ϕ_s	Solid volume fraction	–	–
ρ	Density	–	kg/m^3
ρ_e	Envelope density	–	kg/m^3
ρ_s	Skeletal density	–	kg/m^3
σ	Variance of a distribution	–	–
τ	Characteristic time	h^2/κ	–

13 Mechanical Properties of Aerogels

Numerous daily life materials exhibit a porous structure, e.g., foams made from different polymers (polystyrene, polyurethane), clays, tiles, bricks, oxide ceramics, bones, sponges, wood or diatoms. In many cases, the mechanical properties can be described by simple scaling laws, with the relative or envelope density being the decisive factor. It is generally agreed that similar scaling laws apply to aerogels and xerogels, but the special nanostructured nature of aerogels and the mode by which they are formed out of a solution of monomers or polymers make a difference. A detailed discussion of the conventional approach for closed and open cell foams or honeycombs can be found in the standard textbook of Lorna J. Gibson and Michael F. Ashby [1]. Before we discuss the specialty of aerogels, let us revisit some textbook knowledge about bending and compression of bars, since this is essential for a discussion of mechanical properties of porous materials and aerogels especially.

13.1 Mechanical Testing: A Brief Introduction

We will not derive all the basic equations of linear and non-linear elasticity theory or mechanical testing procedure. The aim of this brief review of mechanical testing is just to give to the reader the important equations (and where possible a short derivation) to understand the aerogel literature if they talk about mechanical properties. An excellent book on mechanical testing is the book *Mechanical Metallurgy* by Dieter [499].

13.1.1 Bending

The three-point bending test is a standard mechanical test procedure which is especially useful for materials that are brittle, such as concrete, ceramics and certain porous materials. For a bending test, typically rectangular bars with a prismatic cross section are used. Figure 13.1 shows schematically such a bending test.

An experiment is performed by putting a sample on two supports separated by a distance ℓ and to position a test indenter in the centre of the bending bar to just touch the sample surface, which defines its reference position (zero). Then the indenter is moved downwards with a preset constant velocity and the force F needed to bend the bar is recorded. The test is finished once the sample is broken. The final output is the force as a function of distance moved by the pistil.

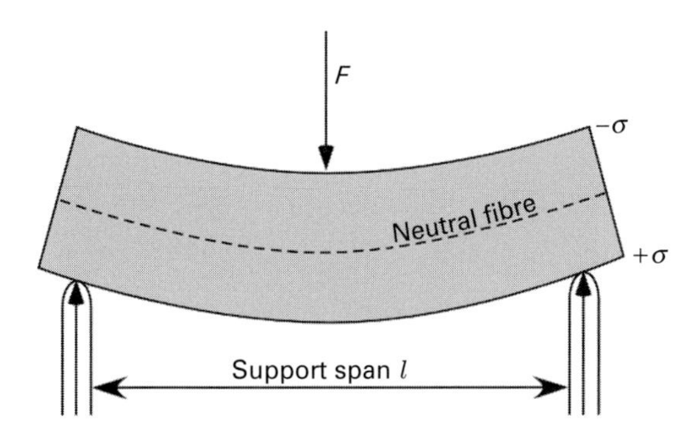

Figure 13.1 Scheme of a bending test. A prismatic bar is loaded with a force F, while it is supported by beam having a curved knife of fixed radius. The support beams span a distance ℓ. The force leads to a bending, which on one side induces tensions in the beam and compressive stresses on the other side.

During a bending test, the material is inhomogeneously stressed. The sample is in a state of compression on the side of the indenter and on the opposite side under tension. Exactly the centreline is at zero stress (neutral plane). The stress distribution is linear in elastically deformed samples.

To evaluate the force-displacement curve $F - d$ of a bending test, we first calculate the bending moment M from the applied force F and the lever arm (= half distance of the support span $\ell/2$). We take as the coordinate origin the position of the indenter. Then the moment induced by the central force F varies linearly from the indenter to the supports:

$$M(x) = F(\ell/2 - |x|) \quad -\ell/2 \leqslant x \leqslant \ell/2. \tag{13.1}$$

The bending moment is maximum in the centre:

$$M_{\text{max}} = \frac{F\ell}{2} \tag{13.2}$$

The bending moment is balanced by the resistance moment or section modulus of the bar W, which is defined by second moment of area I and half the sample thickness. The second moment of area of a bar with width b and thickness h (cross section $A = bh$) is

$$I = \int_A y^2 dA = \int_{-h/2}^{h/2} y^2 b dy = \frac{bh^3}{12} \tag{13.3}$$

and thus the section modulus is

$$W = \frac{I}{h/2} = \frac{bh^2}{6}. \tag{13.4}$$

The stress distribution in the bar will be linear (y will be the coordinate vertical to the x-direction):

$$\sigma(y) = \frac{M}{I} y. \tag{13.5}$$

This again is maximal in the centre, just under the indenter, and reads

$$\sigma(x = \ell/2) = \frac{M_{\max} h}{2I} \tag{13.6}$$

or

$$\sigma_R = \frac{M}{W} = \frac{3F\ell}{bh^2}. \tag{13.7}$$

The compression stress on the side of the applied force has just the opposite sign $-\sigma_R$. The largest deflection of the bar is exactly under the indenter and can be calculated as

$$\delta = \frac{F\ell^3}{48EI} = \frac{F}{4Eb}\left(\frac{\ell}{h}\right)^3. \tag{13.8}$$

The slope m of the deflection force curve determines the elastic or Young's modulus of the material:

$$E = \frac{4bm}{\left(\frac{\ell}{h}\right)^3}. \tag{13.9}$$

Often aerogels are prepared in cylindrical shape and thus a bending test would be performed with a bar of circular cross section. The second moment of inertia for a cylindrical bar with diameter D is

$$I_{circle} = \frac{\pi D^4}{64} \tag{13.10}$$

and the section modulus is

$$W_{circle} = \frac{\pi}{32} D^3. \tag{13.11}$$

The maximum stress is now concentrated in a line and given by

$$\sigma_R = \frac{M}{W} = \frac{16F\ell}{\pi D^3}. \tag{13.12}$$

Let us extend this discussion to larger deformations or larger strains, meaning we extend it to non-linear deformations. To do so, we start with the definition of the bending line. Take a look at the schematic in Figure 13.2, which shows a bending bar deflected such that the curvature of the neutral fibre (dashed line) has a radius ρ. Take a section of arc length ds and move a distance y into the stressed zone. Then the arc length increases by an amount Δds. We then have the simple relation

$$\frac{ds}{\rho} = \frac{ds + \Delta ds}{\rho + y}, \tag{13.13}$$

which can be transformed to

$$\frac{ds + \Delta ds}{ds} = 1 + \frac{y}{\rho}. \tag{13.14}$$

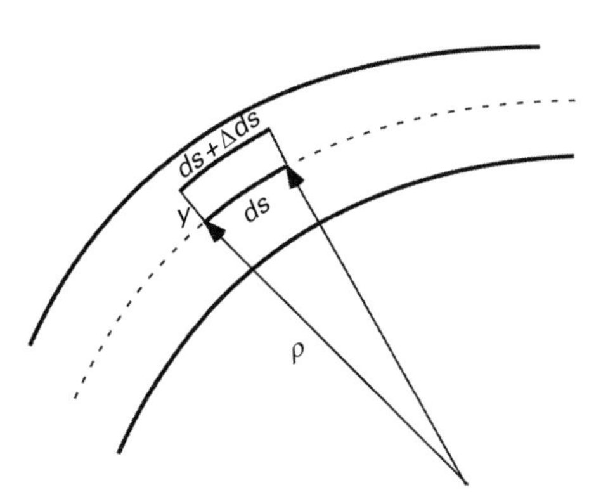

Figure 13.2 Section of a beam such that the neutral fibre is bent by a radius ρ.

By definition the strain is

$$\epsilon_b = \frac{\Delta ds}{ds} = \frac{y}{\rho} = \frac{\sigma}{E} = \frac{M}{IE}y, \tag{13.15}$$

where we have used Eq. (13.5). From this we get that the bending line radius is given as

$$\rho = \frac{EI}{M}. \tag{13.16}$$

To obtain a differential equation for the bending line, let us express the radius of curvature in cartesian coordinates as shown schematically in Figure 13.3. Looking at the figure, we read the following relations:

$$\sigma_R = M/W = \frac{16F\ell}{\pi D^3} \tag{13.17}$$

$$ds = \sqrt{dx^2 + dy^2} = dx\sqrt{1 + y'^2} \tag{13.18}$$

and

$$y'(x) = \frac{dy}{dx} = \tan\alpha(x). \tag{13.19}$$

Taking on the last equation, the second derivative with respect to x leads to

$$\frac{d^2y}{dx^2} = \frac{d\tan\alpha}{d\alpha}\frac{d\alpha}{dx}. \tag{13.20}$$

Using $d\tan\alpha/d\alpha = 1/\cos^2\alpha$ and the identity $1/\cos^2\alpha = 1 + \tan^2\alpha$, we obtain

$$\frac{d^2y}{dx^2} = (1 + y'^2)\frac{d\alpha}{dx}. \tag{13.21}$$

The radius of curvature is defined as

$$\rho = \frac{ds}{d\alpha} = \sqrt{1 + y'^2}\frac{d\alpha}{dx}. \tag{13.22}$$

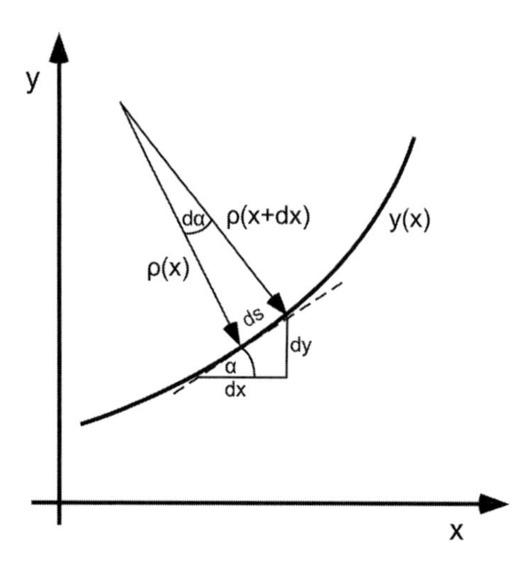

Figure 13.3 On an arbitrary curve with a function given by $y(x)$ choose a point and make from one point a radius $\rho(x)$. At that point the curve has a slope defined by the tangent at that point, dy/dx. If we go an arc length ds further along the curve the radius will have changed to a value $\rho(x + dx)$ by an angle $d\alpha$.

Thus we finally arrive at

$$\frac{1}{\rho} = \frac{y''}{(1 + y'^2)^{3/2}} = \frac{M}{E \cdot I}. \tag{13.23}$$

Only in the case $y' \ll 1$ do we have a simple second-order differential equation for the shape of the beam, otherwise, the equation looks uglier and solutions always involve so-called elliptical integrals (for a discussion, see the classical article by Bisshopp and Drucker on non-linear deformation of beams [500]). We will use Eq. (13.23) later to model the deformation of fibrils or string of pearls in aerogels.

13.1.2 Compression Test

The compression test is used to determine mechanical properties under homogeneous uniaxial loads by recording a stress–strain curve. For a compression test, cylindrical samples are used with a diameter D equal to the height h or prismatic beams with an edge width of w and a height of $h = w$. Other ratios of the diameter or edge width to height are possible, but care has to be taken at large values of h/D because samples may fail through bending or buckling (typically h/d or $h/w \leqslant 2$ are preferred). The sample with an initial cross-sectional area S_0 is fixed between two parallel plates. The compression is raised slowly but steadily charging the sample with an uniaxial load. The corresponding force F is measured while the compression velocity v is kept constant. The compression test is schematically shown in Figure 13.4.

The nominal compression stress σ is given by

$$\sigma = \frac{F}{S_0}. \tag{13.24}$$

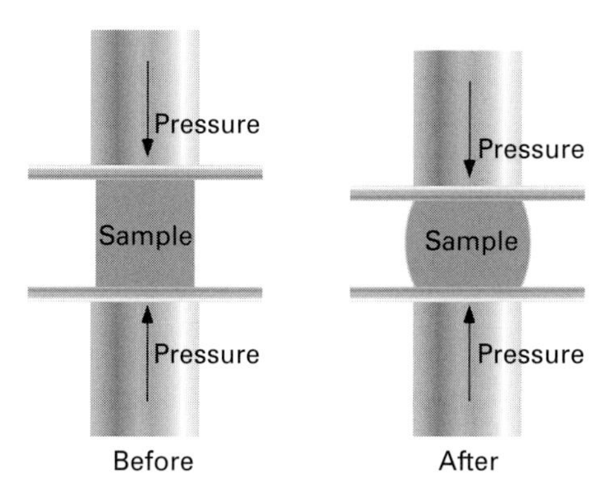

Figure 13.4 Schematic drawing of the compression test. A cylindrical or cubic specimen is fixed between two plates (left), and the upper plate is moved such that a constant compressive strain rate is achieved. The body typically deforms into a barrel due to friction at the plates (right).

In a good setup, the strain or the relative compression ϵ is given by the change of sample length ΔL_d in relation to the initial sample length L_0:

$$\epsilon = \frac{\Delta L_d}{L_0}. \tag{13.25}$$

This strain is measured independently of the force, for instance, by extensometres applied to the sample. Only then a plot of stress against strain can be evaluated to yield the interesting material parameters such as yield stress, fracture strength and Young's modulus. If such a device is not available, one has to assure that the relative compression velocity v/L_0 multiplied with the experiment time t is indeed equal to the reduction in sample length. Typically $\epsilon \neq 1 - v \cdot t/L_0$, since the load cell recording the force required to deform the sample has its own spring constant and also deforms under load, and thus the deformation of the sample is smaller than the calculated value $v \cdot t$. One can nevertheless then detect the fracture stress as a sudden drop in force and even correct the apparent elastic modulus to the true one, if the machine stiffness is known. Figure 13.5 shows schematically a load-compression curve of a porous material. The curve exhibits four regions:

- Compressive response due to irregularities. All curves start with only minor force increase, having often a more or less parabolic shape: sample, sample support and upper plate match in this phase to each other.
- Linear elastic regime. After the initial regime, almost always a linear portion can be measured which characterises the apparent elastic deformation regime: stress and strain are linearly related.[1]

[1] Although in this range stress and strain are linearly related, it is questionable whether in a porous body we can call this a Hookian regime. It is maybe an effective Young's modulus. The linear relation characterises the compliance of the material.

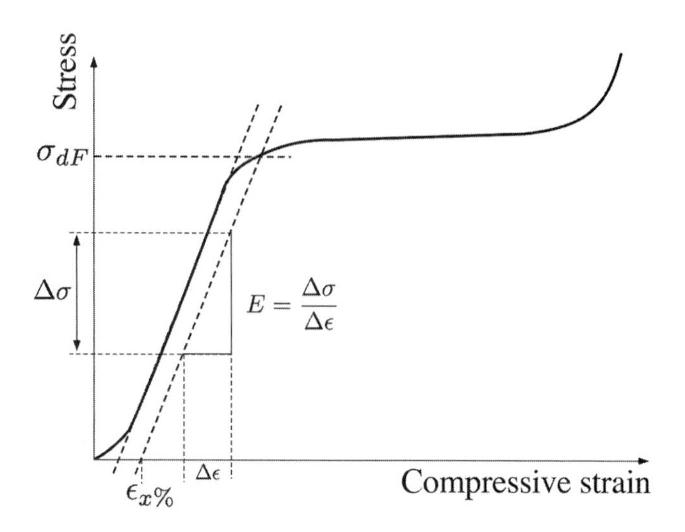

Figure 13.5 Schematic compression stress versus strain curve of a porous body.

- Plateau regime. The linear elastic regime ends either abruptly when the sample brakes or is followed by a third region with a large compressive strain and a smaller increase in stress. Depending on the material, the compressive stress–strain curve is not smooth, but can contain a more or less large scatter around an average value. This is observed in brittle foams, for instance.
- Densification regime. The plateau regime ends when the densification of the porous materials starts to become significant. Then there is only a small increase in deformation strain but a large increase in stress.

Young's modulus is defined as the slope of the linear stress–strain curves at small strains and defined or calculated according to the sketch in the figure. The stress required to leave the region of elastic deformation is called compressive yield stress σ_{dF}. It is defined as the ratio of the force F_f measured when a permanent or irreversible deformation of a certain amount, say x = 0.2%, 0.5% or 1%, is reached and the initial cross-sectional area S_0:

$$\sigma_{dF} = \frac{F_f}{S_0}. \tag{13.26}$$

The value is determined experimentally, by drawing a line parallel to the linear elastic line shifted by an amount $\epsilon_{x\%}$ and then reading on the ordinate the related stress value. A few remarks are necessary:

- If there would be no friction between the front surfaces of the sample and the pressure plates of the testing machine, the sample would retain its cylindrical shape in the region of elastic deformation and the load would be purely uniaxial. The same is true for plastic deformation because it would proceed in the planes of maximum shear stress. These planes are tilted by 45° with respect to the sample axis. However, in reality friction is always present so that the shape of the sample changes more or less to a barrel. Thus, the uniaxial state of stress changes to at

least a biaxial, rotational, symmetric one. On the outer circumference of the sample, tensile stresses are created in circumferential direction. This causes brittle materials to show cracks parallel to the sample axis penetrating into the sample.
- Although the stress state in a compression test is essentially uniaxial, it does not mean that only normal stresses act. On all planes inclined to a certain angle also, shear stresses act. These are maximal in planes tilted 45° with respect to the sample axis. These shear stresses lead to a deformation of the solid network, and it is the combined action of shear and normal stresses that deform porous bodies (in the cited textbook of Gibson and Ashby [1], nice pictures of the deformation of honeycombs are shown).
- Although in many papers on mechanical testing of aerogels under compressive loads the deformation velocity is given, it is preferable to use the strain rate in terms of %/s or %/min, as already suggested by Pekala et al. [425] and used recently by Rege and co-workers [501].

Two final remarks are necessary. First, for the testing of aerogels and similar porous materials, no standards are defined. One can refer to ASTM standard D 695 for compression tests as suggested in the *Aerogel Handbook* [14, chapter 22.2.2]. And second, the ultimate strength of a material in compression is not given by the maximum load read off from the curve in Figure 13.5. That point has no physical meaning. It just tells something about the stiffness of the testing machine used. The ultimate strength of a material always is much smaller than Young's or shear modulus. Typical values, already derived by Frenkel 100 years ago, are that the maximum theoretical strength of a material is only a fraction of Young's modulus, namely $\approx E/10$. Using the maximum point in the load compression curve would always lead to values of strength much higher than Young's modulus, which is unphysical. For a simple discussion, see [502], and a more thorough discussion is given in [503].

13.1.3 Tension and Brazilian Test

Tension tests on aerogels are difficult to perform, especially if they are brittle. In a typical tension test, either a cylindrical rod or a plane sheet is clamped at both ends and a force applied to elongate the specimen. The force is measured and also the elongation, or better directly, the strain. It is similar to a compression test, but in this case the force has opposite direction. Typically, test samples have a threaded shoulder (dog-bone shape) and thus a gauge length in between with a different cross section to avoid, first, that the specimen does not fail in the grips, and second, due to the smaller cross section in the gauge length the specimen will deform there. If the material is brittle, the gripping might directly lead to fracture. Therefore, there are only rare exceptions that aerogels were tested in tension. Cellulose and other biopolymeric aerogels were tested in tension, and we present some data later. There is, however, another method developed in the context of rock mechanics to get tensile strength data. In recent years, the so-called Brazilian test was developed [504] and also applied to aerogels. Figure 13.6 shows the principle. A thin disc of thickness t (like a coin) is taken and put directly between the pistons of a compression testing machine. On applying a force

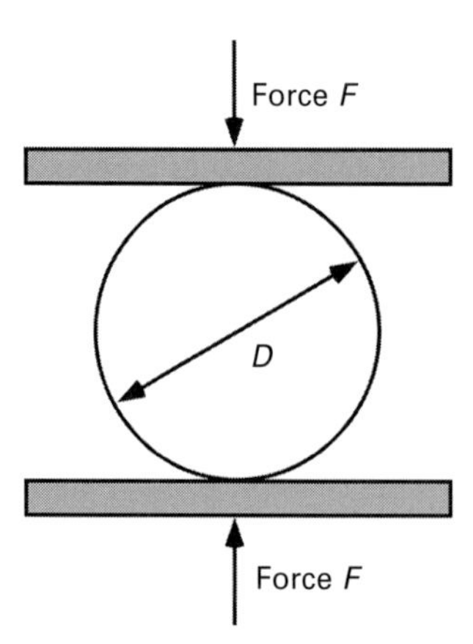

Figure 13.6 Simple specimen configuration to perform a Brazilian test on brittle specimen in compression to allow determination of the tensile fracture strength of a brittle material.

F, the sample deforms and develops tensile stresses in the circumference, leading to failure. Measurement of the force to failure F_f allows us to determine the tensile strength:

$$\sigma_t = \frac{2F_f}{\pi D t}, \tag{13.27}$$

with L the height of the sample, and D its diameter. This test can easily be applied to aerogels.

13.1.4 Flexibility

Since in the last decade the term *flexibilisation* was quite often used in aerogel research, it might be useful to define it with the knowledge of bending of bars, since everyone knows from daily life that brittle materials such as glass, if brought into fibre form, become flexible, and the thinner a fibre is, the better it can be deformed, wrapped around a bobbin etc. How can we understand this? Let us define the term *flexible* more precisely. As mentioned, bending of a circular bar by a bending moment M leads to a curved material, whose deformation can be described by its radius of curvature ρ_c such that the strain in the bar (see Figure 13.7) can be written as

$$\epsilon = \frac{y}{\rho_c} \tag{13.28}$$

and thus the stress distribution is

$$\sigma(y) = \frac{E}{\rho_c} y. \tag{13.29}$$

Using Eq. (13.5) we can write

$$\rho_c = \frac{EI}{M} \tag{13.30}$$

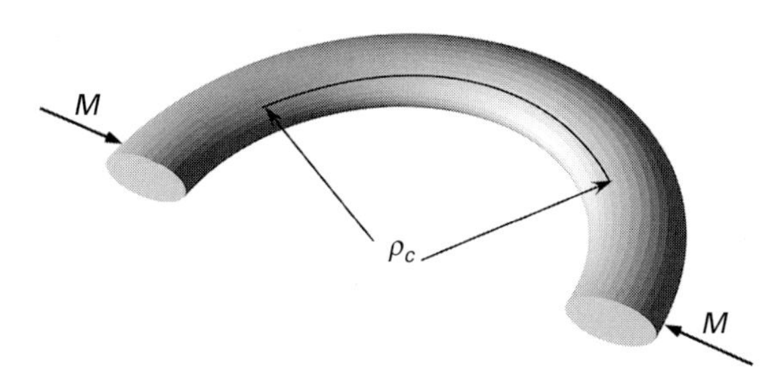

Figure 13.7 Fibres or cylindrical rod subject to the a bending moment M. The radius of curvature depends on the thickness of he rod. A thinner rod has a higher flexibility than a thick one.

We define now flexibility as $\mathcal{F}$ as a term depending only on a material property, Young's modulus and the shape of the material, represented by the moment of inertia I:

$$\mathcal{F} = \frac{1}{\rho_c M} = \frac{1}{EI}. \tag{13.31}$$

The larger Young's modulus, the smaller the flexibility. This agrees with daily experience: a soft material is simple to deform, like a bath sponge. It also agrees with daily life experience, in that thick bars are less flexible and need a much higher force to bend them than thin bars of the same material. Take a bar of a construction steel. A thin rod of a few mm diameter is easy deformable into any shape just by hand, whereas a thick one (>10 mm) needs typically a machine to bring it into a shape needed on a building site. Since $I \propto D^4$, the thinner bar has a higher $\mathcal{F}$ than a thicker one. The same argument holds of course for plate materials. Here the diameter is replaced by the thickness and $\mathcal{F}$ varies as $\mathcal{F} \propto 1/h^3$. Therefore, flexibility is not a simple material property, like Young's modulus, but it is also determined by the shape of the material as it is used.

13.1.5 Compressibility

The compressibility of any type of materials (also gases and liquids) is thermodynamically defined under isothermal conditions as

$$\chi = -\frac{1}{V}\left(\frac{\partial V}{\partial p}\right)_T, \tag{13.32}$$

and thus determines how a material reacts on pressure (typically its volume will decrease and the compressibility defines how much it does). This expression can be written in a different way. Since the mass is defined as $m = \rho V$ via the density and mass on compression remains constant, we can write the compressibility as

$$\chi = \frac{1}{\rho}\left(\frac{\partial \rho}{\partial p}\right)_T. \tag{13.33}$$

Since the density increases on compression, the partial derivative is always positive and thus χ. For isotropic solids, the compressibility is the inverse of the Lamé constant K, namely

$$\chi = \frac{1}{K}, \tag{13.34}$$

and the Lamé constant can be expressed by the shear modulus G and Young's modulus E as

$$K = \frac{1}{3}\frac{GE}{3G - E}. \tag{13.35}$$

Therefore, the compressibility of a solid is

$$\chi = \frac{3(3G - E)}{GE}. \tag{13.36}$$

For porous solids it generally is accepted that Young's modulus E depends in a power-law fashion on the relative density ρ_e/ρ_s. Since the shear modulus and the Young's modulus are related by the Poisson ratio ν

$$\nu = \frac{E}{2G} - 1 \quad \text{or} \quad E = 2G(\nu + 1), \tag{13.37}$$

we can assume that such a power-law description also is valid for the shear modulus, even with the same exponent (provided Poisson's ratio is a constant and does not change with density):

$$E = E_0\left(\frac{\rho_e}{\rho_s}\right)^m \quad \text{and} \quad G = G_0\left(\frac{\rho_e}{\rho_s}\right)^m. \tag{13.38}$$

Insertion into Eq. (13.36) yields a relation for the compressibility as a function of envelope density

$$\chi = \frac{3(3G_0 - E_0)}{G_0 E_0}\left(\frac{\rho_e}{\rho_s}\right)^{-m} = \chi_0\left(\frac{\rho_e}{\rho_s}\right)^{-m} \tag{13.39}$$

or

$$K = K_0\left(\frac{\rho_e}{\rho_s}\right)^m. \tag{13.40}$$

This equation is often used in connection with the description of aerogels or their wet gel precursors.

13.1.6 Young's Modulus from Sound Velocity Measurement

In many cases found in the aerogel literature, Young's modulus is not determined by a static compression test and the slope of the linear region in compression stress against compressive strain but a specimen of a given simple shape (cylinder, rectangular bar) is subject to ultrasound and the speed of sound measured and therefrom the elastic modulus calculated. Elasticity theory (see [505]) proves that in rods, if the wavelength of a sound wave is larger compared to the dimensions of the rod, the velocity of a running wave is given by

$$c^2 = \frac{E}{\rho} \quad \text{rod,} \tag{13.41}$$

with E, the elastic modulus of the material. This relation is true for so-called longitudinal waves in which compressions of a body by the elastic deformation accompanying the wave are in the direction of the wave. If one measures the speed of sound in a thin plate, one has to take into account two types of waves: longitudinal as before and transversal ones, in which the deformation occurs perpendicular to the direction of the travelling wave. The velocity of longitudinal waves in a plate is

$$c_l^2 = \frac{E}{\rho(1 - \nu^2)} \quad \text{plate,} \tag{13.42}$$

with ν as Poisson's ratio and for transversal waves (being nevertheless in the plate)

$$c_t^2 = \frac{E}{2\rho(1 + \nu)} \quad \text{plate.} \tag{13.43}$$

In many papers on sound propagation in aerogels, instead of Eq. (13.41) the equation

$$c_l^2 = \frac{E(1 - \nu)}{\rho(1 + \nu)(1 - 2\nu)} \tag{13.44}$$

is used. This relation is valid for the propagation in an infinitely extended medium. Poisson's ratio for cellulose aerogels is experimentally found to be close to zero, and thus both equations would yield the same result. For silica aerogels, ν is around 0.2. Then the difference between both relations is 11%. Quite often, then, Young's modulus is called the stiffness of the material and abbreviated as c_{11}. In porous bodies, one has to take into account that the pores might be filled with a gas, and this can affect the speed of sound. A simple way to take that into account is to subtract from the expression ρc_l^2 the background air pressure multiplied with the porosity [422, 466]. Experimentally a piece of material (disc, rod) is put between two piezoelectric transducers. One induces a sound wave and the other measures the pressure wave. From the measurement of the run time and the known length or thickness of the sample, the speed of sound can be calculated. A good contact between the piezoelectric transducers and the sample is necessary and typically provided by a small force slightly compressing it. As done before, let us assume that Young's modulus varies in a power-law fashion as written in Eq. (13.38), we then would have for the sound velocity

$$c \propto \frac{E^{m/2}}{\sqrt{\rho_e}} \propto E^{\frac{m-1}{2}} \propto E^{\alpha}, \tag{13.45}$$

from which we have

$$\alpha = \frac{m - 1}{2} \quad \text{or} \quad m = 2\alpha + 1 \tag{13.46}$$

The power-law exponent of Young's modulus and that of the speed of sound are related. Measuring one gives the value of the other. One remark is necessary: sound velocity measurements are almost not affected by the specimen geometry. For example, under compression, as mentioned previously, it is highly important that the specimen surfaces are parallel to each other and to the testing machine cylinders. Minor angles can lead to erroneous stress–strain curves and accordingly influence the property measurements. In the case of sound velocity measurements, it is not necessary to have a perfect cuboid or cylinder.

13.2 Stress–Strain Curves of Aerogels

In many papers, especially older ones, if mechanical properties were determined, the authors looked at the elastic or Young's modulus, or the compression modulus or the fracture strength. In rare cases, the authors looked at, analysed and tried to model the full deformation curve. Such approaches were attempted only recently. Let us first look to some excellent examples of stress–strain curves of aerogels made almost always in compression. An example of the deformation of silica aerogels prepared with different densities is shown in Figure 13.8, which was published by Wong and

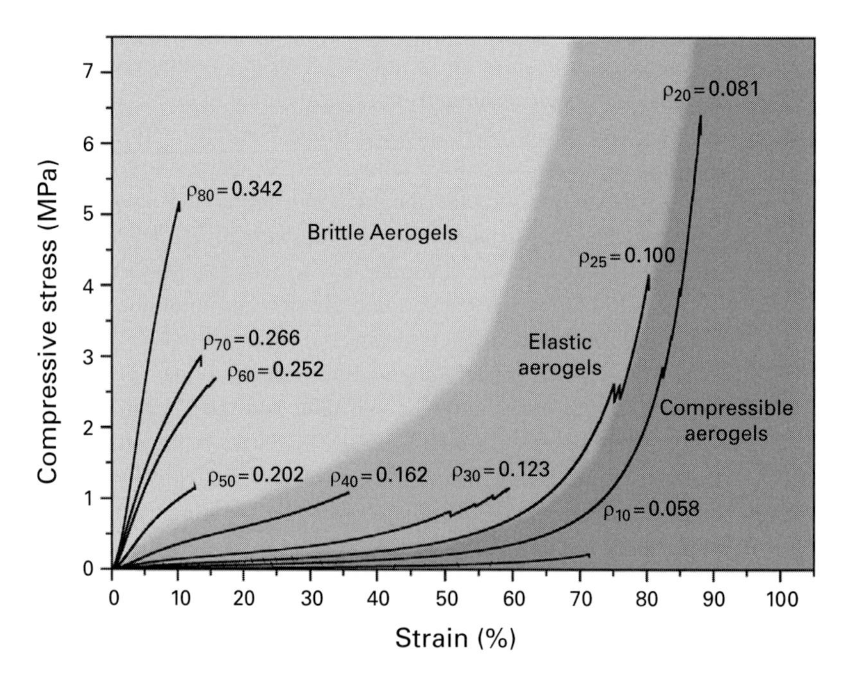

Figure 13.8 Effect of aerogel envelope density on the mechanical response in compression tests. Reprinted from [399] with permission from Elsevier

co-workers [399]. This figure nicely shows the typical behaviour of aerogels under compression. Aerogels with a high envelope density deform almost linearly until they suddenly break. Aerogels with a very low density show the typical behaviour of foams: a linear region is followed by a plateau region in which the stress required to deform the material is almost constant. This phase ends with the densification of the porous material and the stress increases rapidly. Stress–strain curves were also measured for RF aerogels. In principle, they look similar to those shown in Figure 13.8. Figure 13.9 shows an example of an aerogel prepared after the recipe of Pekala [8]. Rege et al. [433] measured the stress–strain curves of cellulose aerogels prepared by different routes and also performed tension tests. Figure 13.10 shows in comparison compression curves for cellulose aerogels prepared by the $ZnCl_2$ hydrate and the rhodanide route (see Section 3.1). It clearly is visible that cellulose aerogels prepared with calcium thiocyanate hydrate melts are much softer than those prepared from a zinc chloride tetrahydrate route, which is, however, also due to a much lower envelope density. Rege and Itskov [506] measured cellulose aerogels prepared with a ZnCl tetrahydrate melt as a solvent also under tension. Figure 13.11 shows the sample in a tensile testing machine before testing and after breakage. The stress–strain curve nicely shows a linear regime to which a straight line at small strain can be applied and Young's modulus extracted. The cellulose aerogels under tension become brittle and show a distinctive maximum, which defines the fracture strength. Similar results were obtained for cellulose aerogel fibres [152].

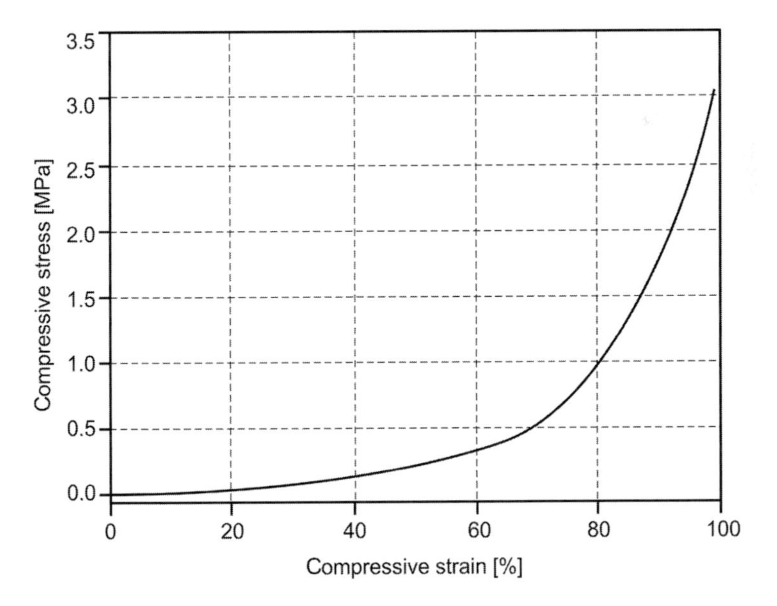

Figure 13.9 Stress–strain curve of a Pekala-type RF aerogel tested in compression (R/F 0.5, R/W 0.008, R/C 200). Courtesy M. Schwan, DLR

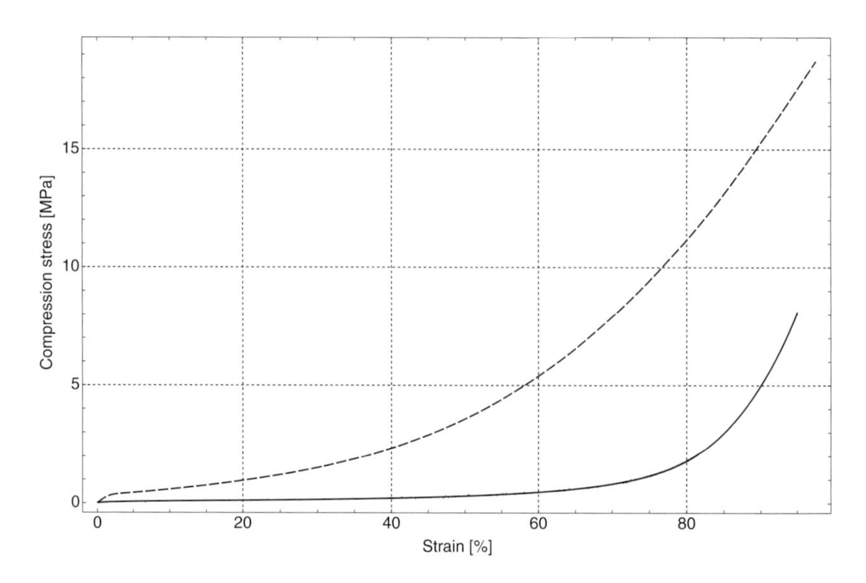

Figure 13.10 Compression curves for cellulose aerogels prepared using either a $ZnCl_2$ tetrahydrate melt to dissolve cellulose (upper curve) or rhodanide melts (lower curve). Courtesy A. Rege, DLR

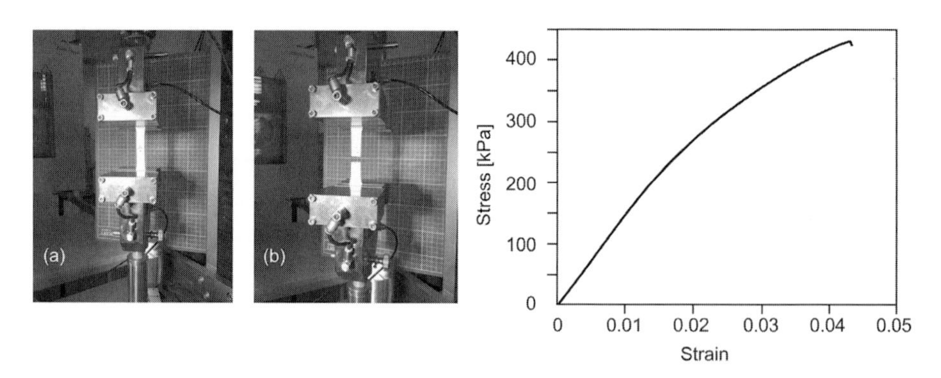

Figure 13.11 Tensile stress–strain curve of a 3 wt% cellulose aerogel prepared using with a $ZnCl_2$–tetrahydrate melt to dissolve cellulose in the right panel. In the two photos at the left, a sample with shoulder is gripped in the tensile testing machine before and after fracture. Reprinted with permission from Springer-Nature

13.3 Young's Modulus of Aerogels

Before we present experimental results on Young's modulus of different aerogels, let us briefly review theoretical model for its dependence on the envelope density. The most popular and well-tested model for porous materials is that of Gibson and Ashby [1].

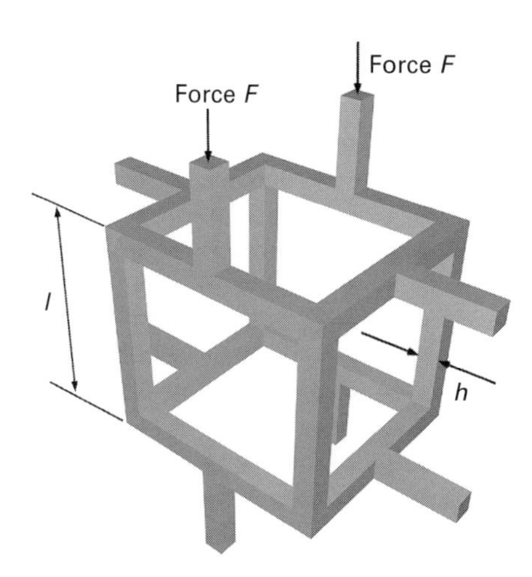

Figure 13.12 Simplified geometry of an open porous foam composed of cells with length l and prismatic slabs with thickness h.

13.3.1 The Gibson and Ashby (GA) Model for Porous Materials

The simplified geometry of an open porous foam (sponge) is depicted in Figure 13.12 and will serve as a model for aerogels. The material is composed of cells with length l and simple prismatic slabs with thickness h. The relative density of these cells can easily be calculated

$$\frac{\rho_e}{\rho_s} \propto \left(\frac{h}{l}\right)^2, \tag{13.47}$$

where ρ_e is the envelope density of the cell and ρ_s the skeletal density. The moment of inertia I of a slab is given by

$$I \propto h^4. \tag{13.48}$$

The Young's modulus is calculated according to the linear elastic deformation theory of a slab. The deformation δ under three-point bending stress is calculated from the force F, the support span l, the Young's modulus of the slab E_s and the moment of inertia I using Eq. (13.8) as

$$\delta \propto \frac{Fl^3}{E_s I} \tag{13.49}$$

The force F is linked to the stress σ acting on the cross section of the cell area $l \times l$ by the expression $F \propto \sigma l^2$, and the strain ϵ is connected to the deformation δ according to $\epsilon \propto \delta/l$. Thus, the Young's modulus of the open porous foam is calculated as follows:

$$E^\star = \frac{\sigma}{\epsilon} = C_E \frac{E_s I}{l^4}, \tag{13.50}$$

where C_E is a constant. Taking Eqs. (13.47) and (13.48) into account, the following relation is obtained:

$$\frac{E^\star}{E_s} = C_E \left(\frac{\rho_e}{\rho_s} \right)^2 . \qquad (13.51)$$

For many sponge structures, there is experimental evidence for this relation with $C_E \simeq 1$ [1].

13.3.2 Simple Extensions of the GA Model for Aerogels

Experiments with aerogels showed that such a power law is not observed. Instead, exponents between 2 and 4 were observed and rarely larger ones are communicated. Gross and Fricke [507] developed a simple model to cover the observed range of exponents. They accept the scaling relation derived in Eq. (13.51) but modify the envelope density and correct it by a term including dead-ends or dead bonds. In this way, the elastically effective density is not ρ_e but $\rho_e \phi_{eff}$, in which ϕ_{eff} is the volume fraction of elastically active bonds or string of pearls or whatever entities make the structure of an aerogel. Assuming that

$$\phi_{eff} \propto \left(\frac{\rho_e}{\rho_s} \right)^\beta , \qquad (13.52)$$

one arrives at

$$E \propto \rho_e^m \quad \text{with} \quad m = 2(1+\beta). \qquad (13.53)$$

Thus an exponent larger than 2 is easily predicted, if β is larger than zero. The experimental results on various aerogels suggest that $m \approx 3.6 \pm 0.4$, as discussed later. Using, for instance, a value of $m = 3.6$, one would have a value for $\beta = 0.8$ or, in other words, the volume fraction of an elastically ineffective structure would be 0.2 or 20%. Gross and Fricke did not, however, provide any idea to theoretically derive a value for β. This approach was later questioned and will be discussed later. One could treat it also in a different, very simple way. The main reason for the higher exponents compared to the model by Gibson and Ashby is that the cells are not all connected. The higher the density, the higher the connectivity. Thus, if not all cells are connected, there is a certain number of cells only partly connected to neighbours. Such cells do not bear any load and therefore effectively reduce the load capacity. One can then assume that the factor C_E is not a constant anymore but a function of the envelope density itself. In the simplest case, C_E is directly proportional to the volume fraction of the solid phase, i.e., the relative density, $C_E \propto \rho_e/\rho_s$, or in a way described earlier by Gross and Fricke. The resulting exponent would be $m = 3$ or dependent on the scaling. Further progress in the understanding of the elastic behaviour of aerogels was made by numerical analysis of both the microstructure using for example DLCA or molecular dynamics and performing in the computer on the simulated structure a compression test with a finite element model. Details are briefly discussed in the following subsection.

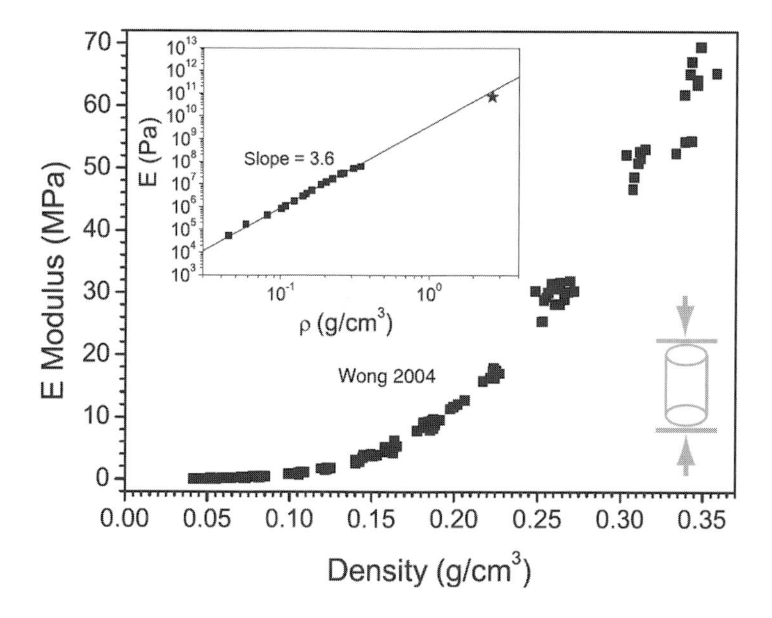

Figure 13.13 Elastic modulus of aerogels as extracted from compression curves, shown in Figure 13.8. The inset shows a double-logarithmic plot with a slope of 3.62±0.05. The star at a high density indicates the value of dense silica ($E_0 = 71$ MPa, density 2648 kg/m^3). Reprinted from [399] with permission from Elsevier

13.3.3 Experimental Results

The mechanical behaviour of aerogels was studied in great detail, especially for silica, carbon aerogels and recently especially biopolymeric ones. We will not review all the literature available but choose only a few representative examples. In a careful study, Wong and co-workers [399] studied the elastic behaviour, the fracture strength and the tensile strength of silica aerogels. One of their results is shown in Figure 13.13. As mentioned, there are numerous experimental studies trying to fit the data to a power-law behaviour. Ma and co-workers collected the results published until 2000 [508]. They found a huge variety for the exponent in the relation $E \propto \rho_e^m$ ranging from 2.57 to 3.8. Most data published suggest an exponent larger than 3. Looking for such a power law is not that simple. It requires that a large density range is covered with exactly the same preparation procedure, such that the microstructure is similar (maybe denser), and one should use a huge number of samples for a given density, such that scatters can be taken into account, and maybe one also should use different methods to measure the elastic modulus, not only static ones from compression curves but also ultrasound methods.

For RF aerogels, similar data as for silica are available either obtained from compression curves or by ultrasonic measurements. Pekala et al. [425] studied the influence of R/C ratio on the mechanical response, and Gross et al. [509] studied it at constant R/C ratio of 200 with a largely varying density. Their combined results are shown in Figure 13.14, in which we have copied the data provided by the authors into

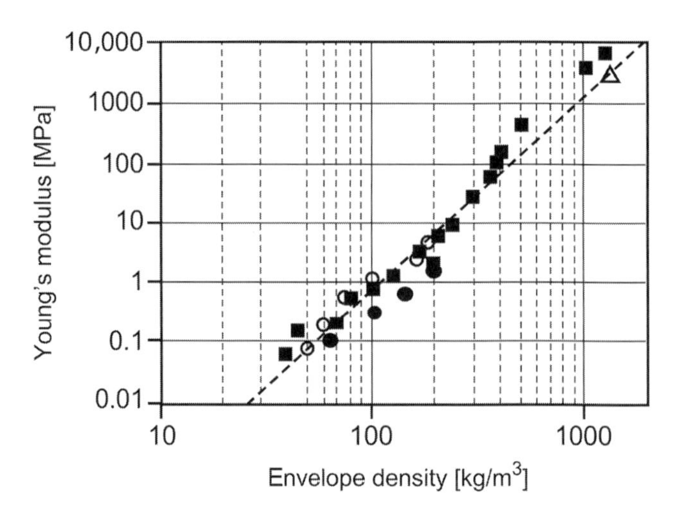

Figure 13.14 Elastic modulus of RF aerogels redrawn according to the data provided by Pekala et al. [425] and Gross et al. [509]. The dashed line corresponds to a power-law fit with an exponent of m = 3.2. The triangle indicates a massive pore-free RF resin.

one figure and drew in a power-law fit with an exponent of $m = 3.2$. For cellulose and other biopolymeric aerogels, several studies looked for a power-law dependence. For cellulose aerogels, a plot of all data available in the literature shows a power-law behaviour with an exponent of m = 1.6 and thus below the value predicted by the classical model. The data, however, exhibit a large scatter, which probably is due to the different solvents used for the cellulose, the different degrees of polymerisation in the starting material (variation from around 150 to 600) and different microstructures (fibrillar, particular and mixed forms). The authors cited in the caption of Figure 13.15 give partly different scaling exponents. Gavillon [155] gives for cellulose dissolved in sodium hydroxide an exponent of 2.33, while Rudaz and co-workers [511] state in their comparison with pectin aerogels that the cellulose aerogels show an exponent of 2.8. Karadagli et al. [149] could fit their data to an exponent of 1.86. As an example that cellulose aerogels prepared with two different salt-melt hydrate routes, namely rhodanide tetrahydrate and zinc chloride tetrahydrate, yield one power-law behaviour, we show Figure 13.16. In their original paper, Schestakow and Rege [433] plotted both data separately. They cover, however, a completely different density range but exhibit the same fibrillar microstructure. A power-law fit, drawn in as a line, follows a behaviour of $\rho_e^{2.34}$ and thus is larger than the value expected by the foam model and larger than the average of all available data shown in Figure 13.15. In a recent review on biopolymeric aerogels, Zhao and co-workers [512] collected many data on various biopolymeric aerogels and plotted the elastic modulus as a function of envelope density. Their result is shown in Figure 13.17. The results show first that for different biopolymers, quite different scalings are observed, with exponents ranging from 1.8 to 4.5. Evaluating all data would give a fit with an exponent of 1.7. One might therefore conclude that the elastic modulus obeys a power-law dependence on

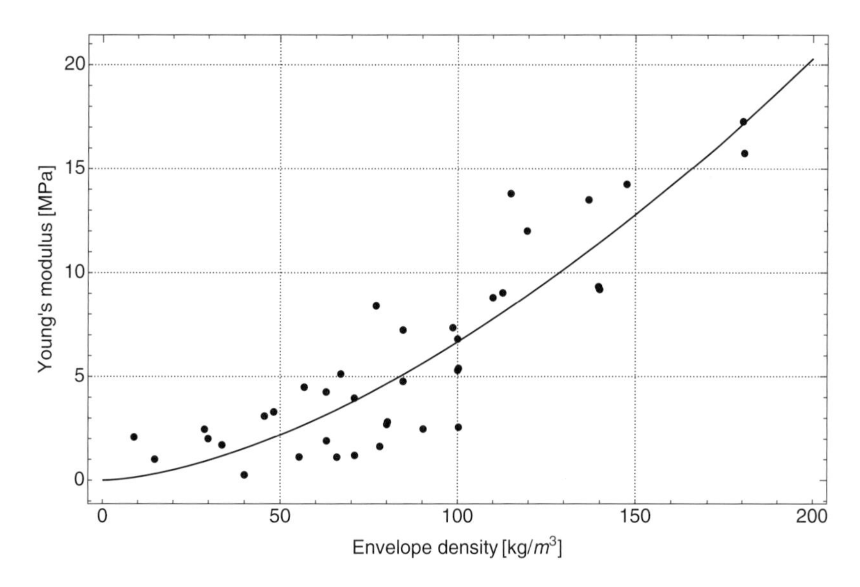

Figure 13.15 Elastic modulus of cellulose aerogels as it depends on the envelope density. Data were collected from various sources, independent of the synthesis route used by the authors [149, 151, 155, 510, 511].

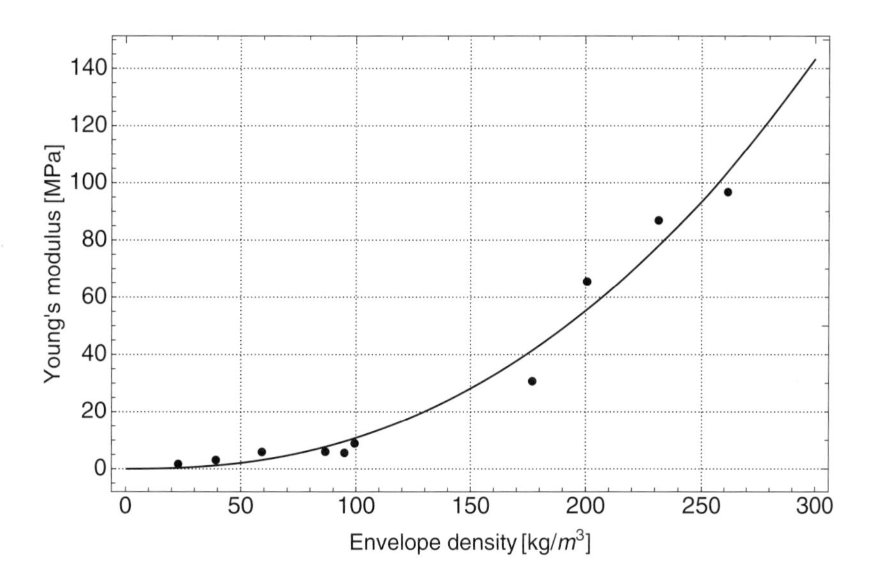

Figure 13.16 Elastic modulus of two cellulose aerogels redrawn after Rege et al. [433]. The density range below 100 kg/m^3 belongs to aerogels prepared by dissolution of microcrystalline cellulose in a calcium thiocyanate–tetrahydrate melt and those above to the same type of cellulose dissolved in zinc chloride–tetrahydrate melts. Both aerogels exhibit the same fibrillar microstructure. Those with the higher density only have a denser network of fibrils.

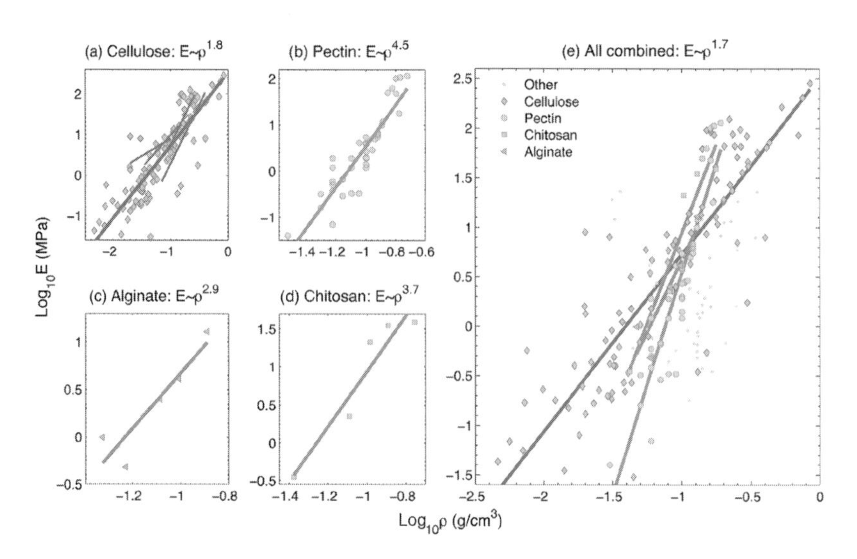

Figure 13.17 Elastic modulus of four different biopolymeric aerogels. Reprinted from [512] with permission of J. Wiley & Sons

envelope density as predicted by classical foam models. In the case of particulate aerogels, typically in the range of 3–4 is observed. Only in the case of biopolymers the power-law exponent is closer to 2 (but with a scatter up to 4.5). This would mean in the simple model by Gross outlined earlier, that in biopolymeric aerogels all structural units are elastically effective, and in particulate aerogels a simple power-law relation does not describe reality. The nanostructure has to be taken into account and calls for improved models.

13.4 Yield and Fracture Strength of Aerogels

Ceramics, glasses and many aerogels show a brittle fracture behaviour. This means that the material suddenly fails when a certain stress σ_{fs} is reached. This situation is demonstrated in Figure 13.18 showing the same open porous structure like in Figure 13.12. The fracture strength (i.e., the compression strength) can be estimated The bending moment M_f at the moment of failure of a prismatic slab is given by

$$M_f = \frac{\sigma_{fs}}{6} h^3. \tag{13.54}$$

The force F acting on a cell wall of length l creates a bending moment proportional to Fl and the stress applied to the cross-sectional area is again F/l^2. Thus, at the moment of failure, the following expression is valid:

$$\sigma_{cr} \propto \frac{M_f}{l^3}. \tag{13.55}$$

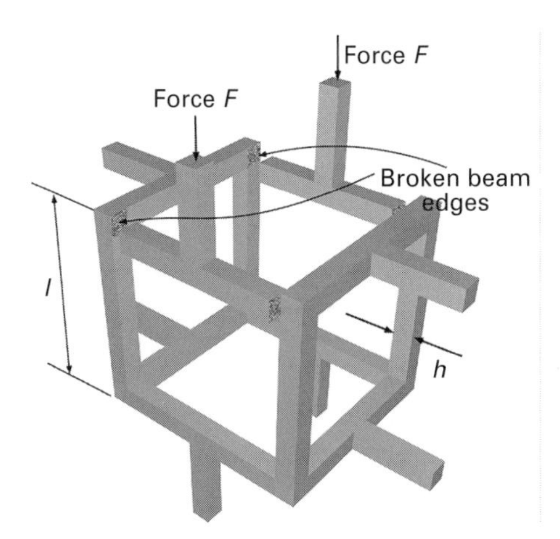

Figure 13.18 The failure of a cell suddenly occurs at the moment when the prismatic slabs breaks, meaning σ_{fs} is reached.

Applying Eqs. (13.55) and (13.47) finally yields

$$\frac{\sigma_{cr}}{\sigma_{fs}} = C_s \left(\frac{\rho_e}{\rho_s}\right)^{3/2}. \tag{13.56}$$

In Eq. (13.56), C_s is a constant with a value of approximately 2 as determined by Gibson and Ashby (G&A) from fitting of experimental data from numerous porous materials [1]. For aerogels, we therefore expect a relation $\sigma_{cr} \propto \rho_e^n$, with $n = 1.5$. As done in the case of elastic modulus, one can easily extend this model to cover other exponents, by assuming that only a fraction of the network structure bears the load and that irreversible deformation sets in locally. This would suggest that the constant C_S in the Gibson and Ashby model is dependent on envelope density leading to an exponent larger than 1.5. Experimental results of almost all aerogels studied so far show that a power law can be observed, but both the onset of irreversible deformation as well as the fracture strength deviate from the G&A model.

The tensile strength of silica aerogels was measured using the Brazilian test method by Wong and co-workers [399]. Figure 13.19 shows their result. As to be expected, the tensile strength increases with envelope density. The inset shows a double logarithmic plot which reveals a power-law dependence with a slope of 2.3 and thus higher than the conventional model outlined previously. Instead of tensile tests, typically compression tests are performed and evaluated as outlined earlier, yielding essentially the onset of irreversible deformation (we do not call that plastic deformation, because especially in the context of metals and alloys, this term implies dislocation movement and in the context of polymers it implies a kind of viscous flow). Woignier and co-workers [513] recently, discussing the mechanical behaviour of silica aerogels, also gave data on the rupture modulus σ_R of silica aerogels prepared either without a catalyst (neutral conditions) or with a base catalyst. The rupture modulus is identical to fracture strength, but it is measured in a three-point bending mode of a rectangular

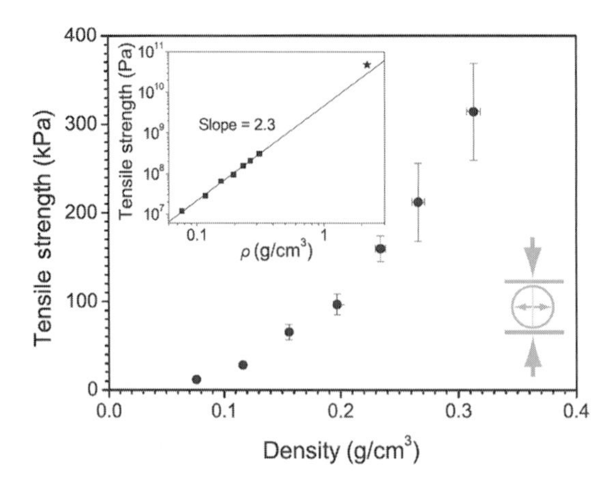

Figure 13.19 Tensile strength of silica aerogels prepared by Wong and co-workers. Note in the inset the double logarithmic representation of the data yielding a power law dependence. Reprinted from [399] with permission from Elsevier

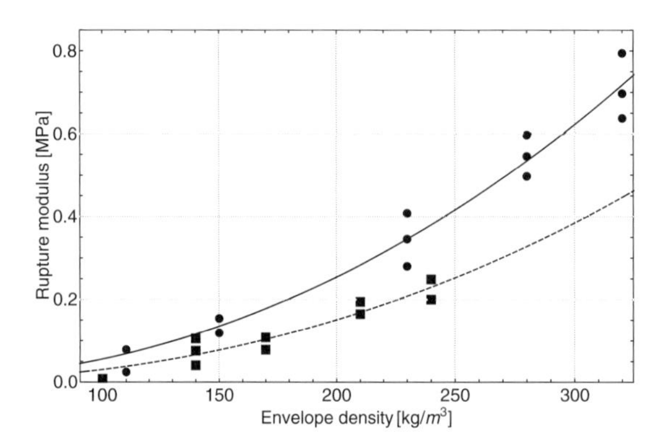

Figure 13.20 Rupture modulus of silica aerogels prepared neutrally or with a base catalyst. The neutral ones have a greater strength than the base one. In both cases, a power-law dependence is observed. Reprinted from [513] under Creative Commons license

bar. For both types of silica aerogels, they found a power-law scaling shown in Figure 13.20. The neutral catalysed silica aerogels obey a relation $\sigma_R \propto \rho_e^{2.2}$, and the base catalysed ones exhibit a power-law dependence of $\sigma_R \propto \rho_e^{2.31}$. The exponents fit well to the value obtained by Wong and co-workers. Pekala et al. measured already in 1990 [425] the compressive strength of RF aerogels and looked for a scaling relation. Their result for three different R/C ratios is shown in Figure 13.21. The compressive strength, as it is called by Pekala et al., is in reality a yield strength, meaning a clear deviation from the linear elastic behaviour. Pekala et al. found three slightly different exponents. For $R/C = 50$, they obtained a value of $n = 2.54$, for $R/C = 200$ $n = 2.07$ and for $R/C = 300$ $n = 2.60$. In any case, there is a clear deviation from the classical foam scaling of $n = 1.5$. For cellulose aerogels, data on the yield strength, measured at an irreversible deformation of 0.2% were obtained by Karadagli et al. [149] in

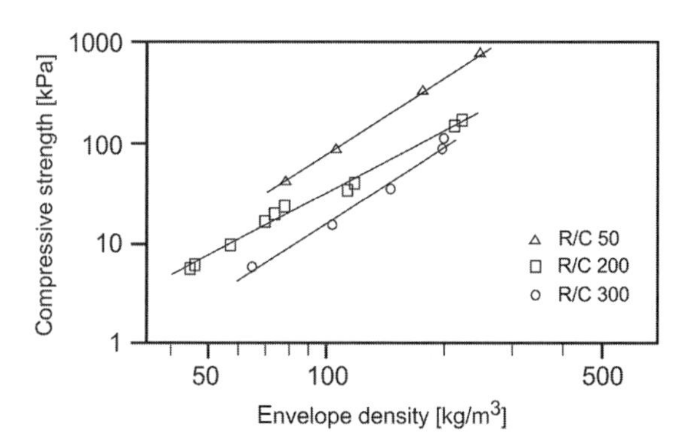

Figure 13.21 Compressive strength of RF aerogels. The aerogels were base catalysed with sodium carbonate using three different R/C ratios. The double logarithmic plot extracted from the authors' diagrams reveals a power-law scaling. Reprinted and adapted from Pekala et al. [425] with permission from Elsevier

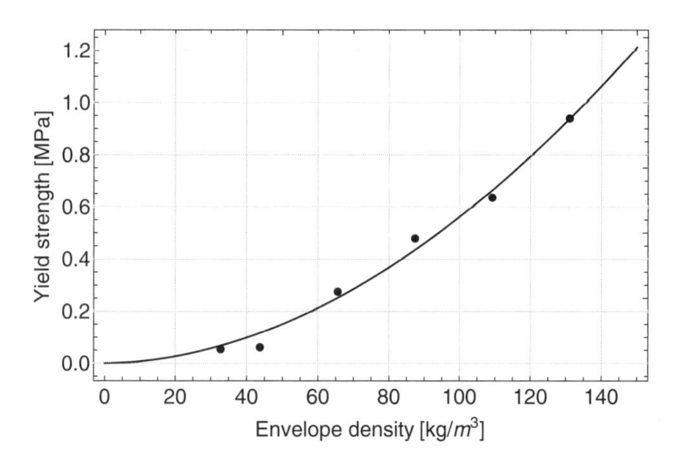

Figure 13.22 Yield strength of cellulose aerogels prepared by the rhodanide route as dependent on the envelope density. The drawn-in line is a least square fit with a power law having an exponent of $n = 1.89$.

compression tests. They analysed cellulose aerogels prepared by the rhodanide route. The scaling law obeyed a relation $\sigma_{yield} \propto \rho_e^{1.89}$ as shown in Figure 13.22.

13.5 Modelling of the Mechanical Response of Aerogels

The brittle behaviour of silica aerogels, as they were produced 30 years ago, made it difficult to handle them, especially tiles which in the 1980s were thought could replace vacuum insulation in windows or could be used for hot water circulators on a roof. Tiles up to a thickness of 6 cm and lateral measures of 60–90 cm were produced, for instance, by Airglass, a Swedish company. The mechanical properties of brittle aerogels were therefore from the beginning interesting, and models were developed to understand the relation to the microstructure. One of the first models was developed by Scherer and co-workers [514].

13.5.1 A Model for Compressive Stress–Strain Curves

Scherer et al. studied the compression of silica aerogels under the action of liquid mercury at varying pressure. Mercury does not enter the pores; it just compresses the aerogel. Using Eqs. (13.33) and (13.40) and the definition of the compressibility in Eq. (13.34), one can derive the following relation in the region, where the compressibility follows a power law:

$$\frac{1}{\rho}\frac{dp}{d\rho} = \frac{1}{K_o \rho^m}. \tag{13.57}$$

Rearranging and integrating from zero pressure to an arbitrary pressure p leads to

$$\int_0^p dp = K_0 \int_{\rho_e^0}^{\rho_e} \rho^{m-1} d\rho \tag{13.58}$$

and thus we have a simple relation:

$$p(\rho_e) = \frac{K_0}{m}(\rho_e^m - (\rho_e^0)^m). \tag{13.59}$$

We can now express the density change on pressurisation using that the mass is constant, meaning we have

$$m(\rho_e) = m(\rho_e^0) \quad \text{or} \quad \rho_e V = \rho_e^0 V_0 \quad \text{or} \quad \frac{\rho_e}{\rho_e^0} = \frac{V_0}{V}. \tag{13.60}$$

Insertion into Eq. (13.59) leads to a simple relation

$$p(\rho_e) = \frac{K_0}{m(\rho_e^0)^m}\left(\left(\frac{V_0}{V}\right)^m - 1\right) \tag{13.61}$$

To derive a stress–strain curve, we go a bit further and relate the volume change on pressurisation to strain. We have

$$\frac{V_0}{V} = \frac{V_0}{V_0 - \Delta V} = \frac{1}{1 - \frac{\Delta V}{V_0}} = \frac{1}{1 - \epsilon_V}. \tag{13.62}$$

In the case of materials with a Poisson ratio of zero, the volumetric strain is identical to the linear strain in compression. We then can rewrite Eq. (13.61) as

$$p(\epsilon_V) = \frac{K_0}{m(\rho_e^0)^m}\left(\left(\frac{1}{1 - \epsilon_V}\right)^m - 1\right). \tag{13.63}$$

At least qualitatively such an equation describes a compressive curve of an aerogel. Initially, at low strain it is linear and at higher strains there is a strong increase in pressure to achieve further deformation. Examples for two different values of m are shown in Figure 13.23, in which we plotted the relative pressure $p(\epsilon_V)/K_0$. Although the behaviour at low strains is linear, it does not include the linear elastic range properly,

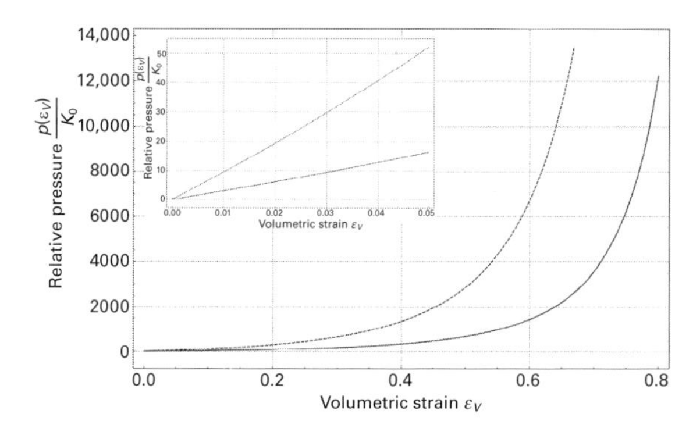

Figure 13.23 Compressive pressure as a function of volumetric strain according to Eq. (13.63). The dashed line is for a power-law exponent $m = 3.6$, the full line for an exponent $m = 3$. The inset shows the linear behaviour at small compressive strains.

but it correctly describes the compression at high strains. Stress–strain curves were not really of interest for quite some time. The power-law exponent and its deviation from the G&A model was astonishing and asked for an interpretation. As mentioned previously, Gross and Fricke already thought this must be related to microstructure of aerogels in which not all bonds existing between particles are really transmitting force. In an excellent approach, Ma et al. [508] analysed already in the year 2000 the elastic and plastic behaviour and developed hypotheses for the strange behaviour of aerogels. Their starting point was a computer model of the aerogel structure. They developed a DLCA algorithm (see Chapter 4) and placed randomly in a cubic lattice particles of unit size as well as clusters onto the lattice and let them move by Brownian motion until aggregation led to a spanning cluster. They developed a special algorithm to burn off all particles which looked disconnected to any cluster and seemed to be isolated. They characterised the structure with the envelope density, the pair correlation function and the mass correlation length. For the analysis of the mechanical behaviour they used the finite element method (FEM). The microstructure obtained from the DLCA algorithm was an input for the FEM calculations. The interparticle bonds imported from the DLCA model were modelled as beam elements undergoing stretching, bending and torsion. The FEM is a standard technique to compute elastic and plastic behaviour of materials with arbitrary structure. From their results, we reproduce only a few which in our opinion give a hint to the origin of the scaling exponent found in experiments.

First of all, they presented the bulk compressive modulus as it varies with density as shown in Figure 13.24. Looking at Figure 13.24, it is obvious that the DLCA-produced microstructure of an aerogel made from particles with a unit size, but randomly and fractally connected by Brownian motion and aggregation, is able to reveal the high exponent of the compression modulus (which also would be the elastic modulus). If the amount of non-connected particles in the structure increases (as shown with the original DLCA structure), the exponent increases even further. Ma and co-workers carefully analysed their results with respect to, for instance, the number of lattice sites or the size of the lattice used and looked for the dependence of the correlation

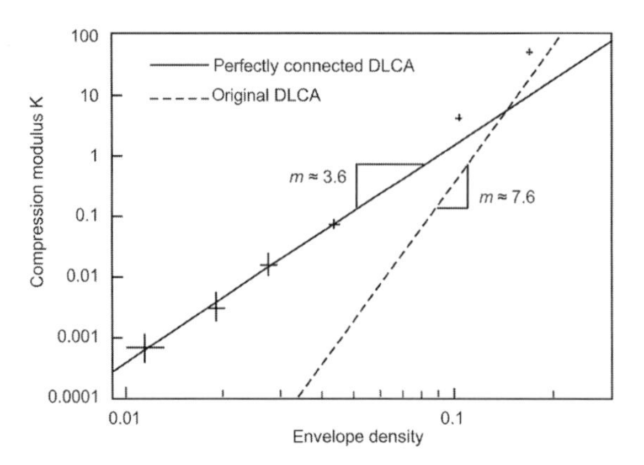

Figure 13.24 Compressive modulus as it depends on simulated envelope density. The full line corresponds to the DLCA microstructure in which all disconnected particles were burnt off, while the dashed line shows the scaling if the not trimmed microstructure was investigated. Reprinted from [508] with permission from Elsevier

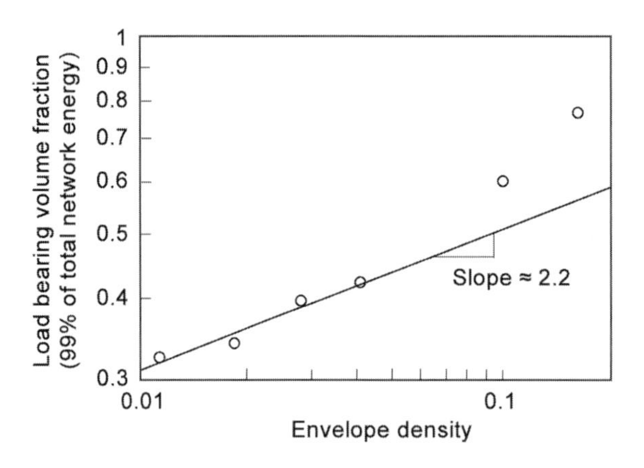

Figure 13.25 The load bearing volume fraction in the network increases with the envelope density. Up to a density below 0.1 less than 40% of the structure bears load, but above this value the load bearing fraction drastically increases. Reprinted from [508] with permission from Elsevier

length on density, etc. Explaining the high exponent, it is interesting that they observe that only a small fraction of connected particles or clusters really bear loads; see Figure 13.25. The load bearing volume fraction in the network increases with the envelope density. Up to a density below 0.1 less than 40% of the structure bears load, but above this value the load bearing fraction drastically increases. These observations are interesting. First, in the simple model outlined one could just say that the scaling approach of Gross and Fricke would suggest exactly this, and second, that the behaviour of very low-density aerogels is different from high-density aerogels. The authors write interestingly in their conclusions: 'In the sol–gel process, density determines the network connectivity, which in turn controls the mechanical stiffness, as illustrated in the simulation result. It is the increase of the mechanical efficiency of the connectivity with respect to the increasing density that accounts for the high exponent. Dangling mass is not a key factor for determining the scaling exponent. If the density of the network changes by varying the network thickness without altering the connectivity, the open-cell foam model, for which the exponent equals 2, is then applicable.' Therefore, the connectivity is not only that of single particles, but more essentially it is the connectivity of the clusters. During aggregation, clusters meet and bond together. The network, especially after trimming, is made up of clusters which are connected, similar to grains in an alloy. The cluster boundaries and the load transmitting abilities (connectivity) of these clusters seem to be essential.

In a so-called coarse-grain model approach, Ferreiro-Rangel and Gelb [515] studied the formation of aerogels and especially the mechanical properties of the dry gels under uniaxial load. Coarse-graining in this context means that they combine a molecular approach with an approach on a larger-scale level in which then averaged properties are used. This method has been well known in statistical physics for decades and allows researchers to model complex systems with simplified models. In their model, they used spherical particles of different sizes (1.5, 1.75 and 2 nm diameter) to represent the entities of an aerogel and allowed them to bond to other particles by simple potentials (a Lennard–Jones or a Morse potential), to move by Brownian motion (exactly applying Langevin-type dynamics, which define the motion of a particle as due three terms: a motion inside a potential gradient, a friction or damping force and a fluctuating random force) such that they can collide and form aggregates and spanning clusters. The simulation of the structure was stopped once all particles set randomly in the simulation volume were part of a cluster (even after the spanning cluster appeared). They varied in the simulations volume fractions from 0.02–0.1, which is quite realistic for silica aerogels. We skip the details of their simulation technique and present a few important results. First of all, they observe that Young's modulus follows a power law for the three different particle diameters used with an exponent around 2.8–2.9, respectively 3 and 3.1, depending in the modelling technique used. Poisson's ratio could also be determined, and they gave a value of around 0.2. The relation between strength and envelope density leads to a power of 1.7–1.91. The exponent increases with decreasing particle diameter. At large tensile strains, they found a strange behaviour: the aerogels behave auxetic, meaning Poisson's ratio becomes negative, a point that needs to be addressed experimentally. Figure 13.26 shows one result of their simulations, namely the density dependence of Young's modulus.

In the last decade, molecular dynamic modelling of aerogels or better aerogel-like structures have been developed and of course used to also describe the mechanical behaviour. We are not going to explain these types of approaches but refer the interested reader to the original literature [36–38] and references cited therein. Instead we briefly summarise and explain a new continuum mechanical model that is able to describe the stress–strain curve of biopolymeric aerogels, both under compressive as under tensile loads and even can cover the effect of humidity on the properties.

Starting point of this continuum model described in detail by Rege and co-workers [433, 506, 516] is a simplifying description of the microstructure of cellulose aerogels.

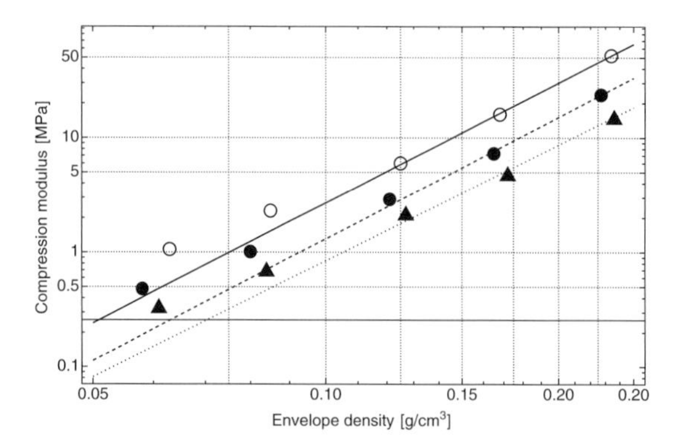

Figure 13.26 Young's modulus increases in a power-law fashion with increasing envelope density and decreases with increasing particle diameter making the network of an aerogel. Drawn from tabulated data given by and Fereiro-Rangel Gelb [515]. The solid line and the open circles belong to a simulation with a particle diameter of 1.5 nm, full circles and the dashed line are for a particle diameter of 1.75 nm and the dotted line and the triangles belong to 2 nm particle size.

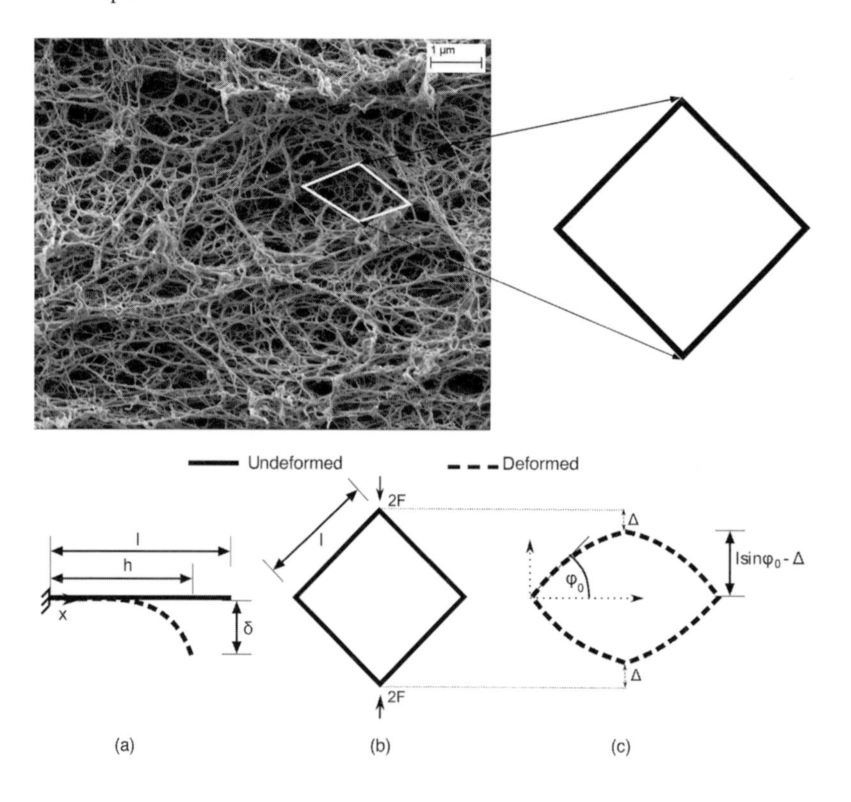

Figure 13.27 Top: in the model of Rege and co-workers [433], a given fibrillar microstructure of a cellulose aerogel is deconstructed into polygonal cells. The cell size is related to the pore size distribution and the cell beams are idealised as cylindrical rods of a diameter taken from image analysis of the SEM pictures and BET adsorption isotherms. Bottom: the beams constructing the square cell in the model deform under load non-linearly. The lateral size of the cell is kept constant, since cellulose aerogels have a zero Poisson ratio, and this is mimicked thereby. Reprinted from [433] with permission of RCS

Figure 13.27 shows the idea. A given fibrillar microstructure of a cellulose aerogel is deconstructed into two-dimensional polygonal cells. The cell size is related to the pore size distribution, determined from adsorption isotherms (BJH analysis; see Chapters 8 and 9), and the cell beams are idealised as cylindrical rods with a diameter taken from image analysis of the SEM pictures and BET desorption isotherms. Each beam in this idealised square cell is allowed to deform under load non-linearly, as described in Section 13.1.1. The idea is shown in the bottom part of Figure 13.27. The beam deformation can be modelled as one with a one-sided fixed support, mimicking that the cellulose aerogels have a zero Poisson ratio. This is an effect of the cancellation of forces exerted by neighbouring cells on the cell under consideration resulting in a near-fixed constraint. This choice of deformation mode makes the model different from other open-cell foam-like models.

Rege and co-workers calculate the strain energy under pure bending and the deflection δ; see Figure 13.27. The micro-stretch applied to the cell in the loading direction can be related to Δ as $\Delta = (1 - \lambda)l \sin \varphi_0$, as demonstrated in Figure 13.27. Only affine deformations are considered meaning the micro-stretches follow the macro-stretches. In the model, as mentioned the cell sizes vary considerably, and this is taken into account. There is no cell interaction at least orthogonal to the loading direction due to the zero Poisson ratio assumption. The model needs, however, a damage criterion. The solid integrity of each fibril is the condition for the fibril to behave as a beam. When the moment exerted on the fibril exceeds some critical value, they loss integrity and collapse according to the well-known Euler buckling model. A cell that has collapsed no longer contributes to the strain energy of the network. The damage criterion is a simple moment criterion. If the bending moment exceeds a critical value, a cell buckles. The model looks like a 2D model and therefore

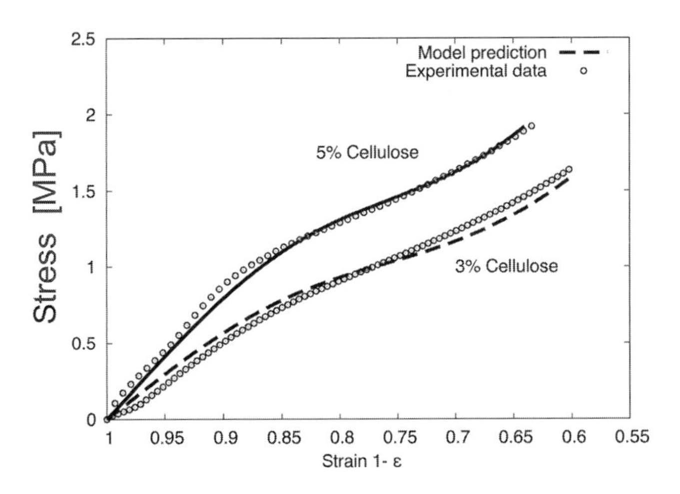

Figure 13.28 Comparison of the calculated stress–strain curves of two cellulose aerogels with different cellulose concentrations. The aerogels were prepared with zinc chloride tetrahydrate as a dissolving agent. Reprinted from [433] with permission of the Royal Chemical Society (RCS)

not suitable to deal with a 3D microstructure. Rege and co-workers use a special technique to convert it from 2D to 3D. We are not going to discuss the details further, but would like to present one result. Figure 13.28 shows the calculated stress–strain curve in comparison with experimental one for two aerogels with a different cellulose concentration. This result and others show that the micro-mechanical model seems to describe the mechanical behaviour very well. The same model was applied for tensile testing of cellulose aerogels, and recently the model was extended to new ways to describe the microstructure, e.g., in Voronoi cells [517]. Combination of such a micro-mechanical model with molecular dynamic modelling of the mechanical behaviour of fibrils could be a good way to link molecular modelling to macroscopic description of aerogel properties.

13.6 Variables and Symbols

Table 13.1 lists many of the variables and symbols used in this chapter.

Table 13.1 Variables and symbols.

Symbol	Meaning	Definition	Units
c	Velocity of a sound wave	–	–
c_l	Longitudinal velocity of a sound wave	–	m/s
c_t	Transversal velocity of a sound wave	–	m/s
c_{11}	Material stiffness	–	Pa
C_E	Constant	–	–
C_s	Constant	–	–
D	Sample diameter	–	m
F	Force	–	N
E	Young's modulus	–	Pa
E_0	Young's modulus of a pore free material	–	Pa
$E^\star$	Young's modulus of a porous material	–	Pa
$\mathcal{F}$	Flexibility	–	1/(N m^2)
h	Thickness of a prismatic slab	–	m
I	Second moment of inertia	$\int_A y^2 dA$	m^4
I_{circle}	Second moment of inertia of a rod	–	m^4
K	Lamé constant	$1/\chi$	–
m	Power-law exponent	–	–
M	Bending moment	–	N m
$M(x)$	Bending moment	–	N m
M_f	Bending moment at failure	–	N m
L_0	Length or height of a sample	–	m
p	Pressure	–	Pa

Table 13.1 Continued

Symbol	Meaning	Definition	Units
S_0	Initial cross section of a sample	–	m^2
t	Sample thickness	–	m
V_0	Sample volume	–	–
W	Section modulus	–	m^3
W_{circle}	Section modulus of a rod	–	m^3
$y(x), y', y''$	Bending line and its derivatives	–	m/s
χ	Isothermal compressibility	$-1/V(\partial V/\partial p)$	–
ℓ	Distance between support beams	–	m
δ	Maximum deflection of a beam in its centre	–	m
ΔL	Length or height change of a sample	–	m
ϵ	Strain	$\Delta L/L_0$	–
ϵ_V	Volumetric strain	$\Delta V/V_0$	–
ν	Poisson ratio	–	–
ϕ_{eff}	Elastically effective volume fraction solid	–	–
ϕ_p	Pore volume fraction	–	–
ϕ_s	Solid volume fraction	–	–
ρ	Radius of curvature	$E \cdot I/M$	1/m
ρ_c	Radius of curvature	–	1/m
ρ_e	Envelope density	–	kg/m^3
ρ_s	Skeletal density	–	kg/m^3
σ	Stress	–	Pa
$\sigma(y)$	Stress field in a beam	–	Pa
σ_{cr}	Crushing strength	–	Pa
σ_{fs}	Fracture strength	–	Pa
σ_{dF}	Compressive yield stress	–	Pa
σ_R	Maximum stress in a beam or rod	M/W	Pa
σ_t	Tensile stress	–	Pa

14 How to Cook Aerogels

Recipes and Procedures

The book would not be complete if anyone who has read it would not be able to make aerogels by themselves. If one is interested in doing so, however, a chemical lab is needed, and anyone doing it should have a bit of experience working in a lab. If supercritical drying is needed, and a lot of aerogels ask for that, a suitable facility should be available. This chapter gives recipes showing how aerogels are made in the chemical lab and the procedures, how they are made.

14.1 Silica Aerogels

14.1.1 Classical Silica Aerogel

Here is a recipe describing how to make silica aerogels from TEOS.

Chemicals Needed

Chemical	Amount (I)	Amount (II)	Amount (III)	Amount (IV)
TEOS	53.4 g	34.2 g	41.7 g	41.7 g
Ethanol	59.1 g	83.1 g	73.7 g	73.7 g
HCl (0.001M)	25.4 g	16.3 g	19.8 g	19.8 g
NH_3 (1M)	2.54 g	1.63 g	1.98 g	1.98 g

Lab Equipment

- 250 mL beaker
- Laboratory balance (accuracy 10 mg)
- Disposable pipettes
- Magnetic stirring bar
- Magnetic stirrer
- Parafilm
- 180 mL screw cap bottles (sample container)
- Cabinet dryer ($T = 50°C$)

Preparation Procedure

The chemicals are added in the following order into the beaker: TEOS, ethanol and hydrochloric acid. The beaker is covered by parafilm and the solution is stirred for 3 h (I and II), 1.5 h (III) or 4 h (IV) at room temperature. Then ammonia solution is added, the resulting solution is stirred for 5 min and equally filled in four screw cap bottles. The screw cap bottles are closed and after the gelation point is determined, the bottles are placed in the cabinet dryer at 50°C for 17–24 h.

Aging:
The gel body is covered with ethanol to 1–2 mm above the gel and stored for 24 h at 50°C.

Washing process:
The gels are transferred into bottles with snap-on caps and covered with 150 mL of ethanol. The solvent is exchanged once per day for five days.

Drying is performed supercritically.

14.1.2 A Superflexible Silica Aerogel

Silica aerogels can be made also flexible and superflexible using instead of TEOS siloxanes in which at least one position at the silicon atoms is a block or even two.

Chemicals Needed

Chemical	Amount
Water	48.97 g
Urea	20.40 g
Acetic acid (100%)	0.02205 g
MTMS	11.67 g
CTAC (25%)	16.32 g
DMDMS	6.87 g

MTMS is methytrimethoxysilane and DMDMS is dimethyldimethoxysilane. CTAC is a surfactant, named hexadecyltrimethylammonium chloride.

Lab Equipment

- 250 mL beaker
- Laboratory balance (accuracy 10 mg)
- Disposable pipettes
- Magnetic stirring bar
- Magnetic stirrer
- Parafilm
- 180 mL screw cap bottles (sample container)

Preparation Procedure

Urea is first dissolved in water and acetic acid under stirring until room temperature is reached. Then add MTMS and CTAC and let the solution stir for 15 min. After that, DMDMS is added and the solution is stirred for another 45 min. The screw cap bottles are filled, closed and placed in the oven at 80°C until the gelation occurs.

Washing process:

The gel body is washed with water first, then with ethanol.

Drying is performed in the fume hood, letting the ethanol evaporate.

14.2 Cellulose Aerogels

Cellulose aerogels need a supercritical drying. Here we give two recipes for cellulose aerogels made with the salt-hydrate route.

Chemicals Needed

Chemical	1 %	5 %
Cellulose	0.5788 g	2.8940 g
Deionised water	7.88 g	7.88 g
$Ca(SCN)_2$	50.0 g	50.0 g

Lab Equipment

- Beaker
- Heating plate
- Glass rod
- Screw cap bottles (sample container)
- Laboratory balance (accuracy 10 mg)
- Ultracentrigue

Preparation Procedure

- Weig the cellulose and water in a beaker.
- Allow the cellulose to soak in the water for 5 min and place it in the oil bath (110 °C).
- Weigh the rhodanite and add it to the soaked cellulose.
- Mix it slowly with the glass rod to avoid air bubbles.
- Go on heating until the solution becomes transparent.
- Take the beaker out of the oil bath and transfer the thickening solution into two screw cap bottles.

- Remove the remaining bubbles by ultracentrifugation and place the samples in the hood to cool down.
- Cover the resulting gel body with ethanol.

Washing process:
The gels are transferred into bottles with snap-on caps and covered with 150 mL of ethanol. The solvent is exchanged once per day for five days. Make a test of the ethanol with silver nitrate to check whether of any thiocyanate is left in the washing fluid. If the test is negative, the ethanol covered gel can be transferred to the supercritical drying unit.

Drying is performed supercritically.

14.3 Alginate Aerogel Beads

The following recipe is adjusted for one specific sodium alginate from AppliChem with the order number A3249. For other sodium alginate sources, one probably must adjust the concentration in order to achieve a certain viscosity for bead production.

Chemicals Needed

Chemical	
Sodium alginate	20 g
Ethanol	40 g
Water	940 mL
$CaCl_2$ solution	1 wt%

Procedure

Take 20 g of sodium alginate. Then disperse the powder in 40 g of pure ethanol. The reason for doing so is the better dispersibility of that mixture in deionised water. Set up a beaker of 2 L volume and fill in 940 g of deionised water. Stir the water with a high torque stirrer. For a stirring tool, you can use any that is designed for stirring highly viscous solutions. A suggestion could be a ViscoJet™. The exact stirring rate depends on the geometry of the chosen stirring tool, beaker and volume of solution. A swirl should be visible, but air dispersing into the water should be prevented at all times. Pour the mixture of sodium alginate with ethanol in the beaker with deionised water. Pour it slowly so the stirring tool has the chance to nicely homogenise the solution. Immediately the viscosity rises. Now the stirring speed can be increased.

Stir the solution until it is clear and all sodium alginate particles are dissolved. If any air is dispersed, an ultrasonic bath can be used for removing those. Evacuating the solution under vacuum works also. Notice that sodium alginate solutions usually are not biologically stable under ambient conditions. Mould can develop after a few days. A bacterial decomposition of sodium alginate can be detected by a sulfuric odor of the solution. In those cases, discard your sodium alginate solution. Store the solution in a fridge at around 4°C. Another possibility is the use of a food preservative.

For the actual bead production, prepare a 1% calcium chloride solution as regeneration medium. Other water-soluble calcium sources such as calcium lactate work as well. As a rule of thumb, a final ratio between regeneration bath and sodium alginate solution of 4/1 should be used. Stir 4 L of 1% calcium chloride solution in a 6 L bowl or beaker. Use a magnetic stirring bar of at least 5 cm length. Now feed any custom-made nozzle system with the sodium alginate solution or use a disposable pipette. Drop from a height between 1 and 20 cm into the regeneration bath. Colouring the sodium alginate solution with a characteristic dispersion colour helps to investigate the bead formation in the regeneration medium. During bead production, pay attention to the stirring speed. Increase it if necessary for continuous flow of beads in the regeneration medium. Prepare beads until all sodium alginate solution is consumed. Stir another 5–15 min to fully regenerate the beads.

Washing process:

Collect the calcium alginate beads with a kitchen sieve. Wash them with deionised water until a silver nitrate test is negative. Exchanging water with an organic solvent has to be performed stepwise starting with 10 vol% ethanol in water. Continue with 30, 50, 70, 90 and three times 100 vol% of ethanol. Exchanging in steps with increasing solvent concentration lowers that stress on the wet gel.

Drying is performed supercritically.

14.4 Resorcinol-Formaldehyde Aerogels/Xerogels

Chemicals Needed

Sample	R/W	R/F	R/C	Chemical	Amount
p-RF[a]	0.019	0.5	200	Deionised water	103.1 g
				Resorcinol-formaldehyde (24 %)	15.0 g 34.1 g
				Na_2CO_3	0.072 g
RF[b]	0.044	0.74	1500	Deionised water	101.93 g
				Resorcinol-formaldehyde (24 %)	40.0 g 61.43 g
				Na_2CO_3	0.026 g

[a] Developed by Richard Pekala [8].
[b] Hard RF aerogel.

Lab Equipment

- 2 beakers (250 mL and 100 mL)
- Magnetic stirrer, stir bar
- Spatula
- Three disposable pipettes
- Parafilm
- Screw cap bottles (sample container)
- Laboratory balances
- Cabinet dryer (T = 80 °C)
- pH electrode
- Weighing paper
- Clock

Preparation Procedure

Caution: Toxic formaldehyde vapour! Work under fume hood only!

- Weigh deionised water in a 250 mL beaker, use a pipette if necessary for the exact amount.
- Put a magnetic stir bar into the beaker.
- Weigh the resorcinol in a 100 mL beaker.
- Add the resorcinol to the water and start stirring at about 200 rpm. Avoid the creation of air blisters; beaker cools down ⇒ endothermic reaction.
- Weigh the formaldehyde solution in the same 100 mL beaker, use a pipette if necessary for the exact amount.
- After complete dissolution of resorcinol in water (5–10 min) add the formaldehyde solution.
- Stir it further for 5 min.
- Weigh the sodium carbonate on analytical balance using weighting paper.
- Add the catalyst after 5 min, carefully considering the low concentration.
- Seal the beaker with parafilm in order to minimise formaldehyde evaporation.
- Mix it further for 30 min.
- Fill it up into the labelled sample container.

Gelation

Place the samples in the cabinet dryer for about 24 h and clean the implements with acetone.

Drying

After 24 h of gelation in the cabinet dryer, the clear solution is gelled and aged. Afterwards the samples need to be dried. The aerogels after the recipe for p-RF have to be dried supercritically, and the aerogel according to the second recipe (RF-aerogel or hard aerogel) can be dried under ambient conditions or mild heating to 80°C.

Appendix A Thermodynamics and Phase Separation in Immiscibles

From daily experience, we know two extreme cases: components which easily mix, like water and alcohol, or those which do not mix at all, like oil and water. There are, however, in between components which mix only at, for instance, a high temperature and at lower temperature the solubility is limited. In a phase diagram plotting the concentration of component B on the x-axis and the temperature on the y-axis one can draw a line, called the binodal line, which marks a boundary between a state of complete dissolution of A in B and a state where two liquid phases co-exist. This is called a miscibility gap. Its thermodynamic origin will be briefly reviewed here, and we will sketch the possible processes of phase separation, which occurs once at a fixed concentration the temperature is lowered and the state point moves from a single-phase region into the two-phase region, the miscibility gap. Concise treatments are given in standard textbooks on thermodynamics of solutions and alloys; see [107, 108, 210, 443, 518, 519] and many others.

A.1 A Bit of Thermodynamics

Let $G_A^0(T)$ and $G_B^0(T)$ be the Gibbs free energy (or free enthalpy) of the pure components A and B, for instance water, ethanol and silica. If we mix them, meaning we put a molar fraction x_B of component B into A, one might make a guess how the Gibbs free energy changes upon mixing. A simple guess would be

$$G_{mix} = x_A G_A^0(T) + x_B G_B^0(T) = (1 - x_B) G_A^0(T) + x_B G_B^0(T), \tag{A.1}$$

where we have used that $x_A + x_B = 1$. This guess is wrong! Why? There are two effects that make the Gibbs free energy of a mixture different from a simple weighted average. The first reason is an entropic one, the second an enthalpic one. This can be understood best if we remember the definition of Gibbs free energy as

$$G = H - T \cdot S. \tag{A.2}$$

The Gibbs free energy is composed of enthalpy H, which essentially at constant pressure is the internal energy of a system, and the entropy. Enthalpy is easy to understand, since it is composed of the sensible heat and of course upon mixing of components they might establish new bonds and thus there can be a heat of mixing or reaction. Entropy, however, is a strange thing. $T \cdot S$ instead has the unit of an energy. Entropy can be visualised as a quantity that measures the order of a system. A disordered state, like a liquid or a gas, has a more disordered state than its crystalline counterpart. One could also understand entropy as a measure of the total number of states a system can take (and a 'state' is here defined as in classical dynamics: if the system is composed of N atoms or molecules, ask for all linear momentums, angular momentums and positions these entities can take. And that is within one mole of a substance a huge number, Avogadro's number multiplied with 6, three for linear momentum and three for position). Thermodynamics shows that entropy always is maximised. (This is easy to understand for everyone: disorder is the natural state of people living in a house, especially with kids. Making order and thus decreasing entropy takes work. Disorder, liked by almost all kids, is the high entropy state and thus preferred by nature.)

On mixing, components A and B will always lead to an increase in entropy. Think of having initially components A and B seperated in two containers. Then take away the wall or lid separating them. Both components will mix (and neglect for the moment any interaction between the components; they will not make a compound) until a uniform concentration of A and B everywhere is achieved. This mixing is irreversible (not really; one can separate both components from each other, but this requires energy and is routinely done in every distillery or chemical lab).

Let us calculate the change in entropy upon mixing of two gases first. As mentioned previously, consider two compartments, 1 and 2, to be separated by a lid. In each compartment is a gas A or B. Entropy is general defined as

$$dS = \frac{dQ}{T}. \tag{A.3}$$

The heat amount of a gas is defined as

$$dQ = dU + pdV, \tag{A.4}$$

in which dU is the internal energy and pdV is the work the system does against the pressure p to achieve a volume change. Since the mixing of the gases happens isothermally, the temperature in both compartment is the same, and we have $dU = c \cdot dT = 0$, with c the specific heat. The ideal gas law relates

$$pV = nR_gT \tag{A.5}$$

pressure, the number of moles n, the volume V and temperature. For both compartments, we therefore have

$$pdV = n_k R_g T \frac{dV}{V}, \tag{A.6}$$

with $k = A, B$. Insertion into Eq. (A.3) using Eq. (A.4) and integrating over the volume leads to

$$\Delta S_1 = \int_{V_1}^{V} n_A R_g \frac{dV}{V} = n_A R_g \ln \frac{V}{V_A} \tag{A.7}$$

and doing the same for compartment 2

$$\Delta S_2 = \int_{V_2}^{V} n_2 R_g \frac{dV}{V} = n_B R_g \ln \frac{V}{V_2}. \quad \text{(A.8)}$$

Note: inasmuch as the components A and B mix, they change their volume in which they extend. The change in entropy of the whole system of both volumes is then

$$\Delta S = \Delta S_1 + \Delta S_2 = R_g \left(n_A \ln \frac{V}{V_1} + n_B \ln \frac{V}{V_2} \right). \quad \text{(A.9)}$$

Since for both gases the ideal gas law holds, we have $pV_1 = n_A R_g T$, $pV_2 = n_B R_g T$ and thus $pV = (n_A + n_B) R_g T = n R_g T$. We therefore also have the relation

$$\frac{V_1}{V} = \frac{n_A}{n_A + n_B} \quad \text{(A.10)}$$

and thus we arrive at

$$\Delta S_m = \frac{\Delta S}{n} = -R_g \left(\frac{n_A}{n} \ln \frac{n_A}{n} + \frac{n_B}{n} \ln \frac{n_B}{n} \right). \quad \text{(A.11)}$$

Defining mole fractions as usual $x_A = n_A/n$ and $x_B = n_B/n$, we can write this also as

$$\Delta S_m = -R_g (x_A \ln x_A + x_B \ln x_B). \quad \text{(A.12)}$$

On mixing, the entropy always increases, even if very small quantities are introduced. The logarithmic dependence shows that a pure material does in reality not exist or needs a huge amount of energy to achieve. Note that the logarithm goes to minus infinity as one concentration goes to zero. We show the concentration dependence of the entropy of mixing in Figure A.1.

Before we discuss systems with a miscibility gap, let us take a look at the most simple kind of solution, one that is only driven by entropy.

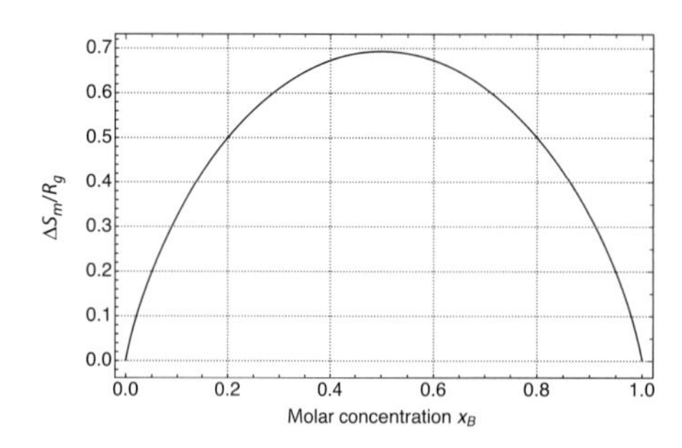

Figure A.1 The figure shows $\Delta S_m/(R_g T)$ as it depends on the concentration of B measured in molar fractions. Note the symmetry around the value $x_B = 1/2$.

A.1.1 Ideal and Regular Solutions

There are many models for solutions developed in the thermodynamic literature. The most simple one is that in which the enthalpy upon mixing is zero, $\Delta H = 0$. This is called an ideal solution. The change in Gibbs free energy on solving B into A is then given as

$$\Delta G_{mix}(x_B, T) = R_g T((1 - x_B)\ln(1 - x_B) + x_B \ln x_B) \tag{A.13}$$

The total Gibbs free energy of a system of components A and B if the y mix ideally is then given as

$$G_{mix} = (1 - x_B)G_A^0(T) + x_B G_B^0(T) + R_g T((1 - x_B)\ln(1 - x_B) + x_B \ln x_B) \tag{A.14}$$

and thus differs from the guess made earlier. We will not further discuss them, since essential for many aerogel building solutions are those which exhibit a miscibility gap. The existence of such gaps can be understood using the so-called regular solution model.

In a regular solution model, the enthalpy of mixing is not zero, but can take negative or positive values. The idea behind this is simply that mixing of the components typically leads to a heat of mixing, which most often is positive, like in the mixture of water and sulphuric acid. A simple model for the concentration dependence is

$$\Delta H_{mix} = W x_A x_B, \tag{A.15}$$

with W as a parameter that can take negative or positive values. This function is also symmetrical around the value $x_B = 1/2$. The essential difference to an ideal solution model comes from adding to this expression the entropy of mixing derived before. We finally have for a regular solution the expression

$$G_{mix} = x_A G_A^0(T) + x_B G_B^0(T) + W x_A x_B + R_g T(x_A \ln x_A + x_B \ln x_B). \tag{A.16}$$

Let us only look at the difference from the linear relation given in Eq. (A.1). We subtract $(1 - x_B)G_A^0(T) + x_B G_B^0(T)$ and obtain

$$\Delta G_{mix} = W(1 - x_B)x_B + R_g T((1 - x_B)\ln(1 - x_B) + x_B \ln x_B). \tag{A.17}$$

If we divide both sides by $R_g T$, we get the following result shown in Figure A.2. The figure shows that at negative values of the mixing enthalpy, the free energy becomes even smaller compared to the ideal solution, but if the sign changes to positive values, two minima appear. These minima are essential to understand the occurrence of a miscibility gap. Let us try to understand the consequences of these minima. Take a look at the dashed line and start to increase the concentration of B in A. On addition, first the free energy decreases until the minimum is reached. The same holds if we start from pure B and add A. What happens if we further increase the concentration of B in A? It looks as if the system again increases the Gibbs free energy. But is this the lowest energy state? Draw a straight line touching both minima. These have a concentration of B given as x_B^1 and x_B^2 (the left one has the superscript 1, the right

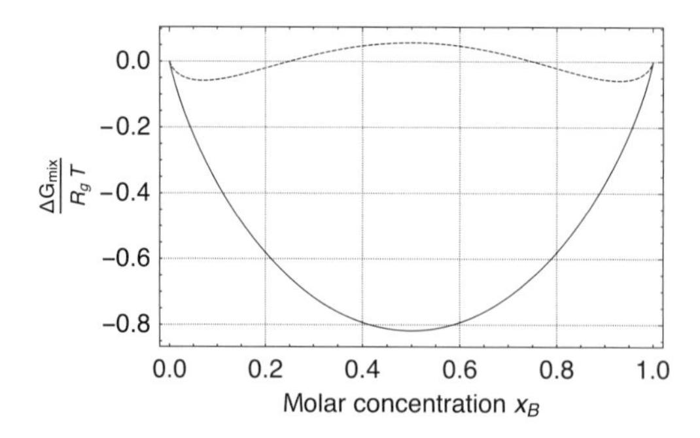

Figure A.2 The figure shows $\Delta G_{mix}/(R_g T)$ defined in Eq. (A.17) as it depends on the concentration of B measured in molar fractions. We arbitrarily have chosen for the solid line a $W/R_g T$ value of -0.5 and for the dashed line a value of 3.

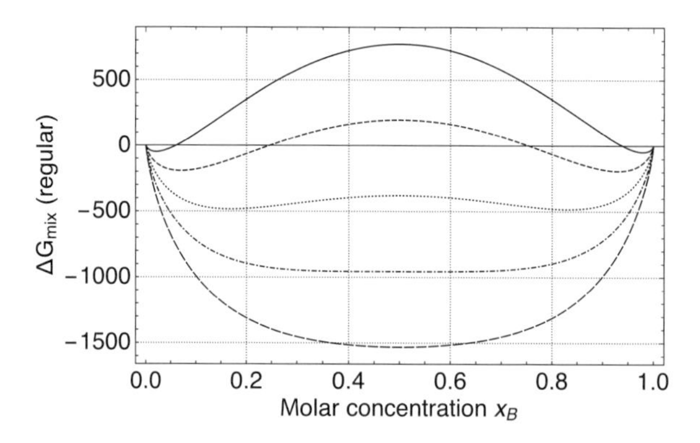

Figure A.3 Variation of the Gibbs free energy in a regular solution model with concentration and temperature according to Eq. (A.17). The interaction parameter W was chosen as 10 kJ/mol. The solid line is for a temperature of 300 K, the dashed line for 400 K, the dotted for 500 K, the dash-dotted for 600 K and the long-dashed one for 700 K.

minimum the superscript 2). Every point of this straight horizontal line has the Gibbs free energy of the minima. If we have a solution of the composition x_0 being between the two minima, the system can lower its free energy by separating into two phases with exact the composition of the two minima. The energy it gains is $\Delta G_{mix}(x_0) - \Delta G_{mix}(x_1) > 0$. Along this line inasmuch as the concentration x_0 varies, the molar fraction of the phases with composition x_B^1 and x_B^2 changes according to the lever rule:

$$f_1 = \frac{x_B^2 - x_0}{x_B^2 - x_B^1} \quad \text{and} \quad f_2 = \frac{x_0 - x_B^1}{x_B^2 - x_B^1}. \tag{A.18}$$

Here f_1 denotes the fraction of the phase with composition x_B^1 and f_2 analogously. The location of the two minima depends strongly on temperature. Figure A.3 shows an example. With decreasing temperature, two minima occur and their position shifts to pure components A and B. The location of the minima as they depend on temperature defines the boundary between a single-phase solution and a two-phase solution. This

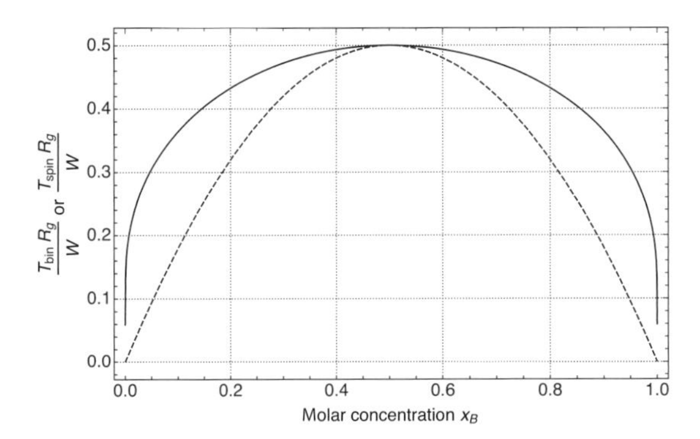

Figure A.4 The solid line shows the binodal line for a regular solution model. The dashed line represents the spinodal. Both touch at the critical point and have there a maximum.

boundary line is called binodal. In the regular solution model, the binodal line can be calculated in a simple way. Since it is defined by the minima of the Gibbs free energy curve, take the first derivative of ΔG_{mix} with respect to x_B, set it zero and solve for temperature of the binodal as a function of x_B. The result is very simple:

$$T_{bin} = \frac{W}{R_g} \frac{1 - 2x_B}{\ln(\frac{1-x_B}{x_B})}. \tag{A.19}$$

Before we plot this function, we have to make a few further calculations. Looking at Figure A.3, there must be a temperature above which both components are miscible to every extent. This temperature is called the critical temperature T_c, and it turns out that this is exactly the temperature at which the binodal has a maximum. Therefore, we take the second derivative and ask for the maximum to obtain

$$T_c = \frac{W}{2R_g}. \tag{A.20}$$

The second derivative is of course a function of concentration. The zeros of the second derivative are the points of inflection of the free energy curve. Their location is given as

$$T_{spin} = \frac{2W}{R_g} x_B(1 - x_B) \tag{A.21}$$

and called the spinodal line. The binodal and the spinodal lines are shown in Figure A.4.

A.2 Phase Separation in a Regular Solution

We had already explained that below the binodal line two phases coexist, in the case of solutions two liquid phases with different compositions. The composition of these two phases vary considerably with temperature: the lower, the more different their composition. There are two possible processes for a transformation of a single-phase liquid to a two-phase one: nucleation and growth and spinodal decomposition. We start with the nucleation process, which seems to be appropriate for all systems which are off-critical, and a quench is just performed to any temperature in the field between binodal and spinodal.

A.2.1 Phase Separation by Nucleation and Growth

Let us think a bit about nucleation (for a deeper understanding, see [520, 521] and a recent critical review on nucleation theory written by Gránásy and co-workers [522]). Assume that we have supercooled a single-phase solution to some temperature below the binodal line. The classical approach of nucleation theory is then that fluctuations of concentration appear locally inside the solution, which have a concentration close to the second-phase liquid droplet that would be there in equilibrium. These fluctuations might disappear if their size is small; they might grow if their size is sufficiently large. The formation, growth and decay of these fluctuations are assumed to happen via a series of single-molecule attachment and detachment events. The work of formation of fluctuations or clusters of second-phase droplets being able to grow depends strongly on the work of their formation. Let R_{gy} be the radius of gyration of such a second-phase liquid-like cluster. Since it is surrounded by a solution having a different composition, there exists an interface tension σ. The clusters may, however, grow, since there is a gain in volumetric free energy ΔG_V, which reflects that a two-phase system is thermodynamically stable. The work of formation is then a sum of a volume term and an interface term, $W_{hom} = -(4\pi/3)R_{gy}^3 \Delta G_V + 4\pi R_{gy}^2 \sigma$. Since for small liquid droplets the surface energy term wins over the volumetric one, small fluctuations or clusters disappear. Once a critical size is reached, the volumetric term wins and the droplets are able to grow. This consideration allows us to calculate the critical size as

$$R_{gy}^* = \frac{2\sigma}{\Delta G_V}. \tag{A.22}$$

In the classical theory of nucleation the distribution of clusters with a radius of gyration R_{gy} is given as a Boltzmann factor

$$n(R_{gy}) = n_{tot} \exp\left(-\frac{\Delta W(R_{gy})}{k_B T}\right), \tag{A.23}$$

with $\Delta W(R_{gy})$ the work of formation of a cluster of size R_{gy} and n_{tot} the number density of single-molecule clusters. From the classical approach of nucleation theory, this can be calculated as

$$\Delta W(R_{gy}) = \Delta W^*(3\xi^2 - 2\xi^3), \tag{A.24}$$

with $\xi = R_{gy}/R_{gy}^*$ and R_{gy}^* the size of a critical cluster radius and

$$\Delta W^* = \frac{16\pi\sigma^3}{3\Delta G_V^2} \tag{A.25}$$

as the critical energy barrier for nucleation. A plot of $\Delta W(R_{gy})/\Delta W(R_{gy})^*$ is shown in Figure A.5. The nucleation rate J_n being the number of nuclei that appear per unit

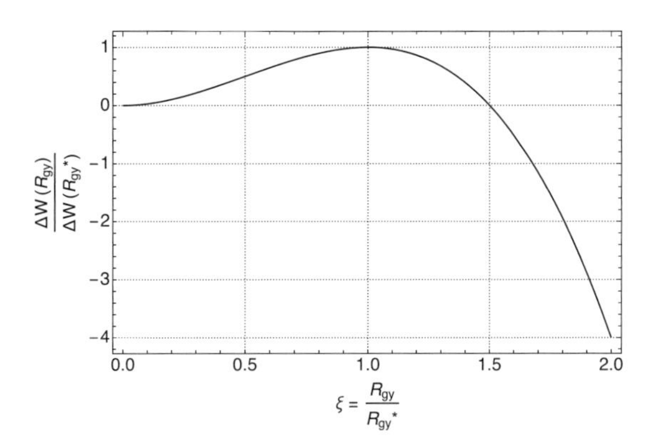

Figure A.5 Plot of $\Delta W(R_{gy})/\Delta W^*$ as a function of $\xi = R_{gy}/R^*_{gy}$. Note the maximum appears at $\xi = 1$.

volume and time is in this classical model expressed as the number density of critical nuclei multiplied with the attachment rate $\Gamma = D/\lambda^2$ of molecules to the interface of the critical cluster

$$J_n \propto n_{tot} \frac{D}{\lambda^2} \exp\left(-\frac{\Delta W^*}{k_B T}\right), \tag{A.26}$$

with D as the diffusion coefficient for the motion of let us say B in the solution and λ the molecular jump distance (which can be set equal to the next nearest neighbour distance in a liquid). If one would like to calculate the nucleation rate, the energy barrier for nucleation is the essential unknown. Two terms make serious problems: the interface energy between particles in aerogel forming solutions and their mother solution is not known and the volumetric gain in free energy requires a phase diagram and the thermodynamics behind it. Let us make a rough estimate. Assume that the free energy difference between the single-phase liquid and the two-phase liquid is proportional to the difference between the binodal temperature T_b and the actual temperature inside the binodal T_a, meaning $\Delta G_V \propto T_b - T_a = \alpha\Delta T$. Then the energy barrier would be proportional to the inverse square of the undercooling below the binodal: the larger the undercooling or the deeper the quench, the more nuclei are able to grow beyond the critical size, and their number density increases and thus the nucleation rate. The interfacial energy term is hard to estimate. From critical point theory, we know that the interface energy close to the critical point of the miscibility gap follows a functional form $\sigma \propto (T - T_c)^\nu$ with $\nu = 1.3$ [521, 523, 524]. Such a behaviour would increase the importance of interfacial tension for the energy barrier: instead of a power of 3 it would be a power of almost 4. If such considerations can be applied to aerogel forming solutions is an open issue, since the polycondensation reaction would have to be taken into account and that changes the regular solution model. A deeper discussion can be found in Appendix B and in Section 4.2.5 on polymerisation-induced phase transformation.

A.2.2 Phase Separation by Spinodal Decomposition

So far we have discussed a quench into the region below the binodal. What, however, is the meaning of the spinodal and what happens if one quenches a solution into this region? Unfortunately, the answer is not so simple. We first have to define the stability of a given phase. Any phase described by a free energy curve $G(x)$ is stable against, let us say, a decomposition in two phases (or more), if such a decomposition would result in an increase of Gibbs free energy of the system. This is exactly the case if the free energy around a given composition has at least locally the shape of an upside parabola or in other words the second derivative of the free energy is greater than zero:

$$\frac{\partial^2 G_{mix}}{\partial x_B^2} > 0. \qquad \text{stability criterion for a solution phase} \qquad \text{(A.27)}$$

Inside the spinodal, however, the second derivative always is negative, and this means the solution is intrinsically unstable against decomposition. More exactly, local fluctuations in composition are able to grow and the system evolves spontaneously to a two-phase system, in which the phases will eventually attain the composition values given by the binodal line (at constant temperature). How can spinodal decomposition be achieved? Looking at Figure A.4, there two ways: first, one can directly from a single-phase region enter the spinodal if the solution has the critical composition and one decreases the temperature from a value above T_c to any value below. A second way is needed if the composition is off-critical. Then the solution must be quenched from a temperature above the binodal line to a temperature below the spinodal. Theoretically, this always is possible practically, heat extraction via the surface of the sample impedes a rapid quench of the whole volume. Thin samples would be required to make such a temperature jump into the spinodal region. In any case, assume a solution is quenched into the spinodal. What happens then? We have derived in Section 10.1 that the mobility of molecules in a diffusional process depends on the the second derivative of Gibbs free energy:

$$M = \frac{D}{\partial^2 G(x_B)/\partial x_B^2}. \qquad \text{(A.28)}$$

Since the diffusion flux is proportional to the mobility and the second derivative inside the spinodal is negative, we have the strange situation called in the literature 'uphill' diffusion. Whereas usually species diffuse from higher concentration to lower ones (downhill), the negative mobility makes it just vice versa. This has consequences: once after a quench in the spinodal region fluctuations appear (and they always appear), the uphill diffusion makes them grow exponentially. However, we have neglected here that there is an interface between a composition fluctuation and the surrounding solution, and that changes the situation. It was shown by Cahn and Hilliard [525–527] that the concentration gradients and the interface energy associated with them leads to a situation that only fluctuations above a certain wavelength can grow

$$\lambda_{crit} = \frac{8\pi\kappa}{-\partial^2 G(x_B)/\partial x_B^2}, \tag{A.29}$$

defined by the second derivative of the free energy curve and a coefficient κ determining the effect of local composition gradients on the free energy. Shorter wavelengths die out. More details on this approach can be found in Section 4.2.5 on polymerisation-induced phase transformation, where the Cahn–Hilliard equation is derived also for polymeric solutions. This leads to interesting microstructures, which easily can be looked at using 'spinodal decomposition' as a search entry and asking for pictures or videos.

Appendix B Flory–Huggins Theory of Polymer Solutions

What happens if not monomers are dissolved in a liquid (solvent), but they have reacted to dimers, trimers, tetramers or pentamers etc. If the monomers are dissolved, such that a good description of the solution is that all molecules constituting it are distributed at random in the liquid, is this also true for all oligomers (k-mers, with $k = 2, 3, 4 \ldots$)? Experience with polymers would tell that this depends on, for instance, the solvent. Polystyrol can easily be brought into solution with acetone, but not with water or a simple alcohol. Thus solvability depends probably on both the polymer and the solvent. How can we attack this problem theoretically?

The classical theory for solutions of polymers goes back to Flory and Huggins and is almost a century old [116]. We describe the Flory–Huggins (FH) model to describe the miscibility gap in a mixture of a polymer with degree of polymerisation DP $= N$ and a solvent. The basic equations can be found almost everywhere in the literature on polymer solutions [54, 108, 528, 529]. The starting point is the Gibbs-free energy of mixing, which reads

$$\Delta G_m = k_B T \left[\frac{\phi}{N} \ln \phi + (1 - \phi) \ln(1 - \phi) + \chi \phi (1 - \phi) \right]. \tag{B.1}$$

The first term represents the entropy change of the solution due to mixing of the polymer with volume fraction ϕ in the solvent. One immediately recognises that with $N = 1$, the well-known regular solution model is recovered [107]. The second term is the latent heat of mixing, characterised by the Flory parameter χ

$$\chi = \frac{Z \Delta \epsilon}{k_B T} \tag{B.2}$$

and being symmetrical in the volume fractions. A derivation of this equation can be found in the textbooks on polymer chemistry cited. The Flory parameter needs some explanation. Since the FH model is a lattice-based model, it considers only nearest neighbour bonds between the polymer and the solvent. Both the solvent molecules and the polymer are strictly located on the points of the lattice (in 2D an sc-lattice with $Z = 4$ bonds, and in 3D a lattice with $Z = 6$ bonds). The bond energies are $\epsilon_{SS}, \epsilon_{PP}$ and ϵ_{PS} between two solvent molecules, two polymer molecules and the solvent and the polymer, respectively. Notice all bond energies are negative. The bond energies themselves are not decisive, but their differences are:

$$\Delta \epsilon = \epsilon_{PS} - (\epsilon_{SS} + \epsilon_{PP})/2. \tag{B.3}$$

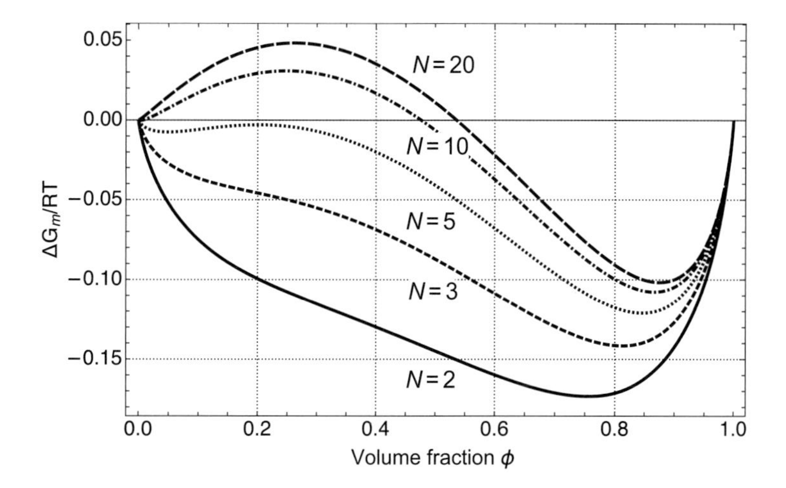

Figure B.1 Free enthalpy of mixing of the Flory–Huggins theory according to Eq. (B.1). From bottom to top, the degree of polymerisation varies from $N = 2, 3, 5, 10, 20$. The FH parameter χ was set to 1.5.

If $\Delta\epsilon < 0$ and thus also $\chi < 0$, the solution of the polymer in the solvent is more favourable compared to the situation, in which $\chi > 0$, which means PP and SS combinations are favoured over PS bonds. A solution with positive χ will decompose into two phases. The dependence of the free enthalpy of mixing on the degree of polymerisation is shown in Figure B.1. One can clearly see that the free energy curves develop with increasing DP two minima, showing that a miscibility gap occurs.

The spinodal decomposition is defined by the zeros of the second derivative of the free mixing enthalpy with respect to the volume fraction:

$$\frac{\partial^2}{\partial\phi^2}\left(\frac{\Delta G_m}{k_B T}\right) = 0 \quad \rightarrow \quad \frac{1}{\phi} + \frac{N}{1-\phi} = 2\chi N. \tag{B.4}$$

If we remember the definition of χ we can see that it is an inverse temperature. Plotting $1/\chi$ against volume fraction with varying degrees of polymerisation reveals the effect of the DP on the shape of the spinodal. This is shown in Figure B.2. The critical point of the miscibility gap depends on the degree of polymerisation only. This can be seen by calculating $\partial\chi/\partial\phi = 0$, leading to

$$\phi_c = \frac{1}{1+\sqrt{N}} \tag{B.5}$$

and the critical temperature (inverse of χ) varies like

$$\frac{k_B T_c}{Z\Delta\epsilon} = \frac{2N}{\left(\sqrt{N}+1\right)^2}. \tag{B.6}$$

Evaluating Eq. (B.5) shows that it needs a large DP for spinodal decomposition to occur at a low volume fraction of polymer typical for aerogels (5–10 vol.%). The degree of polymerisation, however, was never determined and probably is hard

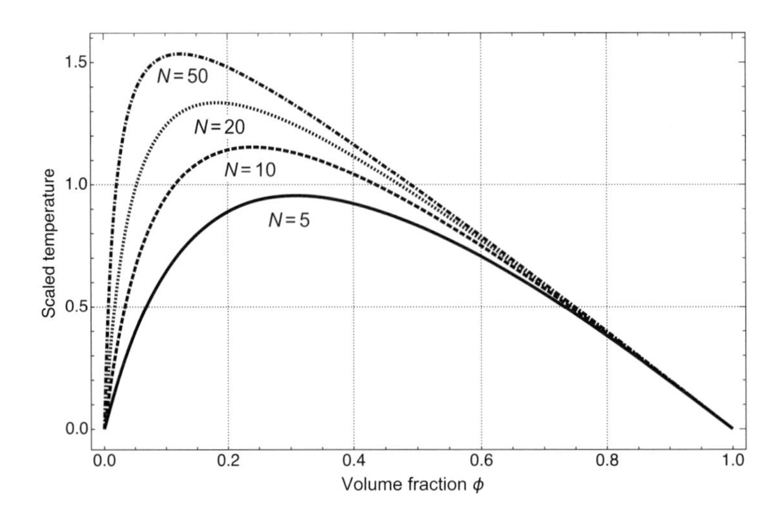

Figure B.2 Spinodal curves in the Flory–Huggins theory according to Eq. (B.4). It is obvious that a miscibility gap appears even at low degrees of polymerisation. The critical point shifts towards the solvent side with increasing DP.

to determine in both a wet gel as well as a dry aerogel, since typically in polymer chemistry the viscosity of a solution is measured and the DP is derived from it. Such a measurement is hard to imagine for, e.g., RF aerogels. We can express the critical temperature also as a function of the critical volume fraction leading to

$$\chi_c = \frac{1}{2(1-\phi_c)^2}. \tag{B.7}$$

Using the definition of the Flory parameter we can also write the critical temperature as

$$T_c = \frac{2Z\Delta\epsilon}{k_B}(1-\phi_c)^2. \tag{B.8}$$

The critical temperature decreases with the critical volume fraction of polymer. Low critical volume fractions, which would be needed in aerogel building solutions, lead to a high critical temperature. The maximum critical temperature is determined by the interaction energy term $\Delta\epsilon$ and the bonds per lattice site Z. The highest critical temperature is just $T_c^{\max} = 2Z\Delta\epsilon/k_B$. The critical temperature changes starting from a solution with monomers ($N = 1$) to the fully developed polymer containing solution. So if we start with a solution at a given temperature $T_{op} < T_c$, such that all monomers are dissolved, the advance of polymerisation might lead to the situation that the critical temperature increases above the operating temperature T_{op} of the solution and a phase separation can occur.

Till now, we have only considered the spinodal. For all phase diagrams exhibiting a miscibility gap, however, it is typical that the single-phase miscible region is limited not by the spinodal but by the binodal line. Both lines merge at the critical point. The binodal is defined as the common tangent to the Gibbs-free energy curve on

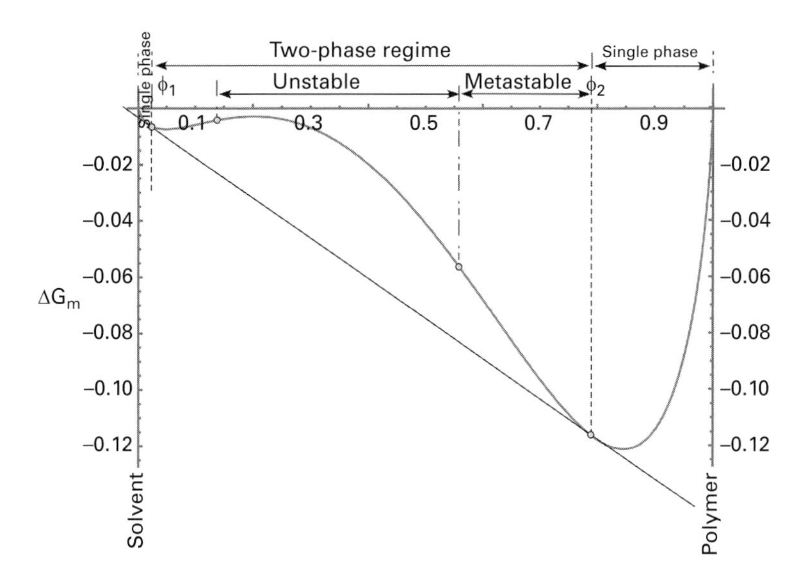

Figure B.3 Free enthalpy of mixing of a polymer-solvent solution with a degree of polymerisation of $N = 5$ and a Flory parameter of 1.5. Touching the curve is the common tangent. The points of touching define the two values of the binodal line, ϕ_1, ϕ_2.

which the chemical potential of the polymer in the solution $\Delta\mu_p$ and the chemical potential of the solvent in the solution $\Delta\mu_s$ are equal (common tangent rule). If the phase diagram exhibits a miscibility gap, the free energy curve has a double-well potential form with two minima and one maximum in a certain temperature range. This is schematically shown in Figure B.3. We can calculate the binodal line from the construction of the common tangent and the equality of the chemical potentials at the point of touching the free enthalpy curve. The chemical potentials are defined as

$$\mu(\phi_1) = \left.\frac{\partial \Delta G_m}{\partial \phi}\right|_{\phi_1} \tag{B.9}$$

and

$$\mu(\phi_2) = \left.\frac{\partial \Delta G_m}{\partial \phi}\right|_{\phi_2}. \tag{B.10}$$

Both derivatives must have the same value. This is not a sufficient condition. The common tangent has the slope m_{ct} determined by the values of the free energy of mixing at the two compositions ϕ_1, ϕ_2 (see [530] for a discussion of phase separation in immiscible melts):

$$m_{ct} = \frac{\Delta G_m(\phi_2) - \Delta G_m(\phi_1)}{\phi_2 - \phi_1}. \tag{B.11}$$

This slope must be equal to both chemical potentials, Eqs. (B.9) and (B.10), meaning $\mu(\phi_1) = \mu(\phi_2) = m_{ct}$. Note: usually the two local minima of the Gibbs-free energy do not define the binodal points (this is only true for $N = 1$). Inserting Eq. (B.1)

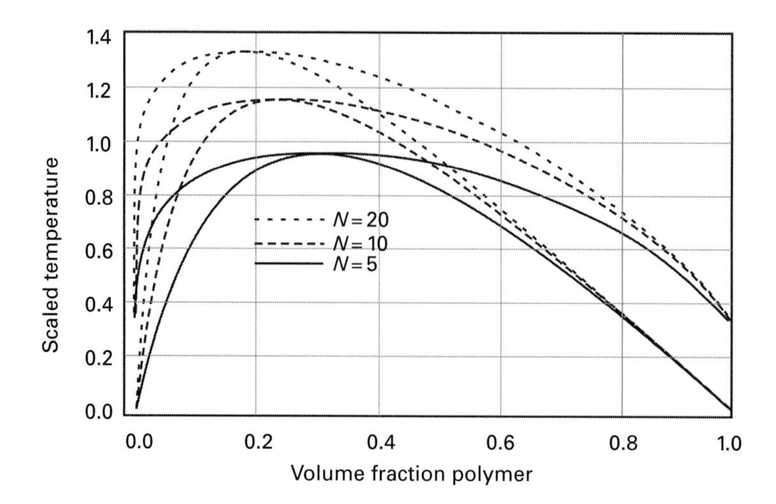

Figure B.4 Calculated spinodal and binodal lines for three different degrees of polymerisation. The critical temperature increases with increasing DP, and the critical volume fraction moves to the solvent side.

and evaluating the preceding equations leads to a set of transcendental equations which have to be solved numerically. A few examples of binodal lines are shown in Figure B.4. The figure shows that the binodal lines are also shifted to the solvent side of the phase diagram, as the spinodal lines are. They also clearly show that the binodals always are even more shifted to the solvent side than the spinodal with increasing DP. This is important with respect to a possible binodal or spinodal decomposition as phase separation mechanisms in aerogel solutions. One has to make a remark here. Looking at Figure B.1, it seems that at a DP of 20 there already is no minimum close to the origin (solvent side), and therefore it would be very difficult to make a common tangent construction. A closer look at the zeros of the first derivative, which defines the local minimum, shows, however, that with increasing DP it rapidly approaches zero, but it never is zero. Fortunately, DPs of 50 and more are probably not necessary to consider for polycondensation reactions, since this already would mean that 98% of the monomers have reacted (see Appendix D).

Let us summarise some results of the classical FH theory. For a given value of the Flory parameter, a miscibility gap with an upper critical solution temperature (UCST) is observed. The critical point shifts towards the solvent side of the phase diagram inasmuch as the degree of polymerisation increases. The volume fraction polymer connected to the critical point decreases in the same direction.

As discussed especially in the context of RF aerogels (see Section 2.2), it might be useful to also discuss another FH model, namely one with a lower critical point, meaning the phase boundaries are inverted to those described for UCST. A lower critical solution temperature (LCST) in a phase diagram is most easily described by a Flory–Huggins parameter χ that depends linearly on temperature:

$$\chi(T) = \chi_1 T, \tag{B.12}$$

with χ_1 a suitable constant. Such a behaviour is, for instance, observed in systems with hydrogen bonds [54, 528, 529]. It would especially describe the polymer solvent interaction being more strongly dependent on temperature. In addition to enthalpic effects, entropic effects are discussed in the literature to explain a linear dependence of χ on temperature, and even complicated extensions and models have been derived to better describe phase diagrams of polymer solutions [270, 531–535]. Here we do not ask for a physical justification of Eq. (B.12) but take it as a model.

The mathematical analysis starts again with the FH expression of the free energy of mixing:

$$\frac{\Delta G_m}{k_B T} = \frac{\phi}{N} \ln \phi + (1-\phi)\ln(1-\phi) + \chi(T)\phi(1-\phi). \tag{B.13}$$

The zeros of the second derivative of this expression defines the locus of the spinodal and reads

$$\frac{1}{N\phi} + \frac{1}{1-\phi} = 2\chi_1 T_{spin}, \tag{B.14}$$

where we have used Eq. (B.12) and denote the spinodal temperature with the subscript 'spin'. Since χ_1 has the unit of 1/K, we plot the scaled spinodal in Figure B.5 for different degrees of polymerisation N. Comparing Figure B.5 with that for an upper critical point, Figure B.2 shows a remarkable difference. In the LCST case, as modelled here, the spinodals are more skew symmetric to the solvent side than in the UCST case. This means that for even a relatively low degree of polymerisation, the spinodal already has its minimum close to the pure solvent. Let us calculate the variation of the critical temperature and the critical volume fraction with the degree

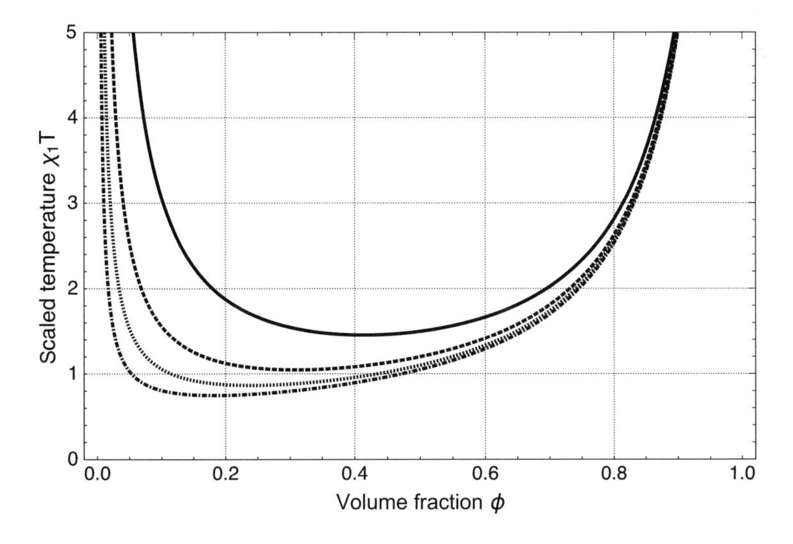

Figure B.5 Variation of the spinodal line in an LCST system with degree of polymerisation. Solid line $N = 2$, dashed line $N = 5$, line with short dashes $N = 10$, dotted line $N = 20$.

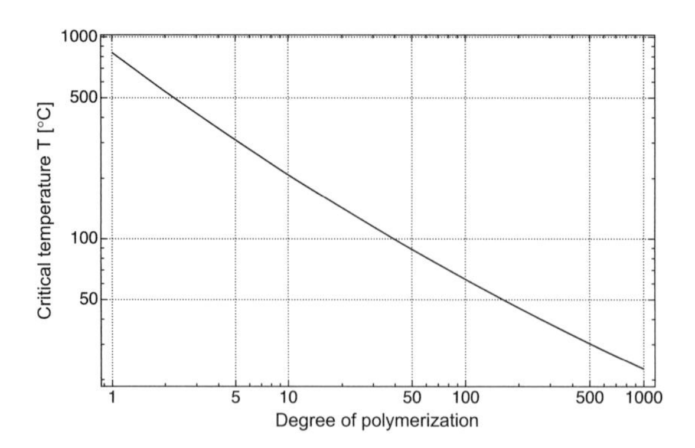

Figure B.6 Double logarithmic plot of the variation of the critical temperature in °C for a system with a lower critical point using $\chi_1 = 0.0018$ as a parameter to rescale temperature. Note that with a DP of around 50–100, the spinodal is below the typical annealing temperature used, for instance, in the gelation process of RF aerogels; see Section 2.2.

of polymerisation. Both are obtained from Eq. (B.14) taking the first derivative with respect to the volume fraction polymer. The two expressions now read

$$T_c = \frac{\left(1+\sqrt{N}\right)^2}{2N\chi_1} \tag{B.15}$$

$$\phi_c = \frac{1}{1+\sqrt{N}}. \tag{B.16}$$

At $N = 1$ the parameter χ_1 is directly related to the critical temperature as $\chi_1 = 2/T_c(1)$. The critical temperature decreases rapidly with the DP. An example is shown in Figure B.6 for $\chi_1 = 0.0018$. This example shows that with a DP of around 50–100 the spinodal is below typical annealing temperature used, for instance, in the gelation process of RF aerogels; see Section 2.2. In the simple expression of Eq. (B.13), the spinodals shift rapidly with increasing DP to the solvent side and become extremely steep. This probably is a bad description of reality. One could, for instance, extend the given description by so-called subregular terms like

$$\frac{\Delta G_m}{k_B T} = \frac{\phi}{N}\ln\phi + (1-\phi)\ln(1-\phi) + \phi(1-\phi)\left[\chi_1 T + \chi_2(1-2\phi)\right], \tag{B.17}$$

which would allow us to shift the miscibility gap by suitable adjusting parameters χ_1, χ_2 a bit more to the polymer side. Although the model used is very simple to describe a system with a lower critical point, it covers at least the essentials. Better thermodynamically motivated models are needed to better describe (aero)gel forming solutions.

Appendix C A Brief Review on Scattering

Scattering of light or X-rays is scattering of electromagnetic waves, and these are described most simply by a wave having a wave vector $\vec{k}$, a frequency ω and an amplitude $\vec{E}_0$, the electric field strength, such that the electrical field can be described as $\vec{E} = \vec{E}_0 \exp(i(\vec{k} \cdot \vec{r} - \omega t)$. If an electromagnetic wave interacts with a material, it interacts with the electrons in it. The beam might be reflected (visible light on many metals) or it may go through (light on a pane of glass). Whenever a beam travels through a material, there also is a scattered beam with a direction different from the incident beam. In order to visualise this, one can take a piece of material and place behind it a suitable detector. If one uses light, let us say a laser beam, the material should be sufficiently transparent such that the incident beam can pass through. As a detector, one then can use any camera system available today, if it is sensitive to the frequency of the laser used. Typically one then will see on the screen in the centre a bright spot stemming from the incident or primary beam, and one might see a radially decaying diffusive light. This can have maxima in the form of rings. In, for instance, a colloidal system, which changes its colloidal size during exposure one might see that the rings move outwards. If one uses X-rays instead, the system under investigation should be sufficiently transparent for X-rays, but this is much easier to achieve compared to visible light and most materials are transparent for X-rays (at least if they are not too thick). The way to measure X-rays passing a material, however, is different from that for light rays. Special detectors are available for X-rays, working similar to a charged coupled device (CCD) sensor (direct photon counting). We are not reviewing all aspects of light or X-ray scattering – this is a discipline in itself and would need at least a book to describe all details (a lot of them can be found in books on classical electrodynamics or quantum mechanics or in special books on scattering [420, 536]). Here we are just concerned with small angle scattering and will only treat a few aspects and try to derive those equations used in the literature about aerogels and their response to light or X-rays. This review is partially based on a review given by Sorensen [537] and uses the original literature of Debye et al. [421] and Porod [418, 419], and of course many publications made in the context of aerogels on small angle X-ray scattering (SAXS) and dynamic light scattering (DLS).

Having an incident beam with a wave vector $\vec{k}_i$, which will be simply described as a wave $\vec{E}_i \propto \exp(i\vec{k} \cdot \vec{r})$, this beam will pass through the sample, but some intensity also goes into a detector being at a distance L away from the sample and inclined at an angle 2θ. The scattered amplitude depends on the scattering ability or scattering

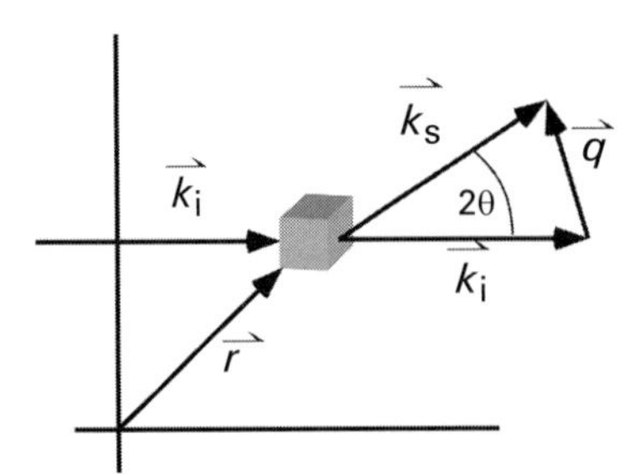

Figure C.1 Schematic diagram of an incident wave hitting a sample volume located at $\vec{r}$ in space having itself a wave vector $\vec{k_i}$ and scattered at an angle θ. The scattered wave has a wave vector $\vec{k_s}$ and we have there $\vec{k_i} + \vec{q} = \vec{k_s}$.

power of the material ρ, and the scattered field strength $d\vec{E}_s$ is proportional to it and the volume element dV it comes from, such that $d\vec{E}_s = \rho dV$. If we have a large particle in which each volume element dV acts like Rayleigh scatterers, we can add them up and the scattered wave will have the electrical field

$$\vec{E}_s = \int_{V_p} e^{-i\vec{q}\cdot\vec{r}} \Delta\rho(\vec{r}) dV, \tag{C.1}$$

in which $\Delta\rho$ is the scattering power difference to, for instance, a surrounding medium of the particle and $\vec{q}$ is the so-called scattering vector. This is defined as the difference between the wave vector of the incident and the scattered beam:

$$\vec{q} = \vec{k_i} - \vec{k_s}. \tag{C.2}$$

Looking at Figure C.1 and performing simple algebra, we obtain for the magnitude of the wave vector if the scattering is elastic

$$|\vec{k_i}| = |\vec{k_s}| = \frac{2\pi}{\lambda} \quad \text{and} \quad q = \frac{4\pi}{\lambda} \sin(\theta), \tag{C.3}$$

with λ the wavelength of the radiation used. Note here that the inverse of the scattering wave vector $\vec{q}$, q^{-1} represents a length scale, and that this length scale first depends on the wavelength of the radiation used but also on the scattering angle. Table C.1 shows some values of q for light (wavelength 400 nm, energy 10 eV) and X-rays (energy 10 keV) as they vary with the angle of the detector. The scattering power difference to a surrounding medium, index 0, of a particle can be expressed for light by the refractive index difference $\Delta n = n_p - n_0$,

$$\Delta\rho \propto \frac{n_p^2 - n_0^2}{n_p^2 + n_0^2} \propto n_p - n_0 = \Delta n, \tag{C.4}$$

or for X-rays the scattering power is proportional to the number density of electrons and the scattering length for X-rays (in 0.282 pm). We should also remark that scattering of either light or X-rays is always due to time or local fluctuations of the electron density or the refractive index in a material. If the material would be without

Table C.1 Values of the scattering vector q as depending on the scattering angle θ.

θ	0.1°	1°	10°	100°	180°
Light q [nm^{-1}]		$0.3\ 10^{-3}$	$3\ 10^{-3}$	0.02	0.03
X-rays q [nm^{-1}]	0.01	0.1	1	10	13

any local fluctuations or variations, there would be no scattering. For example, it is well known that the blue light of the sky is not only due to the frequency dependence of the diffractive index but essentially due to density fluctuations in our atmosphere. The essential difference between small angle X-ray scattering and conventional X-ray diffraction is that in small angle scattering angles up to 5–10° deviation from the primary beam are looked at, and thus small q-values, whereas in wide angle X-ray scattering (WAXS) large angles typically starting from 20° are measured.

Let us now consider a volume of a material which contains an ensemble of N particles each located at a vector $\vec{R}_j$. Then the total scattering is simply the sum of all scattered amplitudes as given in Eq. (C.1):

$$\vec{E}_s = \sum_{j=1}^{N} \vec{E}_{s,j} e^{-i\vec{q}\cdot\vec{R}_j} = \sum_{j=1}^{N} \int_{V_j} \Delta\rho(\vec{r}) e^{-i\vec{q}\cdot\vec{r}} dV e^{-i\vec{q}\cdot\vec{R}_j}. \tag{C.5}$$

We rewrite this equation as

$$\vec{E}_s = \sum_{j=1}^{N} b_j(\vec{q}) e^{-i\vec{q}\cdot\vec{R}_j}, \tag{C.6}$$

introducing the scattering contribution from each particle in the system $b_j(\vec{q})$. Any detector does not measure the scattered field E_s but the intensity, which is the square of the scattered field, and also usually a time or ensemble average is measured (subindex t). The intensity in the detector can then be written as

$$I_s(q) = \langle I_s(q,t) \rangle_t = \langle |E_s(q,t)^2| \rangle_t = \sum_k \sum_j \langle b_j(\vec{q}) b_k(\vec{q}) e^{-\vec{q}\cdot(\vec{R}_j(t) - R_k(t))} \rangle. \tag{C.7}$$

We now make an essential assumption: all particles are identical and their properties are not linked to the position in space. We then rearrange the terms in Eq. (C.7) as

$$I_s(q) = N^2 P(\vec{q}) S(\vec{q}), \tag{C.8}$$

with the definition of a so-called form factor

$$P(\vec{q}) = \langle |b_j(\vec{q})|^2 \rangle \tag{C.9}$$

and the so-called structure factor

$$S(\vec{q}) = \frac{1}{N^2} \sum_k \sum_j \langle e^{-\vec{q}\cdot(\vec{R}_j - R_k(t))} \rangle. \tag{C.10}$$

The decomposition of the scattering curve into a factor stemming from the shape of the scatterers, $P(q)$ and their arrangement in space $S(q)$ shows that scattering of an aerogel would be determined by two things: the shape of the particles making up the aerogel, spheres or cylinders (fibrils) and their networking, which is mirrored in the structure factor. When particles are distributed without any particle–particle correlations as in dilute dispersions of non-interacting particles, then the scattering intensity is entirely from the form factor, or in other words $S(\vec{q}) \to 1$, if the distance between the scatterers becomes very large. When particles are in a well-defined or a correlated structure, the scattering is dominated by the structure factor with, however, a modulating contribution by the form factor. The whole literature on scattering is devoted to the calculation of the form and structure factor, for example, colloidal dispersions, aerogels, porous materials, shape and form of macromolecules suspended in a liquid and so on.

C.1 Form Factor

Since in many aerogels particles make up their structure, it might be useful to calculate the form factor for spherical particles. To start with, we calculate the $b(\vec{q})$ for a sphere. From its definition, we have

$$b(\vec{q}) = \int \Delta\rho e^{-i\vec{q}\cdot\vec{r}} dV = \Delta\rho \int e^{-i\vec{q}\cdot\vec{r}} dV. \tag{C.11}$$

Using polar coordinates, we write the integral over the volume as

$$\int_0^{2\pi} d\phi \int_0^R r^2 \int_0^\pi e^{-iqr\cos\theta} \sin\theta d\theta dr, \tag{C.12}$$

which evaluates $b(q)$ to

$$b(q) = \frac{4\pi\Delta\rho R^3}{3} \frac{3(\sin(qR) - qR\cos(qR))}{(qR)^3} = \Delta\rho V_p F(qR). \tag{C.13}$$

The form factor depends on the scattering vector and the particle size

$$P(q, R) = |b(q)^2| = (\Delta\rho)^2 V_p^2 F(qR)^2. \tag{C.14}$$

The form factor of a sphere is shown in Figure C.2. It is easy to show that $F(qR) = 1$ if $qR = 0$. This result holds independently of the shape of the scatterers. Assume $q = 0$, then the scattering amplitude

$$b(\vec{q}) = \Delta\rho(\vec{r}) \int e^0 dV = \Delta\rho V_p \tag{C.15}$$

and then the form factor is

$$P(0) = \Delta\rho^2 V^2, \tag{C.16}$$

and thus the scattering intensity is always proportional to the square of the particle volume.

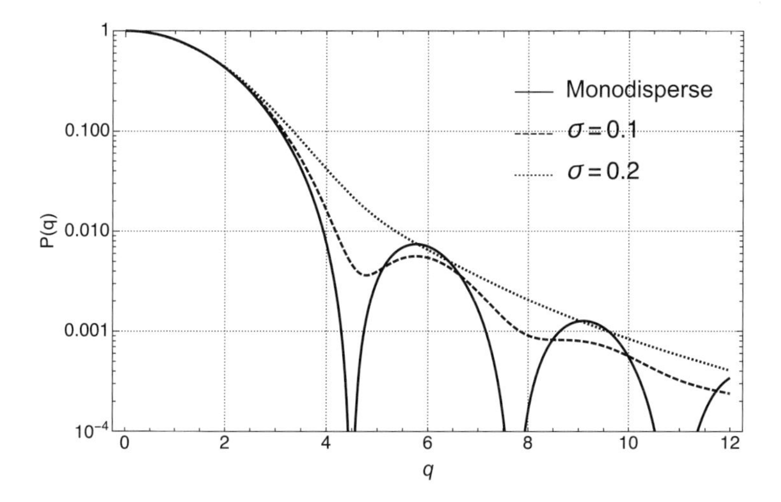

Figure C.2 Form factor of a sphere as a function of scattering vector $q \cdot R$ and that of a dispersion of spheres with a Gaussian size distribution having two different variances of their size. The average particle radius is in all cases set to $\bar{R} = 1$.

Since typically an aerogel or a dispersion of particles does show a size distribution of radii, the form factor might be averaged with the known size distribution $f(R)dR$, denoting the number of particles per unit volume in the size range $R, R + dR$. The average form factor can then be calculated from

$$\langle P(q) \rangle = \frac{\int_0^\infty P(q, R) f(R) dR}{\int_0^\infty f(R) dR}. \tag{C.17}$$

Choosing a Gaussian distribution of particle sizes and performing the calculations leads for two different standard deviations of the particle sizes to a smoothing of the form factor as it depends on qR. The result is shown in Figure C.2 clearly marking that with a narrow distribution the scattering curve exhibits some cusps, while a broad distribution just shows a decrease with increasing qR (note the scattering intensity is exactly the form factor if the structure factor would be unity as in dilute dispersions of the scatterer, e.g., coiled polymers in a solution, nanoparticles or rods in a solvent). Fibrils are the other extreme of structural element making aerogels. Fibrils can be imagined as cylinders. The form factor of cylinders was calculated by Guinier and Fournet [420]. We only state the result here for a cylinder of radius R and a length L:

$$P(q) = \int_0^{\pi/2} \left| \frac{2J_1(qR \sin\alpha}{qR \sin\alpha} \frac{\sin(qL\cos(\alpha/2))}{qL\cos(\alpha/2)} \right]^2 \sin\alpha d\alpha. \tag{C.18}$$

In this equation, $J_1(x)$ is the first-order Bessel function and α is the angle at which the cylinder is oriented in space. The integration can easily be performed numerically. We omit it to show the result and refer to the literature cited on scattering. Today, SAXS or small angle neutron scattering (SANS) software includes a huge number of form factors and models to fit any type of experimental data.

Let us look at some specialties of the form factor for spheres. If one connects in Figure C.2 all maxima with a straight line, one finds that this line has in a double logarithmic plot a slope of −4. At large values of qR, the form factor behaves like

$$P(q) \propto \frac{1}{q^4}\frac{R^2}{R^6} \propto \frac{S_V}{V_p} q^{-4}. \tag{C.19}$$

This behaviour is called Porod's law. The slope of the scattering curve at large scattering vectors is determined by the specific surface area. The exact expression going back to Porod reads

$$V_p P(q) = 2\pi S_V q^{-4} \quad \text{for} \quad \lim q \to \infty. \tag{C.20}$$

Another interesting issue is due to Guinier. If one makes a series expansion of $P(q)$ for small values up to second order, one obtains

$$P(q) \propto \left| \frac{3(\sin(qR) - qR\cos(qR))}{(qR)^3} \right|^2 \cong 1 - \frac{1}{5} q^2 R^2. \tag{C.21}$$

The radius of gyration of a sphere is defined as the second moment of inertia of a sphere and calculates as $R_g = \sqrt{3/5}R$. Insertion yields

$$P(q) \propto 1 - \frac{1}{3} q^2 R_g^2. \tag{C.22}$$

Therefore a lot of $q^2 I(q)$ at very small values of q allow us to determine the radius of gyration of the scatterers. Generally, this Guinier law is expressed as

$$\lim_{q \to 0} I_s(q) = \Delta\rho^2 V_p^2 \exp\left(-\frac{1}{3} q^2 R_g^2\right) \tag{C.23}$$

since the series expansion of the exponential leads exactly to the relation given in Eq. (C.21). Thus we have at very low scattering vectors a Guinier law and at large vectors the Porod law.

C.2 Structure Factor

Aerogels are, however, networked particles or fibrils or a mixture of both simple entities. Therefore, the form factor is not that essential, and the structure factor is more decisive for the scattering intensity. The structure factor $S(q)$ describing the spatial arrangement of the scatterers in space by Eq. (C.10) does not look nice. One can convert the double sum into another expression, which is more useful by introducing a density function of the scatterers in space

$$n(\vec{r}) = N^{-1} \sum_i^N \delta(\vec{r} - \vec{R}_i), \tag{C.24}$$

in which δ denotes Dirac's delta function. We then can write

$$\sum_i^N e^{-i\vec{q}\cdot\vec{R}_i} = N \int e^{-i\vec{q}\cdot\vec{r}} \delta(\vec{r} - \vec{R}_i) d\vec{r}. \tag{C.25}$$

We now use the definition of the density function to calculate the structure factor:

$$S(\vec{q}) = N^{-2} \sum_i^N \sum_j^N e^{-i\vec{q}\cdot(\vec{R}_i - \vec{R}_j)} = \int\int n(\vec{r}) n(\vec{r}') e^{i\vec{q}\cdot(\vec{r}-\vec{r}')} d\vec{r} d\vec{r}'. \tag{C.26}$$

Changing the variables in the integrand by defining $\vec{u} = \vec{r} - \vec{r}'$ leads to

$$S(\vec{q}) = \int\int n(\vec{r}) n(\vec{r} - \vec{u}) e^{-i\vec{q}\cdot\vec{u}} d\vec{r} d\vec{u}. \tag{C.27}$$

Looking at the second integral, one immediately recognises that this has the form of an autocorrelation function, which we write as

$$\gamma(\vec{u}) = \int n(\vec{r}) n(\vec{r} - \vec{u}) d\vec{r}. \tag{C.28}$$

The structure factor can now be written as

$$S(\vec{q}) = \int \gamma(\vec{u}) e^{-i\vec{q}\cdot\vec{u}} d\vec{u} \tag{C.29}$$

and thus the structure factor is a Fourier transform of the autocorrelation function. If one assumes isotropy, meaning there is no preferred orientation, one can perform the integration over the solid angles θ, ϕ, yielding

$$S(q) = 4\pi \int \gamma(u) \frac{\sin(qu)}{qu} u^2 du. \tag{C.30}$$

This can be rearranged into a form quite often used in scattering physics:

$$S(q) - 1 = 4\pi \int (g(r) - 1) \frac{\sin(qr)}{qr} r^2 dr, \tag{C.31}$$

and in that context $g(r)$ is called the radial distribution function. This naming has a simple physical interpretation: from an arbitrary point in a structure, ask how many particles are in a shell $r, r + dr$, and $g(r)$ is the answer to that. By this understanding, we have $g(r = 0) = 1$. The function can have maxima, and they are usually related to the first coordination shell around a particle, the next nearest neighbour shell etc. If one does not believe in that, draw circles and place them either regularly or somehow randomly on a sheet of paper, take an arbitrary central circle, draw around this circle with a given radial distance, count the number of particle centres in each circular shell, divide it by the total number of circles and plot this number as a function of distance from the centre. In the case of a regular arrangement, simple cubic or hexagonal, one yields maxima in that curve. If the circles were placed randomly, one will be more or less pronounced, and you can then find a simple exponential decay with the distance from the origin.

In the application of scattering to aerogels, everything we have discussed so far was just preliminary (and it already contained a lot of equations). Typically the scattering intensity is plotted as a function of (inverse) scattering vector (or angle) and the resulting curve is then interpreted with the help of theoretical models, which make suggestions for the form and the structure factor. One of the models used in the aerogel literature goes back to Porod [418, 419] and Debye et al. [421]. In both references, a model for the autocorrelation function is given:

$$\gamma(r) = \gamma_0 \exp(-r/\xi). \tag{C.32}$$

In this relation, r would be the radial distance from an arbitrary particle in the structure and ξ is a characteristic length, also called the correlation length. Debye gave a derivation of this exponential function, whereas Porod calls it just an exponential model and others call it a two-phase model (TPM). Anyhow, insertion into Eq. (C.30) and performing the integration from $r = 0$ to $r = \infty$ yields

$$S(q) = 2\gamma_0 \frac{\xi^3}{(1 + \xi^2 q^2)^2}. \tag{C.33}$$

The functional dependence is shown in Figure C.3. At larger q-values, the structure factor decays like q^{-4}, which is known as a Porod's law. Fitting such an expression to an intensity curve allows us to derive the characteristic length. One could make here another remark. The radial distribution function is a so-called two-point correlation function. Recently Gommes et al. discussed whether the knowledge of the radial distribution function allows one to unambiguously reconstruct the microstructure from it, which is especially important in the context of scattering [538, 539]. The authors discuss trivial degeneracies of the correlation function, meaning there are space transformations like translation, rigid body rotation and inversion which lead to identical two-point statistics and the more severe case of non-trivial degeneracies

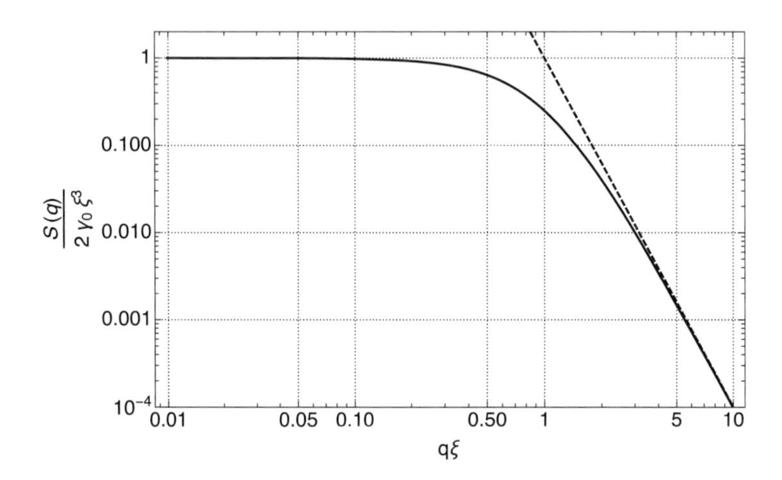

Figure C.3 Structure factor for X-ray scattering with an exponentially decaying autocorrelation function (solid line) and Eq. (C.33) compared to a simple power law q^{-4} (dashed line)

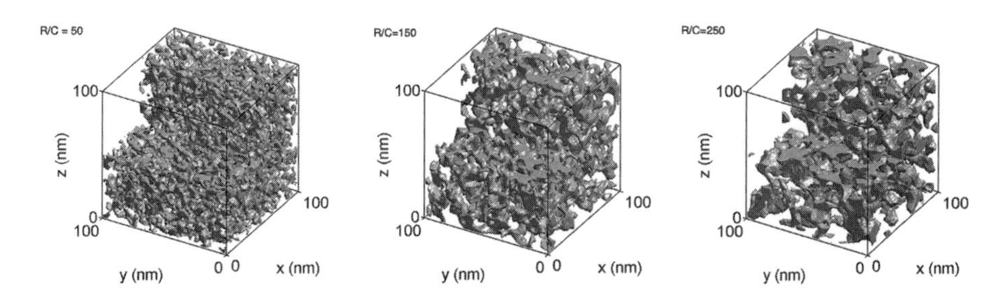

Figure C.4 Three-dimensional reconstruction of three RF aerogels made with different R/C ratios having a fraction solid of 23% after Gommes [540]. Reprinted with permission of APS

as, for instance, observed in random media as constructed by the Debye correlation function. A way out of this dilemma was shown by them. First one constructs a model correlation function $g_m(r)$ and compares it in a certain sense in a least square method with the experimental one $g_e(R)$, called by the authors an energy functional $E = \sum_r |g_m(r) - g_e(r))|^2$, which has to be minimised. Gommes and Roberts [540] discuss in detail the reconstruction of the aerogel structure using as a model Gaussian random waves. Variation of these three parameters allows to minimise the energy functional. They finally could construct 3D models of RF aerogels and reproduce the scattering intensity in agreement with experiments. The structure obtained is shown in Figure C.4. The structure might differ from those shown in Chapter 6, but with respect to the large amount of solid fraction it is quite convincing.

It is also worthwhile to study a so-called invariant. Let us calculate the following integral

$$Q_S = \int_0^\infty q^2 S(q) dq = \gamma_0 \frac{\pi}{2}, \tag{C.34}$$

being a constant or, besides some proportionality factors, the area under the scattering curve multiplied with q^2. A similar result holds for the form factor using the expression derived earlier for spheres:

$$Q_P = \frac{4\pi}{3} R^3 \int_0^\infty q^2 P(q) dq = 2\pi^2, \tag{C.35}$$

which is the invariant given by Porod in his 1951 papers [418, 419] and proved to be correct for various form factors. As done previously with the form factor, one can also make a Guinier approximation of the structure factor being also valid for small q-values. One can expand in Eq. (C.30) the $x^{-1} \sin x$ for small values of x to second order, leading to $\sin(x)/x \approx 1 - 1/6(qu)^2$. Insertion of this approximation leads to

$$S(q) \approx 1 - \frac{q^2}{6} \int \gamma(u) u^2 du. \tag{C.36}$$

Defining as the radius of gyration (which is equivalent to the second moment of inertia)

$$R_g = \frac{1}{2} \int \gamma(u) u^2 du, \tag{C.37}$$

we could express the structure factor as

$$S(q) \cong 1 - \frac{1}{3} q^2 R_g^2 \quad \text{or} \quad S(q) \cong \exp\left[-\frac{1}{3} q^2 R_g^2\right], \tag{C.38}$$

where the last relation holds for small q only. Interestingly, both Guinier's and Porod's laws hold if the scattering is determined by the form or the structure factor alone.

C.3 Specific Surface Area

Debye et al. [421] and Porod [418] have shown that there exists a simple relation between the autocorrelation function and the specific surface area, namely

$$S_V = -4\phi_s(1-\phi_s) \left.\frac{d\gamma}{dr}\right|_{r=0} = 4\frac{\phi_s(1-\phi_s)}{\xi}. \tag{C.39}$$

Thus one needs to extract from measurements the correlation length. As already pointed out by Debye, one can plot $1/\sqrt{I_s(q)}$ against q^2 and the ratio of slope and intercept will allow one to determine ξ as

$$\xi = \sqrt{\frac{\text{slope}}{\text{intercept}}}. \tag{C.40}$$

Another way to determine the specific surface area uses the following expression [102]:

$$S_V = \pi\phi(1-\phi_s)\frac{K}{Q}, \tag{C.41}$$

in which K is the limit of $I(q)q^4$ and Q is defined as $\int q^2 I(q) dq$. In order to understand this relation, assume that the scattered intensity $I_s(q)$ is fully determined by the structure factor and assume also the autocorrelation function is a simple exponential (it must not be; for other approaches, see [537]). We then can write

$$I_s(q) = CS(q) = 2C\frac{\xi^3}{(1+\xi^2 q^2)^2}, \tag{C.42}$$

which allows to determine the constant K

$$K = \lim q^4 I_s(q) = \frac{2C}{\xi} \tag{C.43}$$

and the constant

$$Q = \int_0^\infty q^2 I_s(q) dq = C\frac{\pi}{2}. \tag{C.44}$$

Insertion of both results into Eq. (C.39) to eliminate ξ and C yields Eq. (C.41). The problem to determine the specific surface area is then reduced to integrate the scattering curve and to fit in a suitable q-range the Porod law.

C.4 Dynamic Light Scattering

Dynamic light scattering (DLS) or photon correlation spectroscopy is another scattering technique that uses especially laser light and analyses the scattered intensity. In principle, scattering of light is quite similar to the previously discussed scattering of X-rays. The scattered intensity is also a product of a form factor and a structure factor and the amplitude itself depends on the difference between the refractive index of, for example, particles in a solution and the solution itself. The main difference, however, is that the intensity is typically not analysed as function of the scattering vector (or angle), but the scattered light intensity is measured as a function of time. A sufficiently transparent sample is illuminated with a laser, and the intensity is tracked at a fixed scattering angle. The intensity is, as for X-rays, proportional to the illuminated volume squared. When a dispersion of particles, solid particles, bubbles, droplets, coiled macromolecules, etc. is hit by a laser beam, scattered light waves are generated and spread out in all directions. These scattered waves make interference in the far field region, and the net scattered light intensity I_s is recorded by a detector (photon detector, CCD, complementary metal–oxide–semiconductor [CMOS] chips). If the particles suspended are sufficiently small, around a micron or less, they perform a Brownian, random motion in the illuminated volume which leads to a random scattered intensity. Due to any type of particle motion, the number of particles in the illuminated volume will vary and add to the random scattering signal.

The signal $I_s(t)$ is not analysed itself, but one analyses the autocorrelation function

$$g_2(q,\tau) = \frac{\langle I_s(q,t) I_s(q,t+\tau)\rangle}{\langle |I_s(q,t)|^2\rangle}. \tag{C.45}$$

In this equation, we define as the average

$$\langle I_s(q,t) I_s(q,t+\tau)\rangle = \lim_{T\to\infty} \frac{1}{T} \int_{-T}^{T} I_s(q,t) I_s(q,t+\tau) dt. \tag{C.46}$$

Note any autocorrelation function has a maximum at $\tau = 0$ and is an even function of τ. The autocorrelation function is calculated from the recorded intensity by simply shifting its value with various time lags τ, multiplying the result with $I_s(t)$ and integrating over time (in reality, the integration is over a finite interval). The normalised autocorrelation function $g_2(\tau)$ is related to the scattered electric field correlation function $g_1(\tau)$ by Siegert's relation

$$g_2(q,\tau) = 1 + |g_1(q,\tau)|^2 \tag{C.47}$$

with

$$g_1(q,\tau) = \frac{\langle E_s(q,t)E_s(q,t+\tau)\rangle}{\langle |E_s(q,t)|^2\rangle}. \tag{C.48}$$

Whereas the scattered field autocorrelation function $g_1(\tau)$ is equal to 1 at $\tau = 0$ and decays with increasing time lag, $g_2(\tau)$ starts with a value of 2 and decays after long time lags to 1. Assume a monodisperse dispersion of particles in a solution. Then the autocorrelation function $g_1(q,\tau)$ can be shown to follow a simple exponential decay:

$$g_1(q,\tau) = \exp(-\Gamma\tau) = \exp(-2D_B q^2 \tau). \tag{C.49}$$

Here D_B is the Brownian diffusion coefficient and of course $\Gamma = 2D_B q^2$. The measurement of the autocorrelation function g_2 thus allows us to determine directly the Brownian diffusion coefficient, and using the Stokes–Smoluchowski–Einstein relation, one can determine the hydrodynamic radius R_H of the moving particles:

$$R_H = \frac{k_B T}{6\pi\mu_0 D_B}, \tag{C.50}$$

in which μ_0 is the viscosity of the solvent. Here it is very important to note that the hydrodynamic radius must not be the radius of the particle, not the radius of gyration or so. As an example, assume that nanoparticles are connected to each other in a cluster, and this cluster makes a Brownian motion. The hydrodynamic radius would then be that of the cluster and not that of its constituting entities. If such a cluster would be a fractal aggregate, the hydrodynamic radius might better be identified as the correlation length of the cluster.

With light scattering in solutions which transform to gels, we have to deal with two real problems. First, once particles are nucleated or formed in a solvent, there will be a distribution of particle sizes. Second, the denser a dispersion or suspension of particles is, the less transparent it is and multiple scattering might occur. Standard models for the autocorrelation function always assume first a monodisperse size distribution and second single scattering. Multiple scattering, meaning a photon on its way through the volume is scattered more than once by many particles, poses a real problem for analysing the intensity. The problem with a polydisperse size distribution is typically solved with the assumption that a sum of exponential decays is then acting, meaning

$$g_1(q,\tau) = \sum_{i=1}^{n} G_i(\Gamma_i)\exp(-\Gamma_i\tau). \tag{C.51}$$

Modern DLS equipment uses various methods to extract from measured intensity data the particle size distribution. The problem of multiple scattering is eminent for aerogels, since typically a volume fraction solid is in the range of 1–10%. For typical DLS measurements, the sample should stay transparent all the time. This is definitively not the case in RF aerogels, in which a solution changes from transparent to turbid and then opaque, in contrast to silica and biopolymers. The solid fraction inside a solution is sufficiently high for multiple scattering to occur. One way out of this dilemma is to have the same sample volume illuminated by two beams and then measuring their cross-correlation function, which is the time-averaged product of the scattering intensity from both beams (including a time shift of one beam by τ). Then the single scattering contribution will be identical from both beams, but the multiple scattering contribution will be different, since the light of both beams takes different paths through the volume. The correlation of both with each other results in a cross-correlation function, which contains only the single scattering contribution. Summarising, one would expect for solutions during their transformation to a gel first an exponential decay and a strong deviation from it, once gelation has started, since this suppresses any Brownian motion of the particles.

Appendix D Mathematics of Polycondensation

In almost all texts on aerogels, it simply is stated that the reaction leading to polymeric particles is a polycondensation or polyaddition. Rarely there is mathematical treatment with some predictions which could be tested experimentally. We therefore will give our way of looking at polycondensation reactions. As mentioned, many aerogels are made from solutions of monomers which react under alkaline or acidic conditions, forming dimers, trimers, tetramers, pentamers and so on and eventually polymers mostly by a polycondensation reaction. This is typically described as a bimolecular reaction with a second-order kinetic equation.

D.1 A Simple Model of a Bimolecular Reaction

We will first discuss a bimolecular reaction, meaning a component A reacts with a component B and yields another component C. The reaction rate r is defined as the time rate of change of the concentration of A in the solution c_A:

$$r = -\frac{dc_A}{dt} = kc_A \cdot c_B. \tag{D.1}$$

Assume that $c_A(t=0) \neq c_B(t=0)$. Then one can make use of mass conservation:

$$c_B(t) = c_A(t) + c_B(0) - c_A(0) = c_A(t) + c_B^0 - c_A^0. \tag{D.2}$$

Insertion into Eq. (D.1) and integration with respect to time yields

$$c_A(t) = c_A^0 \frac{c_b^0 - c_a^0}{c_b^0 \exp((c_b^0 - c_a^0)k \cdot t) - c_A^0}. \tag{D.3}$$

In the case of $c_A^0 = c_B^0$ the expression simplifies to

$$c_A(t) = \frac{c_A^0}{1 + c_a^0 k \cdot t}. \tag{D.4}$$

The concentration of A (or B) decrease hyperbolically with time. Let us define a conversion factor of this reaction as

$$p = \frac{c_A^0 - c_A(t)}{c_A(t)} \tag{D.5}$$

and define a degree of polymerisation, DP, as

$$DP = \frac{c_A^0}{c_A(t)}. \tag{D.6}$$

After some manipulations of both equations, we get the so-called Carother's equation:

$$DP = \frac{1}{1-p}. \tag{D.7}$$

For a polycondensation or step- or chain-growth reaction, we have that continuously A and B react to form C (and typically a by-product such as water) and C reacts with either A and B to form the next step and so on until all monomers A and B are consumed. If all subsequent steps can be described by the same reaction laws, Eq. (D.3) holds generally and the DP defined previously would really make sense and then Eq. (D.7) would show that a high degree of polymerisation needs a high conversion rate. At the conversion rate $p = 0.9$, meaning that 90% of A is consumed, one would only have a $DP = 10$. In the next section, we treat the polycondensation or step-growth reaction in more detail. It will turn out that the preceding given relations still hold.

D.2 A More Complex Model of Polycondensation

Instead of following the preceding simple line of thinking, we follow with another approach, which allows us to answer questions like how the volume fraction polymers (and this means everything but the monomers) develop with time and how the average degree of polymerisation develops. Although the last question can be answered by Carother's equation, we will show that the same result can be obtained with a different approach.

D.2.1 A Variant of Smoluchowski's Aggregation Equation

Let us first outline how the reactions proceed in a solution of monomers. We denote the monomers as (1), dimers as (2) and k-mers as (k). The reactions proceeding in a solution can then be described as

$$\begin{aligned}
(1) + (1) &= (2) \\
(2) + (1) &= (3) \\
(3) + (1) &= (4) \\
(3) + (2) &= (5) \\
(2) + (2) &= (4) \\
&\cdots \\
(k) + (1) &= (k+1)
\end{aligned} \tag{D.8}$$

$$(k) + (2) = (k+2)$$
$$\ldots$$
$$(k) + (k) = (2k).$$

These reactions are all bimolecular ones. But one has to note that high-order oligomers can be formed from different starting units: a tetramer can be made from a trimer with a monomer, but also from two dimers. All these reactions may take place concurrently in a solution and therefore at any time there is a distribution of polymers with different degrees of polymerisation (these may also vary locally in a solution, but this is neglected here). The reaction kinetics can be described by an infinite set of first-order differential equations, already done by Marian Smoluchowski 100 years ago for Brownian aggregation of particles [223]. This set reads

$$\begin{aligned}
\frac{dn_1}{dt} &= -K_V n_1^2 - K_V n_1 n_2 - K_V n_1 n_3 - \cdots \\
\frac{dn_2}{dt} &= \frac{1}{2} K_V n_1^2 - K_V n_2^2 - K_V n_1 n_2 - K_V n_2 n_3 - \cdots \\
\frac{dn_3}{dt} &= K_V n_1 n_2 - K_V n_1 n_3 - K_V n_2 n_3 - K_V n_3^2 - \cdots \\
&\ldots
\end{aligned} \tag{D.9}$$

in which the n_k denote the number density of k-mers per unit volume and $K_V = f(x_{cat}, T)$ is the reaction rate constant depending on the concentration of catalyst x_{cat} and temperature T. Note that K_V has the unit volume/second. We implicitly assumed that the reaction rate K_V is a constant and does not depend on the size of the oligomer. The general form of the equation is

$$\frac{dn_m}{dt} = \frac{K_V}{2} \sum_{l=1}^{m-1} n_l n_{m-l} - K_V n_m \sum_{l=1}^{\infty} n_l. \tag{D.10}$$

This set of equations can be solved by iteration and leads to the following solution for the number density of m-mers:

$$n_m(t) = N_0 \frac{(\frac{1}{2} N_0 K_V t)^{m-1}}{(1 + \frac{1}{2} N_0 K_V t)^{m+1}} = \frac{N_0}{(1 + \frac{1}{2} N_0 K_V t)^2} \left(\frac{\frac{1}{2} N_0 K_V t}{1 + \frac{1}{2} N_0 K_V t} \right)^{m-1}, \tag{D.11}$$

with N_0 the total number of monomers per unit volume initially. The behaviour of these oligomeric densities is plotted in Figure D.1. The total number of particles in the system per unit volume is

$$N(t) = \sum_{m=1}^{\infty} n_m. \tag{D.12}$$

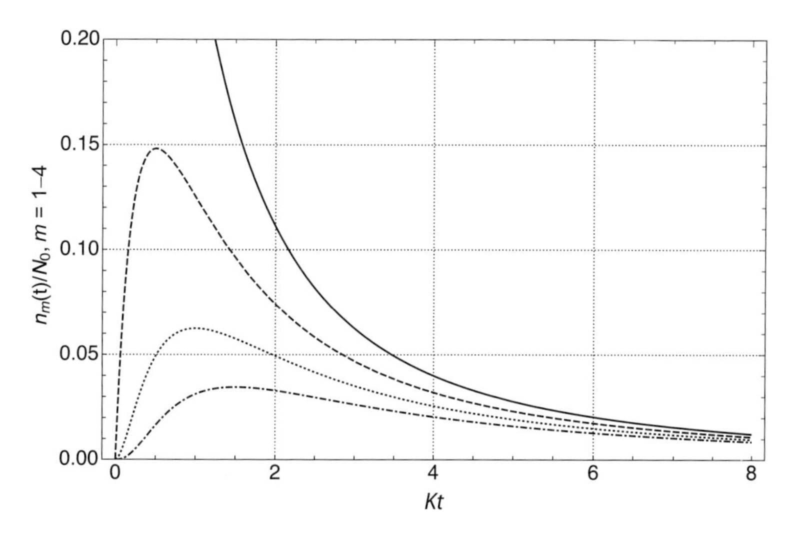

Figure D.1 Number density of oligomers starting from $m = 1$ to $m = 4$, according to Eq. (D.11).

$N(t)$ can be calculated by summing Eq. (D.10) from $m = 1$ to ∞, leading after some manipulations to

$$\frac{dN}{dt} = -\frac{K}{2}N^2, \tag{D.13}$$

which is a differential equation for a second-order reaction describing a bimolecular reaction. The solution is

$$N(t) = \frac{N_0}{1 + \frac{1}{2}K_V N_0 t} = \frac{N_0}{1 + Kt}, \tag{D.14}$$

with $K = \frac{1}{2}K_V N_0$, having the unit 1/s. Let us now calculate the total mass in the system (since all monomers have the same molar volume, this also is the total volume fraction). To do so, imagine that $n_m(t)$ denotes the density of an m-mer as it develops with time. The total mass or the total volume fraction is therefore equal to the expectation value given by

$$E(t) = \sum_{m=1}^{\infty} m \cdot n_m(t) = \frac{N_0}{(1 + Kt)^2} \sum_{m=1}^{\infty} m \left(\frac{Kt}{1 + Kt}\right)^{m-1}. \tag{D.15}$$

Abbreviating

$$q = \frac{Kt}{1 + Kt} \tag{D.16}$$

and understanding that for all t the relation $q < 1$ holds, we can write

$$E(t) = \frac{N_0}{(1 + Kt)^2} \sum_{m=1}^{\infty} mq^{m-1} = \frac{N_0}{(1 + Kt)^2} \frac{d}{dq} \sum_{m=1}^{\infty} q^m = \frac{N_0}{(1 + Kt)^2} \frac{1}{(1 - q)^2}. \tag{D.17}$$

Reinsertion of the definition of q leads to the simple result

$$E(t) = N_0, \tag{D.18}$$

as it should be. The total mass or the total volume fraction is a conserved quantity.

D.2.2 Degree of Polymerisation

The average degree of polymerisation can now be calculated as follows:

$$\overline{m} = \frac{\sum_{m=1}^{\infty} m \cdot n_m(t)}{\sum_{m=1}^{\infty} n_m(t)}. \tag{D.19}$$

Inserting the results from Eq (D.19), we obtain

$$\overline{m} = \frac{N_0}{\frac{N_0}{1+Kt}} = 1 + Kt, \tag{D.20}$$

which is almost Carother's equation, here derived differently and taking into account that all k-mers can react with each other. Let us repeat here the standard found in the chemical literature. For a bimolecular reaction, the extent p of the reaction occurring at a rate K_r follows the relation

$$\frac{dp}{dt} = K_r(1-p)^2. \tag{D.21}$$

This equation can be integrated directly:

$$p(t) = \frac{K_r t}{1 + K_r t}. \tag{D.22}$$

Note, that p varies between 0 and 1. Carother's derived a relation between p and the degree of polymerisation for a polycondensation reaction:

$$DP = \frac{1}{1-p}. \tag{D.23}$$

As usually mentioned in the literature on chemical reaction kinetics, this equation shows that for a polycondensation reaction it is difficult to achieve a high degree of polymerisation. It needs, for instance, a 99% conversion rate to achieve a DP of 100. Insertion of Eq. (D.22) yields

$$DP = 1 + K_r t \tag{D.24}$$

and thus the same relation we derived in a different way in Eq. (D.20); however, our relation is valid for the average DP, and we used $K_r = \frac{1}{2} K_V N_0$. In Section 4.3.3, we already used an experimentally derived reaction rate and converted it to a rate used in Eq. (D.24).

D.2.3 Flory–Schulz Distribution

We will make another check of the derivation by comparing the size distribution of oligomers to the well-established result in polymer science [116]. In polycondensation processes, Flory and Schulz derived the following relation for the probability $w_k = N_k/N_0$ for a polymer to have a degree of polymerisation k if the polycondensation reaction proceeds with a conversion rate p:

$$w_k = p^{k-1}(1-p)^2. \tag{D.25}$$

A derivation of this is given in the textbook of Flory [116, page 319]. Using the preceding results, we insert into Eqs. (D.11), (D.23) and (D.24) and arrive after some rearrangements at

$$\frac{n_k}{N_0} = p^{k-1}(1-p)^2, \tag{D.26}$$

and thus exactly at the same result in a different way. We therefore conclude that our understanding of the polycondensation process is in agreement with classical knowledge in polymer chemistry.

D.2.4 Global Volume Fraction Polymers

For the overall reaction in a dilute solution of monomers leading to a porous gel body, which can be transformed via suitable drying into an aerogel, it is important to have an idea how the volume fraction of polymers develops with time, since the phase separation that might occur depends on the polymers evolving and the miscibility gap evolving with them. We calculate the volume fraction of polymers as follows. Note that the molar volume of a monomer unit is denoted as v_m. Then the volume fraction of polymers (all m-mers with a degree of polymerisation k and higher) is

$$\phi_p = v_m \sum_{m=k}^{\infty} m \cdot n_m(t), \tag{D.27}$$

which leads to a rather simple expression:

$$\phi_p(t) = v_m N_0 \frac{(Kt)^{k-1}(Kt+k)}{(1+Kt)^k}. \tag{D.28}$$

Figure D.2 shows how the fraction of polymers with a DP larger than k develops with time for four different values of k. The curves in this picture show, that the reaction from monomers to dimers is really fast, but to have almost all monomers and k-mers reacted to oligomers with a DP larger than, e.g., 20 takes some time. Assume, for

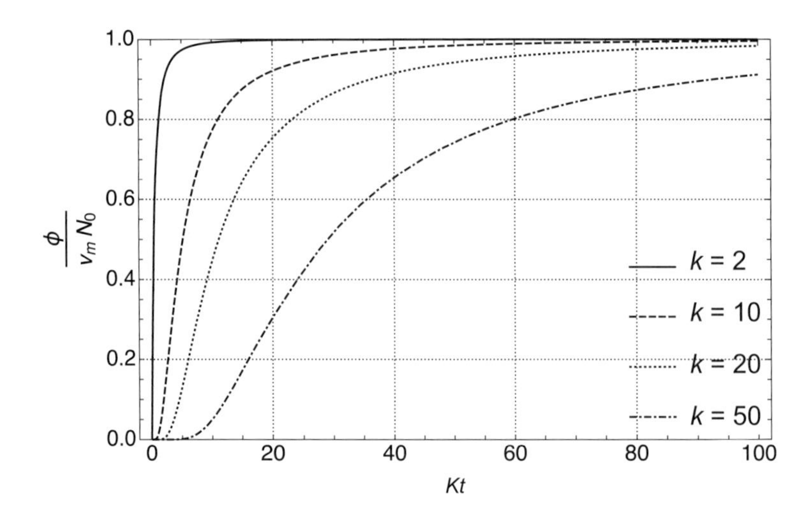

Figure D.2 Volume fraction of polymers with a $DP > k$ as a function of reduced time Kt. Note $Kt = DP - 1$.

instance, that the reaction rate constant would be of the order 10^{-5} per second, so that a value of $Kt = 20$ would mean a time of around two days. At a value of $Kt = 20$, around 80% of the oligomers have a DP larger than 20. It also is worth noting that $v_m N_0$ is the maximum volume fraction polymers in the system, and this is defined experimentally by the initial mixture of, for instance, resorcin, formaldehyde and water.

D.2.5 The Maximum Volume Fraction Polymer

As an example, the maximum volume fraction of polymers will be calculated for RF aerogels. As already stated, RF aerogels are made from resorcinol (R) and formaldehyde (F) dissolved in water and adding a catalyst, such as sodium carbonate or citric acid (for more details, see Section 2.2 and also Chapter 7). Eventually a molecule made of two hydroxymethylresorcinols bonded by a methylene or dimethylenether bridge is formed, and these can react further to produce polymers in either a so-called resol or novolak structure. On reaction of two molecules, hydroxymethylresorcinols always produces one mole of water.

We estimate the polymer volume by calculating the volume of the monomers (R-CH_2-) or (R-CH2-O-CH_2-) from the molar masses of R, methylene and di-methylenether and calculate the volume of the solution from the molar volume of each constituent (R,F, water), before any reaction has happened. We neglect the catalyst, since it usually is used in very low amounts.

Let x_R, X_F, x_W be the molar fraction of the components resorcinol, formaldehyde and water. The volume of the solution before any reaction has happened is calculated as

$$V_{sol} = N_t(x_R V_{mR} + x_F V_{mF} + x_W V_{mW}), \tag{D.29}$$

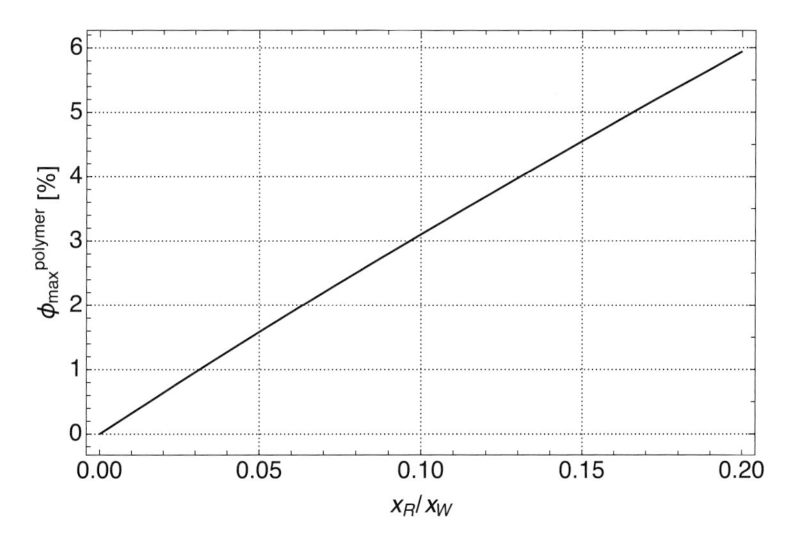

Figure D.3 Estimated maximum volume fraction of polymer in dilute RF water solutions (R/F = 1:2).

with V_{mR}, V_{mW}, V_{mR} the molar volumes of resorcinol, formaldehyde and water and N_t the total number of molecules of the solution. The volume of the polymers if all resorcinol molecules are reacted is

$$V_p = N_t(x_R V_{mR} + \frac{1}{2} x_F (V_{m,CH_2} + V_{m,CH_2OH})), \tag{D.30}$$

with V_{m,CH_2}, V_{m,CH_2OH} the molar volumes of the methyl and the methylenether group. Here we assumed that the base unit of the polymer, the hydroxymethylenresorcinol, is constituted to 50% by methylene and methylenether groups. The maximum volume fraction of polymer is then

$$\phi_p = \frac{V_p}{V_{sol}} = \phi_{\max}. \tag{D.31}$$

Calculating the molar volumes from given tables in the literature and assuming $x_F = 1/2 x_R$, which is typical in many RF aerogel recipes, we can calculate the maximum possible volume fraction as a function of resorcinol to water molar ratio x_R/x_W. The result is shown in Figure D.3 This graph illustrates that at high dilutions with water the maximum volume fraction polymer is around a few percent. We identify the value given in Eq. (D.28) of $v_m N_0$ with $\phi_p = \phi_{max}$. Note: this is a global value, an average value in the whole volume.

Appendix E Time-Dependent Heat Transfer through an Isolated Tube

The equation describing radial transport of heat through a tube is derived in Section 12.6.5. For a hollow cylinder, shown schematically in Figure 12.26, the time- and space-dependent solution is more complicated compared to that in cartesian coordinates, given in Eq. (12.69). Although there is a solution for an even more complicated solution, namely a composite hollow cylinder, given by Li and Lai [541], we will not follow their way,[1] but solve it along lines set by Carslaw and Jaeger. We start with a decomposition of the solution. Let us write

$$T(r,t) = T_{stat}(r) + u(r,t). \tag{E.1}$$

We insert this into Eq. (12.81) and obtain an equation for $u(r,t)$

$$\frac{\partial u}{\partial t} = \kappa \left(\frac{\partial^2 u}{\partial r^2} + \frac{1}{r}\frac{\partial u}{\partial r} \right), \tag{E.2}$$

with the initial and boundary conditions

$$u(r,0) = T_0 - T_{stat}(r) = f(r) \tag{E.3}$$

$$u(a,t) = 0 \tag{E.4}$$

$$\left.\frac{\partial u}{\partial r}\right|_{r=b} + \frac{h}{\lambda} u = 0 \qquad r = b. \tag{E.5}$$

Eq. (E.2) can be solved with a method described in Carslaw–Jaeger [440, page 205ff]. Let us first state that a solution of this equation always will have the form $u(r,t) = w(r)g(t)$, meaning it can be separated into two functions. Knowing that solutions for the heat conduction in cylindrical coordinates always lead to Bessel functions or sums of them (as discussed shortly) we look for a solution of the type

$$u(r,t) = \sum_{n=1}^{\infty} A_n U_0(\alpha_n r) \exp(-\kappa \alpha_n^2 t). \tag{E.6}$$

[1] The solution given by Li and Lai is definitely a complete one to our problem, even including the instationary heat transfer in the inner steel tube, but the final result is so lengthy and complex that it even poses for Mathematica programs utterly complex equations to be solved. Thus it is worthwhile to look for a simpler solution, which we do here, by assuming that the heat conduction in the steel tube is fast enough that it is always at a constant temperature and orders of magnitude faster than the aerogel isolation.

Here α_n are coefficients we have to determine as well as the amplitudes A_n from the boundary and initial conditions. The function $U_0(r)$ can be defined, following the lines set by Carslaw and Jaeger as

$$U_0(\alpha r) = J_0(\alpha r)Y_0(\alpha a) - J_0(\alpha a)Y_0(\alpha r). \tag{E.7}$$

Here J_0, Y_0 are Bessel functions of first and second kind [428, 498] and α is the set of coefficients mentioned previously. At $r = a$, this function automatically fulfils the boundary condition of Eq. (E.4), meaning

$$U_0(\alpha a) = 0 \qquad \text{and therefore} \quad u(a,t) = 0. \tag{E.8}$$

At the boundary $r = b$, the function has to fulfill the boundary condition given in Eq. (E.5). We insert this new function into the boundary condition and obtain

$$\left.\frac{\partial U_0}{\partial r}\right|_{r=b} + \frac{h}{\lambda}U_0 = 0 \tag{E.9}$$

$$\alpha\left[J_0'(\alpha b)Y_0(\alpha a) - J_0(\alpha a)Y_0'(\alpha b)\right] + \frac{h}{\lambda}\left[J_0(\alpha b)Y_0(\alpha a) - J_0(\alpha a)Y_0(\alpha b)\right] = 0. \tag{E.10}$$

Using the relations $J_0' = -J_1$ and $Y_0' = -Y_1$, we arrive at a defining equation for the set of coefficients α, namely

$$-\alpha\left[J_1(\alpha b)Y_0(\alpha a) - J_0(\alpha a)Y_1(\alpha b)\right] + h/\lambda\left(J_0(\alpha b)Y_0(\alpha a) - J_0(\alpha a)Y_0(\alpha b)\right) = 0. \tag{E.11}$$

Using the Nusselt number defined in Eq. (12.88), we can rewrite this as

$$-\alpha b\left[J_1(\alpha b)Y_0(\alpha a) - J_0(\alpha a)Y_1(\alpha b)\right] + \mathrm{Nu}(J_0(\alpha b)Y_0(\alpha a) - J_0(\alpha a)Y_0(\alpha b)) = 0 \tag{E.12}$$

or

$$\mathrm{Nu}(J_0(\alpha b)Y_0(\alpha a) - J_0(\alpha a)Y_0(\alpha b)) = \alpha b\left[J_1(\alpha b)Y_0(\alpha a) - J_0(\alpha a)Y_1(\alpha b)\right] \tag{E.13}$$

$$\frac{J_1(\alpha b)Y_0(\alpha a) - J_0(\alpha a)Y_1(\alpha b)}{J_0(\alpha b)Y_0(\alpha a) - J_0(\alpha a)Y_0(\alpha b)} = \frac{\mathrm{Nu}}{\alpha b} \tag{E.14}$$

$$\frac{U_0'(\alpha b)}{U_0(\alpha b)} = -\frac{\mathrm{Nu}}{b}. \tag{E.15}$$

The last result is useful to later manage the final result. Carslaw and Jaeger (CJ) proved that the functions U_0 are a set of orthogonal functions and that it is possible to expand an arbitrary function $f(r)$ as a series of these functions. We have, however, to consider that here the boundary condition at the outer cylindrical surface $r = b$ is different from that used by CJ. Therefore, we have to calculate the orthogonality relation applicable to our boundary conditions. The general result of orthogonally for the function $U_0(\alpha r)$ reads

$$\int_a^b rU_0(\alpha r)U_0(\beta r)dr = \delta_{\alpha\beta}\int_a^b rU_0^2(\alpha r)dr, \tag{E.16}$$

with $\delta_{\alpha\beta}$ the Kronecker delta, such that, if $\alpha \neq \beta$ the integral on the left side is zero. We cannot use the result of CJ as mentioned. We have to calculate the integral on the right side by ourself. To do so, we follow the lines given by CJ and use especially that the function $U_0(\alpha r)$ is a solution of the differential equation (E.2). This reads

$$\frac{d^2U_0}{dr^2} + \frac{1}{r}\frac{dU_0}{dr} + \alpha^2 U_0 = 0 \quad \text{or} \quad \frac{1}{r}\frac{d}{dr}\left(r\frac{dU_0}{dr}\right) + \alpha^2 U_0 = 0. \tag{E.17}$$

We first multiply the equation by $2r^2\frac{dU_0}{dr}$, which leads after some algebraic manipulations to

$$\frac{d}{dr}\left(r\frac{dU_0}{dr}\right)^2 + \alpha^2 r^2 \frac{dU_0^2}{dr} = 0. \tag{E.18}$$

We now integrate the equation from $r = a$ to $r = b$. For the first term in Eq. (E.18), we immediately get the result

$$\int_a^b \frac{d}{dr}\left(r\frac{dU_0}{dr}\right)^2 dr = \left(r\frac{dU_0}{dr}\right)^2\Bigg|_a^b = b^2U_0'(\alpha b)^2 - a^2U_0'(\alpha a)^2. \tag{E.19}$$

The second term of Eq. (E.18) can be integrated by parts to yield

$$\int_a^b \frac{d}{dr}\left(r\frac{dU_0}{dr}\right)^2 dr = \alpha^2 U_0^2 r^2\Big|_a^b - 2\alpha^2\int_a^b rU_0^2 dr. \tag{E.20}$$

The last integral on the right side is exactly the integral we would like to determine. So collecting terms leads to

$$2\alpha^2\int_a^b rU_0^2 dr = b^2U_0'(\alpha b)^2 - a^2U_0'(\alpha a)^2 + \alpha^2 U_0(\alpha b)^2 b^2, \tag{E.21}$$

using here directly that the boundary condition at $r = a$ means $U_0(\alpha a) = 0$. We are thus left to calculate $U_0'(\alpha a)$. This can be done using a well-known relation for Bessel function, [428], namely the so-called Wronskian determinant of $-J_1 = J_0'$ and $Y_0' = -Y_1$:

$$\frac{dU_0}{dr}\Bigg|_a = \alpha[J_0'(\alpha a)Y_0(\alpha a) - J_0(\alpha a)Y_0'(\alpha a)] = -\frac{2}{\pi a}. \tag{E.22}$$

Using now also the result of Eq. (E.15) leads after some calculations to the final result

$$\int_a^b rU_0(\alpha r)^2 dr = \frac{1}{2\alpha^2}\left(U_0^2(\alpha b)(\mathrm{Nu}^2 + \alpha^2 b^2) - \frac{4}{\pi^2}\right). \tag{E.23}$$

Note that from its definition, α has the dimension of an inverse length, such that αb is dimensionless as the Nusselt number. We can now write down an expression for the amplitudes A_n. We start by taking the general solution, written in Eq. (E.6) at $t = 0$, multiply both sides with $rU_0(\alpha_m r)$ and integrate over r from a to b. This leads to

$$\int_a^b rU_0(\alpha_m r)u(r,0)dr = A_n \int_a^b rU_0(\alpha_m r)U_0(\alpha_n r)dr. \tag{E.24}$$

The right side was calculated in Eq. (E.24) (orthogonality), and we obtain finally for the amplitudes A_n the expression

$$A_n = -\frac{2\alpha_n^2}{(\mathrm{Nu} + \alpha_n^2 b^2)U_0^2(\alpha_n b) - \frac{4}{\pi^2}} \int_a^b rf(r)U_0(\alpha_n r)dr. \tag{E.25}$$

In this equation, $f(r)$ was defined through Eq. (E.3). We are left to perform the integration over the initial temperature field in the sample. To do this, we first rewrite the stationary solution, so that it can be integrated term by term:

$$T_{stat}(r) = C_1 - C_2 \ln r, \tag{E.26}$$

with

$$C_1 = \frac{T_s + \mathrm{Nu}(T_s \ln b - T_0 \ln a)}{1 + \mathrm{Nu}\ln(b/a)} \qquad C_2 = \frac{\mathrm{Nu}(T_s - T_0)}{1 + \mathrm{Nu}\ln(b/a)}. \tag{E.27}$$

Insertion into the integral of Eq. (E.25) leads to two types of integrals to be calculated:

$$\begin{aligned}\int_a^b rf(r)U_0(\alpha_n r)dr = & \\ &= \int_a^b r(T_0 - T_{stat}(r))U_0(\alpha_n r)dr \\ &= \int_a^b r(T_0 - C_1 + C_2 \ln r)U_0(\alpha_n r)dr.\end{aligned} \tag{E.28}$$

We therefore have to think about two types of integrals:

$$H_1 = (T_0 - C_1)\int_a^b rU_0(\alpha_n r)dr \tag{E.29}$$

and

$$H_2 = C_2 \int_a^b r \ln r\, U_0(\alpha_n r)dr. \tag{E.30}$$

Thus the A_n read

$$A_n = -\frac{2\alpha_n^2}{(\mathrm{Nu} + \alpha_n^2 b^2)U_0^2(\alpha_n b) - \frac{4}{\pi^2}}(H_1 + H_2). \tag{E.31}$$

These integrals can be solved using that the U_0 are solutions of the original differential equation. Again, the results presented by Carslaw and Jaeger cannot be used, since they were derived for different boundary conditions; but their method can be used. Since $U_0(a_n r)$ is a solution of the Bessel equation, we have

$$\frac{1}{r}\frac{d}{dr}\left(r\frac{dU_0(\alpha_n r)}{dr}\right) = -\alpha_n^2 U_0(\alpha_n r). \tag{E.32}$$

Multiplication with r and integration over r from a to b leads to

$$\int_a^b rU_0(\alpha_n r)dr = -\frac{1}{\alpha_n^2}\int_a^b \frac{d}{dr}\left(r\frac{dU_0(\alpha_n r)}{dr}\right)dr = -\frac{1}{\alpha_n^2}\left[r\frac{dU_0(\alpha_n r)}{dr}\right]_a^b. \tag{E.33}$$

We have to evaluate this last expression to get the integral H_1. For this, we use Eqs. (E.15) and (E.22) to get the result

$$\int_a^b rU_0(\alpha_n r)dr = -\frac{2}{\pi\alpha_n^2} + \frac{\mathrm{Nu}}{\alpha_n^2}U_0(\alpha_n b). \tag{E.34}$$

The other integral H_2 can be solved in the same way starting with Eq. (E.32), then multiplying both sides with $r \ln r$, integration by parts leads to

$$\int_a^b r\ln r U_0(\alpha_n r)dr = \frac{1}{\alpha_n^2}\left(\frac{-2\ln a}{\pi} + U_0(\alpha_n b)(1 + \mathrm{Nu}\ln b)\right). \tag{E.35}$$

Therefore we obtain

$$H_1 = (T_0 - C_1)\left(-\frac{2}{\pi\alpha_n^2} + \frac{\mathrm{Nu}}{\alpha_n^2}U_0(\alpha_n b)\right) \tag{E.36}$$

and

$$H_2 = \frac{C_2}{\alpha_n^2}\left(-\frac{2\ln a}{\pi} + U_0(\alpha_n b)(1 + \mathrm{Nu}\ln b)\right). \tag{E.37}$$

We now collect all terms and write down the final equation for the temperature field. Let us first add H_1 and H_2 using the definitions of the constants C_1 and C_2. This gives

$$H_1 + H_2 = \frac{2(T_s - T_0)}{\pi\alpha_n^2}. \tag{E.38}$$

We now can write the expression for the amplitudes A_n after some manipulations as

$$A_n = \frac{4\pi(T_s - T_0)}{[\pi^2 U_0^2(\alpha_n b)(\mathrm{Nu}^2 + b^2\alpha_n^2) - 4]} \tag{E.39}$$

and thus for $u(r,t)$, we get

$$u(r,t) = \sum_{n=1}^{\infty} \frac{4\pi(T_s - T_0)}{[\pi^2 U_0^2(\alpha_n b)(\mathrm{Nu}^2 + b^2\alpha_n^2) - 4]} U_0(\alpha_n r)\exp(-\kappa\alpha_n^2 t) \tag{E.40}$$

and for the temperature field, we have

$$T(r,t) = \frac{T_s}{1 + \mathrm{Nu}\ln(b/a)}\left(1 - \mathrm{Nu}\left[\ln\frac{r}{b} - \frac{T_0}{T_s}\ln\frac{r}{a}\right]\right) + \tag{E.41}$$

$$+ 4\pi(T_s - T_0)\sum_{n=1}^{\infty}\frac{U_0(\alpha_n r)}{[\pi^2 U_0^2(\alpha_n b)(\mathrm{Nu}^2 + b^2\alpha_n^2) - 4]}\exp(-\kappa\alpha_n^2 t). \tag{E.42}$$

As expected, the final result is not quite intuitive. However, it might be trivial to note that the solution yields the expected result: after infinite time, the stationary solution is achieved and, of course, that at $t = 0$ the initial temperature is T_0. Another interesting feature: whereas the stationary solution can be written in terms of b/a and thus the thickness of the aerogel shell $b - a$ is important, the time-dependent solution contains b/a or the thickness in the amplitudes of the infinite sum, in the definition of U_0 and of course in the values of the α_n-set.

The surface temperature of the aerogel side is obtained by setting $r = b$ and reads defining $\beta_n = \alpha_n b$

$$T(b,t) = \frac{T_s + \mathrm{Nu}T_0\ln b/a}{1 + \mathrm{Nu}\ln b/a} + 4\pi(T_s - T_0)\sum_{n=1}^{\infty}\frac{U_0(\beta_n)\exp\left(-\frac{\kappa\beta_n^2 t}{b^2}\right)}{\pi^2 U_0(\beta_n)^2(\mathrm{Nu}^2 + \beta_n^2) - 4}. \tag{E.43}$$

This result is visualised solving first Eq. (E.44) to obtain for a preset Nusselt number the dedicated sequence of the β_n, which reads

$$\frac{J_1(\beta_n)Y_0(\beta_n a/b) - J_0(\beta_n a/b)Y_1(\beta_n)}{J_0(\beta_n)Y_0(\beta_n a/b) - J_0(\beta_n a/b)Y_0(\beta_n)} = \frac{\mathrm{Nu}}{\beta_n}. \tag{E.44}$$

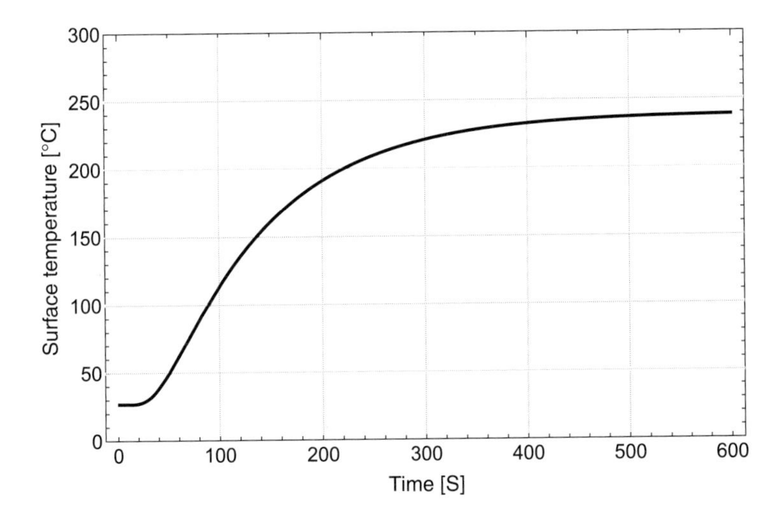

Figure E.1 Time-dependent temperature at the side of the aerogel wrapped around a tube with diameter $a = 25$ mm. The aerogel has a thickness of 5 mm. The hot side of the aerogel is fixed at $T_s = 900$ K and the temperature far away from the colder side of the aerogel at $T_0 = 300$ K.

This equation is solved using a Mathematica^{TM} script. Figure E.1 shows the surface temperature as function of time for a Nusselt number of $\text{Nu} = 10$, an a-value of 25 and a b-value of 30 mm (these values fit to the size of car exhausts). The result shows that in instationary situations the aerogel isolation does a real good job; it isolates the hollow tube interior almost perfectly and thus would bring the hot combustion gas at a high temperature to the catalytic converter and thus increase its efficiency.

References

[1] Gibson L.J., Ashby M.F. *Cellular Solids*. 2nd ed. Cambridge: Cambridge University Press; 1997.

[2] Lunt A.J.G., Chater P., Korsunsky A.M. On the origins of strain inhomogeneity in amorphous materials. *Scientific Reports*. 2018; 8(1):1574.

[3] Brinker C.J., Scherer G.W. *Sol–Gel Science: The Physics and Chemistry of Sol-Gel Processing*. HBJI, editor. San Diego: Academic Press, Inc.; 1990.

[4] Wen D., Eychmüller A. 3D assembly of preformed colloidal nanoparticles into gels and aerogels: function-led design. *Chem Commun*. 2017; 53:12608–12621.

[5] Beier M.G., Ziegler C., Wegner K., et al. A fast route to obtain modified tin oxide aerogels using hydroxostannate precursors. *Mater Chem Front*. 2018; 2:710–717.

[6] Cai B., Sayevich V., Gaponik N., Eychmüller A. Emerging hierarchical aerogels: self-assembly of metal and semiconductor nanocrystals. *Advanced Materials*. 2018; 30(33):1707518.

[7] Du R., Hu Y., Hübner R., et al. Specific ion effects directed noble metal aerogels: versatile manipulation for electrocatalysis and beyond. *Science Advances*. 2019; 5(5).

[8] Pekala R.W. Organic aerogels from the polycondensation of resorcinol with formaldehyde. *Journal of Materials Science*. 1989; 24(9):3221–3227.

[9] Pekala R.W., Alviso C.T., Lu X., Gross J., Fricke J. New organic aerogels based upon a phenolic-furfural reaction. *Journal of Non-Crystalline Solids*. 1995; 188:34–40.

[10] BASF. Slentite aerogel. www.basf.com/global/de/products/plastics-rubber/corpus/ideas-and-solutions/solid_air.html.

[11] Betz M., García-González C.A., Subrahmanyam R.P., Smirnova I., Kulozik U. Preparation of novel whey protein-based aerogels as drug carriers for life science applications. *Journal of Supercritical Fluids*. 2012; 72:111–119.

[12] Kistler S.S. Coherent and expanded aerogels and jellies. *Nature*. 1931; 127:741.

[13] Kistler S.S. Coherent expanded aerogels. *Journal of Physical Chemistry*. 1932; 31: 52–64.

[14] Aegerter M.A., Leventis N., Koebel M.M., editors. *Aerogels Handbook*. New York: Springer Verlag; 2011.

[15] Nicola Hüsing U.S. Aerogels – airy materials: chemistry, structure, and properties. *Angewandte Chemie International Edition*. 1998; 37:22–45.

[16] Pierre A.C., Pajonk G.M. Chemistry of aerogels and their applications. *Chemical Reviews*. 2002; 102:4243–4266.

[17] Teichner S.J., Nicolaon G.A., Vicarini M.A., Gardes G.E.E. Inorganic oxide aerogels. *Advances in Colloid and Interface Science*. 1976; 5(3):245–273.

[18] Fricke J., editor. *Aerogels*. Heidelberg: Springer Verlag; 1988.

[19] Leventis N., Sotiriou-Leventis C., Mohite D.P., et al. Polyimide aerogels by ring-opening metathesis polymerization (ROMP). *Chemistry of Materials*. 2011; 23:2250–2261.

[20] Chidambareswarapattar C., McCarver P.M., Luo H., Lu H., Sotiriou-Leventis C., Leventis N. Fractal multiscale nanoporous polyurethanes: flexible to extremely rigid aerogels from multifunctional small molecules. *Chemistry of Materials*. 2013; 25:3205–3224.

[21] Malakooti S., Rostami S., Churu H.G., et al. Scalable, hydrophobic and highly-stretchable poly(isocyanurate–urethane) aerogels. *RSC Advances*. 2018; 8:21214.

[22] Tan C., Fung B.M., Newman J.K., Yu C. Organic aerogels with very high impact strength. *Advanced Materials*. 2001; 13(9):644–646.

[23] Ganesan K., Ratke L. Facile preparation of monolithic κ-carrageenan aerogels. *Soft Matter*. 2014; 10:3218–3224.

[24] Subrahmanyam R., Gurikov P., Dieringer P., Sun M., Smirnova I. On the road to biopolymer aerogels – dealing with the solvent. *Gels*. 2016; 1:291–313.

[25] Smirnova I., Gurikov P. Aerogels in chemical engineering: strategies toward tailor-made aerogels. *Annual Review of Chemical and Biomolecular Engineering*. 2017; 8(1): 307–334.

[26] Ganesan K., Budtova T., Ratke L., et al. Review on the production of polysaccharide aerogel particles. *Materials*. 2018; 11(11):2144.

[27] Aravind P.R., Mukundan P., P (KP), Warrier K.G.K. Mesoporous silica-alumina aerogels with high thermal pore stability through hybrid sol-gel route followed by subcritical drying. *Microporous and Mesoporous Materials*. 2006; 96(1):14–28.

[28] Liu S., Yu T., Hu N., Liu R., Liu X. High strength cellulose aerogels prepared by spatially confined synthesis of silica in bioscaffolds. *Colloids and Surfaces A: Physicochemical and Engineering Aspects*. 2013; 439:159–166.

[29] Kong Y., Zhong Y., Shen X., Gu L., Cui S., Yang M. Synthesis of monolithic mesoporous siliconcarbide from resorcinol-formaldehyde/silica composites. *Materials Letters*. 2013; 99:108–110.

[30] Demilecamps A., Reichenauer G., Rigacci A., Budtova T. Cellulose-silica composite aerogels from 'one-pot' synthesis. *Cellulose*. 2014; 21:2625–2636.

[31] Demilecamps A., Beauger C., Hildenbrand C., Rigacci A. Cellulose-silica aerogels. *Carbohydrate Polymers*. 2015; 122:293–300.

[32] Zhao S., Malfait W.J., Jeong E., et al. Facile one-pot synthesis of mechanically robust biopolymer-silica nanocomposite aerogel by coagulation of silicic acid with chitosan in aaqueous media. *ACS Sustainable Chemical Engineering*. 2016; 4(10):5674–5683.

[33] Zhao S., Malfait W.J., Demilecamps A., et al. Strong, thermally superinsulating biopolymer-silica aerogel hybrids by cogelation of silicic acid with pectin. *Angewandte Chemie*. 2015; 48:1521–3757.

[34] Corporation C. Aerogel. www.cabotcorp.com/solutions/products-plus/aerogel.

[35] Aerogels A. Aspen Aerogel webpage. www.aerogel.com.

[36] Murillo J.S.R., Bachlehner M.E., Campo F., Barbero E.J. Structure and mechanical properties of silica aerogels and xerogels modelled by molecular dynamics simulation. *Journal of Non-Crystalline Solids*. 2010; 356:1325–1331.

[37] Patil S.P., Rege A., Sagardas, Itskov M., Markert B. Mechanics of nanostructured porous silica aerogel resulting from molecular dynamics simulations. *Journal of Physical Chemistry B*. 2017; 121:5660–5668.

[38] Goncalves W., Morthomas J., Chantrenne P., Perez M., Foray G., Martin C.L. Elasticity and strength of silica aerogels: a molecular dynamics study on large volumes. *Acta Materialia*. 2018; 145:165–174.

[39] Gurav J.L., Jung I.K., Park H.H., Kang E.S., Nadargi D.Y. Silica aerogel: synthesis and applications. *Journal of Nanomaterials*. 2010; 409310.

[40] Koebel M.M., Rigacci A., Achard P. Aerogel-based thermal superinsulation: an overview. *Journal of Sol–Gel Science and Technology*. 2012; 63:315–339.

[41] Maleki H., Durães L., Portugal A. An overview on silica aerogels synthesis and different mechanical reinforcing strategies. *Journal of Non-Crystalline Solids*. 2014; 385:55–74.

[42] Parale V.G., Lee K.Y., Park H.H. Flexible and transparent silica aerogels: an overview. *Journal of the Korean Ceramic Society*. 2017; 54:184–199.

[43] Zsigmondy R. *Kolloidchemie*. vol. II. Spezieller Teil. 5th ed. Verlag Otto Spamer; 1927.

[44] Graham T. Anwendung der Diffusion der Flüssigkeiten zur Analyse. *Justus Liebigs Annalen der Chemie*. 1862; 121:36.

[45] Iler R.K. *The Chemistry of Silica*. Hoboken: John Wiley & Sons, Ltd; 1979.

[46] Grimaux E. *Compt rend.* 1884; 98:1434–1437.

[47] Wang S., Wang D.K., Smart S., Diniz da Costa J.C. Ternary phase-separation investigation of sol-gel derived silica from ethyl silicate 40. *Scientific Reports*. 2015; 5(1):14560.

[48] Schaefer D.W. What factors control the structure of silica aerogels? *Revue de Physique Applique, Colloque C4*. 1989; 24:C4–121–C4–126.

[49] Viscek T. *Fractal Growth Phenomena*. 2nd ed. World Scientific Publishing Co.; 1992.

[50] Ratke L., Voorhees P.W. *Growth and Coarsening: Ostwald Ripening in Material Processing*. Engineering Materials and Processes. Berlin–Heidelberg–New York: Springer Science and Business Media LLC; 2002.

[51] Wagner C. Theorie der Alterung von Niederschlägen durch Umlösen (Ostwald-Reifung). *Z Elektrochemie*. 1961; 65:581–591.

[52] Tiller W.A. *The Science of Crystallization – Microscopic Interfacial Phenomena*. Cambridge: Cambridge University Press; 1991.

[53] Ratke L., Uffelmann D., Bender W., Voorhees P.W. Theory of Ostwald ripening due to a second order reaction. *Scripta Metall et Materialia*. 1995; 33:363–367.

[54] Teraoka I. *Polymer Solutions*. New York: John Wiley & Sons, Inc.; 2002.

[55] Peitgen H.O., Jürgens H., Saupe D. *Fractals for the Classroom*. Part 1. New York: Springer-Verlag; 1992.

[56] Peitgen H.O., Jürgens H., Saupe D. *Fractals for the Classroom*. Part 2. New York: Springer-Verlag; 1992.

[57] Kaye B.H. *A Random Walk through Fractal Dimensions*. Weinheim: Wiley-VCH Verlag GmbH; 1989.

[58] Nichols F.A, Mullins W.W. Surface-(interface-) and volume-diffusion contributions to morphological changes driven by capillarity. *Transactions of the Metallurgical Society of AIME*. 1965; 233:1840.

[59] Nichols F.A. On the spheroidization of rod-shaped particles of finite length. *Journal of Materials Science*. 1976; 11(6):1077–1082.

[60] Einarsrud M.A. Light gels by conventional drying. *Journal of Non-Crystalline Solids*. 1998; 225:1–7.

[61] Einarsrud M.A., Nilsen E., Rigacci A., et al. Strengthening of silica gels and aerogels by washing and aging processes. *Journal of Non-Crystalline Solids*. 2001; 285:1–7.

[62] Haerid S., Dahle M., Lima S., Einarsrud M.A. Preparation and properties of monolithic silica xerogels from TEOS-based alcogels aged in silane solutions. *Journal of Non-Crystalline Solids*. 1995; 186:96–103.

[63] Smitha S., Shajesh P., Aravind P.R., Kumar S.R., Pillai P.K., Warrier K.G.K. Effect of aging time and concentration of aging solution on the porosity characteristics of subcritically dried aerogels. *Microporous and Mesoporous Materials*. 2006; 91:286–292.

[64] Reichenauer G. Thermal aging of silica gels in water. *Journal of Non-Crystalline Solids*. 2004; 350:189–195.

[65] Iswar S., Malfait W.J., Balog S., Winnefeld F., Lattuada M., Koebel M.M. Effect of aging on silica aerogel properties. *Microporous and Mesoporous Materials*. 2017; 241:293–302.

[66] Rao A.V., Bhagat S.D., Hirashima H., Pajonk G.M. Synthesis of flexible silica aerogels using methyltrimethoxysilane (MTMS) precursor. *Journal of Colloid and Interface Science*. 2006; 300(1):279–285.

[67] Guo H., Nguyen B.N., McCorkle L.S., Shonkwiler B., Meador M.A.B. Elastic low density aerogels derived from bis[3-(triethoxysilyl)propyl]disulfide, tetramethylorthosilicate and vinyltrimethoxysilane via a two-step process. *Journal of Materials Chemistry*. 2009; 19:9054–9062.

[68] Aravind P.R., Niemeyer P., Ratke L. Novel flexible aerogels derived from methyltrimethoxysilane/3-(2,3-epoxypropoxy)propyltrimethoxysilane co-precursor. *Microporous and Mesoporous Materials*. 2013; 181:111–115.

[69] Kanamori K., Aizawa M., Nakanishi K., Hanada T. New transparent methylsilsesquioxane aerogels and xerogels with improved mechanical properties. *Advanced Materials*. 2007; 19:1589–1593.

[70] Hayase G., Kanamori K., Kazuki K., Hanada T. Synthesis of new flexible aerogels from MTMS/DMDMS via ambient pressure drying. In: *IOP Conference Series: Materials Science and Engineering*. vol. 18. Ceramic Society of Japan; 2013. p. 032013.

[71] Hayase G., Kanamori K., Nakanishi K. New flexible aerogels and xerogels derived from methyltrimethoxysilane/dimethyldimethoxysilane co-precursors. *Journal of Materials Chemistry*. 2011; 21:17077–17079.

[72] Hayase G., Kanamori K., Hasegawa G., Maeno A., Kaji H., Nakanishi K. A superamphiphobic macroporous silicone monolith with marshmallow-like flexibility. *Angewandte Chemie International Edition*. 2013; 52(41):10788–10791.

[73] Hayase G., Kanamori K., Abe K., et al. Polymethylsilsesquioxane-cellulose nanofiber biocomposite aerogels with high thermal insulation, bendability, and superhydrophobicity. *ACS Applied Materials & Interfaces*. 2014; 6:9466–9471.

[74] Hayase G., Nomura S.M. Large-scale preparation of giant vesicles by squeezing a lipid-coated marshmallow-like silicone gel in a buffer. *Langmuir*. 2018; 34(37):11021–11026.

[75] Durães L., Maia A., Portugal A. Effect of additives on the properties of silica based aerogels synthesized from methyltrimethoxysilane (MTMS). *Journal of Supercritical Fluids*. 2015; 106:85–92.

[76] Maleki H., Durães L., Portugal A. Development of mechanically strong ambient pressure dried silica aerogels with optimized properties. *Journal of Physical Chemistry C*. 2015 04; 119(14):7689–7703.

[77] Maleki H., Durães L., Portugal A. Synthesis of lightweight polymer-reinforced silica aerogels with improved mechanical and thermal insulation properties for space applications. *Microporous and Mesoporous Materials*. 2014; 197:116–129.

[78] Lu X., Caps R., Fricke J., Alviso C.T., Pekala R.W. Correlation between structure and thermal conductivity of organic aerogels. *Journal of Non-Crystalline Solids*. 1995; 188:226–234.

[79] ElKhatat A.M., Al-Muhtaseb S.A. Advances in tailoring resorcinol-formaldehyde organic and carbon aerogels. *Advanced Materials*. 2011; 23:2887–2903.

[80] Reuss M., Ratke L. Subcritically dried RF-aerogels catalysed by hydrochloric acid. *Journal of Sol–Gel Sciences and Technology*. 2008; 47:74–80.

[81] Job N. Matériaux carbonés poreux de texture contrôlée synthétisés par procédé sol-gel et leur utilisation en catalyse hétérogène. University of Liege. Liege, Belgium; 2005.

[82] Chatchawalsaisin J., Kendrick J., Tubleb S.C., Anwar J. An optimized force field for crystalline phases of resorcinol. *CrystEngComm*. 2008; 10:437–445.

[83] Durairaj R.B. *Resorcinol*. Berlin–Heidelberg: Springer Science and Business Media LLC; 2005.

[84] Job N., Panariello F., Crine M., Pirard J.P., Leonard A. Rheological determination of the sol-gel transition during the aqueous synthesis of resorcinol-formaldehyde resins. *Colloids and Surfaces A: Physicochemical and Engineering Aspects*. 2007; 293:224–228.

[85] Barbieri O., Ehrburger-Dolle F., Rieker T.P., Pajonk G.M., Pinto N., Rao A.V. Small-angle X-ray scattering of a new series of organic aerogels. *Journal of Non-Crystalline Solids*. 2001; 285:109–115.

[86] Li T., Cao M., Liang J., Xie X., Du G. Mechanisms of base-catalyzed resorcinol-formaldehyde and phenol-resorcinol-formaldehyde condensation reactions: a theoretical study. *Polymers*. 2017; 9:426.

[87] Mulik S., Sotiriou-Leventis C. Resorcinol-formaldehyde aerogels. In: Aegerter M.F., Leventis N., Koebel M.M., editors. *Aerogels Handbook*. New York: Springer; 2011. pp. 215–234.

[88] Pekala R.W., Kong F.M. A synthetic route to organic aerogels – mechanisms, structure and properties. *Revue de Physique Applique, Colloque C4*. 1989; 24:33–40.

[89] Schwan M. Synthese und Eigenschaften flexibler Resorcin-Formaldehyd- und Kohlenstoffaerogele. RWTH University: Aachen, Germany; 2018.

[90] Yamamoto T., Mukai S.R., Endo A., Nakaiwa M., Tamon H. Interpretation of structure formation during sol-gel transition of resorcinol-formaldehyde solution by population balance. *Journal of Colloid and Interface Science*. 2003; 264:532–537.

[91] Franek J., Nowak A., Turska E. Kinetics of polycondensation. *Acta Polymerics*. 1982; 33:169–174.

[92] Paventi M. Particular solution for any consecutive second-order reaction. *Canadian Journal of Chemistry*. 1987; 65:1987–1994.

[93] Grenier-Loustalot M.F., Larroque S., Grande D., Grenier P., Bedel D. Phenolic resins: 2. Influence of catalyst type on reaction mechanisms and kinetics. *Polymer*. 1996; 37(8):1363–1369.

[94] Tamon H., Ishikaza H. SAXS study on gelation process in preparation of resorcinol-formaldehyde aerogel. *Journal of Colloid and Interface Science*. 1998; 206:577–582.

[95] Horikawa T., Hayashi J, Muroyama K. Controllability of pore characteristics of resorcinol-formaldehyde carbon aerogels. *Carbon*. 2004; 42:1625–1633.

[96] Pekala R.W., Schaefer D.W. Structure of organic aerogels: I. Morphology and scaling. *Macromolecules*. 1993; 26:5487–5493.

[97] Bock V., Emmerling A., Fricke J. Influence of monomer and catalyst concentration on RF and carbon aerogel structure. *Journal of Non-Crystalline Solids*. 1998; 225:69–73.

[98] Czakkel O., Marthi K., Geissler E., Laszlo K. Influence of drying on the morphology of resorcinol–formaldehyde-based carbon gels. *Microporous and Mesoporous Materials*. 2005; 86:124–133.

[99] Gommes C.J., Roberts A.P. Structure development of resorcinol-formaldehyde gels: microphase separation or colloid aggregation. *Physical Review*. 2008; E77:041409.

[100] Gommes C.J., Job N., Pirard J.P., Blacher S., Goderis B. Critical opalescence points to thermodynamic instability: relevance to small-angle X-ray scattering of resorcinol-formaldehyde gel formation at low pH. *Journal of Applied Crystallography*. 2008; 41:663–668.

[101] Gaca K.Z., Sefcik J. Mechanism and kinetics of nanostructure evolution during early stages of resorcinol–formaldehyde polymerisation. *Journal of Colloid and Interface Science*. 2013; 406:51–59.

[102] Brandt R. Sauer katalysierte, unterkritisch getrocknete Resorcin-Formaldehyd-Aerogele und daraus abgeleitete Kohlenstoff-Aerogele. Bayerische Julius-Maximilian Universität Würzburg; 2004.

[103] Mulik S., Sotiriou-Leventis C., Leventis N. Time-efficient acid-catalyzed synthesis of resorcinol-formaldehyde aerogels. *Chemistry of Materials*. 2007; 19:6138–6144.

[104] Milow B., Ratke L, Ludwig S. Ctiric acid catalyzed organic aerogels. In: Smirnova I., Perrut M., editors. *Proceedings 'Seminar on Aerogels'*. Nancy, France: ISASF Society; 2012.

[105] Schwan M., Ratke L. Flexibilisation of resorcinol-formaldehyde aerogels. *Journal of Materials Chemistry*. 2013; 1:13462–13468.

[106] Schwan M., Naikade M., Raabe D., Ratke L. From hard to rubber-like: mechanical properties of resorcinol-formaldehyde aerogels. *Journal of Material Science*. 2015; 50:5482–5493.

[107] Hillert M. *Phase Equilibria, Phase Diagrams and Phase Transformations*. Cambridge: Cambridge University Press; 1998.

[108] Lüdecke D.L. *Thermodynamik*. Heidelberg: Springer-Verlag; 2000.

[109] Atkins P.W., De Paula J., Keeler J. *Atkins' Physical Chemistry*. Oxford University Press; 2018.

[110] Wang F., Ratke L., Zhang H., Altschuh P., Nestler B. A phase-field study on polymerization-induced phase separation occasioned by diffusion and capillary flow – mechanism for the formation of porous microstructures in membranes. *Journal of Sol–Gel Science and Technology*. 2020; 94:356–374.

[111] Stauffer D, Aharony A. *Introduction to Percolation Theory*. 2nd ed. Boca Raton: CRC Press; 1994.

[112] Zugenmaier P. *Crystalline Cellulose and Derivatives: Characterization and Structures*. Springer Series in Wood Science. Berlin, Heidelberg: Springer-Verlag; 2008.

[113] Marsh J.T., Wood F.C. *An Introduction to the Chemistry of Cellulose*. 2nd ed. London: Chapman & Hall Ltd.; 1942.

[114] Klemm D., Phlipp B., Heinze T., Heinze U., Wagenknecht W. *Comprehensive Cellulose Chemistry: Fundamentals and Analytical Methods*. vol. 1 and 2. Weinheim: Wiley-VCH Verlag GmbH; 1998.

[115] Hearle J.W.S. A fringed fibril theory of structure in crystalline polymers. *Journal of Polymer Science*. 1958; 28(117):432–435.

[116] Flory P.J. *Principles of Polymer Chemistry*. Ithaca: Cornell University Press; 1953.

[117] Israelachvili J.N. *Intermolecular and Surface Forces*. Toronto: Academic Press; 1985.

[118] Olsson C., Westman G. Direct dissolution of cellulose: background, means and applications. In: de Wen TGNV, editor. *Cellulose – Fundamental Aspects*. INTECH; 2013. www.intechopen.com/books/howtoreference/cellulose-fundamental-aspects/direct-dissolution-of-cellulose-background-means-and-applications.

[119] Medronho B., Lindmann B. Brief overview on cellulose dissolution/regeneration interactions. *Advances in Colloid and Interface Science*. 2015; 222:502–508.

[120] Budtova T., Navard P. Cellulose in NaOH-water based solvents: a review. *Cellulose*. 2016; 23:5–55.

[121] Hirosi S., Heinz K., Kurt H. Das System Cellulose–Natriumhydroxyd–Wasser in Abhängigkeit von der Temperatur. *Zeitschrift für Physikalische Chemie*. 43B(1): 309–328.

[122] Kihlman M., Medronho B., Romano A., Germgard U., Lindman B. Cellulose dissolution in an alkali based solvent: influence of additives and pretreatments. *Journal of the Brazilian Chemical Society*. 2013; 24(2):295–303.

[123] Cibik T. Untersuchungen am System NMMO/H2O/Cellulose. TU Berlin; 2003.

[124] Walker M., Zimmerman R.L., Whitcombe G.P., Humbert H.H. N-Methylmorpholinoxid (NMMO) – Die Entwicklung eines Lösemittels zur industriellen Produktion von Zellulosefasern. *Lenzinger Berichte*. 1997; 97:76–80.

[125] Wilcox R.J., Losey B.P., Folmer J.C.W., Martin J.D., Zeller M., Sommer R. Crystalline and liquid structure of zinc chloride trihydrate: a unique ionic liquid. *Inorganic Chemistry*. 2015; 54:1109–1119.

[126] Sen S., Losey B.P., Gordon E.E., Argyropoulos D.S., Martin J.D. Ionic liquid character of zinc chloride hydrates define solvent characteristics that afford the solubility of cellulose. *Journal of Physical Chemistry B*. 2016; 120:1134–1141.

[127] Sen S., Martin J.D., Argyropoulos D.S. Review of cellulose non-derivatizing solvent interactions with emphasis on activity in inorganic molten salt hydrates. *ACS Sustainable Chemistry & Engineering*. 2013 08; 1(8):858–870.

[128] Philip B., Schleicher H., Wagenknecht W. Non-aquous solvents of cellulose *Chemtech*. 1977; 7:702–709.

[129] Fischer S. Anorganische Salzhydratschmelzen – ein unkonventionelles Löse- und Reaktionsmedium für Cellulose. TU Bergakademie Freiberg. Freiberg, Germany; 2004.

[130] Deguchi S., Tsujii K., Horikoshi K. Crystalline-to-amorphous transformation of cellulose in hot and compressed water and its implications for hydrothermal conversion. *Green Chemistry*. 2008; 10:191–196.

[131] Hattori M., Koga T., Shimaya Y., Saito M. Aqueous calcium thiocyanate solutions as a cellulose solvent. structure and interactions with cellulose. *Polymer Journal*. 1998; 30:43–48.

[132] Frey M., Theil M.H. Calculated phase diagrams for cellulose/ammonia/ammonium thiocyanate solutions in comparison to experimental results. *Cellulose*. 2004; 11:53–63.

[133] Schestakow M. *Nanostrukturierte Cellulose-Aerogel-Polyesterverbunde*. Aachen, Germany: RWTH Aachen University; 2020.

[134] Gavillon R., Budtova T. Aerocellulose: new highly porous cellulose prepared from cellulose–NaOH aqueous solutions. *Biomacromolecules*. 2008; 9:269–277.

[135] te Nijenhuis K. *Thermoreversible Networks*. vol. 130 of *Advances in Polymer Science*. Berlin, Heidelberg: Springer-Verlag; 1997.

[136] Liu W., Budtova T., Navard P. Influence of ZnO on the properties of dilute and semi-dilute cellulose–NaOH–water solutions. *Cellulose*. 2011; 18:911–920.

[137] Trygg J., Fardim P., Gericke M., Mäkilä E., Salonen J. Physicochemical design of the morphology and ultrastructure of cellulose beads. *Carbohydrate Polymers*. 2013; 93(1):291–299.

[138] Mohamed S.M.K., Ganesan K., Milow B., Ratke L. The effect of zinc oxide (ZnO) addition on the physical and morphological properties of cellulose aerogel beads. *RSC Advances*. 2015; 5:90193–90201.

[139] Cussler E.L. *Diffusion – Mass Transfer in Fluid Systems*. 3rd ed. Cambridge: Cambridge University Press; 2009.

[140] Kuttler C. Reaction–diffusion equations with applications; 2011. Westfälische Wilhelms Universität Münster, online. www.uni-muenster.de/imperia/md/content/physik_tp/lectures/ws2016-2017/num_methods_i/rd.pdf.

[141] Kimura Y.T. The mathematics of patterns: the modeling and analysis of reaction-diffusion equations. Princeton University; 2014.

[142] Innerlohinger J., Weber H.K., Kraft G. Aerocellulose: aerogels and aerogel-like materials made from cellulose. *Macromolecular Symposia*. 2006; 244:126–135.

[143] Liebner F., Potthast A., Rosenau T., Haimer E., Wendland M. Ultralight-weight cellulose aerogels from NBnMO-stabilized lyocell dopes. *Research Letters in Materials Science*. 2007; 2007(ID 7324).

[144] Liebner F., Rosenau T., Wendland M., Potthast A., Haimer E. Cellulose aerogels: highly porous, ultra-lightweight materials. *Holzforschung*. 2008; 62(2):129–135.

[145] Liebner F., Haimer E., Potthast A., et al. Cellulosic aerogels as ultra-lightweight materials. Part 2: Synthesis and properties. *Holzforschung*. 2009; 63:3–11.

[146] Liebner F., Haimer E., Wendland M. et al. Aerogels from unaltered bacterial cellulose: application of scCO2 drying for the preparation of shaped, ultra-lightweighted cellulosic aerogels. *Macromolecular Bioscience*. 2010; 10:349–352.

[147] Perrut M., Smirnova I., editors. Aerogels from bacterial cellulose: a new dimension in preparing shaped cellulose aerogels. ISASF, International Society for Advancement of Supercritical Fluids; 2010.

[148] Hoepfner S., Ratke L., Milow B. Synthesis and characterization of nanofibrillar cellulose aerogels. *Cellulose*. 2008; 15:121–129.

[149] Karadagli I., Schulz B., Schestakow M., Milow B., Gries T., Ratke L. Production of porous cellulose aerogel fibers by an extrusion process. *Journal of Supercritical Fluids*. 2015; 106:105–114.

[150] Schestakow M., Karadagli I., Ratke L. Cellulose aerogels prepared from an aqueous zinc chloride salt hydrate melt. *Carbohydrate Polymers*. 2016; 137:642–649.

[151] Pircher N., Carbajal L., Schimper C., et al. Impact of selected solvent systems on the pore and solid structure of cellulose aerogels. *Cellulose*. 2016; 23:1949–1966.

[152] Schulz B. Cellulose-Aerogelfasern. RWTH Aachen University. Aachen, Germany; 2015.

[153] Pinnow M., Fink H.P., Fanter C., Kunze J. Characterization of highly porous materials from cellulose carbamate. *Macromolecular Symposia*. 2008; 262:129–139.

[154] Gan S., Zakaria S., Chia C.H., Chen R.S., Ellis A.V., Kaco H. Highly porous regenerated cellulose hydrogel and aerogel prepared from hydrothermal synthesized cellulose carbamate. *PLoS One*. 2017; https://doi.org/10.1371/journal.pone.0173743.

[155] Gavillon R. Preparation et caracterisation der materiaux cellulosiques ultra poreux. Ecole des Mines de Paris. Paris; 2007.

[156] Jin H., Nishiyama Y., Wada M., Kuga S. Nanofibrillar cellulose aerogels. *Colloids and Surfaces A: Physicochemical and Engineering Aspects*. 2004; 240:63–67.

[157] Innerlohinger J., Weber H.K., Kraft G. Aerocell: Aerogels from cellulosic materials. *Lenzinger Berichte*. 2006; 86:137–143.

[158] Pircher N., Fischhuber D., Carbajal L., et al. Preparation and reinforcement of dual-porous biocompatible cellulose scaffolds for tissue engineering. *Macromolecular Materials and Engineering*. 2015; 300(9):911–924.

[159] Donati I., Paoletti S. Non-aquous solvents of cellulose. In: Rehm B.H.A., editor. *Material Properties of Alginates*. Berlin, Heidelberg: Springer; 2009. pp. 1–53.

[160] Jones R.G. Dispersity in polymer science. *Polymer International*. 2010; 59(1):22–22.

[161] Aarstad O.A., Tøndervik A., Sletta H., Skjåk-Bræk G. Alginate sequencing: an analysis of block distribution in alginates using specific alginate degrading enzymes. *Biomacromolecules*. 2012 01; 13(1):106–116.

[162] Usov A.I. Alginic acids and alginates: analytical methods used for their estimation and characterisation of composition and primary structure. *Russian Chemical Reviews*. 1999; 68(11):957–966.

[163] Stokke B.T., Smidsroed O., Bruheim P., Skjaak-Braek G. Distribution of uronate residues in alginate chains in relation to alginate gelling properties. *Macromolecules*. 1991 08; 24(16):4637–4645.

[164] Agulhon P., Constant S., Chiche B., et al. Controlled synthesis from alginate gels of cobalt–manganese mixed oxide nanocrystals with peculiar magnetic properties. *Catalysis Today*. 2012; 189(1):49–54.

[165] Grasdalen H., Larsen B, Smisrod O. 13C-n.m.r. studies of monomeric composition and sequence in alginate. *Carbohydrate Research*. 1981; 89(2):179–191.

[166] Pawar S.N., Edgar K.J. Alginate derivatization: a review of chemistry, properties and applications. *Biomaterials*. 2012; 33(11):3279–3305.

[167] Matsushima K., Minoshima H., Kawanami H., et al. Decomposition reaction of alginic acid using subcritical and supercritical water. *Industrial & Engineering Chemistry Research*. 2005 12; 44(25):9626–9630.

[168] Villard P., Rezaeeyazdi M., Colombani T., et al. Autoclavable and injectable cryogels for biomedical applications. *Advanced Healthcare Materials*. 2019; 8(17):1900679.

[169] Pawar S.N. Chemical modification of alginate. In: Venkatesan J., Anil S., Kim S.K., editors. *Seaweed Polysaccharides*. Elsevier; 2017. pp. 111–155.

[170] Pawar S.N., Edgar K.J. Chemical modification of alginates in organic solvent systems. *Biomacromolecules*. 2011 11; 12(11):4095–4103.

[171] Schleeh T., Madau M., Roessner D. Synthesis enhancements for generating highly soluble tetrabutylammonium alginates in organic solvents. *Carbohydrate Polymers*. 2014; 114:493–499.

[172] Budtova T. Cellulose II aerogels: a review. *Cellulose*. 2019 Jan; 26(1):81–121.

[173] Ventura M.G., Paninho A.I., Nunes A.V.M., Fonseca I.M., Branco L.C. Biocompatible locust bean gum mesoporous matrices prepared by ionic liquids and a scCO2 sustainable system. *RSC Advances*. 2015; 5:107700–107706.

[174] Cordeiro T., Paninho A.B., Bernardo M., et al. Biocompatible locust bean gum as mesoporous carriers for naproxen delivery. *Materials Chemistry and Physics*. 2020; 239:121973.

[175] Sun X., Xue Z., Mu T. Precipitation of chitosan from ionic liquid solution by the compressed CO2 anti-solvent method. *Green Chemistry*. 2014; 16:2102–2106.

[176] Aarstad O., Strand B.L., Klepp-Andersen L.M., Skjåk-Bræk G. Analysis of G-block distributions and their impact on gel properties of in vitro epimerized mannuronan. *Biomacromolecules*. 2013 10; 14(10):3409–3416.

[177] Fang Y., Al-Assaf S., Phillips G.O., et al. Multiple steps and critical behaviors of the binding of calcium to alginate. *Journal of Physical Chemistry B*. 2007 03; 111(10): 2456–2462.

[178] Hecht H., Srebnik S. Structural characterization of sodium alginate and calcium alginate. *Biomacromolecules*. 2016 06; 17(6):2160–2167.

[179] Robitzer M., David L., Rochas C., Di Renzo F., Quignard F. Nanostructure of calcium alginate aerogels obtained from multistep solvent exchange route. *Langmuir*. 2008 11; 24(21):12547–12552.

[180] Robitzer M., David L., Rochas C., Di Renzo F., Quignard F. Supercritically-dried alginate aerogels retain the fibrillar structure of the hydrogels. *Macromolecular Symposia*. 2008; 273(1):80–84.

[181] Takeshita S., Sadeghpour A., Malfait W.J., Konishi A., Otake K., Yoda S. Formation of nanofibrous structure in biopolymer aerogel during supercritical CO2 processing: the case of chitosan aerogel. *Biomacromolecules*. 2019 05; 20(5):2051–2057.

[182] Agulhon P., Markova V., Robitzer M., Quignard F., Mineva T. Structure of alginate gels: interaction of diuronate units with divalent cations from density functional calculations. *Biomacromolecules*. 2012 06; 13(6):1899–1907.

[183] Plazinski W., Drach M. Binding of bivalent metal cations by α-l-guluronate: insights from the DFT-MD simulations. *New Journal of Chemistry*. 2015; 39:3987–3994.

[184] Menakbi C., Quignard F., Mineva T. Complexation of trivalent metal cations to mannuronate type alginate models from a density functional study. *Journal of Physical Chemistry B*. 2016 04; 120(15):3615–3623.

[185] Topuz F., Henke A., Richtering W., Groll J. Magnesium ions and alginate do form hydrogels: a rheological study. *Soft Matter*. 2012; 8:4877–4881.

[186] Westmann J.O., Taherzadeh M.J., Franzén C.J. Proteomic analysis of the increased stress *tolerance of saccharomyces cerevisiae encapsulated in* liquid core alginate–chitosan capsules. *PLoS One*. 2012; 7(11):e49335.

[187] Wang Q., Zhang L., Liu Y., Zhang G., Zhu P. Characterization and functional assessment of alginate fibers prepared by metal–calcium ion complex coagulation bath. *Carbohydrate Polymers*. 2020; 232:115693.

[188] Tritz J., Rahouadj R., de Isla N., et al. Designing a three-dimensional alginate hydrogel by spraying method for cartilage tissue engineering. *Soft Matter*. 2010; 6:5165–5174.

[189] Marpani F., Luo J., Mateiu R.V., Meyer A.S., Pinelo M. In situ formation of a biocatalytic alginate membrane by enhanced concentration polarization. *ACS Applied Materials & Interfaces*. 2015 08; 7(32):17682–17691.

[190] Chen D., Lewandowski Z., Roe F., Surapaneni P. Diffusivity of Cu2+ in calcium alginate gel beads. *Biotechnology and Bioengineering*. 1993; 41(7):755–760.

[191] Nestle N.F.E.I., Kimmich R. NMR imaging of heavy metal absorption in alginate, immobilized cells, and kombu algal biosorbents. *Biotechnology and Bioengineering*. 1996; 51(5):538–543.

[192] Kuo C.K., Ma P.X. Ionically crosslinked alginate hydrogels as scaffolds for tissue engineering: Part 1. Structure, gelation rate and mechanical properties. *Biomaterials*. 2001; 22(6):511–521.

[193] Gurikov P., Raman S.P., Weinrich D., Fricke M., Smirnova I. A novel approach to alginate aerogels: carbon dioxide induced gelation. *RSC Advances*. 2015; 5:7812–7818.

[194] Subrahmanyam R., Gurikov P., Meissner I., Smirnova I. Preparation of biopolymer aerogels using green solvents. *JoVE*. 2016; (113):e54116.

[195] Yuguchi Y., Hasegawa A., Padoł A.M., Draget K.I., Stokke B.T. Local structure of Ca2+ induced hydrogels of alginate–oligoguluronate blends determined by small-angle-X-ray scattering. *Carbohydrate Polymers*. 2016; 152:532–540.

[196] Valentin R., Horga R., Bonelli B., Garrone E., Di Renzo F., Quignard F. FTIR spectroscopy of NH3 on acidic and ionotropic alginate aerogels. *Biomacromolecules*. 2006 03; 7(3):877–882.

[197] Mørch Ý.A., Holtan S., Donati I., Strand B.L., Skjåk-Bræk G. Mechanical properties of C-5 epimerized alginates. *Biomacromolecules*. 2008 09; 9(9):2360–2368.

[198] Aarstad O, Strand B.L., Klepp-Andersen L.M., Skjåk-Bræk G. Analysis of G-block distributions and their impact on gel properties of in vitro epimerized mannuronan. *Biomacromolecules*. 2013 10; 14(10):3409–3416.

[199] Lozinsky V. Cryogels on the basis of natural and synthetic polymers: preparation, properties and application. *Russian Chemical Reviews*. 2002; 71:489–511.

[200] Zhao Y., Shen W., Chen Z., Wu T. Freeze-thaw induced gelation of alginates. *Carbohydrate Polymers*. 2016; 148:45–51.

[201] Zhang H., Zhang F., Wu J. Physically crosslinked hydrogels from polysaccharides prepared by freeze–thaw technique. *Reactive and Functional Polymers*. 2013; 73(7): 923–928.

[202] Giannouli P., Morris E. Cryogelation of xanthan. *Food Hydrocolloids*. 2003 July; 17(4):495—501.

[203] Shan L., Gao Y., Zhang Y., et al. Fabrication and use of alginate-based cryogel delivery beads loaded with urea and phosphates as potential carriers for bioremediation. *Industrial & Engineering Chemistry Research*. 2016 07; 55(28):7655–7660.

[204] Ho M.H., Kuo P.Y., Hsieh H.J., et al. Preparation of porous scaffolds by using freeze-extraction and freeze-gelation methods. *Biomaterials*. 2004 January; 25(1):129–138.

[205] Tkalec G., Knez Ž, Novak Z. Formation of polysaccharide aerogels in ethanol. *RSC Advances*. 2015; 5:77362–77371.

[206] Pérez-Madrigal M.M., Torras J., Casanovas J., Häring M., Alemán C., Díaz D.D. Paradigm shift for preparing versatile M2+-free gels from unmodified sodium alginate. *Biomacromolecules*. 2017 09; 18(9):2967–2979.

[207] Gurikov P., Smirnova I. Non-conventional methods for gelation of alginate. *Gels*. 2018; 4:14.

[208] Groot S.R.D., Mazur P. *Non-Equilibrium Thermodynamics*. Amsterdam: North-Holland Publishing Company; 1962.

[209] de Groot S. *Thermodynamics of Irrversible Processes*. Amsterdam: North-Holland Publishing Company; 1952.

[210] Landau L.D., Lifshitz E.M. *Lehrbuch der theoretischen Physik*. vol. X. Berlin: Akademie Verlag; 1983.

[211] Jäckle J. *Einführung in die Transporttheorie*. Braunschweig: Vieweg Verlag; 1987.

[212] Huang K. *Statistical Mechanics*. New York: John Wiley and Sons, Inc.; 1963.

[213] Born M., Green H. A general kinetic theory of liquids. III. Dynamical properties. *Proceedings of the Royal Society A*. 1947; 190:455–474.

[214] Yokoyama I., Tsuchiya S. Excess entropy, diffusion coefficient, viscosity coefficient and surface tension of liquid simple metals from diffraction data. *Materials Transactions*. 2002; 43:67–72.

[215] Christensen R.M. *Theory of Viscoelasticity*. 2nd ed. Mineola: Dover Civil and Mechanical Engineering. Dover Publications; 2013.

[216] Thiel J. *Kinetik des Sol/Gel Übergangs in wässrigen Resorcin/Formaldehyd-Lösungen*. Fachhochschule Aachen and DLR; 2001.

[217] Vogelsberger W., Opfermann J., Wank U., Schulze H., Rudakoff G. A model for viscosity in the early stages of the sol–gel transformation. *Journal of Non-Crystalline Solids*. 1991; 145:20–24.

[218] Buchtová N, Budtova T. Cellulose aero-, cryo- and xerogels: towards understanding of morphology control. *Cellulose*. 2016; 23(4):2585–2595.

[219] Krause J., Lisinski S., Ratke L., et al. Observation of gelation process and particle distribution during sol–gel synthesis by particle image velocimetry. *Journal of Sol–Gel Science and Technology* 2008; 45:73–77.

[220] Raffel M., Willert C., Kompenhans J. *Particle Image Velocimetry*. Berlin, Heidelberg, New York: Springer Verlag; 1997.

[221] Ratke L., Hajduk A. On the size effect of gelation kinetics in RF aerogels. *Gels*. 2015; 1:276–290.

[222] Mandelbrot B. *The Fractal Geometry of Nature*. 3rd ed. New York: W.H. Freeman and Company; 1977.

[223] Smoluchowski M. Drei Vorträge über Diffusion, Brownsche Molekularbewegung und Koagulation von Kolloidteilchen. *Physik Zeitschrift*. 1916; 17:586–599.

[224] Fuchs N. *Aerosol Mechanics*. Oxford: Pergamon Press Ltd.; 1964.

[225] Rogers R.R., Yau M.K. *A Short Course in Cloud Physics*. 3rd ed. Oxford: Pergamon Press Ltd.; 1989.

[226] Ratke L., Diefenbach S. Liquid immiscible allyos. *Materials Science and Engineering Reports*. 1995; R15:263–347.

[227] Christensen K., Moloney N.R. *Complexity and Criticality*. London: Imperial College Press; 2005.

[228] Percolation and thresholds. www.wikiwand.com/en/Percolation_threshold#.

[229] Lorenz C.D., Ziff R.M. Precise determination of the critical percolation threshold for the three-dimensional 'Swiss cheese' model using a growth algorithm. *Journal of Chemical Physics*. 2001; 114(8):3659–3661.

[230] Stauffer D., Coniglio A., Adam M. Gelation and critical phenomena. *Advances in Polymer Science*. 1982; 44:104–158.

[231] Christensen K. *Percolation Theory*; 2002. London: Imperial College London.

[232] Witten, T.A., Sander, L.M. Diffusion-limited aggregation, a kinetic critical phenomenon. *Physical Review of Letters*. 1981; 47:1400–1403.

[233] Meakin P. Formation of fractal clusters and networks by irreversible diffusion-limited aggregation. *Physical Review of Letters*. 1983; 51:1119–1122.

[234] Kolb M., Botet R., Jullien R. Scaling of Kinetically Growing Clusters. *Physical Review of Letters*. 1983; 51:1123–1126.

[235] Halperin, W.P. *Low Temperature Physics Group*. http://spindry.phys.northwestern.edu/agel.htm.

[236] Haard T.M., Gervais G., Nomura R., Halperin W.P. The pathlength distribution of simulated aerogels. *Physica B: Condensed Matter*. 2000; 284-288:289–290.

[237] *Toxiclibs*. Simutils-0001: Diffusion limited aggregation http://toxiclibs.org/2010/02/new-package-simutils/.

[238] Sayama, H. Diffusion-limited aggregation: a real-time agent-based simulation. *Wolfram Demonstration Projects*. https://demonstrations.wolfram.com/DiffusionLimitedAggregationARealTimeAgentBasedSimulation/.

[239] Drake R.L.A. *A General Mathematical Survey of the Coagulation Equation*. edited by Hidy G.M., Brock J.R. Topics in Current Aerosol Research. Oxford, UK: Pergamon Press Ltd.; 1970.

[240] Golovin A.M. The solution of the coagulation equation for cloud droplets in a rising air current. *Bulletin of the Academy of Sciences of the USSR*. 1963; 5:482–487.

[241] Stauffer, D. Percolation and cluster size distribution. In: Stanley H.E. and Ostrowsky, N., editors, *On Growth and Form*. Dordrecht: Martinus Nijhoff Publishers, Dordrecht; 1986. pp 79–100.

[242] Lushnikov A.A. Critical behavior of the particle mass spectra in a family of gelling systems. *Physical Review E*. 2007; 76:011120.

[243] Lushnikov A.A. Postcritical behavior of a gelling system. *Physical Review E*. 2013; 88:052120.

[244] Barenblatt G.I. *Scaling, Self-Similarity, and Intermediate Asymptotics*. Cambridge: Cambridge University Press; 1996.

[245] Friedlander S.K., Wang C.S. The self-preserving particle size distribution for coagulation by brownian motion. *Journal of Colloid and Interface Science*. 1966; 22:126–132.

[246] Pulvermacher B., Ruckenstein E. Similarity solutions of populations balances. *Journal of Colloid and Interface Science*. 1974; 46:428–436.

[247] Family F., Vicsek T. Scaling of the active zone in the Eden process on percolation networks and the ballistic deposition model. *Journal of Physics A: Mathematical and General*. 1985; 18:L75–L81.

[248] Meakin P., Viscek T., Family F. Dynamic cluster-size distribution in cluster–cluster aggregation: effects of cluster diffusivity. *Physical Review B*. 1985; 31:564–569.

[249] Ball R.C., Connaughton C., Stein T.H.M., Zaboronski O. Instantaneous gelation in Smoluchowski's coagulation equation revisited. *Physical Review E*. 2011; 84:011111.

[250] Rezakhanlou F. Gelation for Marcus–Lushnikov process. *Annals of Probability*. 2013; 41:1806–1830.

[251] Fournier N., Laurencot P. Marcus–Lushnikov processes, Smoluchowski's and Flory's models. *Stochastic Processes and Their Applications*. 2009; 119:167–189.

[252] Schaefer D.W., Pekala R., Beaucage G. Origin of porosity in resorcinol-formaldehyde aerogels. *Journal of Non-Crystallline Solids*. 1995; 186:159–167.

[253] Berthon S., Barbieri O., Ehrburger-Dolle F., et al. DLS and SAXS investigations of organic gels and aerogels. *Journal of Non-Crystalline Solids*. 2001; 285:154–161.

[254] Tanaka H. New mechanisms of droplet coarsening in phase-separating fluid mixtures. *Journal of Chemical Physics*. 1997; 107(9):3734–3737.

[255] Tanaka H. Viscoelastic model of phase separation. *Physical Review E*. 1997 Oct; 56:4451–4462.

[256] Tanaka F., Okamura K. Characterization of cellulose molecules in bio-system studied by modeling methods. *Cellulose*. 2005; 12:243–252.

[257] Taniguchi T., Onuki A. Network domain structure in viscoelastic phase separation. *Physical Review of Letters*. 1996; 77:4910–4913.

[258] Glotzer S.C., Marzio E.A.D., Mthukumur M. Reaction-controlled morphology of phase-separating mixtures. *Physical Review of Letters*. 1995; 74:2034–2039.

[259] Glotzer S. Computer simulations of spinodal decompositions in polymer blends. In: Stauffer D, editor. *Annual Reviews of Computational Physics II*. Singapore: World Scientific Publishing Co.; 1995. pp. 1–47.

[260] Oh J., Rey A.D. Theory and simulation of polymerization-induced phase separation in polymeric media. *Macromolecular Theory and Simulations*. 2000; 9(8):641–660.

[261] Zha L., Hu W. Homogeneous crystal nucleation triggered by spinodal decomposition in polymer solutions. *Journal of Physical Chemistry B*. 2007; 111:11373–11378.

[262] Lee K.W.D., Chan P.K., Feng X. A computational study of the polymerization-induced phase separation phenomenon in polymer solutions under a temperature gradient. *Macromolecular Theory and Simulations*. 2003; 12:413–424.

[263] Hara A., Inoue R., Takahashi N., Nishida K., Kanaya T. Trajectory of critical point in polymerization-induced phase separation of epoxy/oligoethylene glycol solutions. *Macromolecules*. 2014; 47:4453–4459.

[264] Luo K. The morphology and dynamics of polymerization-induced phase separation. *European Polymer Journal*. 2006; 42:1499–1505.

[265] Kyu T., Lee J.H. Nucleation initiated spinodal decomposition in an polymerizing system. *Physical Review of Letters*. 1996; 76:3746–3749.

[266] Koyama T., Araki T., Tanaka H. Fracture phase separation. *Physical Review of Letters*. 2009; 102:065701.

[267] Sciortino F., Bansil R., Stanley H.E. Interference of phase separation and gelation: zeroth-order kinetic model. *Physical Review E*. 1993; 47(6):4615–4618.

[268] Muratov C.B. Unusual coarsening during phase separation in polymer systems. *Physical Review of Letters*. 1998 Oct; 81:3699–3702.

[269] Lee D., Huh J.Y., Jeong D., Shin J., Yun A., Kim J. Physical, mathematical, and numerical derivations of the Cahn–Hilliard equation. *Computational Materials Science*. 2014; 81:216–225.

[270] Kontogeorgis G.M., von Solms, N. Thermodynamics of polymer solutions. In: Birdi, K.S., editor, *Handbook of Surface and Colloid Chemistry*. 3rd ed. Boca Raton, New York: Taylor and Francis Group, 2009. pp. 499–537.

[271] Yamakawa H. *Modern Theory of Polymer Solutions*. Electronic edition ed. Kyoto University, original New York: Harper and Row, 1971; 2002.

[272] Courant R., Hilbert D. *Methoden der mathematischen Physik*. Berlin–Heidelberg–New York: Springer-Verlag; 1968.

[273] Morse P.M, Feshbach H. *Methods of Theoretical Physics*. New York: McGraw-Hill Inc.; 1953.

[274] Churchill S.W. *Viscous Flows*. Butterworths Series in Chemical Engineering. Stoneham: Butterworths Publishers; 1988.

[275] Subramanian R.S., Balasubramaniam R. *The Motion of Bubbles and Drops in Reduced Gravity*. Cambridge: Cambridge University Press; 2001.

[276] Levich V. *Physicochemical Hydrodynamics*. Englewood Cliffs: Prentice Hall; 1962.

[277] Wan F., Altschuh P., Ratke L., Zhang H., Selzer M., Nestler B. Progress report on phase separation in polymer solutions. *Advanced Materials*. 2019; pp. 1806733.

[278] Chan P.K., Rey A.D. Polymerization-induced phase separation. 1. Droplet size selection mechanism. *Macromolecules*. 1996; 29:8934–8941.

[279] Anglaret E., Hasmy A., Jullien R. Effect of container size on gelation time: experiments and simulations. *Physical Review of Letters*. 1995; 75:4059–4062.

[280] Vafai K., editor. *Handbook of Porous Media*. 2nd ed. Boca Raton: CRC Press Taylor and Francis Group; 2005.

[281] Metzger T., Kwapinski M., Peglow M., Saage G., Tsotsas E. Modern modelling methods in drying. *Transport in Porous Media*. 2007; 66:103–120.

[282] Prat M. Recent advances in pore-scale models for drying of porous media. *Chemical Engineering Journal*. 2002; 86:153–164.

[283] Prat M. On the influence of pore shape, contact angle and film flows on drying of capillary porous media. *International Journal of Heat and Mass Transfer*. 2007; 50:1455–1468.

[284] Nusselt W. *Heat Transfer, Diffusion and Evaporation*. Washington: National Advisory Committee for Aeronautics; 1954.

[285] Bisson A., Rodier E., Rigacci A., Lecomte D, Achard P. Study of evaporative drying of treated silica gels. *Journal of Non-Crystalline Solids*. 2004; 350:230–237.

[286] Smith D.M., Stein D., Anderson J.M., Ackerman W. Preparation of low-density xerogels at ambient pressure. *Journal of Non-Crystalline Solids*. 1995; 186:104–112.

[287] Reuss M., Ratke L. Drying of aerogel-bonded sands. *Journal of Materials Science*. 2010; 45:3974–3980.

[288] Reuss M., Ratke L. RF-aerogels catalysed by ammonioum carbonate. *Journal of Sol–Gel Sciences and Technology*. 2010; 53:85–92.

[289] Schwerdtfeger F., Frank D., Schmidt M. Hydrophobic waterglass based aerogels without solvent exchange or supercritical drying. *Journal of Non-Crystalline Solids*. 1998; 225:24–29.

[290] Crank J. *The Mathematics of Diffusion*. 2nd ed. Oxford: Clarendon Press; 1975.

[291] Mahadik D.B., Rao A.V., Rao A.P., Waghb P.B., Ingale S.V., Gupta S.C. Effect of concentration of trimethylchlorosilane (TMCS) and hexamethyldisilazane (HMDZ) silylating agents on surface free energy of silica aerogels. *Journal of Colloid and Interface Science*. 2011; 356:298–302.

[292] Zgura I., Moldovsn R., Negrila C.C., Frunza S., Cotorobai V.F., Frunza L. Surface free energy of smooth and dehydroxylated fused quartz from contact angel measuremmets using some particular organics as probe liquids. *Journal of Optoelectronics and Advanced Materials*. 2013; 15:627–634.

[293] Zenkiewicz M. Methods for the calculation of surface free energies of solids. *Journal of Achievements in Materials and Manufacturing Engineering*. 2007; 24:137–145.

[294] Zdziennicka A., Janczuk B. Adhesion of aqueous solution of TX-100 and TX-165 mixture with propanol to quartz. *Journal of Materials Science and Nanotechnology*. 2018; 6. http://www.annexpublishers.co/articles/JMSN/6206-Adhesion-of-Aqueous-Solution-of-TX-100-and-TX-165-Mixture-with-Propanol-to-Quartz.pdf.

[295] Yiotis A.G., Tsimpanogiannis I.N., Stubos A.K., Yortsos Y.C. Pore-network study of the characteristic periods in the drying of porous materials. *Journal of Colloid and Interface Science*. 2006; 297:738–748.

[296] Kohout M., Grof Z., Stepanek F. Pore-scale modelling and tomographic visualisation of drying in granular media. *Journal of Colloid and Interface Science*. 2006; 299:342–351.

[297] Deville S. Freeze-casting of porous biomaterials: structure, properties and opportunities. *Materials*. 2010; 3:1913–1927.

[298] Deville S., Meille S., Seuba J. A meta-analysis of the mechanical properties of ice-templated ceramics and metals. *Science and Technology of Advanced Materials*. 2015; 16:043501.

[299] Glicksman M.E. *Principles of Solidification*. New York: Springer Science and Business Media LLC; 2011.

[300] Dantzig J., Rappaz M. *Solidification*. 2nd ed. EPFL Press; 2018.

[301] Kurz W., Fisher D.J. *Fundamentals of Solidification*. 4th ed. Aedermannsdorf: Trans. Tech. Publisher; 1998.

[302] Rosenberger F. *Fundamentals of Crystal Growth*. Berlin–Heidelberg–New York: Springer; 1979.

[303] Grant P.S., Cantor B. Modelling of droplet dynamic and thermal histories during spray forming – III. Analysis of spray solid fraction. *Acta Metallurgica et Materialia*. 1995; 43(3):913–921.

[304] Grant P.S., Cantor B., Katgerman L. Modelling of droplet dynamic and thermal histories during spray forming – I. individual droplet behaviour. *Acta Metallurgica et Materialia*. 1993; 41(11):3097–3108.

[305] Grant P.S., Cantor B., Katgerman L. Modelling of droplet dynamic and thermal histories during spray forming – II. Effect of process parameters. *Acta Metallurgica et Materialia*. 1993; 41(11):3109–3118.

[306] Kauerauf B., Zimmermann G., Rex S., Mathes M., Grote F. Directional cellular growth of succinonitrile-0.075wt samples Part 1: results of space experiments. *Journal of Crystal Growth*. 2001; 223:265–276.

[307] Kauerauf B., Zimmermann G., Rex S., Billia B., Jamgotchian H., Hunt J.D. Directional cellular growth of succinonitrile-0.075wt samples Part 2: Analysis of cellular pattern. *Journal of Crystal Growth*. 2001; 223:277–284.

[308] Mullins W.W., Sekerka R.F. Stability of a planar interface during solidification of a dilute binary alloy. *Journal of Applied Physics*. 1964; 35(2):444–451.

[309] Grosse K., Ratke L., Feuerbacher B. Solidification and melting of succinonitrle within the porous structure of an aerogel. *Physical Review B*. 1997; 55:2894–2902.

[310] Stefanescu D.M. *Science and Engineering of Casting Solidification*. New York: Springer Science and Business Media LLC; 2009.

[311] Jiménez-Saelices C., Cathala B.S., Grohens Y. Spray freeze-dried nanofibrillated cellulose aerogels with thermal superinsulating properties. *Carbohydrate Polymers*. 2017; 157:105–113.

[312] Ghafar A., Parikka K., Haberthür D., Tenkanen M., Mikkonen K.S., Suuronen J.P. Synchroton microtomography reveals the fine three-dimensional porosity of composite polysaccharide aerogels. *Materials*. 2017; 10:871.

[313] Alkemper J., Sous S., Stöcker C., Ratke L. Directional solidification in an aerogel furnace with high resolution optical temperature measurement. *Journal of Crystal Growth*. 1998; 191:252–260.

[314] Blum J., Wurm G., Kempf S., et al. Growth and form of planetary seedlings: results from a microgravity aggregation experiment. *Physical Review of Letters*. 2000; 85:2426–2429.

[315] Ratke L., Kochan H. Fracture Mechanical Aspects of Dust Emission Processes from a Model Comet Surface. In: *Proceedings of the International Workshop on Physics and Mechanics of Cometary Materials*. ESA SP-302. European Space Agency. Noordwijk: ESA; 1989. pp. 121–128.

[316] Thiel K., Kölzer G., Kochan H., Ratke L., Grün E., Kohl H. Dynamics of crust formation and dust emission of comet nucleus analogues under insolation. In: *Proceedings of the International Workshop on Physics and Mechanics of Cometary Materials*. ESA SP-302. European Space Agency. Noordwijk: ESA; 1989. pp. 221–225.

[317] Newborn R.L., Neugebauer M., Rahe J., editors. *Comets in the Post-Halley Era*. vols. 1 and 2. Dordrecht: Kluwer Academic Publishers; 1991.

[318] Meyer E.F., Meyer T.P. Supercritical fluid: liquid, gas, both, or neither? A different approach. *Journal of Chemical Education*. 1986; 63(6):463.

[319] Tarafder A., Guiochon G. Use of isopycnic plots in designing operations of supercritical fluid chromatography: II. The isopycnic plots and the selection of the operating pressure-temperature zone in supercritical fluid chromatography. *Journal of Chromatography A*. 2011 July; 1218(28):4576–4585.

[320] Pilar F.L. The critical temperature: a necessary consequence of gas non-ideality. *Journal of Chemical Education*. 1967; 44(5):284.

[321] Westwell M.S., Searle M.S., Wales D.J., Williams D.H. Empirical correlations between thermodynamic properties and intermolecular forces. *Journal of the American Chemical Society*. 1995; 117(18):5013–5015.

[322] Joback method https://en.wikipedia.org/wiki/Joback_method.

[323] Moldover M.R. Interfacial tension of fluids near critical points and two-scale-factor universality. *Physical Review A*. 1985; 31:1022–1033.

[324] Domb C. *The Critical Point*. London: CRC Press; 1996.

[325] Liu Y., Lipowsky R., Dimova R. Concentration dependence of the interfacial tension for aqueous two-phase polymer solutions of dextran and polyethylene glycol. *Langmuir*. 2012; 28:3831–3839.

[326] Pierre A.C., Pajonk G.M. Chemistry of aerogels and their applications. *Chemical Reviews*. 2002; 102(11):4243–4266.

[327] Schott J., Dandurand J.L. Prediction of the thermodynamic behavior of aqueous silica in aqueous complex solutions at various temperatures. In: Helgeson H.C., editor. *Prediction of the Thermodynamic Behavior of Aqueous Silica in Aqueous Complex Solutions at Various Temperatures*. Dordrecht: Springer Netherlands; 1987. pp. 733–754.

[328] Kocon L., Despetis F., Phalippou J. Ultralow density silica aerogels by alcohol supercritical drying. *Journal of Non-Crystalline Solids*. 1998; 225:96–100.

[329] Estella J., Echeverría J.C., Laguna M., Garrido J.J. Effect of supercritical drying conditions in ethanol on the structural and textural properties of silica aerogels. *Journal of Porous Materials*. 2008; 15(6):705–713.

[330] Pajonk G.M. A short history of the preparation of aerogels and carbogels. In: Attia Y.A., editor. *Sol-Gel Processing and Applications*. Boston: Springer US, pp. 201–219.

[331] Woignier T., Phalippou J., Quinson J.F., Pauthe M., Laveissiere F. Physicochemical transformation of silica gels during hypercritical drying. *Journal of Non-Crystalline Solids*. 1992; 145:25–32.

[332] Yoda S., Ohshima S. Supercritical drying media modification for silica aerogel preparation. *Journal of Non-Crystalline Solids*. 1999; 248(2):224–234.

[333] Yoda S., Ohshima S., Ikazaki F. Supercritical drying with zeolite for the preparation of silica aerogels. *Journal of Non-Crystalline Solids*. 1998; 231(1):41–48.

[334] Popovici M., Gich M., Roig A., et al. Ultraporous single phase iron oxide – silica nanostructured aerogels from ferrous precursors. *Langmuir*. 2004; 20(4):1425–1429.

[335] Zu G., Shen J., Zou L., et al. Nanoengineering super heat-resistant, strong alumina aerogels. *Chemistry of Materials*. 2013; 25(23):4757–4764.

[336] Ren H., Zhu J., Bi Y., Xu Y., Zhang L. One-step fabrication of transparent hydrophobic silica aerogels via in situ surface modification in drying process. *Journal of Sol–Gel Science and Technology*. 2016; 80(3):635–641.

[337] Lermontov S.A., Malkova A.N., Yurkova L.L., et al. Diethyl and methyl-tert-buthyl ethers as new solvents for aerogels preparation. *Materials Letters*. 2014; 116:116–119.

[338] Lermontov S.A., Straumal E.A., Mazilkin A.A., et al. How to tune the alumina aerogels structure by the variation of a supercritical solvent. Evolution of the structure during heat treatment. *Journal of Physical Chemistry C*. 2016; 120(6):3319–3325.

[339] Wang J., Angnes L., Tobias H., et al. Carbon aerogel composite electrodes. *Analytical Chemistry*. 1993; 65(17):2300–2303.

[340] Szczurek A., Amaral-Labata G., Fierro V., Pizzi A., Masson E., Celzarda A. Porosity of resorcinol-formaldehyde organic and carbon aerogels exchanged and dried with supercritical organic solvents. *Materials Chemistry and Physics*. 2011; 129:1221–1232.

[341] Amaral-Labat G., Szczurek A., Fierro V., Masson E., Pizzi A., Celzard A. Impact of depressurizing rate on the porosity of aerogels. *Microporous and Mesoporous Materials*. 2012; 152:240–245.

[342] Amaral-Labat G., Szczurek A., Fierro V., Pizzi A., Masson E., Celzard A. 'Blue glue': a new precursor of carbon aerogels. *Microporous and Mesoporous Materials*. 2012; 158:272–280.

[343] Feng J., Feng J, Zhang C. Shrinkage and pore structure in preparation of carbon aerogels. *Journal of Sol–Gel Science and Technology*. 2011; 59(2):371–380.

[344] Liang C., Sha G, Guo S. Resorcinol-formaldehyde aerogels prepared by supercritical acetone drying. *Journal of Non-Crystalline Solids*. 2000; 271(1):167–170.

[345] Albert D.F., Andrews G.R., Mendenhall R.S., Bruno J.W. Supercritical methanol drying as a convenient route to phenolic–furfural aerogels. *Journal of Non-Crystalline Solids*. 2001; 296(1):1–9.

[346] Poco J.F., Coronado P.R., Pekala R.W., Hrubesh L.W. A rapid supercritical extraction process for the production of silica aerogels. *MRS Proceedings*. 1996; 431:297.

[347] Gauthier B.M., Bakrania S.D., Anderson A.M., Carroll M.K. A fast supercritical extraction technique for aerogel fabrication. *Journal of Non-Crystalline Solids*. 2004; 350:238–243.

[348] Brown L.B., Anderson A.M., Carroll M.K. Fabrication of titania and titania–silica aerogels using rapid supercritical extraction. *Journal of Sol–Gel Science and Technology*. 2012; 62(3):404–413.

[349] Bono M.S., Anderson A.M., Carroll M.K. Alumina aerogels prepared via rapid supercritical extraction. *Journal of Sol–Gel Science and Technology*. 2010; 53(2):216–226.

[350] Tewari P.H., Hunt A.J., Lofftus K.D. Ambient-temperature supercritical drying of transparent silica aerogels. *Materials Letters*. 1985; 3(9):363–367.

[351] Zhang Z., Scherer G.W. Supercritical drying of cementitious materials. *Cement and Concrete Research*. 2017; 99:137–154.

[352] van Bommel M.J., de Haan A.B. Drying of silica aerogel with supercritical carbon dioxide. *Journal of Non-Crystallline Solids*. 1995; 186:78–82.

[353] Bouledjouidja A., Masmoudi Y., Speybroeck M.V., Schueller L., Badens E. Impregnation of fenofibrate on mesoporous silica using supercritical carbon dioxide. *International Journal of Pharmaceutics*. 2016; 499(1):1–9.

[354] Takenouchi S., Kennedy GC. The binary system H_2 O–CO_2 at high temperatures and pressures. *American Journal of Science*. 1964; 262(9):1055–1074.

[355] Sanz-Moral L.M., Rueda M., Mato R., Martín Á. View cell investigation of silica aerogels during supercritical drying: analysis of size variation and mass transfer mechanisms. *Journal of Supercritical Fluids*. 2014; 92:24–30.

[356] Özbakır Y., Erkey C. Experimental and theoretical investigation of supercritical drying of silica alcogels. *Journal of Supercritical Fluids*. 2015; 98:153–166.

[357] Sahin I., Özbakir Y., Inönu Z., Ulker Z., Erkey C. Kinetics of supercritical drying of gels. *Gels*. 2017; p. gels4010003.

[358] García-González C.A., Camino-Rey M.C., Alnaief M., Zetzl C., Smirnova I. Supercritical drying of aerogels using CO_2: effect of extraction time on the end material textural properties. *Journal of Supercritical Fluids*. 2012; 66:297–306.

[359] Griffin J.S., Mills D.H., Cleary M., Nelson R., Manno V.P., Hodes M. Continuous extraction rate measurements during supercritical CO_2 drying of silica alcogel. *Journal of Supercritical Fluids*. 2014; 94:38–47.

[360] Lebedev A.E., Katalevich A.M., Menshutina N.V. Modeling and scale-up of supercritical fluid processes. Part I: supercritical drying. *Journal of Supercritical Fluids*. 2015; 106:122–132.

[361] Selmer I., Behnecke A.S., Quiño J., Braeuer A.S., Gurikov P., Smirnova I. Model development for sc-drying kinetics of aerogels: Part 1. Monoliths and single particles. *Journal of Supercritical Fluids*. 2018; 140:415–430.

[362] Quiño J., Ruehl M., Klima T., Ruiz F., Will S., Braeuer A. Supercritical drying of aerogel: in situ analysis of concentration profiles inside the gel and derivation of the effective binary diffusion coefficient using Raman spectroscopy. *Journal of Supercritical Fluids*. 2016; 108:1–12.

[363] Bueno A., Selmer I., et al. First evidence of solvent spillage under subcritical conditions in aerogel production. *Industrial and Engineering Chemistry Research*. 2018; 57(26):8698–8707.

[364] Jessop P.G., Jessop D.A., Fu D., Phan L. Solvatochromic parameters for solvents of interest in green chemistry. *Green Chemistry*. 2012; 14:1245–1259.

[365] Silva S.S., Duarte A.R.C., Carvalho A.P., Mano J.F., Reis R.L. Green processing of porous chitin structures for biomedical applications combining ionic liquids and supercritical fluid technology. *Acta Biomaterialia*. 2011; 7(3):1166–1172.

[366] Plappert S.F., Nedelec J.M., Rennhofer H., Lichtenegger H.C., Bernstorff S., Liebner F.W. Self-assembly of cellulose in super-cooled ionic liquid under the impact of decelerated antisolvent infusion: an approach toward anisotropic gels and aerogels. *Biomacromolecules*. 2018; 19(11):4411–4422.

[367] Paraskevopoulou P., Smirnova I., Athamneh T., et al. Mechanically strong polyurea/polyurethane-cross-linked alginate aerogels. *ACS Applied Polymer Materials*. 2020 05; 2(5):1974–1988.

[368] Smidsrød O. Molecular basis for some physical properties of alginates in the gel state. *Faraday Discussions of the Chemical Society*. 1974; 57:263–274.

[369] Di Renzo F., Valentin R., Boissière M., et al. Hierarchical macroporosity induced by constrained syneresis in core-shell polysaccharide composites. *Chemistry of Materials*. 2005; 17(18):4693–4699.

[370] Molvinger K., Quignard F., Brunel D., Boissière M., Devoisselle J.M. Porous chitosan-silica hybrid microspheres as a potential catalyst. *Chemistry of Materials*. 2004; 16(17):3367–3372.

[371] Gurikov P., Raman, S.P., Griffin J.S., Steiner S.A., Smirnova I. 110th anniversary: solvent exchange in the processing of biopolymer aerogels: current status and open questions. *Industrial & Engineering Chemistry Research*. 2019 10; 58(40):18590–18600.

[372] Lindvig T., Michelsen M.L., Kontogeorgis G.M. A Flory–Huggins model based on the Hansen solubility parameters. *Fluid Phase Equilibria*. 2002; 203(1):247–260.

[373] Constantinescu D., Gmehling J. Further development of modified UNIFAC (Dortmund): revision and extension 6. *Journal of Chemical & Engineering Data*. 2016; 61(8): 2738–2748.

[374] Hansen C.M. *Hansen Solubility Parameters: A User's Handbook*. 2nd ed. Boca Raton: CRC Press Taylor and Francis Group; 2007.

[375] Weerachanchai P., Wong Y., Lim K.H., Tan T.T.Y., Lee J.M. Determination of solubility parameters of ionic liquids and ionic liquid/solvent mixtures from intrinsic viscosity. *ChemPhysChem*. 2014; 15(16):3580–3591.

[376] Ravindra R., Krovvidi K.R., Khan A.A. Solubility parameter of chitin and chitosan. *Carbohydrate Polymers*. 1998; 36(2):121–127.

[377] Lehnert R.J., Kandelbauer A. Comments on 'solubility parameter of chitin and chitosan'. *Carbohydrate Polymers*. 2017; 175:601–602.

[378] Rowe R.C. Interaction of lubricants with microcrystalline cellulose and anhydrous lactose – a solubility parameter approach. *International Journal of Pharmaceutics*. 1988; 41(3):223–226.

[379] Pena M.A., Daali Y., Barra J., Bustamente P. Partial solubility parametes of lactose, mannitol and saccharose using the modified extended hansen method and evaporation light scattering detection. *Chemical & Pharmaceutical Bulletin*. 2000; 48(2):179–183.

[380] Pircher N., Veigel S., Aigner N., Nedelec J.M., Rosenau T., Liebner F. Reinforcement of bacterial cellulose aerogels with biocompatible polymers. *Carbohydrate Polymers*. 2014; 111:505–513.

[381] Heath L., Thielemans W. Cellulose nanowhisker aerogels. *Green Chemistry*. 2010; 12:1448–1453.

[382] Heath L., Zhu L, Thielemans W. Chitin nanowhisker aerogels. *ChemSusChem*. 2013; 6(3):537–544.

[383] Watt I.M. *The Principles and Practice of Electron Microscopy*. 2nd ed. New York: Cambridge University Press; 1997.

[384] Reimer L., Pfefferkorn G. *Rasterelektronenmikroskopie*. 2nd ed. Berlin–Heidelberg–New York: Springer-Verlag; 1977.

[385] Stroud R.M., Long J.W., Pietron J.J., Rolison D.R. A practical guide to transmission electron microscopy of aerogels. *Journal of Non-Crystalline Solids*. 2004; 350:277–284.

[386] Liu G., Zhou B., Du A., Shen J., Wu G. Greatly strengthened silica aerogels via co-gelation of binary sols with different concentrations: a method to control the microstructure of the colloids. *Colloids and Surfaces A: Physicochemical and Engineering Aspects*. 2013; 436:763–774.

[387] Despetis F., Bengoura N., Lartigue B., Spagnol S., Olivi-Tan N. Three dimensional reconstruction of aerogels from TEM images. *Journal of Non-Crystalline Solids*. 2012; 358:1180–1184.

[388] Hedge N.D., Rao A.V. Physical properties of methyltrimethoxysilane based elastic silica aerogels prepared by the two-stage sol–gel process. *Journal of Materials Science*. 2007; 42:6965–6971.

[389] Gurav J.L., Rao A.V., Rao A.P., Nadargi D.Y., Bhagat S.D. Physical properties of sodium silicate based silica aerogels prepared by single step sol-gel process dreid at ambient pressure. *Journal of Alloys and Compounds*. 2009; 476:397–402.

[390] Micromeritics Instruments Corporation. Envelope density analyzer. www.micromeritics.com/Product-Showcase/GeoPyc-1365.aspx

[391] Woignier T., Phalippou J. Skeletal density of silica aerogels. *Journal of Non-Crystalline Solids*. 1987; 93:17–21.

[392] Ayral A., Phalipou J., Woignier T. Skeletal density of silica aerogels determined by helium pycnometry. *Journal of Materials Science*. 1992; 27:1166–1170.

[393] Stumpf C., von Gässler K., Reichenauer G., Fricke J. Dynamic gas flow measurements on aerogels. *Journal of Non-Crystalline Solids*. 1992; 145:180–184. Aerogele.

[394] Reichenauer G., Stumpf C., Fricke J. Characterization of SiO_2, RF and carbon aerogels by dynamic gas expansion. *Journal of Non-Crystalline Solids*. 1995; 186:334–341.

[395] Hosticka B., Norris P.M., Brenizer J.S., Daitch C.E. Gas flow through aerogels. *Journal of Non-Crystalline Solids*. 1998; 225:293–297.

[396] Beurroies J., Bourret D., Sempéré R., Duffours L., Phalippou J. Gas permeability of partially densified aerogels. *Journal of Non-Crystalline Solids*. 1995; 186:328–333.

[397] Job N., Théry A., Pirard R., et al. Carbon aerogels, crygels and xerogels: Influence of the drying method on the textural properties of porous cabron materials. *Carbon*. 2005; 43:2481–2494.

[398] Torquato S. Random heterogeneous materials. New York: Springer-Verlag; 2002.

[399] Wong J.C.H., Kaymak H., Brunner S., Koebel M.M. Mechanical properties of monolithic silica aerogels made from polyethoxydisiloxanes. *Microporous and Mesoporous Materials*. 2014; 183:23–29.

[400] Ludwig S. Untersuchungen zur Gelation von RF-Aerogelen mit Zitronensäure und Bestimmung der Eigenschaften der erzeugten Aerogele. Hochschule Bonn-Rhein-Sieg; 2011.

[401] Cai J., Kimura S., Wada M., Kuga S., Zhang L. Cellulose aerogels from aqueous alkali hydroxide-urea solution. *ChemSusChem*. 2008; 1:149–154.

[402] Thommes M., Kaneko K., Neimark A.V., et al. Physisorption of gases, with special reference to the evaluation of surface area and pore size distribution (IUPAC Technical Report). *Pure and Applied Chemistry*. 2015; 87:1051–1069.

[403] Underwood E.E. *Quantitative Stereology*. Reading: Addison-Wesley; 1970.

[404] Cahn J.W. The Significance of average mean curvature and its determination by quantitative metallography. *Transactions of the Metallurgical Society of AIME*. 1967; 239:610–616.

[405] Limodin N., Salvo L., Boller E., et al. In situ and real time 3-D microtomography investigation of dendritic solidification in an Al-10 wt% Cu alloy. *Acta Materialia*. 2009; 57:2300–2310.

[406] Ratke L., Genau A. Evolution of specific surface area with solid fraction during solidification. *Acta Materialia*. 2010; 58:4207–4211.

[407] Neumann-Heyme H., Eckert K., Beckermann C. General evolution equation for the specific interface area of dendrites during alloy solidification. *Acta Materialia*. 2017; 140:87–96.

[408] Heyden S. *Network Modelling for the Evaluation of Mechanical Properties of Cellulose Fibre Fluff*. Sweden: Lund University LTH; 2000.

[409] Karadagli I., Milow B., Ratke L., Schulz B., Seide G., Gries T. Synthesis and characterization of highly porous cellulose aerogels for textile applications. In: Cellular Materials Conference – *Cellmat 2012*, Conventus Congressmanagement, editor. Jena: Marketing GmbH. https://elib.dlr.de/78416/.

[410] Laskowski J., Milow B., Ratke L. The effect of embedding highly insulating granular aerogel in cellulosic aerogel. *Journal of Supercritical Fluids*. 2015; 106:93–99.

[411] Ratke L., Milow B., Lisinski S., Hoepfner S. On an effect of fine ceramic particles on the structure of aerogels. *Microgravity Science and Technology*. 2014; 26:103–110.

[412] Sing K.S.W., Everett D.H., Haul R.A.W., et al. Reporting physisorption data for gas/solid systems with special reference to the determination of surface area and porosity. *Pure and Applied Chemistry*. 1985; 57(4):603–619. IUPAC Recommendations 1984.

[413] Gregg S.J., Sing K.S.W. *Adsorption, Surface Area and Porosity*. London: Academic Press; 1982.

[414] Rouquerol F., Rouquerol J., Sing K.S.W., Llewellyn P., Maurin G. *Adsorption by Powders and Porous Solids*. Oxford: Academic Press; 2014.

[415] Santori G., Luberti M., Ahn H. Ideal adsorbed solution theory solved with direct search minimisation. *Computers and Chemical Engineering*. 2014; 71:235–240.

[416] Langmuir I. *Surface Chemistry*; 1932. www.nobelprize.org/nobel_prizes/chemistry.

[417] Brunauer S., Emmett P.H., Teller E. Adsorption of gases in multimolecular layers. *Journal of the American Chemical Society*. 1938; 60:309–319.

[418] Porod G. Die Röntgenkleinwinkelstreuung von dichtgepackten kolloidalen Systemen. *Kolloidzeitschrift*. 1951; 124:83–114.

[419] Porod G. Die Röntgenkleinwinkelstreuung von dichtgepackten kolloidalen Systemen II. Teil. *Kolloidzeitschrift*. 1951; 125:108–122.

[420] Guinier A., Fournet G. *Small-Angle Scattering of X-Rays*. New York and London: John Wiley & Sons, Inc. and Chapman & Hall Ltd.; 1955.

[421] Debye P., Anderson H.R., Brumberger H. Scattering by an inhomogeneous solid. II. The correlation function and its application. *Journal of Applied Physics*. 1957; 28:679–683.

[422] Weigold L. Ermittlung des Zusammenhangs zwischen mechanischer Steifigkeit und Wärmetransport über das Festkörpergerüst bei hochporösen Materialien. Bayerische Julius-Maximilian Universität; 2015.

[423] Beaumont M., Rennhofer H., Opietnik M., Lichtenegger H.C., Potthast A, Rosenau T. Nanostructured cellulose II gel consisting of spherical particles. *ACS Sustainable Chemistry & Engineering*. 2016; 4(8):4424–4432.

[424] Brandt R., Petricevic R., Pröbstle H., Fricke J. Acetic acid catalyzed carbon aerogels. *Journal of Porous Materials*. 2003; 10:171–178.

[425] Pekala R.W., Alviso C.T., LeMay J.D. Organic aerogels: microstructural dependence of mechanical properties in compression. *Journal of Non-Crystalline Solids*. 1990; 125: 67–75.

[426] Petricevic R., Reichenauer G., Bock V, Emmerling A., Fricke J. Structure of carbon aerogels near the gelation limit of the resorcinol-formaldehyde precursor. *Journal of Non-Crystalline Solids*. 1998; 225:41–45.

[427] Lu B., Torquato S. Chord-length and free path distribution functions for many-body systems. *Journal of Chemical Physics*. 1993; 98(8):6472–6482.

[428] Abramowitz M., Stegun I.A. *Handbook of Mathematical Functions*. 10th ed. Mineola: Dover Publications; 1972.

[429] Chandrasekhar S. Stoachstic problems in physics and astronomy. *Review of Modern Physics*. 1943; 15:1–89.

[430] Steele J.H. *Metallurgical Transactions*. 1976; 7A:1325.

[431] Barrett E.P., Joyner L.G., Halenda P.P. The determination of pore volume and area distributions in porous substances. I. Computations from nitrogen isotherms. *Journal of the American Chemical Society*. 1938; 60:309–319.

[432] Zhang Y., Lam F.L.Y., Yan Z.F., Hu X. Review of Kelvin's equation and its modification in characterization of mesoporous materials. *Chinese Journal of Chemical Physics*. 2006; 19:102–108.

[433] Rege A., Schestakow M., Karadag!i I., Ratke L., Itskov M. Micro-mechanical modelling of cellulose aerogels from molten salt hydrates. *Soft Matter*. 2016; 12:7079–7088.

[434] Boettinger W.J., Kattner U.R., Moon K.W., Perepezko J.H. *DTA and Heat-Flux DSC Measurements of Alloy Melting and Freezing*. Washington: National Institute of Standards and Technology; 2006.

[435] Nedelec J.M., Grolier J.P.E., Baba M. Thermoporosimetry – a powerful tool to study the cross-linking in gels networks. *Journal of Sol–Gel Science and Technology*. 2006; 40:191–200.

[436] Woignier T., Quinson J.F., Pauthe M., Repellin-Lacroix M., Phalippou J. Evolution of the porous volume during the aerogel–glass transformation. *Journal de Physique IV France*. 1992; 02:C2–123,–C2–126.

[437] Reichenauer G., Scherer G.W. Nitrogen adsorption in compliant materials. *Journal of Non-Crystalline Solids*. 2000; 277:162–172.

[438] Reichenauer G., Scherer G.W. Extracting the pore size distribution of compliant materials from nitrogen adsoprtion. *Colloids and Surfaces A: Physicochemical and Engineering Aspects*. 2001; 187–188:41–50.

[439] Reichenauer G., Scherer G.W. Effects upon nitrogen sorption analysis in aerogels. *Journal of Colloid and Interface Science*. 2001; 236:385–386.

[440] Carslaw H., Jaeger J. *Conduction of Heat in Solids*. 2nd ed. Oxford: Clarendon Press; 1959.

[441] Dhont J.K.G.,Briels W.J. Rod-like Brownian particles in shear flow: Sections 3.1–3.9. In: *Soft Matter: Complex Colloidal Suspensions*, Gompper G., Schick M., eds. Berlin: Wiley-VCH Verlag GmbH & Co. KGaA, 2007; pp. 147–216.

[442] Dhont J.K.G., Briels W.J. Rod-like Brownian particles in shear flow: Sections 3.10–3.16. In: Gompper G., Schick M., editors. *Soft Matter: Complex Colloidal Suspensions*, Berlin: Wiley-VCH Verlag GmbH & Co. KGaA, 2007; pp. 216–283.

[443] Hittmair O., Adam G. *Ringvorlesungen zur Theoretischen Physik, Wämetheorie*. Braunschweig: Vieweg Verlag; 1971.

[444] Carrigy N.B., Pant L.M., Mitra S., Secanella M. Knudsen diffusivity and permeability of PEMFC microporous coated gas diffusion layers for different polytetrafluorethylene loadings. *Journal of the Electrochemical Society*. 2013; 160:F81–F89.

[445] Russ S., Zschiegner S., Bunde A., Kärger J. In: Kramer B, editor. *Lambert Diffusion in Porous Media in the Knudsen Regime*. vol. 45 of Adv. in Solid State Phys. Switzerland: Springer-Verlag; 2005. pp. 59–69.

[446] Dammers A.J., Coppens M.O. Anomalous Knudsen diffusion in simple pore models. *Diffusion Fundamentals*. 2005; 2:14.1–14.2.

[447] Zschiegner S., Russ S., Valiullin R., et al. Normal and anomalous diffusion of non-interacting particles in linear nanopores. *The European Physical Journal Special Topics*. 2008; 161:109–120.

[448] Kärger J., Ruthven D.M., Theodoru D.N., editors. *Diffusion in Nanoporous Materials*. John Wiley and Sons, Ltd; 2012.

[449] Scheidegger A.E. *The Physics of Flow through Porous Media*. 3rd ed. Toronto: University of Toronto Press; 1974.

[450] Torrent R.J. A two-chamber vacuum cell for measuring the coefficient of permeability to air of the concrete cover on site. *Materials and Structures*. 1992; 25:358–365.

[451] Tanikawa W., Shimamoto T. Klinkenberg effect for gas permeability and its comparison to water permeability for porous sedimentary rocks. *Hydrology and Earth System Sciences Discussions*. 2006; 3:1315–1338.

[452] Vazquez J.L. *The Porous Medium Equation*. Oxford: Oxford University Press; 2006.

[453] Cunningham R.E., Williams R.J.J. *Diffusion in Gases and Porous Media*. New York: Springer Science and Business Media LLC; 1980.

[454] Kaviany M. *Principles of Heat Transfer in Porous Media*. 2nd ed. New York: Springer-Verlag; 1995.

[455] Gupta N.K., Gianchandani Y.B. Thermal transpiration if zeolites: a mechanisms for motionless gas pumps. *Applied Physics Letters*. 2008; 93:193511.

[456] Woignier T., Primera J., Lamy M., et al. The use of silica aerogels as host matrices for chemical species: different ways to control the permeability and mechanical properties. *Journal of Non-Crystalline Solids*. 2004; 350:299–307.

[457] Woignier T., Anez L., Calas-Etienne S., Primera J. Gas and liquid permeability in nano composites gels: comparison of Knudsen and Klinkenberg correction factors. *Microporous and Mesoporous Materials*. 2014; 200:79–85.

[458] Ganesan K., Barowski A., Ratke L. Gas Permeability of Cellulose Aerogels with a Designed Dual Pore Space System. *Molecules*. 2019; 24:2688.

[459] Ebert H.P. Thermophysical properties of aeorgels. In: Aegerter M.A., Leventis N., Koebel M.M., editors. *Aerogels Handbook*. New York; Springer; 2011. pp. 537 –564.

[460] Cherkaev A. Bounds for effective properties of multi-material two-dimensional conducting composites. *Mechanics of Materials*. 2009; 41:411–433.

[461] Brovelli A., Cassiani G. A combination of the Hashin–Shtrikman bounds aimed at modelling electrical conductivity and permittivity of variably saturated porous media. *Geophysical Journal International*. 2010; 180:225–237.

[462] Hrubesh L.W., Pekala R.W. Thermal properties of organic and inorganic aerogels. *Journal of Materials Research*. 1994; 9:731–738.

[463] Wei G., Liu Y., Zhang X., Yu F., Du X. Thermal conductivities study on silica aerogel and its composite insulation materials. *International Journal of Heat and Mass Transfer*. 2011; 54:2355–2366.

[464] Kittel C. *Introduction to Solid State Physics*. 4th ed. John Wiley and Sons, Inc.; 1971.

[465] Peierls R. *Ann Physik*. 1929; 3:1055–1101.

[466] Gross J., Fricke J., Hrubesh L.W. Sound propagation in SiO_2 aerogels. *Journal of the Acoustical Society of America*. 1992; 91:2004–2006.

[467] Nakayama T., Yakubo K., Orbach R.L. Dynamical properties of fractal networks: scaling, numerical simulations, and physical realizations. *Review of Modern Physics*. 1994; 66:381–443.

[468] Kaganer M.G. *Thermal Insulation in Cryogenic Engineering*; 1969. Jerusalem: Israel Programm for Scientific Translations.

[469] Schwab H. *Vakuumisolationspaneele – Gas und Feuchteeintrag sowie Feuchte- und Wärmetransport*. Würzburg, Germany: Bayerische Julius-Maximilian Universität; 2004.

[470] Dorfmann J.R., van Beijeren H. 3. In: Berne B.J., editor. *Statistical Mechanics*. vol. Part B. New York: Plenum Press; 1977. pp. 65–177.

[471] Zeng S.Q., Hunt A., Greif R. Transport properties of gas in silica aerogel. *Journal of Non-Crystalline Solids*. 1995; 186:264–270.

[472] Swimm K., Reichenauer G., Vidi S., Ebert H.P. Gas pressure dependence of the heat transport in porous solids with pores smaller than 10 μm. *International Journal of Thermophysics*. 2009; 30:1329–1342.

[473] Comings E.W., Nathan M.F. Thermal conductivity of gases at high pressures. *Industrial and Engineering Chemistry*. 1947; 39:964–970.

[474] Guildner L.A. Thermal conductivity of gases. III. Some values of the thermal conductivities of argon, helium, and nitrogen from 0 C to 75 C at pressures of 10^5 to 2.5 10^7 Pascals. *Journal of Research at the National Bureau of Standards*. 1975; 79A:407–413.

[475] Kennard E.H. *Kinetic Theory of Gases*. New York, London: McGraw-Hill Inc.; 1938.

[476] Kleeman R.D. *A Kinetic Theory of Gases and Liquids*. New York: John Wiley and Sons, Inc.; 1920.

[477] Loeb L.B. Kinetic *Theory of Gases*. New York and London: McGraw-Hill Company; 1927.

[478] Ganta D., Dale E.B., Rezac J.R., Rosenberger A.T. Optical method for measuring thermal accomodation coefficients using a whispering-gallery microresonator. *Journal of Chemical Physics*. 2011; 135:084313.

[479] Caps R., Fricke J. Radiative Heat Transfer in Silica Aerogels. In: Fricke J., editor. *Aerogels*. vol. 6 of Springer Proceedings in Physics. Berlin–Heidelberg–New York: Springer-Verlag; 1986. pp. 110–115.

[480] Kubascheweski O., Alcock C.B., Spencer P.J. *Materials Thermo-Chemistry*. Oxford: Pergamon Press Ltd.; 1993.

[481] ASTM. Standard test method for steady state heat flux measurements and thermal transmission properties by mean of the guarded hot plate apparatus. West Conshohocken: ASTM; 1997. ASTM C177.

[482] ISO. *Determination of Steady-state Thermal Resistance and Related Properties Guarded Hot Plate Apparatus*. Geneva: International Organization for Standardization; 1991. ISO 8302.

[483] Dubois S., Lebeau S. Design, construction and validation of a guarded hot plate apparatus for thermal conductivity measurement of high thickness crop-based specimens. *Materials and Structures*. 2013; 46.

[484] Jannot Y., Felix V., Degiovanni A. A centered hot plate method for measurement of thermal properties of thin insulating materials. *Measurement Science and Technology*. 2010; 21:035106.

[485] Eithun C.F. *Development of a Thermal Conductivity Apparatus: Analysis and Design*. Trondheim: NTNU Department of Energy and Process Engineering; 2012.

[486] Gustafsson S.E. Transient plane source techniques for thermal conductivity and thermal diffusivity measurements of solid materials. *Review of Scientific Instruments*. 1991; 62(3):797–804.

[487] Gustafsson M., Karawacki E., Gustafsson S.E. Thermal conductivity, thermal diffusivity, and specific heat of thin samples from transient measurements with hot disk sensors. *Review of Scientific Instruments*. 1994; 65:3856–3859.

[488] He Y. Rapid thermal conductivity measurement with a hot disk sensor: Part 1. theoretical considerations. *Thermochimica Acta*. 2005; 436:122–129.

[489] He Y. Rapid thermal conductivity measurement with a hot disk sensor: Part 2. characterization of thermal greases. *Thermochimica Acta*. 2005; 436:130–134.

[490] Hammerschmidt U., Meier V. New transient hot-bridge sensor to measure thermal conductivity, thermal diffusivity, and volumetric specific heat. *International Journal of Thermophysics*. 2006; 27:840–865.

[491] Dalton M., Nadalini R., Celotti L., et al. The 3 omega transient line method for thermal characterization of superinsulator materials developed for spacecraft thermal control. In: 65th International Astronautical Congress Toronto 2014. IAC-14.C2.7.4. International Astronautical Federation; 2014.

[492] Corbino O. Periodic variation of resistance of metallic filaments on alternating current. *Atti della Reale Accademia Nazionale dei Lincei*. 1911; 20:222–228.

[493] Cahill D.G., Pohl R.O. Thermal conductivity of amorphous solids above the plateau. *Physical Review B*. 1987; 35:4067–4073.

[494] Hemberger F., Ebert H.P., Reichenauer G. *International Journal of Thermophysics*. 2009; 30:1357–1371.

[495] Wiener M., Hemberger F., Reichenauer G., Ebert H.P. *International Journal of Thermophysics*. 2006; 27:1826–1843.

[496] Guo K., Hu Z., Song H., Liang Zhong X.D., Chen X. Low-density graphene/carbon composite aerogels prepared at ambient pressure with high mechancial strength and low thermal conductivity. *RSC Advances*. 2015; 5:5197–5204.

[497] Reichenauer G., Heinemann U., Ebert H.P. Relationship between pore size and the gas pressure dependence of the gaseous thermal conductivity. *Colloids and Surfaces A: Physicochemical and Engineering Aspects*. 2007; 300:204–210.

[498] Oliver F.W.J., Lozier D.W., Boisvert R.F., Clark C.W. *NIST Handbook of Mathematical Functions*. Cambridge: Cambridge University Press; 2010.

[499] Dieter G.E. *Mechanical Metallurgy*. 3rd ed. New York: McGraw Hill; 1986.

[500] Bisshopp K.E., Drucker D.C. Large deflection of cantilever beams. *Quarterly of Applied Mathematics*. 1945; 3:272–275.

[501] Rege A., Schwan M., Chernova L., Hillgärtner M., Itskov M., Milow B. Microstructural and mechanical characterization of carbon aerogels: an in-situ and digital image correlation-based study. *Journal of Non-Crystalline Solids*. 2020; 529:119568.

[502] Wikipedia. https://en.wikipedia.org/wiki/Theoretical_strength_of_a_solid.

[503] Pokluda J., Cerny M., Sandra P., Sob M. Calculations of theoretical strength: state of the art and history. *Journal of Computer-Aided Materials Design*. 2004; 11:1–28.

[504] Li D., Wong L.N.Y. The Brazilian disc test for rock mechanics applications: review and new insights. *Rock Mechanics and Rock Engineering*. 2013; 46:269–287.

[505] Landau L.D., Lifshitz E.M. *Lehbrbuch der theoretischen Physik*. vol. VII. Berlin: Akademie Verlag; 1965.

[506] Rege A., Itskov M. A microcell-based constitutive modeling of cellulose aerogels under tension. *Acta Mechanica*. 2018; 229:585–593.

[507] Gross J., Fricke J. Scaling of elastic properties in highly porous nanostructured aerogels. *Nanostructured Materials*. 1995; 6:905–908.

[508] Ma H.S., Roberts A.P., Prevost J.H., Jullien R., Scherer G.W. Mechanical structure–property relationship of aerogels. *Journal of Non-Crystalline Solids*. 2000; 277:127–141.

[509] Gross J., Scherer G.W., Alviso C.T., Pekala R.W. Elastic properties of crosslinked resorcinol-formaldehyde gels and aerogels. *Journal of Non-Crystalline Solids*. 1997; 211:132–142.

[510] Ganesan K., Dennstedt A., Barowski A., Ratke L. Design of aerogels, cryogels and xerogels of cellulose with hierarchical porous structure. *Materials and Design*. 2016; 92:345–355.

[511] Rudaz C., Courson R., Bonnet L., Calas-Etienne S., Sallee H., Budtova T. Aeropectin: Fully biomass-based mechanically strong and thermal superinsulating aerogel. *Biomacromolecules*. 2014; 15:2188–2195.

[512] Zhao S., Malfait W.J., Guerrero-Alburquerque N., Koebel M.M., Nyström G. Biopolymer aerogels and foams: chemistry, properties and applications. *Angewandte Chemie*. 2018; 57:7580–7608.

[513] Woignier T., Primera J., Alaoui A., Etienne P., Despetis F., Calas-Etienne S. Mechanical properties and brittle behavior of silica aerogels. *Gels*. 2015; 1:256–275.

[514] Scherer G.W., Smith D.M., Wiu X., Anderson J.M. Compression of aerogels. *Journal of Non-Crystallline Solids*. 1995; 186:316–320.

[515] Fereiro-Rangel C.A., Gelb L.D. Computational study of uniaxial deformations in silica aerogel using a coarse-grained model. *Journal of Physical Chemistry C*. 2007; 111:15,792–15,802.

[516] Rege A., Ratke L., Itskov M. Modelling and simulations of polysaccharide and protein based aerogels. In: Sabu Thomas LAP, Mavelil-Sam R, editors. *Biobased Aerogels: Polysaccharide and Protein-Based Materials*. Green Chemistry Series No. 58. London: Royal Society of Chemistry; 2018. pp. 129–150.

[517] Rege A., Hillgärtner M., Itskov M. Mechanics of biopolymer aerogels based on microstructures generated from 2-d Voronoi tessellations. *Journal of Supercritical Fluids*. 2019; 151:24–29.

[518] Haasen P. *Physical Metallurgy*. 3rd ed. Cambridge: Cambridge University Press; 1996.

[519] Hillert M. *Lectures on the Theory of Phase Transformations*. Aaronson H.I., editor. New York: American Institute of Mining, Metallurgy and Petroleum Engineering; 1975.

[520] Kelton K.F., Greer A.L. *Nucleation in Condensed Matter*. 2nd ed. Oxford: Pergamon Press Ltd.; 2010.

[521] Binder K., Stauffer D. Statistical theory of nucleation, condensation and coagulation. *Advances in Physics*. 1976; 25:343–396.

[522] Gránásy L., Tóth G.I., Warren J.A., et al. Phase-field modeling of crystal nucleation in undercooled liquids – a review. *Progress in Materials Science*. 2019; 106:100569.

[523] Merkwitz M. Oberflächen- und Grenzflächenspannung in binären metallischen Entmischungssystemen. TU Chemnitz; 1997.

[524] Kaban I., Köhler M., Ratke L., et al. Interfacial tension, wetting and nucleation in Al-Bi, Al-In and Al-Pb monotectic alloys. *Acta Materialia*. 2011; 59:6880–6889.

[525] Cahn J.W., Hilliard J.E. Free energy of a nonuniform system. I. Interfacial free energy. *Journal of Chemical Physics*. 1958; 28(2):258–267.

[526] Cahn J.W., Hilliard J.E. Free energy of a nonuniform system. III. Nucleation in a two-component incompressible fluid. *Journal of Chemical Physics*. 1959; 31(3):688–699.

[527] Cahn J.W. Phase separation by spinodal decomposition in isotropic systems. *Journal of Chemical Physics*. 1965; 42(1):93–99.

[528] Chandra M. *Introduction to Polymer Science and Chemistry: A Problem-Solving Approach*. Boca Raton: CRC Press Taylor and Francis Group; 2013.

[529] Cowie J.M.G., Arrighi V. *Polymers: Chemistry and Physics of Modern Materials*. CRC Press Taylor and Francis Group; 2007.

[530] Nestler B., Wheeler A.A., Ratke L., Stöcker C. Phase-field model for solidification of a monotectic alloy with convection. *Physica D*. 2000; 141:133–154.

[531] Koningsveld R., Kleintjens L.A., Nies E. Polymers and thermodynamics. *Croatica Chemica Acta*. 1987; 60:53–89.

[532] der Haegen R.V., Kleintjens L.A. Thermodynamics of polymer solutions. *Pure and Applied Chemistry*. 1981; 61:159–170.

[533] Mousavi-Dehghani S.A., Mirzayi B., Vafaie-Sefti M. Polymer solution and lattice theory applications for modeling of asphaltene precipitation in petroleum mixtures. *Brazilian Journal of Chemical Engineering*. 2008; 25:523–534.

[534] Nyström B. *General Polymer Chemistry*. www.uio.no/studier/emner/matnat/kjemi/KJM4550/h08/undervisningsmateriale/KJM5500-Macromol%20in%20solution.pdf.

[535] van Dijk M., Wakker A. *Concepts of Polymer Thermodynamics*. Lancaster: Technomic Publishing Company Inc.; 1997.

[536] Glatter O., Kratky O. *Small Angle X-Ray Scattering*. New York: Academic Press, Inc.; 1982.

[537] Sorensen C.M. Light scattering by fractal aggregates: a review. *Aerosol Science and Technology*. 2001; 35:648–667.

[538] Gommes C.J., Jiao Y., Torquato S. Microstructural degeneracy associated with a two-point correlation function and ist information content. *Physical Review E*. 2012; 85:051140.

[539] Gommes C.J., Jiao Y., Torquato S. Density of states for a specified correlation function and the energy landscape. *Physical Review of Letters*. 2012; 108:080601.

[540] Gommes C.J., Roberts A.P. Structure development of resorcinol-formaldehyde gels: Microphase separation or colloid aggregation. *Physical Review E*. 2008; 77:041409.

[541] Li M., Lai A.C.K. Analytic solution to heat conduction in finite hollow composite cylinders with a general boundary condition. *International Journal of Heat and Mass Transfer*. 2013; 60:549–556.

Index

acetogels, 3
aerocellulose, 72
aerogel classification, 5
aerogel definition, 1
aerogel morphology
    scanning electron microscopy, 203
    SEM pictures of aerogels, 209
    SEM pictures of cellulose aerogels, 213
    SEM pictures of RF aerogels, 212
    SEM pictures of silica aerogels, 210
    TEM pictures of aerogels, 205
    transmission electron microscopy, 205
aerogels
    history, 7
    stress–strain curves, 368
    Young's modulus, 370
alcogels, 3
alcoholysis, 170
alginate
    Ca crosslinking, 84
    cation concentration, 84
    cryogelation, 89
    degradation, 81
    derivatisation, 82
    epimerisation, 82
    fibrillar microstructure, 84
    freeze drying, 85
    gelation, 82, 83
        carbon dioxide induced, 88
        diffusion setting, 86
        metal cations effect, 86
    internal setting, 88
    pyranose ring, 78
    shrinkage, 85
    solvent exchange, 85
    supercritical drying, 85
alginates
    M/G ratio, 81
    molar-mass dispersity, 78
alginic acid, 77
ambient drying, 140
    capillary stress, 145
    drying stages, 147
    evaporation rate, 144
    fluid flow, 147
    mass loss, 141
    observations, 141
    shrinkage, 143
    springback effect, 144
    surfactants, 146
    vapour pressure, 145
aquagels, 3
autoclave, 167
    high-temperature supercritical drying (htscd), 167
autocorrelation function, 417
autocorrelation function (DLS), 421

BET
    specific surface area, 260
BET equation, 259
BET isotherm, 256
BET isotherm derivation, 256
bimolecular reaction, 424
binodal line, 407

Cahn–Hilliard equation, 125
Cahn–Hilliard equation with flow, 127
calcium alginate, 77
capillary pressure
    curvature effect, 146
carbon aerogel, 34
Carother's equation, 425
cellobiose, 60
cellulose, 60
    amphiphilic character, 63
    ballooning, 67
    cellulose I, 61
    cellulose II, 61
    derivatisation, 63
    dissolution, 62
    dissolving agents, 64
    fringe-fibril, 61
    hydrogen bond, 60, 63
    intrinsic gelation, 68
    ionic liquids, 65
    mercerisation, 63

cellulose (Cont.)
 pH inversion, 69
 phase diagram NaOH, 63
 salt hydrate melts, 65
 SEM aerogel, 73
 sodium hydroxide, 64
 TEM aerogel, 73
 temperature-induced gelation, 71
 ZnO addition, 69
cellulose aerogel
 freeze dried, 73
 ionic liquids, 74
 SEM figures, 74
collision volume
 Brownian motion, 120
 homogeneity, 120
 sedimentation, 119
 shear flow, 118
common tangent construction, 407
compressibility, 333
conversion factor polycondensation, 424
Couette flow, 92
critical point, 165
 estimate, 166
 tabulated values, 166
critical pressure, 178
critical temperature, 178, 406
critical temperature LCST, 410
critical volume fraction, 406
critical volume fraction LCST, 410
cryogels, 3

degree of polymerisation, 428
density
 rule of mixtures, 226
depolimarisation
 silica, 168
diffusion, 286
 Fick's first law, 286
 Fick's second law, 287
 gaseous coefficient, 290
 Knudsen diffusion, 290
 measurements, 292
 relaxation time, 288
 scaling solution, 287
 Stokes–Einstein relation, 288
 Zimm model, 289
diffusion limited aggregation, 114
diffusion limited cluster aggregation, 115, 116
diffusive mobility, 127
dynamic light scattering DLS, 421
dynamic low-temperature supercritical drying, 176
dynamic viscosity
 Arrhenius law, 289

entropy of mixing, 396
envelope density, 214
 measurement, 215
envelope density estimates, 227
 cellulose aerogels, 233
 RF aerogels, 231
 silica aerogels, 228
esterification
 silica, 168

Family–Viscek dynamic scaling, 122
Flory–Huggins χ-parameter, 404
Flory–Huggins theory, 404
Flory–Schulz distribution, 429
form factor, 413
 cylinder, 415
 sphere, 414
formaldehyde, 36
fractal
 dimension, 111
 scattering, 110
fractals, 109
 dimension, 110
 measurement, 110
free enthalpy polymer solution, 404
freeze drying, 149
 cells and dendrites, 155
 cells and dendrites equations, 156
 cooling of a liquid, 152
 mushy zone, 155
 pushing and engulfment, 158
 pushing example in biopolymeric aerogels, 159
 setup for controlled freezing, 160
 sublimation, 162
freezing of a gel liquid, 157
freezing of a liquid, 150

G-block, 77
gel time
 size dependence, 135
 spinodal decomposition, 133
 Viscek approach, 132
gelation
 Bethe lattice, 106
 flow damping, 102
 gel time, 130
 gel time Brownian motion, 131
 gel time sedimentation, 131
 gel time shear flow, 130
 light scattering, 103
 light transmission, 103
 percolation, 104
 PIPS, 105
 rheometer, 100
 rotating bob viscosimeter, 100

scaling solutions, 120
storage and loss modulus, 102
tilting meniscus, 99
viscosity change, 97
Gibson–Ashby model, 371
Guinier's law, 416
guluronic acid, 77

heat conduction, 315
Fourier's first law, 315
Fourier's law, 317
Hashin–Shtrikman bounds, 319
parallel arrangement, 319
porous media, 318
series arrangement, 319
heat transfer
characteristic equilibration time, 351
instationary solution plate, 349
Nusselt number, 347
plate, 346
Prandtl number, 349
Reynolds number, 349
stationary solution for a plate, 348
stationary solution tube, 352
tube, 351
heat transfer coefficients, 152
high-temperature supercritical drying, 167
hot supercritical drying, HTSCD, 168
hydrodynamic radius, 422
hydrogels, 3

ideal solution, 397
isopycnic lines CO_2, 164

Knudsen effect, 307
Knudsen flow and diffusion, 306
Knudsen number, 326
Knudsen pump, 307

Langmuir isotherm, 251, 253
loss modulus, 97
loss tangent, 97
low-temperature supercritical drying, 173
depressurisation, 180
flow sheet, 179
lower critical point, LCST, 408
lyogels, 3

M-block, 77
macropores, 2
mannuronic acid, 77
mean field theory aggregation, 117
Brownian motion, 118
mean free path
aerogels, 116
mechanical properties
modelling, 380
modelling cellulose aerogels, 385
modelling stress–strain curves, 381
mechanical testing, 356
bending, 356
Brazilian test, 363
compressibility, 365
compression, 360
flexibility, 364
sound velocity, 367
Young's modulus, 367
mesopores, 2
multilayer adsorption, 256
multilayer adsorption scheme, 258

Navier–Stokes equation, 127
NMMO, 65

oligomer concentration, 426

p-T phase diagram, 163
p-x-diagram, 179
percolation
cluster size, 108
cluster size distribution, 108
finite size scaling, 112
viscosity, 112
permeability, 296
cellulose aerogel, 312
characteristic time, 299, 304
Darcy's law, 297, 300
dynamic measurement, 298
experimental approach, 297
gas flow through an aerogel, 305
Hagen–Poisseulle flow, 296
Knudsen number, 305
porous media equation, 303
Reynolds number, 304
RF and carbon aerogel, 312
silica aerogel, 312
stationary measurement, 300
permeability constant, 297
permeability constant estimates, 308
Karman–Kozeny law, 309, 310
phase diagram CO_2, 163
polycondensation reaction scheme, 425
polymerisation induced phase separation (PIPS), 124
pore shapes, 2
pore size, 268
BJH equation, 276
BJH method, 274
centre-to-centre distance, 272
Kelvin equation, 275
mean free distance, 270

pore size (Cont.)
 randomly distributed spheres, 271
 simple geometrical model, 269
 stereology, 270
pore size distribution, 278
 Gibbs–Thomson effect, 278
 Gibbs–Thomson equation, 279
Porod's law, 416
porosity, 225
porous solids, 1
pressure concentration diagram, 177
pycnometer, 216

radial distribution function, 417
radius of gyration, 420
rapid supercritical extraction
 RSCE, 171
regular solution, 397
resorcinol, 35
 catalyst, 36
 quinonmethide, 37
 reaction, 36
  acidic condition, 38
  base condition, 37
 solubility, 36
RF aerogel, 34
 colour, 34
 di-methylene ether bridge, 37
 methylene bridge, 37
 novolak, 39
 R/C ratio, 44
 R/F ratio, 40
 resol, 39
 synthesis parameters, 40

scattering intensity, 413
seaweed, 77
single-layer adsorption, 252
skeletal density, 214
 effect of adsorbates, 217
 effect of test gas size, 224
 experimental results RF aerogels, 221
 experimental results silica aerogels, 223
 measurement, 215
 non-equilibrium pressures, 219
 pressure evolution, 219
 pycnometer, 216
Smoluchowski equation, 117
sodium alginate, 77
sol–gel process, 3
solvent exchange, 168
solvothermal treatment, 169
sound velocity
 Young's modulus, 367
specific heat, 333
 aerogels, 334
spinodal LCST, 409
spinodal line, 405
springback effect, 143
static low-temperature supercritical drying, 174
Stokes–Einstein relation, 289
storage modulus, 96
structure factor, 413, 417
supercritical drying
 drying time, 183
 mass transfer, 183
 spillage, 184
 theory, 183
supercritical fluid, 163
 virial expansion, 165
superinsulation, 345
surface area
 cellulose aerogels, 246
 composites, 247
 definition, 236
 embedded particles, 248
 fibrillar structures, 244, 245
 Halsey equation, 262
 Harkins–Jura equation, 262
 irregular shaped bodies, 241
 linear intercept, 242
 Magee equation, 262
 measurement, 249
 neck formation, 238
 packaging of spheres, 239
 rule of mixtures, 247
 SAXS, 420
 simple shapes, 237
 small angle X-ray scattering, 265
 t-plot, 262
 t-plot conversion, 262
 truncated spheres, 238
 volume fraction, 243
surface tension
 liquids, 168

thermal conductivity, 315
 3-omega method, 344
 aerogels, 332
 fractons, 322
 gas phase, 325
 guarded hot plate, 337
 hot wire method, 344
 instationary methods, 339
 Knudsen effect, 326
 Laser Flash method, 345
 mean free path, 328
 phonons, 321
 pore size spectrum, 329

pressure dependence, 328
radiation transport, 331
solid backbone, 325
sound velocity, 321
stationary methods, 336
Transient Hot Bridge Method, 343
transient plane source method, 341
thermal conductivity aerogels, 320
solid backbone, 321
thermal diffusivity, 317, 332
relaxation time, 333
thermoporosimetry, 280
time-dependent heat transfer tube, 432
tortuosity, 309

upper critical point, 405

viscoelasticity, 94
viscosity, 92
dynamic, 93
Kelvin model, 95
Maxwell model, 95
Richardsson–Zaki relation, 94
volume fraction, 94
Zimm model, 94
volume fraction polymers, 429

xerogel, 3

yield and fracture strength
cellulose aerogel, 379
model, 376
RF aerogel, 378
silica aerogel, 377
Young's modulus
biopolymeric aerogels, 374
RF aerogel, 374
silica aerogel, 373